CHEMICALS BY ENZYMATIC
AND MICROBIAL PROCESSES

CHEMICALS BY
ENZYMATIC AND
MICROBIAL PROCESSES

Recent Advances

Edited by J.I. Duffy

NOYES DATA CORPORATION

Park Ridge, New Jersey, U.S.A.

1980

Published in the United States of America by
Noyes Data Corporation
Noyes Building, Park Ridge, New Jersey 07656

Library of Congress Cataloging in Publication Data

Duffy, Joan Irene, 1950–
    Chemicals by enzymatic and microbial processes.

    (Chemical technology review ; no. 161)
    "Supersedes ... Chemicals by fermentation,
published in 1973."
    Includes index.
    1. Chemicals--Patents--United States.
2. Fermentation--Patents--United States.
I. Gutcho, Sidney. Chemicals by fermentation.
II. Title. III. Series.
TP210.D83   660'.6   80-16150
ISBN 0-8155-0805-0

FOREWORD

The detailed, descriptive information in this book is based on U.S. patents issued since January 1975 that deal with recent advances in the manufacture of chemicals by enzymatic and microbial processes.

This book is a data-based publication, providing information retrieved and made available from the U.S. patent literature. It thus serves a double purpose in that it supplies detailed technical information and can be used as a guide to the patent literature in this field. By indicating all the information that is significant, and eliminating legal jargon and juristic phraseology, this book presents an advanced commercially oriented review of the manufacture of chemicals by enzymatic and microbial processes. This title supersedes our previous title *Chemicals by Fermentation,* published in 1973.

The U.S. patent literature is the largest and most comprehensive collection of technical information in the world. There is more practical, commercial, timely process information assembled here than is available from any other source. The technical information obtained from a patent is extremely reliable and comprehensive; sufficient information must be included to avoid rejection for "insufficient disclosure." These patents include practically all of those issued on the subject in the United States during the period under review; there has been no bias in the selection of patents for inclusion.

The patent literature covers a substantial amount of information not available in the journal literature. The patent literature is a prime source of basic commercially useful information. This information is overlooked by those who rely primarily on the periodical journal literature. It is realized that there is a lag between a patent application on a new process development and the granting of a patent, but it is felt that this may roughly parallel or even anticipate the lag in putting that development into commercial practice.

Many of these patents are being utilized commercially. Whether used or not, they offer opportunities for technological transfer. Also, a major purpose of this book is to describe the number of technical possibilities available, which may open up profitable areas of research and development. The information contained in this book will allow you to establish a sound background before launching into research in this field.

Advanced composition and production methods developed by Noyes Data are employed to bring these durably bound books to you in a minimum of time. Special techniques are used to close the gap between "manuscript" and "completed book." Industrial technology is progressing so rapidly that time-honored, conventional typesetting, binding and shipping methods are no longer suitable. We have bypassed the delays in the conventional book publishing cycle and provide the user with an effective and convenient means of reviewing up-to-date information in depth. The table of contents is organized in such a way as to serve as a subject index. Other indexes by company, inventor and patent number help in providing easy access to the information contained in this book.

16 Reasons Why the U.S. Patent Office Literature Is Important to You

1. The U.S. patent literature is the largest and most comprehensive collection of technical information in the world. There is more practical commercial process information assembled here than is available from any other source. Most important technological advances are described in the patent literature.

2. The technical information obtained from the patent literature is extremely comprehensive; sufficient information must be included to avoid rejection for "insufficient disclosure."

3. The patent literature is a prime source of basic commercially utilizable information. This information is overlooked by those who rely primarily on the periodical journal literature.

4. An important feature of the patent literature is that it can serve to avoid duplication of research and development.

5. Patents, unlike periodical literature, are bound by definition to contain new information, data and ideas.

6. It can serve as a source of new ideas in a different but related field, and may be outside the patent protection offered the original invention.

7. Since claims are narrowly defined, much valuable information is included that may be outside the legal protection afforded by the claims.

8. Patents discuss the difficulties associated with previous research, development or production techniques, and offer a specific method of overcoming problems. This gives clues to current process information that has not been published in periodicals or books.

9. Can aid in process design by providing a selection of alternate techniques. A powerful research and engineering tool.

10. Obtain licenses—many U.S. chemical patents have not been developed commercially.

11. Patents provide an excellent starting point for the next investigator.

12. Frequently, innovations derived from research are first disclosed in the patent literature, prior to coverage in the periodical literature.

13. Patents offer a most valuable method of keeping abreast of latest technologies, serving an individual's own "current awareness" program.

14. Identifying potential new competitors.

15. It is a creative source of ideas for those with imagination.

16. Scrutiny of the patent literature has important profit-making potential.

CONTENTS AND SUBJECT INDEX

Contents and Subject Index

INTRODUCTION

Fermentation processes can probably rank as man's first attempt at controlling chemical reactions to produce useful products. Although the mechanisms involved were not understood until relatively modern times (indeed the role of microorganisms was not even recognized until the work of Pasteur), the benefits derived from microbial processes have been widespread through the ages, notably in the winemaking and brewing industries.

With the increasing sophistication and progress of the chemical industry, man is turning more and more to the use of microorganisms to produce an ever-growing list of chemical products. Utilizing microbes and their enzymes, reactions can be carried out with a speed and efficiency that could not be attained in their absence, if in fact the reaction could be accomplished at all.

Progress in developing new microbial processes has proceeded along a number of pathways. A better understanding of the mechanisms involved, new strains of microorganisms, and easier identification and separation of the desired product are only a few ways in which techniques have improved in the last years.

The food industry has probably been the major recipient of recent advances in this field. Amino acid and protein supplementation has been the focus of much attention, and the large number of patents on this subject attest to this fact. New strains of microorganisms have been developed by chemical and physical mutations of the parent strains which produce these substances in higher yield and more efficiently than was heretofore possible.

The large number of processes to produce a wide variety of carbohydrates, from simple sugars to the high molecular weight polysaccharides, also reflect the desire of the industry to use to advantage the wide spectrum of microorganisms available, mainly through isomerization techniques to convert one form of sugar to another, or the use of enzymes to break down polysaccharides and starches to their components.

1

The information presented in this volume covers a broad spectrum of microbial processes, ranging from the production of a long list of different organic acids, to the nucleic acids and amino acids which are the fundamental building blocks of our life processes, to carbohydrates and vitamins which perpetuate those processes. Microbes and their enzymes are now an integral part of the chemical industry in which the technology is improving at a rapid rate.

ORGANIC ACIDS

CITRIC ACID

Organic Acid as Carbon Source

Citric acid is widely used in the food and pharmaceutical industry as, for example, an acidulant in beverages or in the preparation of jams. Citric acid is also used as an antioxidant or as a stabilizer in various food products and has been employed in various detergent compositions as a detergent builder.

R.I. Leavitt; U.S. Patent 4,178,211; December 11, 1979; assigned to Ethyl Corporation describes a process for producing citric acid which comprises aerobically fermenting a mixed carbon source nutrient medium in a submerged culture of *Candida lipolytica*, ATCC-20510, the process being further characterized in that the mixed carbon source is a mixture of a carbohydrate and a straight chain monocarboxylic acid or salt thereof having from 2 to about 20 carbon atoms, the fermentation liquor having an initial concentration of from about 1 to 5% by weight of the monocarboxylic acid or salt thereof and, as a fermentation initiator, from 0.5 to 2.5% by weight of the carbohydrate with no further addition of the carbohydrate to the fermentation mixture and a total of from about 10 to 20% by weight of the monocarboxylic acid or salt thereof being added over the total fermentation period of from about 1 to 15 days whereby the process produces a ratio of citric to isocitric acid of at least 4:1, respectively.

Preferably, the carbon source is a lower organic acid or salt thereof such as those having from 2 to 6 carbon atoms. The most preferred carboxylic acid is acetic acid because of its wide availability and low cost. The initiating carbohydrate is preferably a monosaccharide, such as glucose.

The microorganism employed in the process is a member of the genus *Candida* and a stable mutant of the species *Candida lipolytica*, ATCC-8661. It has been found that this organism produces species of a different colonial and microscopic morphology detectable in plate cultures 48 hours or more old. This variant is distinguished by a smooth textured colonial surface, is stable in that it does not

3

Ingredient	Grams Per Liter
Glucose	20.0
Thiamine hydrochloride	0.001
$MgSO_4 \cdot 7H_2O$	0.20
$CaCO_3$	17.0
Urea	4.05
KH_2PO_4	0.75

Fermentation is allowed to proceed for 24 hours at which time 100 ml of lique-fied slack wax (Moore & Munger, Inc.) is then added to one of the fermentors. Similarly, 100 ml of n-octane is added to a second fermentor, and to a third fermentor is added 100 ml of n-octane and 100 ml of slack wax. Identical additions are made to each fermentor at 48 hours, 72 hours and 96 hours. The fermentation in each fermentor is allowed to proceed for a total of 137 hours.

Additions	Grams of Citric Acid Per Fermentor
n-Octane	None
Slack wax	None
n-Octane + slack wax	5.44

Production from Olefins

T. Nagata, S. Satoh, and T. Matsumoto; U.S Patent 4,180,626; December 25, 1979; assigned to Showa Oil Co., Ltd., Japan describe a method to provide a manufacturing process of citric acid by culturing olefin-assimilable microorganisms in culture medium containing olefins under aerobic conditions.

The above object is accomplished by cultivating of the microorganism selected from the group of *Candida tropicalis, Candida intermedia* and *Candida brumptii* in the culture medium containing any olefin of C_{8-40} as carbon source. Especially, the most suitable microorganism utilized in the process is the strain belonging to *Candida tropicalis* and its auxotrophic mutants and its variants. Also, mixture of olefins may be used as carbon source. The concentration of the olefins in the culture medium is 1 to 20% weight, preferably 5 to 15% weight.

As the nitrogen source of the culture medium, inorganic and organic ammonium salts, such as ammonium chloride, ammonium acetate, and various nitrogen compounds may be used individually or in mixture.

Ordinary inorganic salts such as phosphate, sulfate, hydrochloric acid salts, potassium salts, sodium salts, magnesium salts, iron salts, manganese salts, copper salts and zinc salts may be used as inorganic nutrients in culture medium. Calcium carbonate and alkali compounds may be used to regulate pH of the medium. Biotin and thiamine as organic nutrients may be used in a trace amount individually or in mixture, and also natural substances such as yeast extract or corn steep liquor containing biotin and thiamine may be used. When biotin and thiamine are used individually, an amount less than 1,000 $\mu g/l$ of biotin and thiamine suffices to increase the production amount of citric acid. A preferable amount of the organic nutrients is from 50 to 100 $\mu g/l$. In cases where both biotin and thiamine are added in the medium, the total amount of biotin and thiamine should be less than 1,000 $\mu g/l$, preferably 50 to 100 $\mu g/l$. α-Olefin or mixture of olefins may be used as carbon source. These substrates are in liquid state at

ORGANIC ACIDS

CITRIC ACID

Organic Acid as Carbon Source

Citric acid is widely used in the food and pharmaceutical industry as, for example, an acidulant in beverages or in the preparation of jams. Citric acid is also used as an antioxidant or as a stabilizer in various food products and has been employed in various detergent compositions as a detergent builder.

R.I. Leavitt; U.S. Patent 4,178,211; December 11, 1979; assigned to Ethyl Corporation describes a process for producing citric acid which comprises aerobically fermenting a mixed carbon source nutrient medium in a submerged culture of *Candida lipolytica*, ATCC-20510, the process being further characterized in that the mixed carbon source is a mixture of a carbohydrate and a straight chain monocarboxylic acid or salt thereof having from 2 to about 20 carbon atoms, the fermentation liquor having an initial concentration of from about 1 to 5% by weight of the monocarboxylic acid or salt thereof and, as a fermentation initiator, from 0.5 to 2.5% by weight of the carbohydrate with no further addition of the carbohydrate to the fermentation mixture and a total of from about 10 to 20% by weight of the monocarboxylic acid or salt thereof being added over the total fermentation period of from about 1 to 15 days whereby the process produces a ratio of citric to isocitric acid of at least 4:1, respectively.

Preferably, the carbon source is a lower organic acid or salt thereof such as those having from 2 to 6 carbon atoms. The most preferred carboxylic acid is acetic acid because of its wide availability and low cost. The initiating carbohydrate is preferably a monosaccharide, such as glucose.

The microorganism employed in the process is a member of the genus *Candida* and a stable mutant of the species *Candida lipolytica*, ATCC-8661. It has been found that this organism produces species of a different colonial and microscopic morphology detectable in plate cultures 48 hours or more old. This variant is distinguished by a smooth textured colonial surface, is stable in that it does not

3

revert to stock type after repeated transfer, breeds true while the parent does not and produces levels of total citric acid from acetate equal to or higher than its parent.

Example: In a 7-liter stirred tank fermenter, there was added 3 liters of an aqueous solution containing: 1.0 g/l $MgSO_4 \cdot 7H_2O$, 6.0 g/l NH_4CL, 2.0 g/l KH_2PO_4, and 0.5 g/l yeast extract. To this solution was then added 2% by weight of glucose and 4 wt % of sodium acetate. The temperature was controlled at 30°C, the pH at 7.2, and the stirrer set at 1,000 rpm. The air rate through the fermentation tank was 2,000 cc/min. Then, 225 ml of a 24-hour inoculum of *Candida lipolytica* ATCC-20510 was added and allowed to run for 24 hours. At this time, and subsequently every 24 to 36 hours, additional acetate was added to a total level of 12 wt %.

At the end of 6 days, there was produced 8 wt % of citric acid. The maximum rate of production of citric acid during this run was over a period of 48 hours in which 2.1 wt % of citric acid was produced every 24 hours.

Slack Wax as Carbon Source

R.C. Nubel; U.S. Patent 4,014,742; March 29, 1977; assigned to Pfizer Inc. describes a process for producing citric acid which comprises aerobically propagating a citric acid-accumulating strain of yeast of the genus *Candida* in an aqueous nutrient medium containing enough readily assimilable source of carbon to promote growth but insufficient to permit the accumulation of citric acid; introducing slack wax as the principal source of assimilable carbon together with a solubilizing agent into the aqueous nutrient medium after at least 50% of the readily assimilable source of carbon has been utilized, continuing the aerobic propagation until a level of at least about 1 g of citric acid has been accumulated per liter of aqueous medium and recovering the citric acid, the solubilizing agent being selected from the group consisting of alkanols having 4 to 10 carbon atoms in each alkyl moiety, lower alkyl esters of alkanoic acids having 2 to 6 carbon atoms, alkenes and alkanes having 8 to 19 carbon atoms, turpentine, mineral oil and mixtures thereof.

Petroleum wax, known generically in the petroleum industry as slack wax, is a relatively homogeneous material consisting of long chain hydrocarbons having 20 to 30 carbon atoms. It is a solid substance at room temperature and has a freezing point of about 45°C

The initial source of carbon may be glucose, a vegetable oil, a fatty acid ester or mixtures thereof, a normal paraffin having 9 to 19 carbon atoms or mixtures thereof or other suitable carbon sources known to those skilled in the art. The preferred carbon source is glucose at a concentration of 5 to 10% w/v, preferably 7.5% w/v. In addition, the medium contains a source of assimilable nitrogen such as ammonium nitrate, ammonium sulfate, ammonium chloride, wheat bran, soybean meal, urea, amino acids and peptones. It is, of course, well known that such vitamins as biotin and thiamine and such mineral cations and anions as sodium, potassium, cobalt, phosphate and sulfate are also beneficial to the growth of yeasts.

The amount of readily assimilable source of carbon in the production medium is limited to that amount which is adequate for the growth of the added *Candida* cells but is insufficient to permit the accumulation of citric acid (less than 0.5 g/l). The upper concentration limit is fixed as that amount to which any incremental addition will permit the accumulation of citric acid (greater than 0.5 g/l). When glucose is used in the production medium, the concentration is 1 to 3% w/v, preferably 2% w/v.

The production fermentation is allowed to proceed for a period of time until at least 50%, and preferably 80%, of the glucose is assimilated (approximately 24 hours). An amount of slack wax is then added to provide a concentration of at least 3% by weight of the medium, preferably 5 to 20% by weight, along with a solubilizing agent. The addition of slack wax and solubilizing agent may be accomplished by adding all of the combination at one time or by additions at various times during the fermentation. For example, the first addition can be followed by similar additions at 24-hour intervals during the fermentation cycle.

The usual temperatures known in the art for growing yeasts, e.g., about 20° to 37°C, may be employed, a range of from 25° to 29°C being preferred with fermentation times of 4 to 7 days. The production fermentation is continued until at least about 1 g of citric acid per liter of medium has been accumulated. The initial growth period of the yeast cells for preparation of the inoculum is preferably 24 to 48 hours. These general conditions of growth and fermentation are well known in the art, as are also methods for the recovery of the citric acid produced as the free acid, sodium salt or calcium salt by centrifugation, filtration, concentration under vacuum, etc.

Any of the known citric acid-accumulating strains of *Candida* will provide varying levels of accumulated citric acid as the art appreciates. Illustrative strains include citric acid-accumulating strains of *Candida lipolytica*, *Candida guilliermondii*, *Candida tropicalis*, *Candida parapsilosis* and *Candida brumptii*. The preferred citric acid-accumulating strains are those belonging to the species *Candida lipolytica*.

Example: Cells of *Candida lipolytica* NRRL Y-1094 grown on a potato dextrose agar slant are inoculated into a Fernbach flask containing 800 ml of sterile medium having the following composition:

Ingredient	Grams Per Liter
Glucose	75.0
$MgSO_4 \cdot 7H_2O$	0.25
KH_2PO_4	0.50
Calcium phytate	0.50
$(NH_4)_2SO_4$	4.0
$CaCO_3$	5.0
Yeast extract	1.0

The inoculated flask is incubated at 26°C on a rotary shaker for 48 hours after which a 100 ml portion of the grown culture (first stage inoculum) is aseptically transferred to a 4-liter fermentor containing 2 liters of sterile medium having the same composition as that used in the Fernbach flask. The fermentor is stirred at 1,750 rpm and air is introduced through a sparger at the rate of 4 standard cubic feet per hour per gallon of medium. The temperature is maintained at 26°C. The fermentation in the fermentor is continued for 72 hours at which time 100 ml portions (second stage inoculum) are transferred to a series of identical fermentors, each containing 2 liters of sterile medium of the following composition.

Ingredient	Grams Per Liter
Glucose	20.0
Thiamine hydrochloride	0.001
$MgSO_4 \cdot 7H_2O$	0.20
$CaCO_3$	17.0
Urea	4.05
KH_2PO_4	0.75

Fermentation is allowed to proceed for 24 hours at which time 100 ml of liquefied slack wax (Moore & Munger, Inc.) is then added to one of the fermentors.
Similarly, 100 ml of n-octane is added to a second fermentor, and to a third fermentor is added 100 ml of n-octane and 100 ml of slack wax. Identical additions are made to each fermentor at 48 hours, 72 hours and 96 hours. The fermentation in each fermentor is allowed to proceed for a total of 137 hours.

Additions	Grams of Citric Acid Per Fermentor
n-Octane	None
Slack wax	None
n-Octane + slack wax	5.44

Production from Olefins

*T. Nagata, S. Satoh, and T. Matsumoto; U.S Patent 4,180,626; December 25,
1979; assigned to Showa Oil Co., Ltd., Japan* describe a method to provide a
manufacturing process of citric acid by culturing olefin-assimilable microorganisms in culture medium containing olefins under aerobic conditions.

The above object is accomplished by cultivating of the microorganism selected
from the group of *Candida tropicalis, Candida intermedia* and *Candida brumptii*
in the culture medium containing any olefin of C_{8-40} as carbon source. Especially, the most suitable microorganism utilized in the process is the strain belonging to *Candida tropicalis* and its auxotrophic mutants and its variants. Also,
mixture of olefins may be used as carbon source. The concentration of the olefins in the culture medium is 1 to 20% weight, preferably 5 to 15% weight.

As the nitrogen source of the culture medium, inorganic and organic ammonium
salts, such as ammonium chloride, ammonium acetate, and various nitrogen compounds may be used individually or in mixture.

Ordinary inorganic salts such as phosphate, sulfate, hydrochloric acid salts, potassium salts, sodium salts, magnesium salts, iron salts, manganese salts, copper
salts and zinc salts may be used as inorganic nutrients in culture medium. Calcium carbonate and alkali compounds may be used to regulate pH of the medium.
Biotin and thiamine as organic nutrients may be used in a trace amount individually or in mixture, and also natural substances such as yeast extract or corn
steep liquor containing biotin and thiamine may be used. When biotin and thiamine are used individually, an amount less than 1,000 μg/l of biotin and thiamine
suffices to increase the production amount of citric acid. A preferable amount
of the organic nutrients is from 50 to 100 μg/l. In cases where both biotin and
thiamine are added in the medium, the total amount of biotin and thiamine
should be less than 1,000 μg/l, preferably 50 to 100 μg/l. α-Olefin or mixture
of olefins may be used as carbon source. These substrates are in liquid state at

fermentation temperature. Thus, any surface active agent which hinders the cultivation activity of the microorganism is not necessary.

Details of the process are as follows. The microorganisms are cultivated under aerobic conditions. The fermentation temperature is in the range of 25° and 40°C, preferably about 30°C. pH value of the medium is in the range of 3 to 10, preferably 4 to 6. pH value of the medium is regulated by adding alkalis or salts such as sodium carbonate or calcium carbonate. The cultivation is ordinarily carried out for 50 to 150 hours, preferably 80 to 100 hours. Citric acid accumulated in the medium is isolated in the form of calcium citrate by normal method, for example, by filtration or by centrifugal method and then purified by chromatographic or ion exchange method.

Carbon Source as By-Product of Fructose Production

R. Uchio, K. Kikuchi, S. Asai, and K. Yarita; U.S. Patent 4,040,906; August 9, 1977; assigned to Ajinomoto Co., Inc., Japan have found that when fructose is recovered from a sucrose hydrolyzate in the form of an addition product of calcium hydroxide, and when the mother liquor, wherein the calcium ions are eliminated, is neutralized, the mother liquor is suitable as a carbon source of a fermentation of citric acid by using a strain belonging to species *Aspergillus niger* or *Aspergillus awamori*.

Raw materials include various products and intermediates produced in the sugar manufacturing process from cane and beet, such as juices of cane and beet, crude sugar, and molasses.

Hydrolysis of sucrose in the raw material may be carried out by known methods, such as by the use of mineral acid or an enzyme. For example, if the pH of the raw material is adjusted to 1.5 to 2 with hydrochloric acid and is heated to 60° to 100°C for 0.5 to 4 hours, most of the sucrose may be hydrolyzed to hexose, i.e., glucose and fructose.

When the raw material is hydrolyzed with mineral acid, the hydrolyzate is neutralized with an alkali, such as calcium hydroxide, preferably in an amount of 0.7 to 1.5 times the molar quantity of the hexose. Instead of calcium hydroxide, calcium oxide may be employed. In this instance, calcium oxide is first converted to calcium hydroxide in the hydrolyzate, and then reacted with fructose.

The mixing must be carried out cautiously. First, the neutral hydrolyzate is cooled to below 10°C, preferably below 5°C, and calcium hydroxide, in an amount of 1.2 to 1.6 times the molar quantity of the fructose, is added to the cold hydrolyzate. Then, the seed adducts of fructose and calcium hydroxide are preferably added, and the seeded mixture is aged for 15 to 60 minutes with moderate stirring. The adduct crystals may also crystallize without seeding. The remaining calcium hydroxide is added gradually over a period of about 1 to 2 hours. During the reaction of calcium hydroxide with fructose, the temperature of the reactant is preferably kept below 5°C in order to decrease the decomposition of fructose and glucose. The above procedure produces large crystals of the adduct which are especially suitable for separation on an industrial scale.

The crystals so produced are recovered by filtering or centrifuging. The recovered crystals are preferably washed with chilled water.

Fructose crystals or a fructose solution containing a small amount of glucose can be prepared from the recovered crystals by known methods. The yield of fructose from the hydrolyzate in the above solution may be as high as 70%, usually 50 to 65%.

The mother liquor is neutralized, and calcium ions are removed from the liquor. Suitable neutralizing agents include carbon dioxide, sulfuric acid, phosphoric acid, cation exchange resins, or the like. When the neutralization is carried out by use of one of the exemplified agents, calcium ions can be removed simultaneously. Among the above agents, carbon dioxide is the most preferable because recovered calcium carbonate can be used again by calcination. The other superiority of carbon dioxide is decolorization effect during neutralization.

Since both fructose and glucose are not stable in alkali solution, it is necessary that fructose and glucose are kept cold throughout the above procedure. It is also necessary that, after fructose adduct is separated, both the adduct and the mother liquid are quickly neutralized.

The calcium-eliminated mother liquid is suitable as a carbon source for citric acid fermentation by using a strain belonging to species *Aspergillus niger* or *Aspergillus awamori*. Examples of the strains are: *Aspergillus niger* ATCC 6275 and *Aspergillus awamori* FERM-P 3512.

Fermentation is carried out according to the conventional manner. The yield of citric acid from the mother liquid of the process is much higher than the same from the raw material which is the conventional carbon source for citric acid fermentation. Particularly in case of molasses, in spite of lowering carbon purity by extracting fructose, the yield of this method is also much higher than the yield of the conventional method.

Addition of Monofluoroacetic Acid or Alcohol

K. Kimura and T. Nakanishi; U.S. Patent 3,873,424; March 25, 1975; assigned to Kyowa Hakko Kogyo Co., Ltd., Japan describe a process for producing citric acid, isocitric acid and/or microbial cells by fermentation which may be carried out advantageously on an industrial scale at low cost to give a high yield of product.

It has been found that the yield of isocitric acid is greatly increased by the addition of potassium ferrocyanide to the medium, whereas the yield of citric acid is considerably increased by the addition of monofluoracetic acid to the medium. The addition of these compounds to the media not only effects the yield and accumulation of isocitric acid and citric acid, but also has a considerable effect upon the variation in formation ratio of these organic acids by means of the increase in the productivities thereof.

The addition of an effective amount up to about 5% by weight of potassium ferrocyanide to the medium provides the benefits discussed hereinabove. The amount of monofluoroacetic acid to be added to the medium ranges from an effective amount up to about 20 millimols.

When citric acid and microbial cells are produced by fermentation using an alcohol as the additive in the medium, it has been found that alcohols in general may be used in this connection. However, organic alcohols having 1 to 20 carbon atoms are particularly effective, and the preferred alcohols to be employed are alkanols having 1 to 20 carbon atoms. These alcohols themselves are effectively assimilated by the microorganisms employed and also show an inhibiting action. Thus, it is necessary to properly select the concentration thereof.

No citric acid is formed at all in the culture liquor when an alcohol is not added to the medium. However, a small amount of citric acid is formed and accumulated when various alcohols are added thereto. Furthermore, the microorganism employed is capable of assimilating alcohols having 10 to 20 carbon atoms and forms microbial cells in a yield (based on carbon) which is almost as good as that for n-paraffins or ethanol. In addition, the protein content is somewhat increased by the addition of an alcohol to the medium.

Hydrocarbon-assimilating microorganisms, including bacteria, yeasts, molds, etc. in a wide range, can be employed in the process. Preferred bacteria are those belonging to the genera *Arthrobacter, Corynebacterium, Brevibacterium* or *Nocardia*. The preferred yeasts are those belonging to the genera *Candida, Torulopsis, Endomyces, Pichia, Hansenula, Mycoderma* or *Endomycopsis*. Preferred molds are those of the genus *Aspergillus* or the genus *Penicillium*.

Either a synthetic culture medium or a natural nutrient medium is suitable in the fermentation process as long as it contains the essential nutrients for the growth of the microorganism strain employed. Such nutrients are well known in the art and include substances such as a carbon source, a nitrogen source, inorganic compounds and the like which are utilized by the microorganism employed in appropriate amounts.

The fermentation or culturing of the microorganisms is conducted under aerobic conditions, such as aerobic shaking of the culture or with stirring and aeration of a submerged culture, at a temperature of, for example, about 15° to 45°C and at a pH of, for example, about 1.0 to 9.0 in the case of the embodiment wherein an alcohol is added to the medium, and at a temperature of, for example, about 20° to 37°C and at a pH of, for example, about 1.0 to 7.5 in the case of the embodiment wherein potassium ferrocyanide or monofluoroacetic acid is added to the medium.

Example 1: 18.9 liters of a fermentation medium having the following composition is placed into a 30-liter volume jar fermentor:

n-Paraffins (C_{12}, C_{13}, C_{14}) (equal volume mixture)	2.5% (v/v)
NH_4Cl	20% (w/v)
KH_2PO_4	0.1% (w/v)
K_2HPO_4	0.1% (w/v)
$MgSO_4 \cdot 7H_2O$	0.1% (w/v)
$MnSO_4 \cdot 4H_2O$	0.001% (w/v)
$FeSO_4 \cdot 7H_2O$	0.002% (w/v)
$ZnSO_4 \cdot 7H_2O$	0.001% (w/v)
$CuSO_4 \cdot 5H_2O$	0.0005% (w/v)
$(NH_4)_6Mo_7O_{24} \cdot 4H_2O$	0.0065% (w/v)
H_3BO_4	0.0002% (w/v)
Nonion LP-20R (surfactant)	500 γ/ml

Soybean oil	1 ml/l
Thiamine hydrochloride	200 γ/l
Cornsteep liquor	0.05% (w/v)

1 liter of a microbial seed liquor of *Torulopsis famata* ATCC 15586 is added to the above fermentation medium. Then, 0.1 liter of ethanol is also added thereto. Culturing is carried out at a cultivation temperature of 34°C with aeration at the rate of 20 l/min and with stirring at the rate of 500 rpm. The pH is kept at 6.50 with 18% aqueous ammonia during the culturing. After 42 hours, the culturing is completed, and the concentration of microbial cells in the culture liquor reaches 20 mg/ml. The yield of citric acid is 3.0 mg/ml.

As a comparison, culturing is carried out in the same manner in a medium containing no ethanol. The concentration of formed cell bodies is 14 mg/ml, and no by-product citric acid is observed at all.

The microbial cells obtained as a result of adding ethanol to the medium are repeatedly washed several times in a Shaples centrifugal separator and subjected to centrifugal separation. After freeze-drying, 365 g of cell bodies are obtained. The protein content per cell body is 61%. However, when cultivation is conducted in a medium containing no ethanol, only 246 g of cell bodies are obtained with a protein content of 59%.

After removing the microbial cells from the culture liquor, 200 g of calcium chloride is added, and the pH is adjusted to 8.0 with aqueous ammonia. The resultant liquor is heated at 100°C for 5 minutes, and the obtained crystals of calcium citrate are filtered and 2 liters of water is added to the crystals. The pH is adjusted to 2.0 with hydrochloric acid, thereby dissolving the calcium citrate. Then, by repetition of this operation, 65 g of calcium citrate crystals is obtained.

Example 2: *Arthrobacter paraffineus* ATCC 15591 is cultured at 30°C for 24 hours with aerobic shaking in 20 ml of a medium containing 0.25% yeast extract, 0.5% meat extract, 0.5% peptone and 0.25% sodium chloride at a pH of 7.0 in a 250-ml conical flask. The resulting culture is transferred, in a proportion of 10%, to 20 ml of a fermentation medium contained in 500-ml Sakaguchi flasks and having the following composition:

	Percent
KH_2PO_4	0.1
$Na_2HPO_4 \cdot 12H_2O$	0.1
$MgSO_4 \cdot 7H_2O$	0.05
$FeSO_4 \cdot 7H_2O$	0.01
$MnSO_4 \cdot 4H_2O$	0.002
Cornsteep liquor	0.05
NH_4NO_3	2.0
n-Paraffins (C_{12}, C_{13}, C_{14})	
(equal volume mixture)	5
$CaCO_3$	5

Culturing is conducted at 30°C with reciprocating shaking. Sodium monofluoroacetate is added thereto 6 hours after the initiation of culturing in a concentration of 0.1%. Then, the cultivation is continued for 4 days, whereby 18.2 mg/ml of calcium citrate (in terms of citric acid) is found to be accumulated in the culture liquor. Using the same medium in a control experiment, only 9.3 mg/ml of citric acid is obtained when no sodium monofluoroacetate is added to the medium.

Addition of Sodium Hydroxide

A.J. Kabil; U.S. Patent 3,886,041; May 27, 1975; assigned to Aktiengesellschaft Jungbunzlauer Spiritus- und Chemische Fabrik, Austria describes a process for the production of citric acid by submerged fermentation of a carbohydrate-containing material by means of the fungus *Aspergillus niger*, with the addition of an alkaline substance.

The process comprises adding to the fermenting medium (containing at least partially purified carbohydrates, for example, sugar, purified by ion exchange) sodium hydroxide as an alkaline substance, optionally in portions, whereby the first addition is made, preferably 24 to 36 hours, after inoculation, in order to control the mycelium growth.

The addition of sodium hydroxide after inoculation causes a stimulation of the citric acid production, and at the same time, allows exact control of the process. It was also found that addition of sodium hydroxide after inoculation makes it unimportant to adjust the alkali metal/phosphate ratio to specific values. The first addition of sodium hydroxide is preferably made 30 hours after inoculation.

The addition at the abovementioned time promotes the development of a mycelium, which is capable of an increased citric acid production, thereby decreasing the time of fermentation. Further, it was shown to be an advantage to add sodium hydroxide throughout the fermentation period in order to maintain the pH of the fermenting medium between 1.5 and 2.0, preferably between 1.8 and 2.0. Thereby the formation of undesirable acids is almost completely suppressed.

Example: A decationized sugar solution with a sugar content of 25.2% (w/v) was sterilized for 30 minutes in streaming steam and cooled down to 35°C. Six sterilized glass fermenters with a diameter of 150 mm and a height of 1,000 mm were each charged with 11 liters of the abovementioned solution, and the temperature adjusted to 30°C. To each fermenter was added a piece of stainless steel of the type DIN 4586 of the following composition: $C \leqslant 0.07\%$; $Si \leqslant 1\%$; Mn, 2%; Cr, 18%; Mo, 2.75%; Ni, 22%; Cu, 2%; $Nb \geqslant 0.56\%$. Before the fermentation was started, the following nutrient salts were added to the aerated sugar solution:

	g/l
$(NH_4)_2SO_4$	6
$MgSO_4 \cdot 7H_2O$	1.1
$CaCl_2 \cdot 2H_2O$	0.55
NaCl	0.15
KH_2PO_4	0.15
$ZnSO_4 \cdot 7H_2O$	0.0015

After initial adjustment of the pH to 2.90 and addition of $K_4[Fe(CN)_6] \cdot 3H_2O$ as an inhibitor, the solution was inoculated with spores of *Aspergillus niger*. Throughout the fermentation period, the temperature was kept at 30°C, and further portions of $K_4[Fe(CN)_6] \cdot 3H_2O$ were added to a total concentration of 10 ppm. After 30 hours, 40 ml NaOH (16% by weight) were added to three fermenters, and further amounts of NaOH were added as required. The pH is adjusted to 1.9. The amount of citric acid formed, the fermentation yield and the fermentation period were measured. The improved results as compared to the samples without addition of sodium hydroxide are shown in the following table:

Number of Sample	Addition of NaOH	Fermentation Period (days)	Citric Acid-Monohydrate (g)	Yield (%)
106	–	12	2,320	83.7
107	–	12	2,299	82.9
108	–	11	2,318	83.6
112	+	9½	2,475	89.3
113	+	10	2,414	87.1
114	+	9	2,489	89.8

Addition of Refined Steel

A.J. Kabil; U.S. Patent 3,936,352; February 3, 1976; assigned to Aktiengesellschaft Jungbunzlauer Spiritus- und Chemische Fabrik, Austria describes a process for preparing citric acid by subjecting a carbohydrate-containing material to a submerged fermentation with a citric-acid-producing strain of *Aspergillus niger* in the presence of refined steel.

Suitable for the process are steels identified as No. 1.4505 and 1.4586 in *Nachschlagewerk Stahlschlussel*, 8, 159, 151 (1968). Two other steels listed in the *Stahlschlussel* publication identified as Standard No. 1.4571 and 1.4541 serve as comparisons. These steels will be referred to hereafter by their numbers 1.4505, 1.4586, 1.4541 and 1.4571.

The steels 1.4505 and 1.4586 contain as components, in addition to iron and carbon, silicon, manganese, chromium, molybdenum, nickel, copper and niobium. The components in steel 1.4505 have values of 2% manganese, about 0.07% carbon, about 1% silicon, 17.5% chromium, 2.25% molybdenum, 20% nickel, 2% copper and niobium in an amount eight times greater than that of the carbon component. Steel 1.4586 contains about 0.07% carbon, about 1% silicon, 2% manganese, 18.0% chromium, 2.75% molybdenum, 22.0% nickel, 2% copper and an amount of niobium eight times greater than that of carbon. Steel 1.4541 contains about 0.10% carbon and about 1% silicon, as well as manganese, chromium, nickel and titanium. Steel 1.4571 contains about 0.10% carbon, about 1% silicon, as well as manganese, chromium, molybdenum, nickel and titanium. The titanium component of steels 1.4571 and 1.4541 is supposed to be present in an amount five times greater than that of carbon.

It is of advantage to adjust the pH value to approximately 2.8 after addition of the nutrient solution. It is useful to follow this up with sterilization at temperatures of 100°C or above and, after cooling off, the addition of, e.g., the refined steel and its coming into contact with the nutrient solution, respectively, if parts of the construction of the fermenter, feeder lines, cooling coils or the entire fermenter are made of this steel.

Then, the solution is inoculated with the spores of *Aspergillus niger* and fermentation is carried out at temperatures of approximately 30° to 32°C. The improvement in yield in submerged fermentation in the presence of refined steel as compared to fermentation without the addition of metal is illustrated by the following examples.

Example 1: 11 liters of a sugar solution partially purified by decationization with a content of 25% by weight per volume of sugar were adjusted to a pH value of 2.8 after the addition of nutrient salts, sterilized in streaming steam at a temperature of 100°C, cooled off to 30°C and then transferred to a glass fermenter with a diameter of 150 mm and a height of 1,000 mm under sterile conditions. Ventilation was effected by a distributor pipe at the bottom of the fermenter. The fermenter contained a piece of refined steel of the type 1.4541 with a surface area of 131 cm^2. The fermentation solution was inoculated with *Aspergillus niger* at a temperature of 30°C. After a fermentation period of 14 days, 614 g citric acid crystals were obtained, which corresponds to a yield of 22.3%.

Example 2: The fermentation solution treated as described in Example 1 was put into a fermenter as described in Example 1 containing a piece of refined steel of the type 1.4571 with a surface area of 131 cm^2. The fermentation solution was inoculated at a temperature of 30°C with spores of *Aspergillus niger* and, at the same time, 11 mg of potassium hexacyanoferrate-II were added. On the third day, 6.6 mg, on the fifth day, 4.4 mg and on the eighth day, 3.3 mg of potassium hexacyanoferrate-II were added. After a fermentation period of 12 days, 1.254 g of citric acid crystals were obtained, which corresponds to a yield of 45.6%.

Example 3: The fermentation solution treated as described in Examples 1 and 2 was put into a fermenter as described in Example 1 containing a piece of refined steel of the type 1.4505 with a surface area of 131 cm^2. The inoculation with *Aspergillus niger* was effected. The fermentation period was terminated after 10 days. 2.456 g of citric acid crystals were obtained, which corresponds to a yield of 89.3%.

Production Using Ferrous Salts

It is known that the production of isocitric acid in a fermentation process is accelerated by the presence of iron ion or iron-containing ion such as $Fe(CN)_6^{-4}$ [*Journal of the Agricultural Chemical Society of Japan*, Vol. 44, p. 562 (1970), Japan and U.S. Patent 3,773,620]. Therefore, in order to prevent the by-production of isocitric acid, it is necessary that iron ion or iron-containing ion be absent from the medium. However, from a commercial standpoint, contamination of the medium by iron ion is unavoidable due to the fermentation raw materials and equipment. For example, in the commercial production of citric acid, usually an apparatus made of stainless steel or one having a glass lining is used. In either case, contamination of the medium with iron ion is unavoidable. Even the slight contamination of the medium due to the apparatus is considered to be sufficient to cause a significant production of isocitric acid and, therefore, renders the process impractical.

K. Takayama, T. Adachi, M. Kohata, K. Hattori, and T. Tomiyama; U.S. Patent 3,926,724; December 16, 1975; assigned to Kyowa Hakko Kogyo Co., Ltd., Japan have found that mutant yeast strains which require a higher amount of iron for growth as good as that of the parent strains by-produce no substantial amount of isocitric acid and produce citric acid in a yield equal to or even higher than the yield of citric acid produced by the parent strain. Such mutants, therefore, have a nutritional requirement for iron.

The mutants usually exhibit a growth comparable to that of the parent strains in the presence of 0.1 mg/l, and preferably 0.2 mg/l or more of iron. Particularly preferred mutants are *Candida zeylanoides* T-15, ATCC 20391; T-20, ATCC 20392; T-57, ATCC 20393 and IC 142, ATCC 20367, all of which are derived from *Candida zeylanoides* No. 19-5, ATCC 20347. These mutants are deposited with American Type Culture Collection, Rockville, Md., and are freely available to the public. The aforementioned strains require a higher amount of iron for growth than the parent strain. *Candida zeylanoides* IC 142, ATCC 20367 requires glycerin for growth in addition to an iron requirement.

Any culture medium normally used for the culturing of yeasts is suitable as long as it contains an assimilable carbon source, a nitrogen source, inorganic materials and other growth promoting factors which may be required by the specific yeast strain used.

As the carbon source, hydrocarbons, such as n-paraffins and kerosene, carbohydrates such as glucose and blackstrap molasses and acetic acid are suitable. As the nitrogen source, inorganic compounds such as ammonium chloride, ammonium sulfate, ammonium nitrate, ammonium phosphate and ammonium acetate, urea and nitrogen-containing natural substances such as peptone, meat extract and corn steep liquor may be used. Additionally, as inorganic materials, potassium dihydrogen phosphate, magnesium sulfate, manganese sulfate, ferrous sulfate, ferric chloride and zinc sulfate may be used.

Although ferrous sulfate and ferric chloride are preferred sources of iron ion in the culture medium, any soluble iron salt which does not prove toxic to the microorganism is appropriate.

As the source of iron-containing ion, potassium ferrocyanide, sodium ferrocyanide, potassium ferricyanide, sodium ferricyanide and iron alum may be used.

Culturing is carried out under aerobic conditions at 20° to 40°C, and at a slightly acidic to neutral pH of about 3 to 7 for 2 to 5 days, at which time a considerable amount of citric acid is formed in the culture liquor. The pH may be adjusted with calcium carbonate, sodium hydroxide or an aqueous ammonia.

After the completion of culturing, the microbial cells are removed from the culture liquor by, for example, filtration and the filtrate is concentrated. By adding calcium hydroxide to the filtrate, citric acid is readily recovered as calcium citrate. The calcium citrate is converted to citric acid by the addition of sulfuric acid, thus precipitating out calcium sulfate. Of course, the citric acid may also be recovered by any other of the usually used purification techniques.

Addition of Potassium Ferrocyanide

H. Hustede and H. Rudy; U.S. Patent 3,940,315; Feb. 24, 1976; and U.S. Patent 3,941,656; Mar. 2, 1976; both assigned to Joh. A. Benckiser GmbH, Germany describe the production of citric acid by submerged fermentation which is carried out in the presence of ferrocyanide or ferricyanide ions added to the fermentation at a specific time, by reference to the evolution and growth of the citric-acid microorganism. In accordance with the process, there are used in the main fermentation for the production of citric acid special pellets of *Aspergillus niger* or other suitable citric acid-producing microorganism, which have already been preactivated or pre-treated in a separate preculturing method. The preculturing method comprises activating by a treatment with cyanide ions a young inoculum of the microorganism during its highly intensive physiological development and labile phase, which takes place during the transition period from the spore-swelling stage to the beginning of the germination stage of the microorganism by means of an aimed or selective shock of the enzyme system of the microorganism. The method comprises treating, at the particular time defined, the inoculum of the microorganism with potassium ferrocyanide or other equivalent ferri- or ferrocy-anide salt with a selected amount such as of about 0.5 to 3 g, particularly 1.5 to 2 g of potassium ferrocyanide per liter of fermentation medium. This treat-ment apparently causes a selective shock of the microorganism's enzyme system.

It has been found that by timing the addition of the cyanide ion, like potassium ferrocyanide, selectively so as to take place during this intensive and labile growth stage, it has been possible to cause a remarkably high development of acid-stable pellets which have a remarkable propensity for producing citric acid.

In the process, there may be used any of the citric-acid-producing microorganisms such as the genera *Aspergillus, Penicillum* or *Mucor*. Examples of useful species of these genera are *A. niger, A. wentii, A. clavatus, P. citrinum, Mucor piriformis* and *Trichoderma viride* (ATCC No. 1323). The species which has been found most useful is *A. niger*. Among these such strains at ATCC 10577, ATCC 1015 or Wisconsin 72-4, also named NRCA-1-233 (National Research Council Publica-tion No. 2359), and mutants thereof are quite suitable. Other suitable microor-ganisms are disclosed in the scientific literature.

In practice, the spores of the inoculum of the citric-acid-producing microorganism reach the stage of intensive physiological development approximately 7 to 10 hours after inoculation. The desired morphological development stage of the spores is determined by periodic macro- and microscopic observations of the cul-ture on samples removed from the medium. Depending on the particular condi-tions selected, the stage of intensive physiological development of the spores may be controlled to be reached earlier or later.

When the spores show evidence of swelling and prior to germination, they are subjected to a ferrocyanide treatment, like with potassium ferrocyanide. The addition of the potassium ferrocyanide insures the formation of citric-acid-form-ing activated pellet of mycelium and essentially eliminates the adaptation period in the subsequent fermentation of the sugar. Moreover, the inoculum so treated develops a pellet from at least 90% of the spores present, generally from each labile inoculated spore. Accordingly, the prefermentation medium only needs to be inoculated with the number of labile spores that correspond to the number of pellets necessary for the main fermentation. In contrast, in accordance with known methods, the growth of pellet mycelium requires 500 to 600 spores to develop each pellet, so that the requirement for spores by far exceeds that of this process.

It is advantageous to carry out the preparation of the activated pellet in about a period of 20 to 28 hours. At that time, the individual activated pellets reach a diameter of about 0.15 to 0.2 mm, and the pH of the prefermentation medium drops by about 0.5 to 1 pH unit from an initial pH range of about 4.3 to 6.2, which is substantially similar to that of the fermentation medium. Thus, the pellet is adequately activated when a distinct pH drop is noted, thereby evidencing the initiation of the acid-producing stage and the termination of the spore-activating stage. Conveniently, this later stage can be considered terminated when the pH drops within the range of 4.0 to 4.5.

In the second phase of the process, the resulting activated pellets are used as inoculum for the main fermentation to citric acid, for instance, of a fermentation of a black-strap molasses having a suitable sugar concentration as, for instance, a 15% concentration. The size of inoculum can be varied as desired, for instance, from 2 to 20% or more, commonly 8 to 12% of inoculum per volume of fermentation medium may be used. The inoculum of pellets remains uninterruptedly in the acid-forming stage so that already after 24 hours a pH drop of about 1.5 to 2 pH units takes place.

The fermentation medium may contain any carbohydrate source which can be converted to citric acid by the microorganism, like a sugar from conventional crude sugar sources, as sugar beet, sugar cane molasses or citrus molasses or other carbohydrates and suitable nutrient salts such as, for instance, phosphates, nitrates, and so on in suitable amounts and conditions of growth, as is known from the prior art.

Example 1: In a suitable fermentation vessel, there are fed 320 liters of fermentation medium of black-strap molasses. The molasses is diluted with tap water to an approximate concentration of 15% of sugar. There is then added to the medium 80 g of ammonium dihydrogen monophosphate. The pH is adjusted to 5.0 and the medium then sterilized. Aeration provided by an air sparger feeds from the beginning a total of about 6 to 8 m^3/hr of air into the medium. The air is distributed by stirring with a stirrer rotating at the speed of 300 rpm.

The medium is inoculated at 32°C with 4×10^{10} spores of *Aspergillus niger.* Eight hours after the inoculation, there are added 480 g of potassium ferrocyanide. Within a total of 18 to 19 hours, there is formed from each spore a mycelium of a length of about 0.4 to 0.7 mm. Each one of the extremeties of the mycelium is club-like thickened and has a branched cauliflower-like appearance. This structure develops into pellets which have an approximate diameter of 0.2 mm in the next 4 hours and is accompanied by a sharp pH drop. When the pH value drops to about 4.3, enough pellets in medium are collected to serve as inoculum.

Example 2: A fermenter tank of a capacity of 3 m^3 is charged with a fermentation medium of the same composition and sugar content as was described in Example 1. Its pH is likewise adjusted to about 5.0 and then the medium is sterilized in a similar manner. After cooling the fermentation medium to a temperature range of about 28° to 32°C, there are added 3,000 g of potassium ferrocyanide.

Prior to inoculation, the medium is aerated by a sparger so as to feed 0.2 volume of air per volume of medium per minute in an amount equivalent to about 35 to 40 m^3/hr. Stirring by means of an appropriate stirrer rotating at a rate of 100 rpm distributes the air throughout the medium. An inoculant of pellets prepared as in Example 1 is used to inoculate the medium. After the inoculation, samples are removed to determine the pH which has been lowered as a result of the inoculation. After 8 further hours of uninterrupted fermentation by the activated pellet mycelium, the pH has already dropped by 0.1 to 0.2 unit. After about 18 hours of fermentation, there are added 750 g of ammonium nitrate and the potassium ferrocyanide concentration is determined. When it is found to fall to a concentration below 0.3 g/l, it is raised again to a concentration of at least 0.4 g/l by addition of potassium ferrocyanide. In this matter, the concentration of potassium ferrocyanide does not fall below 0.2 g/l throughout the entire fermentation. After 48 hours of fermentation, there are again added 750 g of ammonium nitrate.

After 3 days of fermentation, the pH value of the fermentation medium has dropped to a pH of 2.45 and the citric acid concentration is 10.5%. This corresponds to a yield of 70% based on the initial amount of sugar present. After a further 15 hours, the fermentation is terminated. The citric acid concentration is 12.3% corresponding to a yield of 82% based on the initial amount of sugar present.

Addition of Nitrogenous Heterocyclic Organic Compounds

J.-M. Charpentier, G. Glikmans, and P. Maldonado; U.S. Patent 3,966,553; June 29, 1976; assigned to Institut Francais du Petrole, des Carburants et Lubrifiants et Entreprise de Recherches et d'Activites Petrolieres Elf, France describe a method of producing citric acid by aerobic culture of yeast strains in a medium containing at least one n-paraffin and an aqueous nutrient phase, the method being characterized in that at least one nitrogenous organic heterocyclic compound is added to the medium.

Among these compounds are preferred those which contain at least one carboxylic group in α position of the nitrogen atom of the heterocycle, such as 2-pyridine carboxylic acid (picolinic acid), 2,6-pyridine dicarboxylic acid (2,6-dipicolinic acid), or quinaldic acid. These acids may be used in the form of their nontoxic salts, for example, their alkali metal salts. However, other heterocyclic nitrogenous compounds can be used with success, such as orthophenantroline, dipyridyl, 5-hydroxy quinoxaline, and 8-hydroxy quinoline.

The inoculum may be prepared by suspending yeast cells, preferably *Candida*, in aerobic conditions, in an aqueous medium containing an assimilable carbon source, generally an industrial n-paraffinic hydrocarbon cut of from C_{10-24}, and a source of assimilable nitrogen. The nitrogenous heterocyclic compounds may be added to the inoculum or may be introduced subsequently. The medium is stirred at a temperature of 25° to 30°C, for example, 36 hours.

After having obtained a sufficient proportion of cells of *Candida* or another yeast in the inoculum culture, at least a portion thereof is added to the fermentation medium which contains a n-paraffinic carbon source, a source of assimilable nitrogen, as well as different cations, anions and vitamins known as favoring the growth. The nitrogenous heterocyclic compound used in the process is added either at the beginning of the fermentation step or preferably when the growth of the yeast has been sufficient (end of the exponential phase) or by fractions at each of the stages.

As the nitrogen source in the fermentation medium, inorganic ammonium salts and preferably ammonium nitrate, ammonium sulfate and ammonium carbonate may be employed.

The following cations and anions are also favorable to the growth of the yeasts of the *Candida*-type: potassium, sodium, zinc, magnesium, manganese, iron, phosphate, carbonate. It is also well known that the addition to the yeast cultures of traces of oligo-elements and vitamins such as thiamine and biotine has a favorable action on the growth of the cells.

After the inoculation, the fermentation takes place in a stirred and strongly ventilated medium at a temperature usually from 25° to 35°C and preferably close to 30°C. Sterile compressed air is diffused through the fermentation medium at a rate of about 1 to 2 liters of air per minute and liter of the medium. During the first hours of fermentation, i.e., during the growth phase, the pH is preferably so adjusted that the multiplication rate of the yeast cells be optimum, i.e., preferably in the range of from 4.5 to 5.5. This pH adjustment may be carried out, for example, by addition of a basic aqueous solution such as a solution of ammonia, sodium hydroxide, potassium hydroxide, sodium or potassium carbonate.

After a sufficient growth of the yeasts (end of the growth exponential phase), heterocyclic nitrogenous organic compounds are added at a concentration usually from 0.05 to 10 g/l and preferably from 0.1 to 5 g/l, and the fermentation continues without the necessity of a pH adjustment by addition of a basic solution, since the pH progressively decreases down to a value of about 3 to 3.5 at the end of the fermentation stage.

When sufficient amounts of citric acid have been accumulated, this compound may be separated by conventional methods, for example, in the form of a calcium salt.

Example 1: *Comparative Example* — Cells of *Candida lipolytica* IFP 29 are inoculated on gelose, into a flask of a 100 ml capacity, containing 20 ml of a preculture liquid medium having the following composition:

KH_2PO_4	3.4 g/l
$Na_2HPO_4 \cdot 12H_2O$	1.5 g/l
$MgSO_4 \cdot 7H_2O$	0.7 g/l
$(NH_4)_2SO_4$	4 g/l
$CaCl_2$	0.1 g/l
$FeSO_4 \cdot 7H_2O$	2 mg/l
$CuSO_4 \cdot 5H_2O$	5 µg/l
H_3BO_3	10 µg/l
$MnSO_4 \cdot H_2O$	10 µg/l
$ZnSO_4 \cdot 7H_2O$	10 µg/l
$(NH_4)_6M_7O_{24} \cdot 4H_2O$	100 µg/l
$Co(NO_3)_2 \cdot 6H_2O$	10 µg/l
Yeast extract	100 mg/l
Tap water complemented to	1 liter
C_{12-19} n-paraffin cut	15 g/l

The n-paraffin cut has the following composition:

Hydrocarbon	% by Wt
C_{12}	0.07
C_{13}	2.05
C_{14}	15.71

(continued)

Hydrocarbon	% by Wt
C_{15}	32.00
C_{16}	30.83
C_{17}	17.90
C_{18}	1.38
C_{19}	0.06

The *Candida* cells are incubated at 30°C after the flask containing the inoculum has been secured to a stirring table driven at a speed of 140 rpm. After incubation for 36 hours, 10 ml of the inoculum are inoculated into a Fernbach flask of 1.5 liter capacity containing 200 ml of sterile nutrient medium A having the following composition:

KH_2PO_4	2 g/l
$MgSO_4 \cdot 7H_2O$	1 g/l
NH_4NO_3	2.5 g/l
$CaCO_3$	20 g/l
$FeSO_4 \cdot 7H_2O$	0.2 g/l
$MnSO_4 \cdot 7H_2O$	0.026 g/l
Yeast extract	100 mg/l
Tap water complemented to	1 liter
C_{12-19} n-paraffin cut	25 g/l

The Fernbach flask is secured onto a stirring table and the medium is stirred for 6 days at 30°C. The citric acid concentration is 10.1 g/l.

Example 2: The fermentation described in Example 1 is repeated after addition of 1.2 g/l of α-picolinic acid to medium A. The citric acid concentration is 26 g/l.

Production Using Cyanoacetic Acid

P. Maldonado, M. Charpentier, and G. Glikmans; U.S. Patent 3,996,106; December 7, 1976; assigned to Institut Francais du Petrole, des Carburants et Lubrifiants et Entreprise de Recherches et d'Activities Petrolieres Elf, France describe a process for the production of citric acid and/or isocitric acids, which comprises cultivating a citric and/or isocitric acid-producing strain of a yeast in a culture medium therefor, the medium comprising at least one source of carbon, at least one source of nitrogen, at least one inorganic compound, and cyanoacetic acid or an organic or inorganic derivative thereof in an amount sufficient to increase the production of citric and/or isocitric acids by the yeast. The amount of cyanoacetic acid or a derivative thereof added to the culture medium is preferably between an effective amount and about 0.1% by weight.

Although the cyanoacetic acid or a derivative thereof can be added to the nutrient medium, all at once or intermittently, it is usually preferred to add the particular compound at between 24 and 48 hours after the commencement of culture, corresponding to the end of the exponential growth phase of the yeast. The following are examples of derivatives of cyanoacetic acid which can be used: salts such as sodium and potassium cyanoacetates; organic esters such as methyl and ethyl cyanoacetates; and organic derivatives such as cyanoacetamide.

Yeasts which assimilate hydrocarbons are preferably chosen from the genus *Candida*. The culture medium can be synthetic or a natural nutrient medium. The medium will generally include at least one source of carbon, at least one source of nitrogen and at least one inorganic compound. Suitable aqueous culture media usually contain a hydrocarbon or a mixture of hydrocarbons as the principal carbon source. These hydrocarbons are preferably straight-chain alkanes containing from 12 to 20 carbon atoms (n-paraffins).

Suitable nitrogen sources include various organic or inorganic compounds, such as urea, ammonium salts, for example, ammonium chloride, ammonium sulfate, ammonium nitrate, ammonium phosphate, or ammonium acetate, one or more amino acids, and natural substances containing nitrogen, for example, maize maceration liquor, yeast extract, meat extract, fish meal, or peptone, etc.

Inorganic compounds that can be present in the culture medium include magnesium sulfate, potassium dihydrogen phosphate, phosphoric acid, potash, iron sulfate, manganese sulfate, manganese chloride, calcium carbonate, calcium chloride, sodium carbonate, and potassium carbonate.

Moreover, it may be necessary to add one or more nutrients to the culture medium depending upon the yeast used. Examples of such nutrients are amino acids such as aspartic acid, threonine, methionine, iso-leucine, valine glutamic acid, and lysine, etc., and vitamins such as thiamin, nicotinic acid, and biotin, etc. In certain cases, the addition of an emulsifying agent can increase the yield of citric and/or isocitric acids.

Culture of the yeast will generally be effected under aerobic conditions by stirring and aerating the culture medium, preferably at a temperature of from 20° to 40°C, and advantageously at from 25° to 35°C, conveniently at a pH of from 2 to 7, and preferably from 3 to 6. The pH of the culture medium can be adjusted by the addition of a nonnitrated basic aqueous solution, for example, of soda, potash, basic sodium carbonate or basic potassium carbonate.

The citric and/or isocitric acid-producing yeast mutants can be obtained by physical techniques such as the use of x-rays or ultraviolet rays, or by chemical techniques such as the use of nitrosomethylurethane or nitrosoguanidine, which are well known for their mutagenous action.

When citric and/or isocitric acids have accumulated in the culture medium, they can be isolated therefrom by classical methods, for example, as the insoluble calcium salts. This can be effected as follows: after elimination of the yeast cells, a stoichiometric amount of calcium chloride, for example, is dissolved in the culture medium, corresponding to the amounts of citric and isocitric acids present. The medium can then be neutralized with aqueous ammonia, and the calcium citrate precipitated by heating.

Production from Lysine-Enriched Yeasts

P. Maldonado, C. Gaillaridin, G. Sylvestre, and G. Glikmans; U.S. Patent 3,986,933; October 19, 1976; assigned to Institute Francais du Petrole et Entreprise de Recherches et d'Activities Petrolieres E.R.A.P., France describe a method of preparing a yeast enriched in L-lysine and which are capable of excreting an organic acid, comprising the steps of subjecting a yeast of the species *Candida lipolytica* to the action of a mutagenic agent, cultivating the thus-treated yeast in a culture medium containing a lysine analogue in a concentration which is about 50 to 100 times greater than that to which the cells of the original yeast are sensitive to obtain a mutated resistant yeast, and cultivating the resistant yeast on a medium containing at least one hydrocarbon.

The mutagenic agent used in the first step of this method can be either a chemical agent, such as, for example, nitrosomethylurethane (NMU) or nitrosomethylguanidine (NMG), or radiation, such as, for example, x-rays or ultraviolet rays.

During the second step of this method, cultivation can be carried out in the presence of any lysine analogue. As is known, strains of Candida yeast are sensitive to all lysine analogues, which are, therefore, called toxic analogues by specialists, examples of such toxic analogues including: aminocyclohexylalanine, transdehydrolysine (TDL), S-(β-aminoethyl)cysteine, 2,6-diaminoheptanoic acid, 5-methyl lysine, and α-amino ∈-hydroxycaproic acid. The sensitivity to lysine analogues is such that if a lysine analogue is present in the culture medium for a Candida yeast, the growth of the yeast is completely stopped. In the second step of this method, therefore, the yeast obtained in the first step is cultivated in a medium containing a lysine analogue at a relatively high concentration, followed by selection of the yeasts which can grow in the presence of a lysine analogue.

In the third step of this method, the resistant yeast selected during the second step is cultivated on a hydrocarbon-containing substrate, so as to obtain a biological material which has been enriched in L-lysine and excretes metabolic products such as citric acid, isocitric acid and/or glutamic acid.

The inoculum can be prepared by suspending Candida yeast cells made resistant to a toxic lysine analogue under aerobic conditions in an aqueous medium containing a source of assimilable carbon (usually an industrial fraction of n-paraffin C_{10-20} hydrocarbons) and a source of assimilable nitrogen and agitating the medium at a temperature of 25° to 30°C, for e.g., 36 hours.

After Candida cells have been obtained at sufficient density in the inoculum culture, a part thereof is added to the fermentation medium, which contains a source of n-paraffin carbon, a source of assimilable nitrogen and various cations, anions and vitamins which are known to favor growth.

After seeding, fermentation is brought about in an agitated, strongly aerated medium at a temperature which is usually from 25° to 35°C, and preferably 30°C, by diffusing sterile compressed air into the fermentation medium at a rate of 0.5 to 1.5 liters of air per minute and per liter of medium. The pH of the medium containing the growing cells is preferably adjusted so that the multiplication rate of the yeast cells is at the optimum value, i.e., from 3 to 6 and preferably from 4 to 5.

After the yeasts have grown sufficiently and adequate quantities of citric and isocitric acid have accumulated, fermentation is stopped and the medium is centrifuged. The resulting cake of lysine-enriched cells is washed and dried in a manner conventional for processing yeasts for nutritional use, and is then stored.

The concentrated citric acid in the supernatant fluid is isolated by conventional methods, e.g., in the form of its calcium salt. For example, the stoichiometric quantity of calcium chloride required for chelating the citric acid in the solution can be added to the supernatant liquid, followed by neutralization, e.g., by adding an aqueous ammoniacal solution, after which calcium citrate is precipitated by heating to boiling for 1 hour with agitation.

The citric acid can subsequently be recovered from its calcium salt by processing with sulfuric acid. For example, the salt can be washed in cold water and suspended in water cooled to 5° to 10°C, the suspension being agitated and kept at the same temperature, after which the stoichiometric quantity of sulfuric acid is added for completely releasing the citric acid, i.e., until the pH of the suspension has stabilized at 1.75 to 1.85.

The resulting calcium sulfate is recovered by filtration. The filtrate is conveyed through basic, then acid ion-exchange resin columns to eliminate the mineral impurities remaining in solution, and is then evaporated to dryness at reduced pressure and at a temperature not greater than 35°C, thus recovering the entire crystalline mass of citric acid formed during fermentation.

Production Using *Saccharomycopsis lipolytica*

T. Furukawa and H. Kaneyuki; U.S. Patent 3,902,965; September 2, 1975; assigned to Mitsui Petrochemical Industries, Ltd., Japan describe a method for producing citric acid by the steps of aerobically culturing a yeast belonging to the genus *Saccharomycopsis* and possessing an ability to produce citric acid by assimilating, as a main carbon source, at least one hydrocarbon, in a culture medium which uses the assimilable compound as the carbon source and contains nitrogen sources, inorganic salts and, if necessary, other suitable nutrient sources thereby accumulating citric acid in the culture medium and isolating the formed citric acid from the culture medium.

Microorganisms usable in the process include not only *Saccharomycopsis lipolytica* MT 1002 and mutants thereof but also all yeasts belonging to the genus *Saccharomycopsis*, so far as they possess an ability to produce citric acid from hydrocarbons.

Mutation can be effected with high energy irradiation, such as with ultraviolet ray, ^{60}Co ray or x-ray, or chemicals such as N-methyl-N'-nitro-N-nitrosoguanidine, sodium nitrite, hydroxylamine, nitrogen mustard, acenaphthene, etc.

Hydrocarbons desirably used as raw materials for this process are saturated aliphatic hydrocarbons and unsaturated aliphatic hydrocarbons, with those normal paraffins having from about 12 to 19 carbon atoms such as, for example, tetradecane, pentadecane and hexadecane being especially advantageous.

A typical composition of a culture medium usable in this process will be shown herein below by way of illustration. To distilled water, city water or well water, there are added 15 to 150 ml/l of normal paraffin, 1 to 10 g/l of a nitrogen source (such as ammonium chloride), 0.05 to 2 g/l of potassium dihydrogen phosphate, 0.05 to 2 g/l of potassium hydrogen phosphate or sodium hydrogen phosphate, 0.1 to 2 g/l of magnesium sulfate heptahydrate, 0.1 to 1.0 g/l of yeast extract or corn steep liquor, 0 to 2 mg/l of ferrous sulfate heptahydrate, 1 to 50 mg/l of zinc sulfate heptahydrate, 0.1 to 10 mg/l of manganese sulfate tetrahydrate, 2 to 50 mg/l of calcium chloride and 0 to 0.4 mg/l of copper sulfate.

A medium of a composition like the one shown above is adjusted to pH 6 to 7, sterilized at 120°C for 10 minutes and subsequently inoculated with a separately cultured citric acid-producing yeast belonging to the genus *Saccharomycopsis* such as *Saccharomycopsis lipolytica* MT 1002. The microorganism is subjected to culture under aeration either by means of agitation or shaking of the culture (with a reciprocating or rotary shaker) or other similar aerobic conditions at 25° to 37°C for 5 to 10 days, with the culture medium maintained at pH 3 to 7.

As the amount of citric acid accumulated in the culture medium increases to the extent of lowering the pH of the medium to a value not more than 3, the citric acid-producing ability of the yeast is degraded. At a proper time, such alkaline substance as calcium carbonate, calcium hydroxide, barium carbonate, potassium carbonate, potassium hydroxide, sodium carbonate, sodium hydroxide or aqua ammonia, particularly aqua ammonia, is added to the medium in a proper amount not so large as to turn the pH value of culture medium alkaline. The alkaline substance thus added serves to neutralize the greater part of citric acid and, at the same time, keep the culture medium at a neutral to weakly acidic pH value. The yield of citric acid can be improved by adding a proper surfactant (such as Tween, Noigen, Span and Plysurf).

Citric acid is accumulated in the form of a salt or partly in a free form in the culture medium. In isolating citric acid from the culture medium, any of the means available for the separation and purification of citric acid may be employed. In case where the greater part of citric acid occurs in the form of calcium salt or barium salt, for example, the medium is acidified with sulfuric acid so as to have the citrate dissolved and then subjected to centrifugation to effect separation therefrom of cells and other precipitates (such as calcium sulfate or barium sulfate).

Then, the resultant supernatant is neutralized with calcium carbonate, heated to educe precipitates and, while still hot, filtered to separate calcium citrate. To obtain citric acid from the calcium citrate, the calcium citrate is suspended in water of a volume 10 to 15 times as large and then sulfuric acid is gradually added to the suspension. In this case, sulfuric acid is added in an amount only slightly larger than is required to free calcium citrate and never in any large excess. The mixture is heated for some time and, while still hot, filtered and the resultant filtrate is concentrated. If calcium sulfate is educed in the early stage of the concentration, it is removed from the filtrate, after which renewed concentration is made. Then, the concentrate is cooled to educe citric acid crystals, which are separated by filtration and dried to produce citric acid.

Diploid Strain of *Candida lipolytica* and Thiamin

P. Maldonado and M. Charpentier; U.S. Patent 3,997,399; December 14, 1976; assigned to Institut Francais du Petrole, des Carburants et Lubrifiants et Entreprise de Recherches et d'Activities Petrolieres Elf, France describe a process for the production of citric and isocitric acids by a culture of a diploid *Candida lipolytica* yeast in a nutrient medium containing substances which are sources of carbon, nitrogen, and mineral compounds, the medium containing thiamin in a concentration in excess of 200 µg/l.

The yeast used in the process is a diploid *Candida lipolytica* yeast. A yeast of this kind can be obtained by means of the following operations: in a first stage, two haploid germ strains of *Candida lipolytica* of opposite signs are cultivated separately, this culture being effected in media rich in assimilable hydrocarbon substrate. In a second stage, the two cultures are brought together and cultivated in a medium poor in assimilable hydrocarbon substrate; this culture brings about the appearance of colonies of diploids. In a third stage, these colonies of diploids are treated with a mutagen in order to stabilize them.

The diploid *Candida lipolytica* yeast is cultivated as described hereinbelow. The culture medium has a hydrocarbon substrate. This hydrocarbon substrate is preferably composed of paraffinic hydrocarbons, that is, either an n-paraffin alone or a n-paraffinic cut. Although the use of an n-paraffin by itself makes it possible to obtain better results, it is generally preferred, essentially for reasons of cost, to work with an n-paraffinic cut usually containing from 5 to 22 carbon atoms per molecule.

Fermentation is preferably effected at a temperature between 24° and 34°C and with an acid pH between 2 and 7, preferably between 3 and 6. The culture is effected in the presence of air, and the culture medium is vigorously agitated by any suitable means in order to disperse the hydrocarbon substrate as much as possible. Before the actual fermentation process is effected, the selected strains of diploid *Candida lipolytica* yeast are subjected to preculture intended to bring about growth, thus enabling the fermenter to be seeded under good conditions of cellular concentration (for example, 10^7 cells/ml).

When the concentration of thiamin in the culture medium is low (between 0.01 and 10 μg/l), the culture does not excrete citric or isocitric acid, but only alpha-ketoglutaric acid. When the concentration of thiamin is between 10 and substantially 200 μg/l, no excretion of acids, either citric acids or alpha-ketoglutaric acid, is observed. When the concentration of thiamin exceeds 200 μg/l, the excretion of citric and isocitric acids is observed, but no production of alpha-ketoglutaric acid takes place.

Direct Production of Free Acid

R. Nubel, R. Fitts, and G. Findlay; U.S. Patent 4,155,811; May 22, 1979; assigned to Pfizer Inc. describe a process for producing citric acid by aerobically fermenting *Candida lipolytica* ATCC No. 20,228 in a nutrient medium containing at least one n-alkane or n-alkene hydrocarbon of from 9 to 19 carbon atoms, which is intimately mixed with an aqueous phase containing an assimilable source of nitrogen, minerals and other usual nutrients. The accumulated free citric acid is recovered directly by concentrating the filtered fermentation broth after removal of the metal ions of the buffer.

Example: A potato dextrose agar slant containing cells of *Candida lipolytica* ATCC 20,228 is transferred to a liquid medium prepared from 3 g of NZ Amine YTT, a commercial source of assimilable nitrogen comprising peptones from the degradation of casein, 34.7 g of C_{14-16} n-paraffins, and 600 ml of tap water. The medium is first sterilized for 30 minutes at 120°C. The *Candida* cells are then incubated aerobically in the medium with agitation at room temperature (27°C) for 48 hours, using a rotary shaker. At the end of that time, a 5% inoculum of the *Candida* growth is transferred to an aqueous sterilized nutrient medium containing, per liter of medium, 5.0 g of corn steep liquor, 4.0 g of ammonium sulfate,

15.0 g of calcium carbonate and 155 g of C_{14-16} n-paraffins. The inoculated medium is stirred for 48 hours and aerated at the rate of 4.0 scfhg (standard cubic feet of air per hour per gallon) at a temperature of 26°C. During this 48-hour propagation period, the pH is maintained at about 5 to 6 to permit optimum cell development, adding more calcium carbonate buffer in small increments as needed. This limited amount of buffer is consumed as the cell mass produces citric acid, so that by the end of the 48 hours the pH is already slightly below 4 and falling quite rapidly.

A 5% aliquot of this actively growing inoculum is then added to a large fermenter containing the following ingredients per liter of sterilized medium: 4.7 g of urea, 0.001 g of thiamine hydrochloride, 180 g of C_{14-16} n-paraffins and 0.375 g of KH_2PO_4 (sterilized separately). The fermentation medium is stirred for 144 hours (6 days) at 1,725 rpm, 4.0 scfhg, and a temperature of 26°C. The fermentation yield of citric acid monohydrate is 225 g/l.

Early in this 6-day run, the pH levels off of its own accord at about 2 to 3 and no adjustments are necessary. After the 6 days, the solids and mycelium are removed by filtration and the filtered fermentation broth concentrated under vacuum at 45°C to 40°Bé. The crystals of citric acid which are formed after standing overnight are removed by centrifugation. Second and third crops of citric acids are recovered by concentrating the mother liquors under vacuum to 45°Bé at 75°C. The total recovery of free citric acid is approximately 50% by weight. Residual product in the final mother liquor is lastly recovered as the monosodium salt by neutralizing with sodium hydroxide and concentrating under vacuum.

Use of Two Air Supply Systems

T. Messing and K.H. Wamser; U.S. Patent 4,052,261; October 4, 1977; assigned to Standard-Messo Duisburg Für Chemietechnik GmbH, Germany describe a method of producing citric acid whereby a reduction in the susceptibility to contamination and infection during the germination phase is achieved by a process carried out with an apparatus having a fermentation chamber and an air supply system, the process being characterized in that: highly sterile air is preheated through coarse-fine-absolute filters, UV gate, and overheating to 120°C, and fed into the fermentation chamber during the germination phase of a fermentation of carbohydrate-containing nutrient substrate in a quantity sufficient to maintain the chamber at a slightly positive pressure; the fermentation chamber is heated separately and independently of the air supply during the germination phase; the feeding of the highly sterile air and the heating of the fermentation chamber are discontinued during the fermentation phase; and fresh, normally filtered, unconditioned air is introduced in a quantity sufficient to maintain the fermentation solution at an optimum fermentation temperature.

Figure 1.1 schematically shows a fermentation chamber of a citric acid production system. A liquid carbohydrate-containing substance such as molasses is used as the starting material for the citric acid fermentation. The molasses is provided with fermentation-promoting additives and sterilized after the adjustment of the most favorable pH value. This sterilized fermentation solution reaches the fermentation chamber 2, through sterile conduits 1, wherein flat vessels 3 are arranged over each other for accepting the fermentation solution. The fermentation chamber 2 and the vessels 3 have been sterilized prior to filling.

Figure 1.1: Fermentation Chamber of a Citric Acid Production System

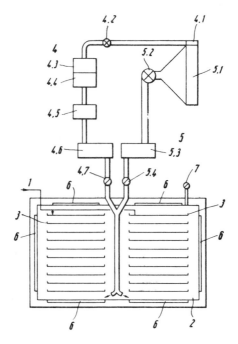

Source: U.S. Patent 4,052,261

Two air supply means or air systems **4** and **5**, which are separate from each other, are connected to the fermentation chamber **2**. Air system **4** is provided for the germination phase, and air system **5** is operative during the fermentation phase. In addition, fermentation chamber **2** is equipped with a wall heating unit or element **6**. Through these measures, the strict separation of the germination and fermentation phases, which is essential for this process, is made possible. Air system **4** for the germination phase consists of coarse filter **4.1**, ventilator **4.2**, fine filter **4.3**, absolute filter **4.4**, UV gate **4.5**, air overheater **4.6** and valve **4.7**, as well as the associated conduits. The wall heating unit **6** consists of conventional heating elements which are provided in walls, ceiling and floor.

In addition, air system **5** for the fermentation phase is connected to the fermentation chamber **2** and consists of coarse filter **5.1**, ventilator **5.2**, absolute filter **5.3**, and valve **5.4**, as well as the associated conduits through which unconditioned, normally filtered air can be fed into the fermentation chamber **2**.

For extreme temperature conditions, an apparatus must be so constructed, for example, in winter to preheat air which is too cold, or in tropical countries, to cool air which is too warm and too humid. This is carried out through a partial flow, which is not illustrated in the figure, which is installed between the coarse filter **5.1** and the absolute filter **5.3**. The discharge of the air takes place through an air discharge valve **7**.

Through the described air systems **5** and **4**, as well as the separate wall heating unit **6**, the processes of air supply and chamber heating are separated during the germination phase. The fermentation chamber **2** is kept warm during the germination phase in that the walls, which cause the significant dissipation losses, are provided with heating elements, which compensate for the heat losses. In addition, an air supply with an extremely small quantity of air takes place during the germination phase, because the oxygen requirement during this phase is so low that the quantity of air which is in the chamber is basically sufficient. For reasons of maintaining sterile conditions, a slightly positive pressure is therefore produced with the sterile air which is fed in during the germination phase, thus avoiding the penetration of impure normal air through leakages of the fermentation chamber **2** during the germination phase.

Depending on the size of the fermentation chamber **2**, a quantity of air in the order of magnitude of 100 to 300 m^3/h is introduced into the fermentation chamber **2** during the germination phase. This quantity of air is sterilized in air system **4** for the germination phase. The supply takes place through the opening of the open/closed valve **4.7** in a simple manner.

In this manner, the conditions for the germination phase for the optimum course of the germination process are fulfilled. The fermentation chamber **2** is kept sufficiently warm through the wall heating **6**, and is kept with a slight positive pressure through the supply of the sterile air through the air system **4**.

With the beginning of the development of heat, when the fermentation phase starts, the fermentation solution which is in the vessels **3** is heated by the exothermic processes. At a certain temperature in the fermentation solution, which indicates the end of the germination phase, and the beginning of the fermentation phase, air system **4** and the wall heating **6** are turned off and air system **5**, which is configured as cooling air system, is turned on. For this purpose, valves **4.7** and **5.4** are switched over. Normally, filtered air reaches the fermentation chamber **2** through air system **5**. By known means, the temperature in the fermentation solution is automatically controlled during the fermentation phase, by means of the quantity of air which is blow in.

L-TARTARIC ACID

Production from *Acetobacter* and *Corynebacteria* Microorganisms

Heretofore, L(+)-tartaric acid has solely been manufactured by a process utilizing as the raw material the raw tartar which is by-produced in the production of wine. The amount of supply by this process has its limit. It is observed that the supply of this compound has become increasingly tighter on the worldwide scale as a result of the increase of demand. In the circumstances, the price thereof is continuing to increase. The tartaric acid, which is obtained by the technique of organic synthetic chemistry, is generally an optically inactive DL-tartaric acid. This DL-tartaric acid has a lower solubility than L(+)-tartaric acid, therefore, is disadvantageous for commercial uses. As a food additive, such as for refreshing beverages, for example, it is still susceptible to doubts in terms of safety, palatability, etc.

Y. Tsurumi and T. Fujioka; U.S. Patent 4,092,220; May 30, 1978; assigned to Mitsubishi Gas Chemical Co., Inc., Japan describe a process for the manufacture of L(+)-tartaric acid or salts thereof, which comprises causing a hydrolase to react upon cis-epoxysuccinic acid or salts thereof, thereby producing specifically L(+)-tartaric acid or salts thereof and collecting the products.

In the culture of the microorganism, the culturing temperature is generally in the range of 20° to 35°C, preferably in the range of 28° to 31°C and the pH value of the medium is generally in the range of 4.5 to 7.5, preferably in the range of 5.5 to 7.0. The culture is incubated aerobically. The growth of the microorganism reaches its stationary phase after 2 to 5 days. When the culture is carried out outside the tolerable range of culture conditions, the microorganism fails to obtain sufficient growth and, in the worst case, ends up in total destruction. When the pH value of the medium rises in consequence of gradual consumption of the tartaric acid or salts thereof, it is desirable to continue the culture while controlling the pH value by addition of an inorganic acid such as sulfuric acid or hydrochloric acid or an organic acid such as acetic acid.

This method has great commercial significance in that the cis-epoxysuccinic acid-hydrolase can be obtained by effectively utilizing DL-tartaric acid, for example, which occurs as a by-product in the chemical synthesis of cis-epoxysuccinic acid, the very compound constituting the substrate of cis-epoxysuccinic acid-hydrolase.

Desired induction of the enzyme can also be obtained by using *Acetobacter curtus* No. 4, *Acetobacter curtus* No. 10, *Acetobacter curtus* No. 21 or *Corynebacterium* S-13, etc. utilizing cis-epoxysuccinic acid or salts thereof as the substrate, similarly to tartaric acid or salts thereof.

As the raw material, not merely cis-epoxysuccinic acid but metallic salts or nonmetallic salts of the acid or a substance containing the acid or salts may be used. Examples of the metallic moieties in the metallic salt of cis-epoxysuccinic acid include sodium, potassium, calcium, magnesium, iron, aluminum, zinc, manganese and cobalt which have no inhibitive action to the conventional enzymes. Examples of the base in the nonmetallic salts of the acid include ammonium and amines. Concrete examples of the salts of cis-epoxysuccinic acid include disodium cis-epoxysuccinate, calcium cis-epoxysuccinate, sodium calcium cis-epoxysuccinate, potassium sodium cis-epoxysuccinate, calcium hydrogen cis-epoxysuccinate, and ammonium cis-epoxysuccinate.

A sample of cis-epoxysuccinic acid-hydrolase is added to the aqueous solution of the raw material or to the suspension of the raw material, with the resultant mixture adjusted to pH 4 to 10, preferably pH 7.5 to 8.5. The mixture is greatly agitated at temperatures below 65°C, desirably in the range of from 20° to 45°C, and preferably around 38°C. There is consequently obtained an enzymatic reaction mixture.

The development of the reaction can be followed by sampling the reaction mixture at required intervals and determining the cis-epoxysuccinic acid concentration. For the determination of the cis-epoxysuccinic acid concentration, there may be adopted a method which resorts to the reaction between hydrochloric acid and the epoxy group, for example.

The cis-epoxysuccinic acid-hydrolase sample may be suitably selected from among a culture broth, live microbic cell, dry microbic cell, gel-entrapped microorganism, cell free extract, crude enzyme solution, purified enzyme, immobilized enzyme, etc.

Use of a gel-entrapped microorganism is recommended. The method, which proves advantageous for the entrapment of the aforementioned microorganisms, comprises causing an acrylic acid amide-type monomer to undergo polymerization in a liquid containing therein a microorganism capable of exhibiting an enzymatic activity of producibility of L(+)-tartaric acid or salts thereof.

The enzymatic reaction involving the use of the gel-entrapped micoorganism can be carried out by a batchwise method, a continuous method or a semicontinuous method, whichever may prove advantageous to the occasion.

In the batchwise method, an enzymatic reaction product mixture containing L(+)-tartaric acid or salts thereof can be obtained by mixing the aqueous solution of cis-epoxysuccinic acid or salts thereof with the gel-entrapped microorganism, then shaking or agitating the mixture and subsequently separating the gel-entrapped microorganism from the resultant mixture by an ordinary solid-liquid separating technique such as centrifugation or filtration. The gel-entrapped microorganism thus separated and recovered may be reused in the next cycle of enzymatic reaction.

As the continuous method, the so-called fixed-bed method is available which comprises allowing the aqueous solution of cis-epoxysuccinic acid or salts thereof to flow through a column packed with the gel-entrapped microorganism.

As the semicontinuous method, there can be cited the so-called suspended-bed method, which comprises continuously feeding the gel-entrapped microorganism and the aqueous solution of cis-epoxysuccinic acid or salts thereof at the same time to a reaction tank, agitating the charge, withdrawing from the reaction tank the reaction mixture of an amount corresponding to the combined amount of the charge and separating and recovering the gel-entrapped microorganism from the withdrawn reaction mixture as by the technique of decantation.

For the collection of the produced L(+)-tartaric acid or salts thereof, there is employed a procedure which comprises, for example, adding the aqueous solution of calcium chloride, for example, to the product mixture of the enzymatic reaction or, as occasion demands, to the solution obtained by removing the enzyme sample used from the product mixture of enzymatic reaction, thereby inducing the precipitation of sparingly soluble or totally insoluble salts of L(+)-tartaric acid such as calcium L(+)-tartrate separating the salts by filtration, again suspending the salts in water and, while under agitation, adding dilute sulfuric acid, etc. thereto, thereby adjusting the pH value of the suspension to 1.8, removing from the suspension the precipitated sparingly soluble or totally insoluble salts such as calcium sulfate and finally removing the water content from the remaining liquid such as by, for example, reduced-pressure concentration to afford crude crystals of L(+)-tartaric acid.

Required refining of the crude crystals is accomplished by dissolving the crude crystals in water, passing the aqueous solution through a column packed with a strongly acidic cation exchange resin such as, for example, Amberlite IR 120 (H$^+$-type cation exchange resin), thereby causing adsorption of the acid on the resin, eluting the adsorbed acid with water, again passing the eluate through a column packed with a strongly basic anion exchange resin such as, for example, Amberlite IRA400 (formic acid-type anion exchange resin) and eluting the adsorbed acid with 9 N formic acid. Purified L(+)-tartaric acid is obtained by concentrating the resultant eluate.

Example: In a 100-ml portion of a culture medium containing 0.5% of dipotassium phosphate, 0.2% monopotassium phosphate, 0.3% of ammonium sulfate, 0.05% of heptahydrated magnesium sulfate, 0.001% of heptahydrated ferrous sulfate, 0.01% of yeast extract and 1.0% of disodium cis-epoxysuccinate and adjusted to pH 6.5, the strain of *Acetobacter curtus* No. 4 (Ferm-P No. 2879) was inoculated and cultured at 30°C for 24 to 48 hours. The disodium cis-epoxysuccinate present in the culture broth was wholly consumed to produce 0.45 g of sodium L(+)-tartrate (yield 40% based on disodium cis-epoxysuccinate). The reaction mixture was centrifuged to separate the microbic cells and to obtain a supernatant. By adding to this supernatant about 25 ml of an aqueous calcium chloride having a concentration of 0.1 mol/l, there was obtained a precipitate of calcium L(+)-tartrate. The precipitate was separated by filtration and again suspended in about 50 ml of water. While the suspension was kept under agitation, 0.1 N sulfuric acid was added thereto until the pH of the suspension became 1.8.

The produced precipitate of calcium sulfate was separated by filtration. The filtrate was concentrated to dryness under reduced pressure to afford about 0.28 g of L(+)-tartaric acid in the form of crystals. The L(+)-tartaric acid thus obtained was found to have a specific rotatory power of $[\alpha]_D^{30} = +15.5°$ (10% in concentration). Comparison of the results with those of the extra pure reagent grade L(+)-tartaric acid available on the market revealed the product to have an optical purity of 100%.

Production from *Achromobacter* and *Alcaligenes* Microorganisms

E. Sato and A. Yanai; U.S. Patent 3,957,579; May 18, 1976; assigned to Toray Industries, Inc., Japan describe a method for producing d-tartaric acid by microbiological conversion of cis-epoxysuccinic acid to d-tartaric acid. According to this process, the hydrolyzing activity of the microorganisms belonging to the genera *Achromobacter* and *Alcaligenes,* and the enzymes produced by the microorganisms, brings about the conversion of cis-epoxysuccinic acid to d-tartaric acid via a selective hydrolytic step, whereby the desired product is readily obtained in substantially pure form and in quantity.

Cis-epoxysuccinic acid for use in this process can be easily and inexpensively prepared from maleic anhydride. The reaction is represented as the following equation in which the abovementioned previous steps starting with maleic anhydride are included.

(I)   (II)   (III)   (IV)

In the above equation (1) is a reaction easily conducted by dissolving maleic anhydride in water, (2) is a reaction producing cis-epoxysuccinic acid from maleic
acid, for example, by reacting maleic acid with hydrogen peroxide in the presence of an epoxidizing catalyst such as sodium tungstate, and (3) is a reaction
achieved according to this process.

Examples of microorganisms employed in this process include the following bacterial strains: *Achromobacter tartarogenes* nov. sp. Toray 1246, Ferm-P 2507,
Achromobacter epoxylyticus nov. sp. Toray 1270, Ferm-P 2508, *Achromobacter
acinus* nov. sp. Toray 1366, Ferm-P 2509, *Achromobacter sericatus* nov. sp.
Toray 1190, Ferm-P 2510, *Alcaligenes epoxylyticus* nov. sp. Toray 1128, Ferm-
P 2511, *Alcaligenes margaritae* nov. sp. Toray 1110, Ferm-P 2512. (Ferm-P is
the official deposit number of each organism in the Fermentation Research Institute in Japan.)

Regarding the composition of the culture medium when cultivating such microorganisms, either a synthetic or a natural culture medium is suitable so long as
it contains the essential nutrients for the growth of the microorganisms employed, and a small amount of cis-epoxysuccinic acid, its salt, and d-tartaric acid
or its salt is used as the inducer for the enzyme. Such an inducer can be added
either before or during the cultivation.

The cultivation is preferably carried out aerobically at 20° to 35°C, more preferably 26° to 30°C. The pH of the culture medium is preferably in the range of
6 to 8. The enzyme, d-tartrate epoxidase, accumulates in the cells as the cultivation proceeds.

In connection with the reaction step of this process, preferred embodiments are
as follows. The reaction mixture is composed of a proper amount of cis-epoxysuccinic acid-hydrolyzing agent and an aqueous solution of cis-epoxysuccinic acid.
Cis-epoxysuccinic acid is added to the reaction mixture as the free acid or as its
salt. Essentially any kind of salt can be used, but from the economical point of
view, alkali metal or alkaline earth metal salts such as sodium, potassium, calcium
or ammonium salts are preferred. Essentially, there is no restriction on the range
of concentration of the cis-epoxysuccinic acid. Usually, 10 to 25% (w/v) of cis-
epoxysuccinic acid (or its salt) is preferably added in the reaction mixture at the
beginning of the reaction.

The pH of the reaction mixture is preferably maintained between 6.5 to 9.0,
more preferably 7.0 to 8.0. For the adjustment of the pH of the reaction mixture, alkaline metal hydroxide and the like can be employed. The reaction is
conducted at any temperature conducive to satisfactory stability of the enzyme,
for example, about 20° to 50°C. The temperature may be 50°C in the case of 2
to 3 hours of reaction, or 40°C in the case of the reaction longer than 20 hours
at which cis-epoxysuccinic acid is substantially stoichiometrically converted to
d-tartaric acid.

A variety of well known procedures can be employed in the recovery of d-tartaric acid from the reaction mixture. For example, the reaction mixture or the
filtrate resulting from its filtration is passed through a column filled with an
anion exchange resin to absorb d-tartaric acid. After washing the column, absorbed d-tartaric acid is eluted by a suitable acid such as formic acid. d-Tartaric

acid can be obtained from the resulting eluate by a simple evaporation and crystallization procedure. Recovery of d-tartaric acid is also effected by the addition of a calcium salt to the reaction mixture. Calcium sulfate, calcium chloride and calcium carbonate are preferred. After adding one of these calcium salts to the reaction mixture, d-tartaric acid is recovered from the reaction mixture as calcium d-tartrate by simple filtration (usually with $4H_2O$). The salt is preferably converted to the free acid by conventional methods.

Example: *Achromobacter tartarogenes* nov. sp. Toray 1246 (Ferm-P 2507) was inoculated into a medium of 0.5% disodium cis-epoxysuccinate, 0.2% ammonium sulfate, 0.14% potassium dihydrogen phosphate, 0.31% disodium hydrogen phosphate (12 hydrate), 0.05% magnesium sulfate (7 hydrate) and 0.05% yeast extract (Difco). The pH of the medium was adjusted to 7.2 before sterilization at 120°C for 15 minutes. Cells were grown aerobically at 30°C for 36 hours and were harvested by centrifugation when a stationary phase was reached. After being washed with distilled water, cells were frozen at –20°C and thawed at room temperature. About 108 mg (dry weight) of cells with 1,270 units of d-tartrate epoxidase activity were obtained from 120 ml of the culture broth.

To the aqueous suspension of the cells, 3.0 g of disodium cis-epoxysuccinate and 75 mM tris (hydroxy-methyl) amino methane-HCl buffer (pH 8.0) were added to make the total volume 20 ml. This mixture was allowed to stand at 37°C for 21.0 hours. After the reaction was completed, 18.2 ml of 1 M calcium chloride were added to the mixture. The resulting precipitates of calcium tartrate were collected on a glass filter after being kept overnight at 5°C. The precipitates were suspended in a small volume of distilled water and the suspension was mixed with about 40 ml of Amberlite IR 120 B (H^+-type) and stirred until it became clear at room temperature. The resin was filtered off and washed with water. Combined filtrates were evaporated to dryness below 60°C under decreased pressure.

About 50 ml of 99.5% ethanol was added to the resulting solid material and the insoluble material was removed by filtration. The filtrate was again dried to produce 2.33 g of d-tartaric acid crystal with a yield of 91%. The following characteristics were observed: rotation value, $[\alpha]_D^{20}$ (c = 20, H_2O); for authentic d-tartaric acid, +12.4; for reaction product, +12.4.

Production from *Nocardia* Microorganisms

Y. Miura, K. Yutani and Y. Izumi; U.S. Patent 4,010,072; March 1, 1977; and Y. Miura, K. Yutani, H. Takesue, and K. Fujii; U.S. Patent 4,017,362; April 12, 1977; both assigned to Tokuyama Soda, KK, Japan have found that *Nocardia* microorganisms have the ability to produce an enzyme that can enzymatically convert readily available cis-epoxysuccinic acid to L-tartaric acid with superior activity.

According to the process, a microorganism of the genus *Nocardia*, preferably *Nocardia tartaricans* ATCC 31190 and ATCC 31191, having the ability to produce cis-epoxysuccinate hydrolase, or an enzyme isolated from the microorganism is contacted with cis-epoxysuccinic acid or its derivatives in an aqueous medium. The cis-epoxysuccinate hydrolase is obtained by cultivating the above microorganism in a customary manner. The cultivation of the above microorganism is carried out usually in an aqueous medium. It can also be carried out on a solid surface. Natural culture media such as nutrient broth and synthetic culture media composed of a suitable carbon source such as glucose, sucrose, glyc-erol, ethanol, isopropanol or n-paraffin, a suitable nitrogen source such as ammonium salts, nitric acid salts and urea and other inorganic salts can be used for the cultivation. It is desirable to add cis-epoxysuccinic acid or its derivatives to the culture medium for inducing the required enzyme.

The pH of the culture medium is adjusted to 5.5 to 11, preferably about 6 to 10. The cultivation is carried out aerobically at about 10 to 45°C, preferably about 25° to 40°C. In batchwise operations, the cultivation is carried out usually for ½ to 10 days.

The reaction to produce the L-tartaric acid can be carried out by a batch method, a continuous method, or a semibatch method in which the substrate and other required materials are successively fed during the reaction. It is also possible to employ a flow reaction technique in which the microorganism or the enzyme is fixed to a suitable support and packed in a column. The reaction can be carried out at a temperature of about 20° to 60°C, preferably about 25° to 45°C, at a pH of about 5 to 10, preferably about 6 to 10, more preferably 7 to 9.

Example: Each 500 ml shaking flask was charged with 50 ml of a culture medium (pH 7.0) containing 1% of meat extract, 1% of peptone and 0.3% of sodium chloride, and the culture medium was sterilized at 121°C for 15 minutes. Sterilized sodium cis-epoxysuccinate was added in a concentration of 1% to the culture medium, and *Nocardia tartaricans* (Strain ESI-T: Ferm-P 2882: ATCC 31191) was inoculated, and cultivated shakingly at 30°C for 24 hours. The cell concentration was 3.5 g/l. The cells were separated centrifugally from the resulting culture broth, washed once with a physiological saline solution and further centrifuged to collect the resulting cells.

The collected cells, 25 mg as a dry weight, were added to 25 ml of a 1 mol/l aqueous solution of sodium cis-epoxysuccinate (pH 7.0), and the reaction was carried out at 30°C. The results are shown in the table below:

Time, hr	Tartaric Acid Formed, g/l	Tartaric Acid Yield, %
24	39.5	26.3
48	90.0	60.0
72	139.1	92.7
90	145.7	97.1

The maximum rate of tartaric acid formation was 2.04 g/l/hr which was 2.04 g/g of cell per hour. This corresponded to 47.6 units as the enzymatic activity of 1 ml of the original culture broth.

Production from *Pseudomonas, Agrobacterium* **or** *Rhizobium* **Microorganisms**

Y. Kamatani, H. Okazaki, K. Imai, N. Fujita, Y. Yamazaki, and K. Ogino; U.S. Patent 4,011,135; March 8, 1977; assigned to Takeda Chemical Industries, Ltd., Japan describe a method for producing L(+)-tartaric acid which comprises having the culture of microorganisms or the processed matter thereof, which is capable of hydrolyzing cis-epoxysuccinic acid belonging to genus *Pseudomonas, Agrobacterium,* or *Rhizobium,* thereby producing L(+)-tartaric acid, brought in contact with cis-epoxysuccinic acid, and recovering so-produced L(+)-tartaric acid. The identifying numbers are assigned to each of the microorganisms as listed below:

Microorganism	IFO No.	FERM Accession No.	ATCC No.
Pseudomonas (KB-86)	13645	2855	31106
Agrobacterium aureum (KB-91)	13647	2857	31108
Rhizobium validum (KB-97)	13648	2858	31109
Agrobacterium viscosum (KB-105)	13652	2862	31113
Rhizobium validum (KB-106)	13653	2863	31114

The method is to be carried out by bringing the cultures of above microbes or the treated material thereof into contact with cis-epoxysuccinic acid which is the raw material to be employed.

Good results can be obtained by making incubation under aerobic conditions, at around 20° to 40°C, and for 1 to 7 days, meanwhile maintaining the culture broth at around pH 5 to 9.

In the case where it is intended to produce L(+)-tartaric acid by the method of bringing the raw materials into contact with the culture of the microorganism, or with the treated substance thereof, ordinarily it is preferable to make the inoculation in an aqueous medium. In this instance, a more advantageous result may be obtained by making the concentration of cis-epoxysuccinic acid in the reactant liquid as high as possible, provided that the concentration be kept within the bounds not to impede the activity of the microbial culture or the processed matter thereof. If circumstances so require, it is feasible that the cis-epoxysuccinic acid be added in parts at intervals of a certain definite period of time.

Reaction may be initiated by the methods of standing, shaking or agitating. In the case where entrapped microbial cells or the insolubilized enzyme is utilized, the cis-epoxysuccinic acid solution is let flow through the column. As for the reaction temperature, in ordinary cases, 5° to 50°C is used.

L(+)-tartaric acid having been formed in the culture broth or in the reaction medium through the mechanisms as have been described above can be easily isolated by a proper combination of various methods, as based on the specific chemical characteristics of L(+)-tartaric acid. By way of examples, the precipitation as calcium salts or the like, or the impurities elimination method by means of ion exchange resin, activated charcoal and the like is used, each of which produces effective results.

Example: A strain of *Rhizobium validum* KB-97 is used to inoculate 500 ml of liquid culture medium (pH 7.0) contained in two sets of 2-liter-capacity Sakaguchi flasks, the medium being composed of glucose (0.5%), ammonium nitrate (0.5%), dipotassium phosphate (0.1%), plus magnesium sulfate (0.05%), and this liquid culture medium is subjected to incubation under reciprocating-shaking culture at 30°C for 24 hours, obtaining therefrom around 1 liter of culture broth. The broth thus obtained is transferred into a tank of 50 liter-capacity which contains 30 liters of liquid culture medium (pH 7.0) being composed of disodium cis-epoxysuccinate (0.6%), ammonium nitrate (0.5%), dipotassium phosphate (0.1%), and magnesium sulfate (0.05%); and the tank contents are subjected to aerated agitation culture at 30°C for 30 hours. The culture broth obtained from the above (15 liters) is transfused into a 200-liter-capacity tank which contains 100 liters of culture medium (pH 7.0) which is composed of ammonium nitrate (0.5%), dipotassium phosphate (0.1%), potassium chloride (0.05%), magnesium sulfate (0.05%), and ferrous sulfate (0.001%), and simultaneously 2.7 kg of the crystals of disodium cis-epoxysuccinate is added. The mixture is subjected to aerated agitation culturing at 30°C. This culturing is done continuously for 7 running days, during which 2.0 kg each of the crystals of disodium cis-epoxy-succinate is added on the third and fifth day, respectively. Around 95 liters of the culture broth obtained as above is filtered by means of filterpress and, while agitating the filtrate thoroughly well, 5.8 kg of the crystals of calcium chloride (dihydrate) is added by piecemeal. The result of the above is left standing for 1 night, following which filtration by means of filterpress is carried out, and thus the crystals of calcium L(+)-tartarate are recovered. Amount of yield is 5.7 kg (as anhydride).

Addition of Nonionic Surfactant

Y. Kamatani, H. Okazaki, K. Imai, N. Fujita, Y. Yamazaki, and K. Ogino; U.S. Patent 4,013,509; March 22, 1977; assigned to Takeda Chemical Industries, Ltd., Japan describe a method for producing L(+)-tartaric acid which comprises: (1) incorporating calcium cis-epoxysuccinate as the raw material in the culture medium; (2) incubating a microorganism which is capable of hydrolyzing cis-epoxysuccinate to L(+)-tartaric acid; and thereby converting the calcium cis-epoxysuccinate into calcium L(+)-tartrate.

A further improved feature of the method comprises incorporating a nonionic type surfactant, in combination with the calcium cis-epoxysuccinate, in the culture medium.

As for the microorganisms to be employed, any sort of microbe is employable as long as it is capable of hydrolyzing cis-epoxysuccinic acid and of forming L(+)-tartaric acid. For example, the below itemized ones may be employed, i.e., *Acinetobacter tartarogenes* KB-82 (IFO 13644, Ferm No. 2854, ATCC 31105); the same species KB-99 (IFO 13650, Ferm No. 2860, ATCC 31111); the same species KB-111 (IFO 13656, Ferm No. 2866, ATCC 31117); the same species KB-112 (IFO 13657, Ferm No. 2867, ATCC 31118); *Agrobacterium aureum* KB-91 (IFO 13647, Ferm No. 2857, ATCC 31108); *Agrobacterium viscosum* KB-105 (IFO 13652, Ferm No. 2862, ATCC 31113); *Rhizobium validum* KB-97 (IFO 13648, Ferm No. 2858, ATCC 31109); the same species KB-106 (IFO 13653, Ferm No. 2863, ATCC 31114); *Pseudomonas species* KB-86 (IFO 13645, Ferm No. 2855, ATCC 31106).

The incubating conditions involving culturing temperature, duration of culturing time, acidity-alkalinity of liquids prevalent in the culture medium, and the like factors are subject to variation according to the kind of microorganisms employed, or to the composition and elements of the culture medium. In many cases, good results can be obtained by making incubation under aerobic conditions at around 20° to 40°C for 1 to 7 days, meanwhile maintaining the culture medium at around pH 5 to 9.

As the starting raw material, calcium cis-epoxysuccinate may be in whichever state of normal salt, acid salt or the mixture of them but, in the case of employing the raw material which contains an acid salt, it is commonly a favored practice to neutralize beforehand by calcium chloride, calcium carbonate and the like.

The total amount of calcium cis-epoxysuccinate employable during the incubation of microorganisms may be not less than 5% (w/v) and it is possible to raise the amount up to as high as 50% (as free acid).

As for the nonionic surfactant to be employed, there are a wide range of applicable ones, which are effectively used, such as sorbitan fatty ester (e.g., sorbitan monooleate, sorbitan trioleate, and the like); polyoxyethylene sorbitan fatty ester (e.g., polyoxyethylene sorbitan monolaurate, polyoxyethylene sorbitan monooleate, and the like); etc.

In ordinary cases, the surface active agents are used within the concentration range of 0.05 to 5.0% (w/v) and more preferably at around 0.05 to 2.0%. Addition of the total amount of surface active agents may be effected all at one time before starting the incubation, or fractionally in the course of the incubation.

Calcium L(+)-tartarate having been formed in the culture broth or in the reaction medium can be readily recovered by means of filtration or centrifugation.

Example: A strain of *Rhizobium validum* KB-97 is used to inoculate 500 ml of a liquid culture medium (pH 7.0) contained in a 2-liter-capacity Sakaguchi flask, which is composed of corn steep liquor (2.0%), glucose (0.5%), and this liquid culture is incubated at 28°C for 24 hours under reciprocating-shaking culture. Thus obtained culture broth is transferred to a tank of 50 liter-capacity which contains 30 liters of the culture medium, the same in composition as the one described above and the liquid culture medium in the tank is subjected to incubation at 28°C for 24 hours under aerated agitation culturing. About 15 liters of the culture broth obtained therefrom is transferred into a tank of 200 liter-capacity which contains 100 liters of a liquid culture medium composed of corn steep liquor (0.5%), ammonium nitrate (0.1%), sodium dihydrogen phosphate (0.2%), magnesium sulfate (0.05%), ferrous sulfate (0.001%), polyoxyethylene lanolin derivative (0.1%) (pH 7.0) and simultaneously 40 kg of calcium cis-epoxysuccinate (as free acid) is added.

The whole is subjected to aerated agitation culturing at 30°C for 30 hours. The resultant cultured broth and the washing water of the tank, about 150 liters in total, are subjected to decantor type centrifuge (Sumitomo Heavy Industries, Ltd. TS-210F type) to separate calcium L(+)-tartarate as crystals. The crystals are suspended in about 100 liters of water. After sufficiently stirring, the crystals are separated by the decantor type centrifuge and again suspended in about 70 liters of water and stirred well. The suspension is subjected to a filterpress to give 54 kg of calcium L(+)-tartarate (purity 98%) as anhydride.

In an improvement on the previous process, *Y. Kamatani, H. Okazaki, K. Imai, N. Fujita, Y. Yamazaki, and K. Ogino; U.S. Patent 4,028,185; June 7, 1977; assigned to Takeda Chemical Ind., Ltd., Japan* describe a similar method for producing L-tartaric acid where the particle size of the calcium cis-epoxysuccinate does not exceed 100 μ.

POLY-(β-HYDROXYBUTYRIC ACID)

Selection of Bacterial Strains

R.M. Lafferty; U.S. Patent 4,138,291; February 6, 1979; assigned to Agroferm AG, Switzerland describes a method of obtaining selected bacterial strains which can substantially convert an assimilable carbon source selected from carbohydrates, methanol, ethanol, glycerin, carbon dioxide and spent lyes from caprolactam synthesis containing monocarboxylic and dicarboxylic acids into poly-(D-3-hydroxybutyric acid). This method comprises selecting from poly-(D-3-hydroxybutyric acid)-producing bacterial strains those which form colonies of milky-white appearance on an agar nutrient medium. These colonies have dome-shaped elevations above the agar surface or attain large dimensions. The selected bacterial strains are bred on constantly increasing concentrations of the assimilable carbon source and are then subjected at least once, either before, after or during selection and/or breeding, to the action of mutagenic agents. In this way, the desired bacterial strains are obtained. The bacterial strains thus obtained may then be used to produce poly-(D-3-hydroxybutyric acid) by culturing them on the assimilable carbon source.

Polyesters of D(−)-3-hydroxybutyric acid (abbreviation:poly-β-hydroxybutyric acid or PHB) have the formula:

$$HO-CH(CH_3)-CH_2-C(=O)-\left[O-CH(CH_3)-CH_2-C(=O) \right]_n-O-CH(CH_3)-CH_2-C(=O)-OH$$

in which n is approximately 6,000 to 8,000.

In addition to the selection procedure described above, it is desirable that the bacterial strains are further subjected to one or both of the following two selection procedures, which two procedures may be carried out after the abovedescribed selection procedure in either order: (a) selection of bacterial strains having the lowest specific weight under conditions which hinder or do not help the microbiological synthesis of poly-(D-3-hydroxybutyric acid), (b) selection of bacterial strains having the strongest coloring after brief flooding with a solution of Sudan black B.

Preferred bacterial strains include strains derived from *Alcaligenes eutrophus* ATCC 23440, e.g., the mutant CBS 388.76 (GD-5); from *Bacillus megatherium* ATCC 32, e.g., the mutants CBS 389.76 (GB-1003) and CBS 390.76 (GBM-13); from *Zoogloea ramigera* ATCC 19623, e.g., the mutant CBS 391.76 and from *Mycoplana rubra*, e.g., the mutant CBS 385.76.

Method of Extraction

The ability, widespread among microorganisms to synthesize poly-(D-β-hydroxybutyric acid) as an energy store has been known for a long time. The hitherto known methods of producing and recovering fermentation polyester (PHB) have proved to be impractical and uneconomic because extremely large amounts of solvents are consumed and complex and costly precipitation steps are necessary. In addition, the solvents can only be recycled in the known methods by using expensive, inefficient regeneration methods since the solvents either have too low boiling points to be separated by distillation or are miscible with water or the precipitants, forming azeotropic mixtures. For this reason, the PHB-containing fermentation mass has to be subjected to an expensive drying step before the extraction takes place. Furthermore, the hitherto used solvents and precipitants are damaging to health and their use involves a high explosion risk.

R.M. Lafferty and E. Heinzle; U.S. Patent 4,101,533; July 18, 1978; assigned to Agroferm AG, Switzerland describe a process for the extraction of racemic or optically active poly-(β-hydroxybutyric acid) from a fermentation mass which comprises contacting a fermentation mass containing poly-(β-hydroxybutyric acid) with a liquid cyclic carbonic acid ester of the formula:

$$
\begin{array}{c}
R_1 \\
| \\
R_2-C-O \\
| \qquad\qquad C=O \\
R_3-C-O \\
| \\
R_4
\end{array}
$$

wherein R_1, R_2, R_3 and R_4, which are the same or different, are hydrogen or alkyl with 1 to 6 carbon atoms, as a solvent for the poly-(β-hydroxybutyric acid).

Example 1: *Extraction of PHB from Native Cell Masses with Ethylene Carbonate* — 12 g of fermentation mass sediment containing 2.9 g PHB were taken up in 25 ml ethylene carbonate and heated with stirring up to about 118° to 130°C. After a 44-minute extraction time, the hot suspension was suction filtered to remove the cells. An aliquot portion A of the filtrate was then cooled to about 40°C and the precipitated PHB filtered off, washed with alcohol and dried in vacuo . 1.56 g of PHB were obtained, 5 to 10% by volume of water was added to an aliquot part B of the filtrate. PHB was precipitated immediately and 2.1 g of PHB, which was identical with an authentic sample, was isolated by filtering at 90° to 100°C in the usual way. The resulting filtrate was used three more times to extract PHB from native cell masses with virtually the same result.

Example 2: *Extraction of PHB from Native Cell Masses with 1,2-Propylene Carbonate* — 11.5 g of fermentation mass were taken up in 25 ml 1,2-propylene carbonate and heated with continuous stirring to 119° to 135°C and held for about 40 to 50 minutes at 119° to 135°C. After suction filtering the hot suspension, 1 to 5 ml of water were added with stirring to the filtrate. PHB precipitated immediately. The resulting suspension was filtered at room temperature and the filter cake washed well with ethanol and dried in vacuo. 95% of the PHB in the cell mass was extracted. The melting point of the product obtained lay between 174° and 178°C and was unaltered by two recrystallizations from chloroform-ether.

Analysis of the product—Calculated for $(C_4H_6O_2)_n$: C, 55.8%; H, 6.96%; N, 0.0%; ash, 0.0%. Found: C, 55.7%; H, 6.28%; N, 0.0%; ash, 0.0%.

The 1,2-propylene carbonate obtained as the last filtrate was subsequently used three more times without any kind of preparation to isolate pure PHB from native cells with the same success.

Related work is described in *R.M. Lafferty and E. Heinzle; U.S. Patent 4,140,741; February 20, 1979; assigned to Agroferm AG, Switzerland.*

L-MALIC ACID

Production from *Paracolobactrum aerogenoides*

L. Degen, N. Oddo, and R. Olivieri; U.S. Patent 3,980,520; September 14, 1976; assigned to Snam Progetti SpA, Italy describe a process for the production of L-malic acid which comprises inoculating a culture with the organism *Paracolobactrum aerogenoides,* the culture comprising a salt solution that contains a fumarate, a nitrogen source and a carbon source, and fermenting the culture under aerobic conditions at a temperature ranging from 20° to 40°C for a period of 18 to 36 hours at a pH of 5.5 to 8, and thereafter recovering the L-malic acid. These strains, isolated from agricultural ground samples, are marked by the number 743 (CBS 405.73) and have been deposited at Central Bureau Voor Schimmelcultures in Holland.

Example: A culture medium was prepared having the following composition:

	g/l
Fumaric acid	20
Glucose	20
Yeast extract	1
NaNO$_3$	5
K$_2$HPO$_4$	2
MgSO$_4$·7H$_2$O	0.5
Tween 80	1

The aforesaid products were dissolved in demineralized water brought to a pH of 7 by sodium hydroxide. It was sterilized for 20 minutes at 110°C. The medium was put into 500 ml flasks having a broad neck by adding 100 ml of medium to each flask. They were inoculated by a suspension of strain 743 grown for 24 hours at 30°C on slants of nourishing agar. The optical density of the bacterial suspension used as inoculant, diluted 1/10, was 0.200 at 550 nm.

The incubation was carried out under an orbital stirring at 3°C for 88 hours. 4.5 g/l fumaric acid remained in the fermentation medium and 17.25 g/l of L-malic acid were formed corresponding to a 75% yield.

Immobilization of Microorganism

I. Chibata, T. Tosa, T. Sato, and K. Yamamoto; U.S. Patent 3,922,195; November 25, 1975; assigned to Tanabe Seiyaku Co. Ltd., Japan describe a process whereby L-malic acid can be prepared by the steps of polymerizing at least one acryloyl monomer in an aqueous suspension containing a fumarase-producing microorganism, and subjecting the resultant immobilized fumarase-producing microorganism to enzymatic reaction with fumaric acid or a salt thereof.

Preferred examples of the fumarase-producing microorganisms, which are employed in the process, include *Brevibacterium ammoniagenes* IAM (Institution of Applied Microbiology, Japan) 1641 (ATCC 6871), *Brevibacterium ammoniagenes* IAM 1645 (ATCC 6872), *Corynebacterium equi* IAM 1038, *Escherichia coli* ATCC 11303, *Microbacterium flavum* IAM 1642, *Proteus vulgaris* IFO (Institute for Fermentation, Japan) 3045, *Pichia farinosa* IFO 0574. All of these microorganisms are publicly available from the abovementioned depositories. A suitable amount of the fumarase-producing microorganism is 0.1 to 5 g, especially 1 to 3 g, per gram of the acryloyl monomer or monomers used. The polymerization reaction serves to tightly entrap each of the microorganisms into the lattice of the polymer thereby affording high enzymatic activity for a long period of time.

The polymerization reaction can be carried out in the presence of a polymerization initiator and a polymerization accelerator. Potassium persulfate, ammonium persulfate, vitamin B_2 and methylene blue are suitable as polymerization initiators. On the other hand, β-(dimethylamino)-propionitrile and N,N,N',N'-tetramethyl-ethylenediamine are employed as polymerization accelerators. A suitable amount of the polymerization initiator to be added to the aqueous suspension of the fumarase-producing microorganism is 1 to 100 mg/g of the acryloyl monomer or monomers. A suitable amount of the polymerization accelerator to be added is 5 to 10 mg/g of the acryloyl monomer or monomers. It is preferred to carry out the reaction at 5° to 50°C, especially 10° to 30°C. The reaction may be completed within 5 to 60 minutes. The acryloyl monomers, which are suitable for use, include acryloylamide, N,N'-lower alkylene-bis-acryloylamide, bis(acryloylamidomethyl)ether and N,N'-di-acryloylethyleneurea(N,N'-diacryloyl-imidazolidine-2-one).

After the polymerization reaction is completed, the resultant immobilized fumarase-producing microorganism is granulated by passing it through a sieve to form granules of 0.5 to 30 mm, especially 1 to 5 mm in diameter. The fumarase activity of the immobilized preparation thus obtained can be enhanced by treating it with an aqueous solution containing a surfactant and fumaric acid or a salt thereof, and allowing the suspension to stand at 20° to 40°C. Any one of a cationic surfactant (e.g., cetyl pyridinium chloride, cetyl trimethyl ammonium chloride), an anionic surfactant (e.g., sodium dodecyl sulfate) and a nonionic surfactant (e.g., glyceryl monostearate) is employed as the surfactant. On the other hand, sodium, potassium, ammonium, calcium, magnesium and barium fumarates are employed as the salt of fumaric acid. A preferred concentration of the surfactant in the solution is about 0.005 to 0.5% (w/v). A preferred concentration of fumaric acid or a salt thereof is about 0.5 to 1.0 mol/l. Furthermore, it is preferred to treat the immobilized preparation with an organic solvent or a mixture of an organic solvent and water.

The treatment with the organic solvent can be carried out by soaking the immobilized preparation in an organic solvent or a mixture of an organic solvent and water, preferably under stirring. It is preferred to carry out the treatment at 20° to 40°C at a pH of 6 to 9. Preferred examples of the organic solvent include an alkanone having 3 to 6 carbon atoms (e.g., acetone, methyl ethyl ketone), an alkanol having 1 to 6 carbon atoms (e.g., methanol, ethanol, propanol, butanol), a dialkyl ether having 2 to 6 carbon atoms (e.g., diethyl ether), dioxane, toluene, ethyl acetate and chloroform.

L-malic acid can be prepared by enzymatic reaction of the immobilized fumarase-producing microorganism with fumaric acid or a salt thereof. Sodium, potassium, ammonium, calcium, magnesium and barium fumarates are suitable as the salt of fumaric acid. The enzymatic reaction can be carried out at 5° to 60°C, especially at 10° to 50°C. It is preferred to carry out the reaction at a pH of 5 to 10, especially at pH 7 to 7.5. The enzymatic reaction can be accelerated by conducting it in the presence of a surfactant. For this purpose, the same surfactant as used for activation of the granules of the immobilized fumarase-producing micro-organism can be employed. A preferred concentration of the surfactant in the reaction solution is about 0.005 to 0.5% (w/v). The concentration of substrate is not critical. For example, when sodium fumarate is employed as the substrate, it is dissolved in water at any concentration.

The abovementioned immobilized microorganism is suspended in the aqueous so-lution of sodium fumarate, and the suspension is stirred. After the reaction is completed, the mixture is filtered or centrifuged to remove the immobilized mi-croorganism for subsequent use. An aqueous solution containing sodium L-mal-ate is obtained as the filtrate or supernatant solution. L-malic acid can be re-covered from the filtrate or supernatant.

On the other hand, when a slightly water-soluble fumarate, for example, calcium fumarate is employed as the substrate, it is suspended in water. The immobilized microorganism is added to the aqueous suspension of calcium fumarate and the mixture is stirred. After the reaction is completed, calcium L-malate is obtained as the crystalline precipitate. The optimum condition for conversion of fumaric acid or a salt thereof to L-malic acid can be readily obtained by adjusting the re-action time.

Alternatively, the enzymatic reaction can be performed by a column method. The column method enables the reaction to be carried out in a successive man-ner. For example, the immobilized microorganism is charged into a column, and an aqueous solution of fumaric acid or a salt thereof is passed through the col-umn. An aqueous solution containing L-malic acid or a salt thereof is obtained as the effluent. Recovery of L-malic acid from the effluent can be carried out in a conventional manner.

MISCELLANEOUS ORGANIC ACIDS

Hydrolyzation of Nitriles

A. Commeyras, A. Arnaud, P. Galzy, and J.-C. Jallageas; U.S. Patent 3,940,316; February 24, 1976; assigned to Agence Nationale de Valorisation de la Recherche (ANVAR), France describe a technically simple process by which it is possible to hydrolyze almost all the nitriles under mild conditions.

In the process, an organic acid is produced from the corresponding nitrile by sub-jecting that nitrile, in aqueous solution, to the action of bacteria which show ni-trilase activity, and by subsequently separating the bacterial mass from the acid solution. The bacteria showing nitrilase activity used for the process are preferably selected from the species *Bacillus, Bacteridium* as defined by Prevot, *Micrococcus* and *Brevibacterium* as defined by Bergey.

More particularly, these bacteria are selected from the strains lodged in the Col-lection de la Chaire de Genetique de l'Ecole Nationale Superieure Agronomique de Montpellier (France), under the numbers R 332, R 340, R 341, A 111, B 222, A 112, A 13, A 141, A 142, B 211, B 212, B 221, C 211, R 21, R 22, R 311, R 312, R 331.

The nitrile solution is preferably adjusted to a slightly basic pH value, for example, with potash or ammonia, before being exposed to the action of the bacteria.

The strains with nitrilase activity, after culture in a nutrient medium, are suspended in an aqueous solution of the nitrile to be hydrolyzed for a period of a few hours. The pH value of the solution is kept slightly basic (for example, pH 8) or neutral. In some cases, it is necessary in order to complete hydrolysis slightly to acidify the pH value of the solution after about 1 hour.

On completion of hydrolysis, i.e., generally after a few hours, the bacteria are eliminated by any method known in the science of biology, for example, by centrifuging. The acid is extracted from the solution by known methods, for example, by extraction or precipitation.

Example: *Preparation of Racemic Lactic Acid* — The strain R 312 is cultured on a medium containing glucose as carbon source. After culture, the cells are centrifuged, washed with a physiological salt solution and then suspended in the reaction medium consisting of a 10% by weight solution of lactonitrile obtained by chemical synthesis. The pH value is adjusted to 8 with potash or ammonia. The bacterial cells, representing approximately 20 to 40 g of dry material per liter, completely hydrolyzed the nitrile over a period of 2 to 3 hours with stirring at a temperature of 25°C. They are then eliminated by centrifuging.

The supernatant liquid contains ammonium lactate which can be recovered in a quantitative yield by drying. This product may be used as such, because its applications are numerous, for example, as an antiscaling agent in washing solutions. The lactic acid can also be recovered in a quantitative yield by methods known per se. For example, acidification may be followed by continuous extraction with ethyl ether or with any other suitable organic solvent. The lactic acid thus recovered is suitable for use, for example, in the food industry and in the chemical or pharmaceutical industry.

Dicarboxylic Acids from *Torulopsis candida*

H. Kaneyuki and K. Ogata; U.S. Patent 3,912,586; October 14, 1975; assigned to Mitsui Petrochemical Industries, Ltd., Japan describe a process for producing the dicarboxylic acids, pimelic acid, adipic acid, and sebacic acid by use of yeasts belonging to the species of *Torulopsis candida*, cultured in media containing hydrocarbons, nitrogen sources and the like, or resting cells of the yeasts are brought into contact with hydrocarbons in proper solutions containing carbon sources mainly composed of the hydrocarbons but no nitrogen source, thereby accumulating dicarboxylic acids in the media or solutions.

According to a preferred embodiment of the process, a yeast belonging to the species of *Torulopsis candida*, e.g., *Torulopsis candida* NRRL No. Y-7506, is inoculated and cultured in a medium containing a hydrocarbon, a nitrogen source, and other necessary nutriment sources, whereby dicarboxylic acids, identical in number of carbon atoms with, or less in number of carbon atoms than the starting hydrocarbon, are obtained. n-Decane gives sebacic acid with smaller amounts of succinic acid and adipic acid, and n-paraffins having 9 to 16 carbon atoms give corresponding dicarboxylic acids identical in number of carbon atoms with, or less in number of carbon atoms than, the n-paraffins.

A preferable medium is composed of 10 to 100 g of n-paraffin, 1 to 40 g of a nitrogen compound (e.g., ammonium sulfate, corn steep liquor), 0.2 to 3 g of monopotassium hydrogen phosphate, 0.2 to 3 g of dipotassium hydrogen phosphate, 0.1 to 1 g of magnesium sulfate heptahydrate, each 0.1 to 20 mg of at least one member selected from iron sulfate heptahydrate, zinc sulfate heptahydrate, manganese sulfate tetrahydrate, calcium chloride, sodium chloride and copper sulfate, and 100 to 1,500 mg of yeast extract or corn steep liquor, which have been dissolved in 1 liter of distilled water, city water or well water.

A medium of such a composition as mentioned above is sterilized at 120°C for 5 to 15 minutes, and then a dicarboxylic acid-producing yeast belonging to the species of *Torulopsis candida*, e.g., *Torulopsis candida* NRRL No. Y-7506, is inoculated into the medium. The pH at the time of cultivation is preferably from 2 to 9, more preferably from 5 to 8. The cultivation is carried out under aerobic conditions according to shaking culture or aerobic stirring culture at a temperature of 20° to 35°C, preferably 20° to 30°C.

According to another preferred embodiment of the process, cells of a yeast belonging to the species *Torulopsis candida*, which has previously been cultured in a proper medium, are collected by centrifuge or the like means, the collected cells are dispersed in a suitable solvent, e.g., phosphate buffer or the like, which contains a hydrocarbon but no nitrogen source, and the resulting suspension is stirred or shaken to contact and react the cells with the hydrocarbon (this method will be referred to as resting cell method hereinafter), whereby a dicarboxylic acid having carbon atoms equal to or fewer than the carbon atoms of the hydrocarbon can be obtained.

A preferable medium for the resting cell method is prepared by dissolving 3 to 5 g of monoammonium hydrogen phosphate, 1 to 3 g of dipotassium hydrogen phosphate, 0.5 to 2 g of monopotassium hydrogen phosphate, 0.2 to 1 g of magnesium sulfate heptahydrate, 0.1 to 2 g of yeast extract, and as a carbon source, 20 to 60 g of at least one member selected from various saccharides, sugar alcohols, organic acids such as citric acid, alcohols such as ethanol, and hydrocarbons such as n-decane, in 1 liter of water (distilled water or city water), and adjusting the resulting solution to pH of 6.5 to 7.5. The effect of the thus prepared medium varies depending on the kind of carbon source, and the use of xylose, sorbitol or citric acid, for example, gives favorable results.

The reaction solution used for contacting and reacting resting cells with a hydrocarbon is prepared, for example, by adjusting to a pH of 5 to 9, a 0.5 M phosphate buffer solution or a solution formed by removing the nitrogen and carbon sources from the aforementioned medium. To this solution are added precultured, separated and washed cells and a suitable hydrocarbon, and then the solution is stirred or shaken under aerobic conditions at 20° to 35°C, preferably 20° to 30°C, whereby dicarboxylic acids are formed.

Dicarboxylic Acids from *Torulopsis bombicola*

D.O. Hitzman; U.S. Patent 3,975,234; August 17, 1976; assigned to Phillips Petroleum Company describes a process for the production of dicarboxylic acids which comprises:

(A) Growing the microorganisms of mutant strains of *Torulopsis*

        bombicola on a hydrocarbon-free growth medium containing
        a carbohydrate source;

(B) Separating at least a portion of the microorganisms from the
        growth medium;

(C) Introducing at least a portion of these microorganisms into a
        converting medium under nongrowth conditions;

(D) Contacting these microorganisms in this nongrowth converting
        medium with n-alkane or the corresponding alcohol such as to
        partially convert this n-alkane or alcohol into the correspond-
        ing dicarboxylic acid having the same number of carbon atoms
        as the n-alkane or the corresponding alcohol; and

(E) Recovering the dicarboxylic acid from the converting medium.

The parent culture of *Torulopsis bombicola,* η sp, PRL 319-67, is obtainable from
Prairie Regional Laboratory, National Research Council of Canada. The yeast
mutant is produced in accordance with the following preferred procedure.

A 0.5 ml sample of a 3-day culture of *Torulopsis bombicola* grown on a sucrose-
containing nutrient is added to 100 ml of glucose-containing nutrient and the
mixture is agitated for a period of about 5 hours. Thereafter, 5 ml of a 300 mg
per 100 ml solution of N-methyl-N-nitroso-N'-nitroguanidine is added to the culture. Agitation is continued for about ¼ hour and a 1 ml sample is removed
and plated on plates at a 10^5 dilution.

After 6 days of incubation at 25°C, the plates, averaging 26 colonies per plate,
are replicated on nonglucose-based media containing 0.2 wt % methyl laurate
and on the glucose-containing nutrient.

After 6 days of incubation at 25°C, colonies, which develop on the glucose-containing nutrient but not on the methyl laurate-containing media, are selected and
tested to determine if they produce long-chain dicarboxylic acids from n-paraffins.
The glucose-containing nutrient, employed in the abovedescribed procedure for
producing the mutant, was of the following composition per liter of aqueous solution: commercial yeast extract, 3 g/l; malt extract, 3 g/l; peptone, 5 g/l; and
glucose, 10 g/l.

The time for the conversion of C_{6-22}, preferably C_{16-22}, n-alkanes or alcohols into
the corresponding dicarboxylic acids according to this process is about 20 to 50
hours. The preferred reaction or converting time is about 30 hours. The temperature of the conversion is usually in the range of about 15° to 45°C. Preferably, this temperature is between about 20° and 30°C.

In accordance with one preferred embodiment, the microorganism is grown on
a growth medium comprising at least one carbohydrate and yeast extract or
other nitrogen source such as peptones, ammonia, urea or ammonium nitrate.
The preferred yeast extract constitutes a nitrogen source necessary for the growth
of the organism. Good results are achieved by including into the growth medium
a mineral solution, as well as small amounts of KH_2PO_4 and $MgSO_4$. The growth
medium preferably contains these ingredients per liter of water in the following
ranges: carbohydrate, 20 to 50 g/l; yeast extract, 2 to 10 g/l; KH_2PO_4, 0.5 to
2.0 g/l; $MgSO_4$, 1.0 to 5.0 g/l; and trace mineral solution, 0.5 to 5 ml/l.

Whereas a wide variety of carbohydrates can be used to grow the microorganism, it is preferred to grow them in a growth medium containing at least one carbohydrate selected from the group consisting of sucrose, glycerol, molasses, glucose, and starch.

The nongrowth converting medium can comprise various ingredients, the important limitation being, however, that the nongrowth converting medium does not contain a nitrogen source such as yeast extracts, peptone, or other nitrogen sources as ammonia or derivatives thereof such as urea or ammonium nitrate which can be utilized by the organism for cell reproduction, i.e., growth. In other words, the nongrowth converting medium lacks at least one essential growth ingredient.

The preferred converting medium comprises a carbohydrate, KH_2PO_4, $MgSO_4$ and a mineral solution. The ranges for the quantities of these ingredients are the same as given above for the corresponding ingredients in the growth medium.

While any carbohydrate can be used, the preferred carbohydrate to be included in the nongrowth converting medium is one selected from the group consisting of sucrose, glycerol, starch or mixtures of these. The nongrowth converting medium may also contain a small quantity of an alkali metal citrate, particularly sodium citrate, in order to reduce the production of the monoacids.

The microorganisms are separated from the growth medium by standard sampling techniques. The diacid is recovered by standard technologies well known in the art such as solvent extraction and fractional distillation. At the end of the process, the microorganisms, having produced the diacids, are no longer viable. These microorganisms are recovered and used as a protein source, e.g., in animal feed materials.

L-2-Amino-4-Methoxy-trans-3-Butenoic Acid

T.C. Demny and J.P. Scannell; U.S. Patent 3,859,171; January 7, 1975; assigned to Hoffman-La Roche Inc. describe a compound of the formula:

$$CH_3O-\underset{\underset{H}{|}}{C}=\underset{\underset{H}{|}}{C}-\underset{\underset{H}{|}}{\overset{\overset{NH_2}{|}}{C}}-COOH$$

and pharmaceutically acceptable cationic salts thereof. The compound of the formula is prepared by culturing a known microorganism *Pseudomonas aeruginosa* ATCC 7700 in a fermentation medium containing nutrient sources under submerged aerobic conditions, until activity against bacteria occurs and then isolating the so-obtained compound of the formula from the fermentation broth.

The conditions of fermentation are generally the same as conventional methods for producing a substance by fermentation. The fermentation medium contains the usual nutrient and mineral sources supplying carbon, nitrogen, and energy to the developing culture. Suitably, the fermentation is permitted to proceed for from about 3 to 10 days.

The compound of the formula can be isolated from the fermentation broth in which it is prepared by adsorption into an anion exchange column, followed by elution with any suitable eluent preferably trimethylammonium bicarbonate or any other equivalent volatile buffered medium. The isolation procedure can be illustrated diagrammatically as follows:

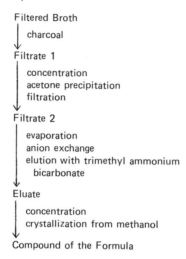

```
              Filtered Broth
                  │ charcoal
                  ▼
              Filtrate 1
                  │ concentration
                  │ acetone precipitation
                  │ filtration
                  ▼
              Filtrate 2
                  │ evaporation
                  │ anion exchange
                  │ elution with trimethyl ammonium
                  │   bicarbonate
                  ▼
              Eluate
                  │ concentration
                  │ crystallization from methanol
                  ▼
              Compound of the Formula
```

The compound of the formula forms cationic salts with pharmaceutically acceptable bases, such as sodium hydroxide, potassium hydroxide, calcium hydroxide, whereby respectively sodium, potassium and calcium salts are obtained. Such addition salts, prepared by admixture of the compound of the formula and base as the case may be, are equivalent to each other and to the compound of the formula in nonsalt form.

The compounds of the formula and their pharmaceutically acceptable cationic salts are useful as antibacterial agents and thus can combat gram-positive and gram-negative bacteria. They are also useful as antitrichomonal agents. They may be administered orally or parenterally, in conventional dosage forms with dosage adjusted to the individual needs of the one treated.

The compound of the formula and its salts can be used in the form of conventional pharmaceutical preparations. For example, the compounds can be mixed with conventional organic and inorganic inert pharmaceutical carriers for parenteral or enteral administration such as, for example, water, gelatin, starch, magnesium stearate, petroleum jelly or the like. They can be administered in conventional pharmaceutical forms, for example, solid forms such as capsules, tablets, suppositories or liquid forms such as solutions and emulsions.

L-2-Amino-4-(2-Aminoethoxy)-Butanoic Acid

J. Berger, D. Pruess, and J.P. Scannell; U.S. Patent 3,865,694; February 11, 1975; assigned to Hoffman-LaRoche Inc. describe the preparation of the L-antipode of the compound of the formula:

$$H_2NCH_2CH_2OCH_2CH_2CH-COOH$$
$$|$$
$$NH_2$$

by the fermentation of the known microorganism *Streptomyces sp.* X-11085.

A viable culture of the organism labelled with the laboratory designation *Streptomyces sp.* X-11085 has been deposited in the Northern Utilization Research and Development Division, Agriculture Research Service, U.S. Department of Agriculture in Peoria, Illinois, where this culture has been added to the NRRL collection under Registration No. NRRL 5331 and has been made available to the public.

Cultivation of the organism *Streptomyces sp.* X-11085 to produce the desired compound of the formula may be carried out utilizing a variety of fermentation techniques. In general, the following basic techniques can be employed in both flask and tank procedures.

In the flask fermentation, a loopful of spores from an agar slant of the culture is inoculated into 100 ml of nutrient medium in a 500 ml Erlenmeyer flask and incubated at about 28°C on a rotary shaker for up to 3 days. The inoculum nutrient medium contains a nitrogen source, preferably selected from an acid or enzyme hydrolyzed protein source such as enzyme hydrolyzed milk products, enzyme hydrolyzed bean meal products and the like, a carbohydrate source such as glucose, and inorganic salts such as phosphates, sodium chloride, and the like. Trypticase soy broth (pancreatic digest of casein, 15 g/l; enzymatic digest of soybean protein, 5 g/l; dipotassium phosphate, 2.5 g/l; dextrose, 2.5 g/l; and sodium chloride, 5 g/l) prepared by the Baltimore Biological Laboratories is the preferred inoculum medium.

After incubation in the inoculum medium for up to 3 days, small samples of the broth are transferred to the culture medium where they are incubated at about 28°C on a rotary shaker for from about 1 to 5 days. Whole broth samples are aseptically removed periodically for determination of the course of fermentation, usually every second day. For preparation of larger volumes of broth, inoculum is first prepared in 6-liter Erlenmeyer shake flasks or in 5-gallon pyrex bottles, fitted for aeration, sampling, etc. This broth is then transferred to the tank fermentors. Aeration in bottles and tanks is provided by forcing sterile air through the fermenting medium. In tanks, further agitation is provided by mechanical impellers. Antifoam agents such as lard oil, soybean oil, silicone surfactants, etc. are added as needed to control foam.

One of the preferred media for production of the compound of the formula in large fermenters contains: glucose, 10.0 g/l; enzyme hydrolyzed protein such as Bacto peptone prepared by Difco, 5.0 g/l; yeast extracts such as Bacto yeast extract prepared by Difco, 3.0 g/l; and $Fe(NH_4)_2(SO_4)_2 \cdot 6H_2O$, 0.031 g/l.

After fermentation is complete, a variety of procedures can be employed for the isolation and purification of the compound of the formula. Suitable isolation and purification procedures include ion exchange chromatography, partition chromatography and adsorption chromatography.

The compound of the formula above either alone or in combination with ascorbic acid enhances ethylene production in fruits and, therefore, is useful as an abscission agent.

Example: *Fermentation of Streptomyces Species X-11085* — A spore suspension of *Streptomyces species* X-11085 from a nutrient agar test tube slant was inoculated into a 500-ml Erlenmeyer flask containing 100 ml of Trypticase Soy Broth (Baltimore Biological Laboratories). The flask was incubated at 28°C for 72 hours on a rotary shaker (240 rpm with a 2" stroke). 2 ml of this inoculum was then added to 100 ml of fermentation medium containing: glucose, 20 g/l; Bacto peptone (Difco), 5.0 g/l; Bacto yeast extract, 3.0 g/l; and ferrous ammonium sulfate hexahydrate, 0.03 g/l.

The pH of the medium was about 7.0 before sterilization. The culture was incubated in a 500-ml shake flask at 28°C on a rotary shaker (240 rpm with a 2" stroke). After 4 days, the contents of the shake flask were filtered by centrifugation.

The undiluted broth (pH 2 to 2.2) (10 μl) was applied to the fluorometric amino acid analyzer [A. Felix and G. Terkelsen, *Arch. Biochem. Biophys.*, 157, 177 (1973)] and eluted with standard buffers.

Long-Chain α-Hydroxyalkanoic Acids

K. Hachikubo and S. Suzuki; U.S. Patent 4,059,488; November 22, 1977; assigned to Bio Research Center Co., Ltd., Japan have found that a microorganism of the *Candida lipolytica* species, which itself is known to assimilate n-paraffins as a carbon source in an aqueous nutrient medium under aerobic conditions, can be adapted to produce more selectively and in higher yields long-chain α-hydroxyalkanoic acids as the major metabolite of the assimilation of α-olefins having from 10 to 18 carbon atoms. The resulting α-hydroxyalkanoic acids, which are recovered from the culture broth have the same number of carbon atoms as the α-olefin precursor.

In the adaption procedure, the microorganism is first cultivated in an aqueous nutrient medium containing as the carbon source, one or more hydrocarbons boiling in the kerosene range to obtain a culture of the microorganism, then cultivating the resulting culture in a second separate aqueous nutrient medium containing an α-olefin having from 10 to 18 carbon atoms as the carbon source to obtain a second separate microorganism culture, then repeating a series of separate successive cultivation steps in aqueous nutrient media each containing the above α-olefin as the carbon source, starting with the above second separate microorganism culture as inoculum in the first of the successive cultivation steps and using as inoculum in succeeding cultivation steps the microorganism culture obtained in the next preceding cultivation step.

The above repetitive culturing is continued until the microorganism culture is capable of assimilating the α-olefin to produce as the major metabolite thereof the desired α-hydroxyalkanoic acid. The resulting culture is used as inoculum in a final cultivation in an aqueous nutrient medium containing an α-olefin having from 10 to 18 carbon atoms and there is recovered from the culture broth as the major metabolite of the α-olefin an α-hydroxyalkanoic acid having the same number of carbon atoms as the α-olefin. All the above cultivations are conducted under aerobic conditions.

The product acids or esters thereof have such uses as oil-soluble metal complexing agents, emulsifiers, plasticizers for vinyl polymers, lubricant additives, etc.

Example: There is first prepared a basal nutrient medium which is used in the adaptation cultivation and in the final cultivation to obtain the desired α-hydroxy-alkanoic acid. The basal medium is prepared by dissolving in 1 liter of distilled water the indicated amounts of the following compounds:

Component	Grams
$(NH_2)_2CO$	2.0
$(NH_4)_2SO_4$	2.2
$Na_2HPO_4 \cdot 12H_2O$	3.0
KH_2PO_4	1.5
$FeCl_3 \cdot 6H_2O$	0.01
$MgSO_4 \cdot 7H_2O$	0.5
NaCl	0.5
Yeast extract	0.2
Malt extract	0.2

Adaptation of Candida Lipolytica — Step 1: To 1 liter of the above basal medium in a shake flask there was added 100 ml of kerosene having a boiling range of 180° to 230°C, and the resulting mixture was sterilized. The sterilized mixture was then inoculated with two platinum loopfuls of a culture of *Candida lipolytica* ATCC 20496 from a malt extract agar slant. The mixture was then cultured on a reciprocal shaker at 110 oscillations per minute (65 mm amplitude) at 30°C for 4 days.

Step 2: To a separate 1-liter portion of the above basal medium in a shake flask there was added 100 ml of 1-decene. The mixture was sterilized, inoculated with 5 ml of the culture obtained in Step 1, and then cultured under the identical conditions set forth in Step 1.

Step 3: To another separate 1-liter portion of the above basal medium in a shake flask there was added 50 ml of 1-decene. The mixture was sterilized, inoculated with 5 ml of the culture obtained in Step 2, and then cultured under the identical conditions set forth in Step 1.

Step 4: To still another separate 1-liter portion of the above basal medium in a shake flask there was added 20 ml of 1-decene. The mixture was sterilized, inoculated with 5 ml of the culture obtained in Step 3, and then cultured under the identical conditions set forth in Step 1.

Step 5: To yet another separate 1-liter portion of the above basal medium there was added 20 ml of 1-decene. The mixture was sterilized, inoculated with 5 ml of the culture obtained in Step 4, and then cultured under the identical conditions set forth in Step 1. In order to obtain a pure culture isolate, one platinum loop of the resulting culture was transferred to a nutrient agar medium in a Petri dish, a sheet of filter paper soaked in 1-decene was placed on the inoculated agar which was then incubated at 30°C for 72 hours to form a colony of the microorganism.

Final Cultivation: To a separate 1-liter portion of the above basal medium there was added 10 ml of 1-decene. The mixture was sterilized, inoculated with one platinum loopful of microorganism cells from the pure culture colony obtained in Step 5 and cultured on a reciprocal shaker under conditions identical to those in Step 1, except that the temperature was 30°±3°C and cultivation was stopped after 60 hours. The culture was removed from the flask, the yeast cells were separated from the broth, and the remaining broth was extracted with ether. After evaporating the ether from the extract, there was obtained 6.2 g of crystalline α-hydroxydecanoic acid having a purity of 95%. The yield of α-hydroxydecanoic acid, based on the 1-decene, was approximately 62 mol %.

Itaconic Acid

T. Kobayashi, I. Nakamura, and M. Nakagawa; U.S. Patent 3,873,425; March 25, 1975 describe a process for the production of itaconic acid which comprises:

(a) Culturing an itaconic-acid-producing mold in a culture broth containing a sugar source, nitrogen source, and inorganic salts under aerobic conditions;

(b) Removing high molecular substances having the property to prevent electrodialysis of itaconic acid salts from this broth;

(c) Contacting the filtrate thus obtained with a cation-exchange resin of an alkali metal salt form which is subsequently regenerated through the process to the acid form;

(d) Neutralizing the thus produced cation-exchanged liquor;

(e) Feeding the neutralized liquor into the diluting compartment of an electrodialyzer which is composed of ion-exchange membranes, while an aqueous solution of alkali metal salt of a member of the group consisting of strong acid, itaconic acid, and mixtures thereof is fed into the concentrating compartment of the same, and then electrodialyzing the itaconic acid salts contained in the neutralized liquor;

(f) Incorporating the concentrate recovered in the electrodialysis with the retentate from step (b) which is refined prior to incorporation with the concentrate;

(g) Crystallizing and isolating itaconic acid or alkali metal salt of itaconic acid from the resulting incorporated liquor, while contacting the dialyzate recovered in the electrodialysis with about an equivalent quantity of cation-exchange resin of acid form regenerated in step (c); and

(h) Recycling the resulting cation-exchange liquor to form the raw material of the next fermentation medium.

Example: *Aspergillus terreus* K26 (ATCC No. 10020) capable of producing itaconic acid is cultured in a high test molasses medium in a routine manner, and the resultant fermented broth is filtered to give 2,727.8 g of filtrate containing 99.2 g in calculated quantity as free itaconic acid (g in calc. itaconic acid) and 36.5 g of remaining sugar. To the above filtrate of the fermented broth are added 258.2 g of high test molasses containing 176.9 g of sugar, and the resulting mixture is subjected to ultrafiltration for 6.4 hours under a pressure of 4 kg/cm^2 using a filter membrane (Diaflo UM-10), whereby 2,292.4 g of filtrate containing 74.9 g in calc. itaconic acid and 144.4 g of sugar as well as 693.6 g of retentate containing 24.3 g in calc. itaconic acid and 69.0 g of sugar are obtained.

2,230.9 g of the filtrate (containing 72.9 g in calc. itaconic acid and 140.5 g of sugar) are caused to flow through a column of cation-exchange resin of potassium salt form (Amberlite XE232) which has been produced through the later step of this example, at a space velocity below 1, thereby the resin is regenerated to acid form and 2,166.9 g of cation-exchange liquor containing 72.1 g in calc. itaconic acid and 138.6 g of sugar are obtained. 345.0 g of water are caused to flow through the resin column and 360.9 g of washing containing 0.8 g in calc. itaconic acid and 1.9 g of sugar flow out. 119.7 g of potassium bicarbonate are added to the cation-exchanged liquor so as to adjust the pH thereof to 7, thereby 2,237.8 g of neutralized liquor containing 114.1 g of potassium itaconate and 138.6 g of sugar are obtained.

The above neutralized liquor is placed into the diluting compartments of an electrodialyzer, which is so composed that eleven sheets of cation-exchange membrane (Selemion CMV) and ten sheets of anion-exchange membrane (Selemion AMV), each dialysis area of these membranes being constructed as to be 6 x 6 cm^2, are alternately arranged in parallel and at a distance of about 10 mm from each other, between two electrode compartments, and 2,150.2 g of aqueous solution containing 42.8 g of potassium chloride are placed into the concentrating compartments of the same, so as to electrodialyze potassium itaconate contained into the neutralized liquor.

The liquor placed in the diluting compartments and the solution placed into the concentrating compartments are both circulated in routine manner in the respective compartments. As the result, 2,005.0 g of the dialyzate containing 28.2 g of potassium itaconate and 129.2 g of sugar as well as 2,428.4 g of the concentrate containing 81.8 g of potassium itaconate and 42.1 g of potassium chloride are respectively recovered from the diluting and concentrating compartments. Starting electric current of 770 mA and tension of 25 V have become respectively 205 mA and 61 V at the end of the dialysis.

Thereafter, the diluting compartments are washed with 1,265.0 g of water, causing 1,546.2 g of washing liquor containing 2.1 g of potassium itaconic acid and 9.4 g of sugar to flow out. The concentrating compartments are washed with 1,202.0 g of water, causing 923.3 g of washing liquor containing 1.3 g of potassium itaconate and 0.7 g of potassium chloride to flow out.

To 2,005.0 g of the abovementioned dialyzate (containing 28.2 g of potassium itaconate and 129.2 g of sugar) are added 160 ml of cation-exchange resin of acid form (Amberlite XE232), which has been regenerated through the aforementioned step in this example and the quantity of which is about equivalent to that of potassium itaconate contained in the dialyzate, and the resulting mixture is agitated periodically for an hour, whereafter the resin is separated through filtration and further washed with 3,570 g of water. The filtrate is incorporated with washing and concentrated at the temperature of 50°C to give 2,005.0 g of concentrated liquor (containing 16.0 g of free itaconic acid, 2.86 g of potassium itaconate and 129.2 g of sugar).

To the above mother liquor are added 5.71 g of ammonium nitrate, 1.96 g of magnesium sulfate, 1.96 g of corn steep liquor and 0.14 g of concentrated nitric acid so as to prepare 2,014.8 g of culture medium with pH 2.5. This medium is placed 118 ml by 118 ml into seventeen 500-ml flasks and sterilized at 120°C for 7 minutes. After 4.00 g of sterilized aqueous solution containing 0.12 g of surface-active agent (Emulgen 910) is added to each flask, the medium in each flask is inoculated with 6.11 g of seed of *Aspergillus terreus* K26, which has previously been prepared through cultivation and contains about 0.08 g of itaconic acid and about 0.29 g of sugar. The inoculated mold is cultured at 36°C on a reciprocating shaker for 4 days. As the result, 1,942.0 g of fermented liquor containing 75.15 g of itaconic acid, 2.86 g of potassium itaconic, 24.33 g of sugar and 12.13 g of mycelium are obtained in totalizing the seventeen flasks. The yield of itaconic acid to the consumed sugar amounts to 55.6% and no reduction in the yield appears.

On the other hand, 693.6 g of the retentate come out through the ultrafiltration and are caused to flow through a column of anion-exchange resin of weak basic form (Amberlite IRA93) at a space velocity below 1, so as to cause itaconic acid to be absorbed on the resin and, after being washed with water, the resin is subjected to elution with concentrated solution of potassium bicarbonate to give 198.2 g of eluate containing 34.5 g of potassium itaconate.

The eluate is incorporated with 2,428.4 g of the concentrate having been recovered in the electrodialysis and containing 81.8 g of potassium itaconate together with 42.1 g of potassium chloride, and concentrated in vacuo at the temperature of 50°C to 437.3 g in amount. 118.5 g of hydrochloric acid having concentration of 11.35 N are added to the resulting concentrated liquor, and after being left to stand at the temperature of 10°C overnight so as to crystallize itaconic acid, the crystals are centrifuged. The yield amounts to 53.8 g which corresponds to 73.2% of the calculated quantity as free itaconic acid in the solution.

502.0 g of the resulting mother liquor containing 19.7 g of itaconic acid and 126.1 g of potassium chloride are subjected to electrolysis using diaphragm type electrolyzer, thereby potassium ions in the liquor are recovered as potassium hydroxide at the cathode and chlorine ions are recovered as chlorine gas at the anode. The potassium hydroxide is converted to potassium bicarbonate through contacting with carbonic acid gas, and the potassium bicarbonate is used for the abovementioned neutralization step of the cation-exchanged liquor. The chlorine gas is changed into hydrochloric acid by causing it to react on hydrogen having formed at the cathode, and the resulting hydrochloric acid is used for the removal of potassium from the concentrate recovered in the electrodialysis. 398.3 g of the liquor which has been electrolyzed and contains 19.7 g in calc. itaconic acid and 42.8 g of potassium chloride are, after addition of 1,751.9 g of water, recycled to the concentrating compartments of the electrodialyzer.

Urocanic Acid

I. Chibata, T. Tosa, T. Sato, and K. Yamamoto; U.S. Patent 3,898,127; August 5, 1975; assigned to Tanabe Seiyaku Co. Ltd., Japan describe the production of urocanic acid or a mixture of urocanic acid and D-histidine by enzymatic reaction of an immobilized L-histidine ammonialyase-producing microorganism with L- or DL-histidine.

According to the process, urocanic acid or a mixture of urocanic acid and D-histidine can be prepared by the steps of polymerizing at least one acrylic monomer in an aqueous suspension containing an L-histidine ammonialyase-producing microorganism, and subjecting the resultant immobilized L-histidine ammonialyase-producing microorganism to enzymatic reaction with L-histidine, DL-histidine or an acid addition salt thereof.

The polymerization reaction can be carried out in the presence of a polymerization initiator and a polymerization accelerator. Potassium persulfate, ammonium persulfate, vitamin B_2 and methylene blue are suitable as the polymerization initiator. On the other hand, β-(dimethylamino)propionitrile and N,N,N',N'-tetramethylethylenediamine are employed as polymerization accelerators. It is preferred to carry out the reaction at 5° to 80°C, especially at 10° to 50°C. The reaction may be completed within 10 to 60 minutes. In some cases, in order to carry out the subsequent enzymatic reaction advantageously, it may be preferred to heat the L-histidine ammonialyase-producing microorganism at an elevated temperature such as 60° to 80°C for 30 minutes prior to the immobilization reaction thereof, or to heat the immobilized L-histidine ammonialyase-producing microorganism at 60° to 80°C for about 30 minutes. The acrylic monomers which are suitable for use in this process include acrylamide, N,N'-lower alkylene-bis-acrylamide and bis(acrylamidomethyl) ether.

Preferred examples of L-histidine ammonialyase-producing microorganisms include *Achromobacter aquatilis* OUT (Faculty of Technology, Osaka University, Japan) 8003, *Achromobacter liquidum* IAM (Institute of Applied Microbiology, Tokyo University, Japan) 1667, *Agrobacterium radiobacter* IAM 1526, *Flavobacterium flavescens* IFO (Institute for Fermentation, Osaka, Japan) 3085, and *Sarcina lutea* IAM 1099. All of these microorganisms are publicly available from the abovementioned depositories.

Urocanic acid can be prepared by enzymatic reaction of the resultant immobilized microorganism with L-histidine or an organic or inorganic acid addition salt thereof. Alternatively, urocanic acid and D-histidine can be prepared by using DL-histidine or an organic or inorganic acid addition salt thereof instead of L-histidine. Suitable examples of the organic or inorganic acid addition salt of L- or DL-histidine include hydrochloride, sulfate, nitrate, acetate, etc. It is preferred to carry out the enzymatic reaction at 0° to 60°C, especially at about 37°C. The enzymatic reaction can be accelerated by carrying it out in the presence of a surfactant.

Preferred concentration of the surfactant in the reaction solution is about 0.005 to 0.5% (w/v). Moreover, the enzymatic activity of the immobilized microorganism can be stabilized effectively by adding a metal ion to the reaction solution. For this purpose, it is preferred to use 10^{-4} to 10^{-1} M of the metal ion such as magnesium, calcium, zinc and ferric ion.

The concentration of a substrate employed is not critical. For example, L- or DL-histidine is dissolved in water at any concentration. The aforementioned immobilized microorganism is suspended in the solution of L- or DL-histidine, and the suspension is stirred. After the reaction is completed, the mixture is filtered or centrifuged to recover the immobilized microorganism for subsequent use. Urocanic acid or a mixture of urocanic acid and D-histidine is recovered from the filtrate or supernatant solution.

Example: An aqueous nutrient medium (pH 7.0) containing the ingredients shown on the following page is prepared.

	Weight Per Volume (%)
Glucose	1
Dipotassium phosphate	0.2
Monopotassium phosphate	0.05
Ammonium chloride	0.1
Magnesium sulfate heptahydrate	0.02
Yeast extract	0.1
L-histidine hydrochloride	0.02

Achromobacter liquidum IAM 1667 is inoculated into 200 ml of the medium. The medium is cultivated at 30°C for 24 hours under shaking. Then, the medium is centrifuged. The microbial cells thus collected are suspended in 12 ml of a physiological saline solution, and the suspension is heated at 70°C for 30 minutes. 2.25 g of acrylamide, 0.12 g of N,N'-methylene-bis-acrylamide, 1.5 ml of 5% β-(dimethylamino)propionitrile and 1.5 ml of 2.5% potassium persulfate are added to the suspension. Then, the suspension is allowed to stand at 25°C for 10 minutes. The insoluble product is ground and washed with a physiological saline solution. 25 g of an immobilized preparation of *Achromobacter liquidum* IAM 1667 are obtained.

25 g of the immobilized preparation of *Achromobacter liquidum* IAM 1667 are charged into a 1.6 x 25.5 cm column, and 500 ml of an aqueous 0.25 M L-histidine hydrochloride solution (pH 9.0) are passed through the column at 37°C at the flow rate of 6 ml/hr. 500 ml of the effluent is adjusted to pH 4.7 with concentrated sulfuric acid. Then, the effluent is allowed to stand at 5°C overnight. The crystalline precipitate is collected by filtration, washed with ice water, and dried. 21.5 g of urocanic acid dihydrate is obtained (MP 225°C).

Pyruvic Acid

R. Uchio, K. Kikuchi, H. Enei, and Y. Hirose; U.S. Patent 3,993,543; November 23, 1976; assigned to Ajinomoto Co., Inc., Japan have found that remarkably improved yields of pyruvic acid may be obtained compared with the conventional methods by culturing pyruvic-acid-producing mutants of the genus *Candida* which require thiamine and methionine for growth.

The methionine and thiamine requiring mutants are derived from the parent strains, which may or may not require thiamine for growth, by known mutagenic processes such as exposure to mutagenic doses of ionizing radiation (ultraviolet light, x-rays) or chemical agents (nitrous acid or N-methyl-N'-nitro-N-nitrosoguanidine) and thereafter screening the parent strains to select those mutants having the described properties.

An example of an effective pyruvic-acid-producing mutant derived by this process is *Candida lipolytica* AJ 14353 (Ferm P-2628) and strains thereof. The microorganism, identified by Ferm P-number, is available from the Fermentation Research Institute, Agency of Industrial Science & Technology, Japan.

The culture media in which the mutants produce pyruvic acid are largely conventional. They will contain sources of assimilable carbon, nitrogen, inorganic salts, vitamins and methionine. Suitable carbon sources include saccharides such as glucose, fructose, sucrose, starch hydrolysates; organic acids such as acetic acid, propionic acid; and alcohols such as methanol or ethanol. Several organic or inorganic nitrogen-containing materials such as ammonium sulfate, ammonium nitrate, sodium nitrate, potassium nitrate, urea, peptone, meat extract may be advantageously employed as nitrogen sources.

Such inorganic salts as KH_2PO_4, K_2HPO_4, $MgSO_4$, $MnSO_4$, and $FeSO_4$ may be used. It is not necessary to employ pure methionine and thiamine. Sources of methionine and thiamine such as their biologically transformable analogs, or compositions known to contain the amino acids such as yeast extract, corn steep liquor may be utilized.

During cultivation, the pH of the culture media is preferably held at from 3 to 8 using appropriate alkaline agents such as sodium or potassium hydroxide. The fermentation is advantageously carried out at from 25° to 35°C.

The pyruvic acid accumulated in the fermentation broth is recovered by conventional methods. One useful procedure is to remove the cells by filtration or centrifugation, acidify the broth, extract with ether, concentrate under reduced pressure, take up the residue in water, and crystallize by the addition of a miscible liquid such as an alkanol.

Aldonic Acids

T. Miyake and Y. Sato; U.S. Patent 3,862,005; January 21, 1975; assigned to Hayashibara Company, Japan describe a commercial and economical method of production of aldonic acids which comprises cultivation of microorganisms that form mono-, di-, and trisaccharide dehydrogenase on mediums containing reducing sugars, and oxidation of highly concentrated solutions which contain only disaccharides or mixtures of mono-, di- and trisaccharides with utilization of the abovementioned enzyme-containing cells with agitation and aeration to obtain highly pure aldonic acids or their salts.

In this process, only the cells of the microorganism, which have been previously cultivated, are used. No nutrients or nitrogen sources necessary for growth of the cells are added. Much higher concentrations of sugar solutions may be converted than is possible in fermentation processes. Furthermore, it has been discovered that the reaction can be attained by mere aeration with air or oxygen and no addition of hydrogen acceptor is necessary.

The process can not only be performed with aqueous solutions of reducing disaccharides such as maltose, lactose, etc., but may also be performed in the reactive solutions from which these disaccharides form. Thus, great simplification can be obtained not only by the direct high yield production of aldobionic acid directly from starch, but also in production time cutting.

Another object of the process is the preparation of syrups which combine the sourness of organic acids and sweetness and viscosity of starch syrups by using various starch syrups as sweetener for foods and oxidizing low-molecular-weight oligosaccharides contained in those starch syrups, that is, oxidizing glucose, maltose, maltotriose, etc. to aldonic acids, thereby obtaining mixed syrups of various aldonic acids, oligosaccharides and dextrins.

It has also been discovered that trisaccharides may also be fermentatively oxidized to trionic acids. When the oxidizing effects of certain *Pseudomonas* bacteria were studied, it was found that, when cultured on starch syrups as media, satisfactory oxidation of disaccharides and trisaccharides to bionic and trionic acids occurred. Similarly using only cells of microorganisms productive of such a dehydrogenase without hydrogen acceptor and in conditions which will not sustain growth of such cells, enzymatic oxidation is possible. Such enzymatic oxidation can also be attained on a commercial scale by adding the enzymes during the process of producing the starch syrups.

Dehydrogenase-forming strains that can be used include gluconic-acid-producing strains, fungi such as *Penicillium* genus, *Aspergillus* genus, and bacterias such as *Pseudomonas* genus, *Acetobacter, Gluconobacter*, etc., while the presence of amylases that hydrolyze disaccharides and trisaccharides are unsuitable to produce aldonic acids as well as the presence of enzymes that produce 2-ketogluconic acid, 5-ketogluconic acid, etc. It has been found that *Pseudomonas* species, especially *Pseudomonas fragi* and *Pseudomonas graveolens* exhibited the best result. The latter is particularly suitable for the production of bionic acids. The latter two microorganisms may be found in many type culture collections and are catalogued in the Institute for Fermentation, Osaka, Japan, respectively, as IFO 3458 and IFO 3460.

Production of High Yield, High Purity Bionic Acids: Oxidative reaction is performed by adding the enzyme-containing cells either in an aqueous disaccharide solution or in a reactive solution productive of disaccharides, with aeration or charging oxygen with agitation. This process can be performed with relatively high concentration of sugars of 10 to 30%, The reaction is also practiced under normal temperature and atmospheric pressure, preferably at 20° to 50°C. Reaction at temperatures over 50°C involves the risk of enzyme inactivation. The most suitable initial pH in the case of *Pseudomonas* is 5.0 to 8.0.

As the pH of the reactive solution and reaction speed decrease naturally on formation of aldobionic acids, desirable results can be obtained by adding suitable counteractive or counteragents such as sodium carbonate or ammonia water. Accordingly, it is preferable to maintain pH around the neutral zone during reaction progress with additions of these counteragents.

Example: 1-liter medium containing 50 g of maltose, 10 g of corn steep liquor, 2 g of urea, 0.6 g of monobasic potassium phosphate, 0.25 g of magnesium sulfate (heptahydrate), 20 g of calcium carbonate was sterilized and inoculated with *Pseudomonas graveolens* IFO 3460. Cultivation was carried out at 30°C for 50 hours with mechanical agitation (400 rpm) and aeration 1 vvm. After completion of cultivation, cells were harvested from the culture broth by centrifugation and used as the enzyme source. Cells obtained from above culture broth had enzyme activity of 131 units/mg dried matter.

Maltose solution was oxidized using the above cells. Enzyme solution was adjusted to pH 6.50 and subjected to reaction under the conditions listed in the table below. The reactive solution with a sugar concentration of 10% was agitated with aeration at 30°C for 21 hours.

No.	Maltose Monohydrate (g)	Calcium Phosphate (g)	Amount of Cells (mg)	Total Enzyme Activity	Reducing Sugar After 21 Hours (%)
1	10	1.5	0	0	5.7
2	10	1.5	43	5,633	2.1
3	10	1.5	86	11,266	1.3
4	10	1.5	215	28,125	0.1

When the reactive solution stated in No. 4 was identified by paper chromatogram with aniline hydrogen phthalate, no presence of residual sugar was observed and that is evidence of complete oxidation. The reaction mixture formed into free acid with cationic ion exchange resin and identified by paper chromatogram showed formation of only maltobionic acid, and no other presence of acids besides gluconic acid or other acid was observed. Paper chromatogram of the hydrolyzed solution exhibited the presence of only glucose as sugar, gluconic acid was the only acid detected. These determinations are proof that the obtained aldobionic acids are of extremely high purities. Calcium was removed from the mixture obtained by filtration, decoloration and concentration with cationic ion exchange. It was found by titration method that 95% of the theoretical yield was converted into maltobionic acid.

AMINO ACIDS, PEPTIDES AND PROTEINS

ALANINE

L-Alanine

I. Chibata, T. Tosa, T. Sato and K. Yamamoto; U.S. Patent 3,898,128; August 5, 1975; assigned to Tanabe Seiyaku Co., Ltd., Japan describe a process for preparing L-alanine or L-alanine and D-aspartic acid by enzymatic reaction of an immobilized L-aspartic acid β-decarboxylase-producing microorganism with L-aspartic acid, DL-aspartic acid or a salt thereof.

According to the process, L-alanine can be prepared by polymerizing at least one monomer selected from the group consisting of acrylamide, N,N'-lower alkylene-bis(acrylamide) and bis(acrylamidomethyl) ether in an aqueous suspension of an L-aspartic acid β-decarboxylase-producing microorganism to produce an immobilized L-aspartic acid β-decarboxylase-producing microorganism, and subjecting the immobilized L-aspartic acid β-decarboxylase-producing microorganism to enzymatic reaction with L-aspartic acid, DL-aspartic acid or a salt thereof.

The polymerization reaction is preferably carried out in the presence of a polymerization initiator and a polymerization accelerator. Potassium persulfate, ammonium persulfate, vitamin B_2 and methylene blue are suitable as the polymerization initiator. On the other hand, β-(dimethylamino)propionitrile and N,N,N',N'-tetramethylethylenediamine are employed as the polymerization accelerator. It is preferred to carry out the reaction at 5 to 60°C, especially at 10 to 40°C. The reaction may be completed within 10 to 60 minutes.

Examples of L-aspartic acid β-decarboxylase-producing microorganisms include *Acetobacter rancens,* OUT (Faculty of Technology, Osaka, Japan) No. 8300; *Achromobacter pestifer,* IAM (Institute of Applied Microbiology, Tokyo University, Japan) No. 1446; *Achromobacter pestifer* ATCC No. 23584; *Alcaligenes faecalis* ATCC No. 25094; and *Pseudomonas dacunhae* IAM No. 1152. All of these microorganisms are publicly available from the abovementioned collections.

L-alanine or a mixture of L-alanine and D-aspartic acid can be prepared by contacting the resultant immobilized microorganism with L-aspartic acid, DL-aspartic acid or a salt thereof. Suitable examples of the salts of L- and DL-aspartic acid include the ammonium, potassium, sodium and magnesium salts.

Pyridoxal phosphate, a divalent metal ion, e.g., cobaltous or nickelous ion, and/or a surface-active agent, e.g., polyethylene sorbitan monolaurate, polyoxyethylene stearate, may be added to the enzymatic reaction to keep the enzymatic activity of the immobilized microorganism at a high level during the reaction. The preferred concentrations of pyridoxal phosphate, the divalent metal ion and the surface-active agent are respectively about 0.05 to 10 mmol/l, 0.1 to 10 mmol/l, and 0.05 to 1.0% w/v.

The concentration of substrate employed is not critical. That is, L-aspartic acid, DL-aspartic acid or a salt thereof is dissolved in water in any concentration. The solution is then adjusted to a pH of 4 to 9. The aforementioned immobilized microorganism is suspended in the solution, and the mixture is incubated at 5 to 55°C, especially at 30 to 45°C, with stirring, until the reaction is completed. When the reaction is completed, the mixture is filtered or centrifuged. Thus, an aqueous solution containing L-alanine or a mixture of L-alanine and D-aspartic acid is obtained as the filtrate of supernatant liquid. L-alanine and/or D-aspartic acid are recovered from the filtrate or supernatant liquid in the conventional manners, for example, by applying the method of direct crystallization, the treatment with an ion-exchange resin, or the combination of these operations to the filtrate or supernatant liquid.

Example: *Pseudomonas dacunhae* IAM No. 1152 is inoculated into 200 ml of a nutrient medium (7.0 pH) containing 0.5% w/v ammonium fumarate, 1.0% w/v sodium fumarate, 0.55% w/v corn steep liquor, 1.8% w/v peptones, 0.05% w/v potassium dihydrophosphate and 0.01% w/v magnesium sulfate 7 hydrate. The medium is cultivated at 30°C for 24 hours under shaking. After the cultivation, the microbial cells of *Pseudomonas dacunhae* IAM No. 1152 are collected by centrifugation. The microbial cells are suspended in 20 ml of physiological saline solution. 3.75 g acrylamide, 0.2 g of N,N'-methylenebis(acrylamide), 2.5 ml of 5% β-(dimethylamino)-propionitrile and 2.5 ml of 2.5% potassium persulfate are added to the suspension. Then, the suspension is allowed to stand at 25°C for 10 minutes. The immobilized microbial cells are ground into pieces and then washed with physiological saline solution. 40 g of an immobilized preparation of *Pseudomonas dacunhae* IAM No. 1152 are obtained.

40 g of the immobilized preparation of *Pseudomonas dacunhae* No. 1152 are charged into a 1.6 x 19 cm column. 500 ml of a 1 M-ammonium L-aspartate aqueous solution (5.5 pH) containing 10^{-4} M concentration of pyridoxal phosphate are continuously passed through the column at 37°C at a flow rate of 6 ml/hr. 500 ml of the effluent are concentrated to 150 ml; 150 ml of methanol are added to the concentrated effluent. The crystalline precipitate thus formed is collected by filtration and then washed with 30 ml of cold methanol. 40.1 g of L-alanine are obtained [270°C MP (dec.); $[\alpha]_D^{23}$ + 14.4°(c = 6.46, 1N HCl)].

D-Alanine

I. Chibata, S. Yamada, M. Wada, H. Maeshima and N. Izuo; U.S. Patent 3,871,959; March 18, 1975; assigned to Tanabe Seiyaku Co., Ltd., Japan describe a process whereby D-alanine can be produced by cultivating a D-alanine-producing strain of *Corynebacterium fascians* in an aqueous nutrient medium under aerobic conditions, and recovering the accumulated D-alanine from the fermentation broth.

The fermentation of a D-alanine-producing strain of *Corynebacterium fascians,* such as *Corynebacterium fascians* ATCC No. 21950, may be conducted by either shaking cultivation or submerged fermentation under aeration. It is preferred to carry out the fermentation at a pH of 5.0 to 9.0, especially at a pH of about 7.0. When the pH of the nutrient medium is adjusted within the abovementioned range prior to the fermentation, readjustment of the pH of the medium may be unnecessary because it scarcely varies during the fermentation. The preferred temperature range for the fermentation is 20 to 37°C, especially 25 to 30°C. The aqueous nutrient medium should contain a carbon source and a nitrogen source. Suitable sources of carbon include polyalcohols, e.g., glycerol; sugar alcohols, e.g., sorbitol; and monosaccharides, e.g., glucose, sorbose and fructose.

Inorganic ammonium salts, such as ammonium phosphate and ammonium sulfate, are suitable as the sources of nitrogen. In some cases, sodium or potassium phosphate, calcium carbonate and the like may be further added to the nutrient medium. The preferred amount of the carbon source which is added to the medium is within the range of 1 to 14%, especially 3 to 10%. On the other hand, the preferred amount of the nitrogen source is 0.3 to 10%, especially 1 to 8%. In carrying out the fermentation, D-alanine productivity of the abovementioned microorganism may be enhanced by addition of corn steep liquor. For this purpose, it is preferred to add 0.01 to 1.0%, especially 0.05 to 0.5%, of corn steep liquor to the nutrient medium.

Alternatively, a mixture of an organic nitrogen source, vitamins and minerals may be used instead of corn steep liquor in order to enhance the D-alanine productivity of the microorganism. The fermentation can be accomplished in about 24 to 120 hours. D-alanine is accumulated in the fermentation broth.

After fermentation is completed, cells and other solid compositions are removed from the fermentation broth by conventional procedures such as filtration or centrifugation. Known procedures may be used in the recovery and/or purification of D-alanine from the filtrate or supernatant solution. For example, the filtered fermentation broth is passed through or treated with a strong cation-exchange resin. Then, the resin is eluted with a dilute alkaline solution such as aqueous ammonia. D-alanine is readily recovered by evaporating the eluate.

Example: A loopful of *Corynebacterium fascians* ATCC No. 21950 is inoculated into 100 ml of an aqueous nutrient medium (7.0 pH) containing 1% glycerol, 1% ammonium phosphate and 0.15% corn steep liquor. The medium is cultivated at 30°C for 2 days under shaking. A seed culture is obtained.

100 ml of an aqueous nutrient medium (7.0 pH) containing 5% glycerol, 4% ammonium phosphate and 0.4% corn steep liquor are charged into a 500 ml shaking flask, and the medium is sterilized at 120°C for 10 minutes by autoclaving; 5 ml of the seed culture are added to the medium. The medium is then cultivated at 30°C for 5 days under shaking. The fermentation medium thus obtained contains 7 mg/ml of D-alanine. 1 liter of the fermentation medium is filtered by centrifugation. The supernatant solution is treated with an active carbon and then passed through a column of cation-exchange resin (H-form)

(Amberlite IR-120). After washing with water, the column is eluted with an aqueous 2 N-ammonia solution. The eluate is evaporated to dryness under reduced pressure. Ethanol is added to the residue obtained, and the crystalline precipitate is collected by filtration. The precipitate is recrystallized from an aqueous ethanol. 5.6 g of D-alanine are obtained $[\alpha]_D^{25} = -14°(c = 2, 5N$ HCl).

ARGININE

Bacillus **Mutant Having Nutritional Requirement**

L-arginine is an amino acid classified as essential with respect to its growth effect in rats. The amino acid is also useful inter alia as a starting compound in the preparation of arginine glutamate which is an adjunct in management of ammonium intoxication due to hepatic failure.

K. Nakayama, K. Araki and H. Yoshida; U.S. Patent 4,086,137; April 25, 1978; assigned to Kyowa Hakko Kogyo Co., Ltd., Japan have found that increased yields of L-arginine are produced by culturing a strain of microorganism belonging to the genus *Bacillus* capable of producing L-arginine and having a nutritional requirement for at least one member of the group consisting of methionine, histidine, threonine, proline, isoleucine, lysine, adenine, guanine and uracil (or its precursor) in a nutrient medium; accumulating L-arginine in the culture liquor and recovering the L-arginine therefrom.

The strains listed below are examples of mutants suitable for the process, which are obtained by mutating the arginine-producing parent strain, *Bacillus subtilis* 110M-59 (prototrophic arginine hydroxamate-resistant strain) (ATCC 31193) derived from *Bacillus subtilis* ATCC 15244 (wild strain) by conventional methods for inducing mutation. The following specific exemplary strains have been deposited with the American Type Culture Collection, Rockville, Md., U.S. (ATCC) and have been accorded the noted accession numbers. These strains are freely available to the public.

Strain	ATCC
Bacillus subtilis BA-22	
(Requiring methionine)	31185
Bacillus subtilis BA-26	
(Requiring histidine)	31186
Bacillus subtilis BA-10	
(Requiring threonine)	31184
Bacillus subtilis BA-9	
(Requiring proline)	31183
Bacillus subtilis BA-32	
(Requiring isoleucine or lysine)	31187
Bacillus subtilis BA-43	
(Requiring adenine or guanine)	31188
Bacillus subtilis 59PL-1	
(Requiring uracil or orotic acid)	31189

Media usually used in the production of amino acids by fermentation are suitable for the process; that is, either a synthetic medium or a natural medium may be used so long as it contains a carbon source, a nitrogen source, inorganic materials and other nutrients. As the carbon source, carbohydrates such as glucose, sucrose, fructose, mannose, starch, starch hydrolyzate, blackstrap molasses, etc.; glycerol, polyalcohol, organic acids such as pyruvic acid, fumaric acid, lactic acid, acetic acid, etc.; alcohols such as ethanol, methanol, etc; amino acids such as glutamic acid, aspartic acid, etc.; and n-paraffins and other hydrocarbons may be used. As the nitrogen source, ammonia, inorganic and organic ammonium salts such as ammonium chloride, ammonium sulfate, ammonium carbonate, ammonium acetate, etc.; urea; and other nitrogen-containing compounds; and natural substances such as peptone, meat extract, yeast extract, corn steep liquor, casein hydrolyzate, fish meal, digest of fish meal, defatted soybean cake, digest of defatted soybean cake, chrysalis hydrolyzate, etc. may be used. As inorganic materials, potassium phosphate, magnesium sulfate, sodium chloride, ferrous sulfate, manganese sulfate, calcium carbonate, etc. are appropriate.

The medium must, of course, be supplemented with an appropriate amount of the nutrients required by the particular microorganism for growth. In some cases, these nutrients are supplied as components of natural substances which are used as the nitrogen source.

The productivity of L-arginine by the auxotrophic strain can be further enhanced by adding L-glutamic acid to the fermentation medium either at the start of the fermentation or during the growth phase of the cells. In either case, it is preferable to add a total amount of L-glutamic acid to make up 0.1 to 3% w/v of the medium.

Culturing is carried out under aerobic conditions, for example, by shaking culture, agitation submerged culture or the like. The preferred temperature for culturing is generally 20 to 40°C, but culturing can be carried out at a temperature which is out of this range so long as the particular microorganism can grow. In order to obtain a high yield of the product, it is desirable that the pH of the medium be maintained at around neutral during culturing. Usually after culturing for 1 to 5 days under these conditions, a considerable amount of L-arginine is produced and accumulated in the culture liquor.

After the completion of culturing, the microbial cells and any precipitate are removed from the culture liquor by conventional methods. Then the L-arginine is recovered from the culture liquor by known methods such as an ion exchange resin treatment.

Pyrimidine-Metabolic-Antagonist-Resistant Bacterial Strains

I. Chibata, M. Kisumi and J. Kato; U.S. Patent 3,902,967; September 2, 1975; assigned to Tanabe Seiyaku Co., Ltd., Japan describe a process for producing L-arginine by cultivating a pyrimidine-metabolic-antagonist-resistant mutant of a microorganism being capable of producing L-arginine in a medium and harvesting the produced L-arginine from the broth.

Examples of the resistant mutant used in the process may be 6-azauracil-resistant mutant, 2-thiouracil-resistant mutant and 6-azauridine-resistant mutant which may be derived from an arginine hydroxamate-resistant mutant of *Bacillus subtilis.* The representative mutants are deposited with ATCC and with Fermentation Research Institute, Agency of Industrial Science and Technology, Japan (FERM). They are *Bacillus subtilis* AHr.AUr-9 (arginine hydroxamate-resistant and 6-azauracil-resistant mutant: ATCC No. 31002; FERM-P No. 1998), *Bacillus subtilis* AHr.TUr-61 (arginine hydroxamate-resistant and 2-thiouracil-resistant mutant: ATCC No. 31003; FERM-P No. 1999) and *Bacillus subtilis* AHr.AUDr-62 (arginine hydroxamate-resistant and 6-azauridine-resistant mutant: ATCC No. 31004; FERM-P No. 2000).

The medium used for cultivation of the resistant mutants may include 5 to 15% by weight of saccharides, e.g., glucose or starch hydrolysates, as a carbon source; 0.5 to 4% by weight of an inorganic ammonium salt, e.g., ammonium chloride or ammonium sulfate; urea or the like as a nitrogen source; and 0.02 to 2% by weight of peptone, yeast extract, corn steep liquor or the like as an organic nutrient; and optionally a small amount of inorganic salt, e.g., potassium phosphate, magnesium sulfate, manganese sulfate or ferric sulfate. It is also preferable to add calcium carbonate or the like to maintain the pH value of the medium at 6 to 9. Furthermore, it is preferable to add 1 to 5% by weight of glutamic acid and/or aspartic acid to the medium for enhancing the accumulated amount of the desired L-arginine.

The resistant mutant may be inoculated into a medium containing the above components and it is cultivated under an aerobic condition such as by strongly shaking at 25 to 37°C for 1 to 4 days and thereby the desired L-arginine is accumulated in the broth. The L-arginine is accumulated in a large amount in the broth and, on the other hand, the by-produced amino acids other than L-arginine are included in extremely slight amount. The L-arginine thus produced can be easily harvested by a conventional method, for instance, by using an ion exchange resin.

Example: To a 500 ml shake flask is added a medium (7.6 pH, 30 ml) containing 8% glucose, 2.5% ammonium chloride, 3.5% L-aspartic acid, 0.1% peptone, 0.1% yeast extract, 0.5% potassium dihydrogen phosphate, 0.05% magnesium sulfate, and 2% calcium carbonate; the mixture is sterilized under pressure. The glucose and calcium carbonate added to the medium are previously sterilized. The medium is inoculated with one platinum loop of *Bacillus subtilis* AHr.AUr-9 (ATCC No. 31002; FERM-P No. 1998) and subjected to shake culture at 30°C for 72 hours. L-arginine is produced in an amount of 31.2 mg/ml in the broth.

The broth (1 liter) thus obtained is harvested, heated and then filtered. The filtrate is passed through a column packed with Amberlite IR-120 (H type) and then the column is washed with water. The adsorbed L-arginine is eluted with 5% aqueous ammonia. The eluate is concentrated under a reduced pressure. To the residue are added hydrochloric acid and methanol, and the precipitated crystals are separated by filtration. The crystals are recrystallized from water-containing methanol to give L-arginine hydrochloride (29 g).

Arginine-Analogue-Resistant Bacterial Strains

K. Kubota, T. Onoda, H. Kamijo and S. Okumura; U.S. Patent 3,878,044; April 15, 1975; assigned to Ajinomoto Co., Inc., Japan have found large amounts of arginine in generally conventional culture media on which guanine-requiring mutant strains of *Brevibacterium* resistant to feedback inhibition and/or repression by arginine or arginine analogs had been cultured. The arginine is recovered as the L-form.

Mutant strains which differ from the parent by resisting feedback inhibition and/or repression by arginine or arginine analogs are obtained by exposing the parent strain to mutagenic agents and screening the treated microorganisms for their ability of growing on culture media containing arginine or arginine analogs in amounts of 5 mg/ml or more. Suitable arginine analogs include canavanine, homoarginine, α-methylarginine, 2-thienylserine, D-serine, ethionine, 2-thiazolalanine, α-amino-β-hydroxyvaleric acid, 6-chloropurine, and sulfa drugs such as sulfaguanidine, sulfamerazine, sulfisomezole, and sulfisoazole.

Mutants of *Brevibacterium* capable of growth of media containing the arginine analogs, such as *B. flavum*, ATCC 21493, produce arginine to some extent even if they do not require specific organic nutrients. However, substantially enhanced arginine productivity is found in strains derived therefrom by further mutation and requiring guanine for their growth. The most effective arginine-producing mutant of genus *Brevibacterium* found so far is the guanine-requiring strain *B. flavum* AJ 3401, FERM-P 1639, which is resistant to 2-thiazolalanine and canavanine.

Example: 20 ml batches of an aqueous medium containing 10% glucose, 6% ammonium sulfate, 0.1% potassium dihydrogen phosphate, 0.04% magnesium sulfate hepahydrate, 2 ppm each of Fe and Mn ions, 50 μg/l biotin, 20 μg/l thiamine hydrochloride, 1% soy protein hydrolyzate (2.4% total nitrogen), 0.015% guanine were sterilized in respective 500 ml shaking flasks and adjusted to pH 7 with 5% separately sterilized calcium carbonate.

Inocula of *B. flavum* AJ 3401 prepared on bouillon agar slants were added to each flask which was thereafter held at 31°C with aeration and agitation for 72 hours. The combined broth contained 2.5 g/dl arginine and was centrifuged to remove the cells. 1 liter of the supernant liquid was passed over a column packed with an ion exchange resin (Amberlite C-50, NH$_4$ type) and the arginine adsorbed by the resin was eluted with 2 N ammonium hydroxide solution. The eluate was partly evaporated to precipitate crude, crystalline arginine which, when dried, weighed 16.2 g.

ASPARTIC ACID

Production from Fumaric Acid

N.-C. Duc; U.S. Patent 3,933,586; January 20, 1976; assigned to Les Produits Organiques du Santerre Orsan, France describes a method of making L-aspartic acid by bioconversion of fumaric acid. The *Pseudomonas* PO 7111 strain is confirmed as being by far the most productive. Consequently, samples thereof have been deposited at the ATCC (ATCC No. 21973).

The most appropriate nutritive medium for the amino-transferase production and, consequently, for the conversion of the fumaric acid is a medium which has substantially the following composition:

> 15 to 60 g/l of molasses (in terms of total sugar), preferably 40 g/l expressed in total sugar;
>
> 1 to 4 g/l of phosphoric acid or mineral phosphates (expressed in phosphoric acid quantities), preferably 2 to 2.6 g/l;

0.1 to 1 g/l of Mg (expressed in MgSO$_4$), preferably 0.4 to 0.6 g/l of MgSO$_4$; and

0.5 to 10 g/l of fumaric acid, preferably 0.9 to 1.1 g/l.

The pH of the resulting aqueous medium is adjusted between 7 and 8.5, preferably at about 7. During the culture, the duration is between 9 and 15 hours, preferably between 10 and 12 hours, the temperature should not be higher than 38°C. In the same way, the growth of the microorganism is prevented if the temperature becomes lower than 25°C; an optimal density of population is obtained at 30°C. Traces of oligo-elements can be added within a range of 0 to 10 ppm, either by incorporating same to the culture medium in the form of mineral salts, or by using tap water. Sufficient water is added to adjust the volume of the medium to 1 liter. The bioconversion is performed at a pH between 8.5 and 10, preferably between 9 and 9.5.

According to a preferred embodiment, the production of the amino-transferase necessary to the bioconversion is carried out by aerobian incubation of the PO 7111 strain in a culture medium containing sugar beet or sugar cane molasses as the only carbon-providing material, and a limited fumaric acid amount as an effective agent for the amino-transferase accumulation by this microorganism.

When the optimum percentage is obtained, the culture is stopped and then put into contact with the substrate, so that the latter would be transformed into aspartic acid.

The enzyme activity is not affected by the presence of other microorganisms; as a consequence it is not necessary to work in a sterile medium at the time of the bioconversion of fumaric acid into L-aspartic acid. It is then possible to use a substrate without special treatment thereof and in particular pure crystallized fumaric acid such as found on the market; a fumarate can also be employed. The solid is then added, without previous sterilization, to the bioconversion medium so as to maintain therein a concentration of about 10%. This bioconversion phase is carried out without aeration in order to avoid undesirable by-products formation, especially oxidation products of ethylenic acid. The implied reagent is the diammonium fumarate when fumaric acid is poured into the ammoniacal syrupy residual medium. The reaction is performed in the presence of an inert gas such as nitrogen.

During the conversion, the temperature is preferably kept to 55 to 57°C and the pH adjusted to 9 to 9.5 by addition of gaseous ammonia. At the end of the reaction, namely when the residual fumaric acid percentage has been reduced to values such that the yield is maximum, the reactive mixture is submitted to a separation by usual and known means, such as centrifugation, filtration, etc. in order to remove therefrom microorganisms and any other insoluble compounds.

Thus a concentrated solution of aspartic acid is obtained and the amino acid is extracted from same by using conventional means such as insolubilization when the pH takes a value equal to that of the isoelectric point.

The resulting crystallized product has a sufficiently high purity for industrial uses, but can be further purified to obtain a product satisfying pharmaceutical standards by a mere recrystallization treatment after discoloration.

Strains Resistant to 6-(Dimethylamino)Purine

T. Tsuchida, K. Kubota and Y. Hirose; U.S. Patent 4,000,040; December 28, 1976; assigned to Ajinomoto Co., Inc., Japan have found that a mutant derived from a microorganism which belongs to the genus *Brevibacterium* and the genus *Corynebacterium* and which is resistant to 6-(dimethylamino)purine which suppresses the growth of the parent strain is especially effective for the production of L-aspartic acid.

The best L-aspartic-acid-producing mutants known are listed below with their accession numbers.

Brevibacterium flavum AJ 3859 (FERM-P 2799), parent strain: *Brevibacterium flavum* (ATCC 14067);

Brevibacterium lactofermentum AJ 3860 (FERM-P 2800), parent strain: *Brevibacterium lactofermentum* (ATCC 13869);

Corynebacterium acetoacidophilum AJ 3877 (FERM-P 2803), parent strain: *Corynebacterium acetoacidophilum* (ATCC 13870); and

Corynebacterium glutamicum (Micrococcus geritamicus) AJ 3876 (FERM-P 2802), parent strain: *Corynebacterium glutamicum (Micrococcus glutamicus)* (ATCC 13032).

The culture media are conventional, and contain a carbon source, a nitrogen source, inorganic ions and, when required, minor organic nutrients. Suitable carbon sources include carbohydrates such as glucose and sucrose, alcohols such as ethanol, and organic acids such as acetic acid. As the nitrogen source, gaseous or aqueous ammonia, ammonium ions, and urea are preferably used. As the minor organic nutrients, preferably 2 to 8 μg/l are added to the culture medium.

Especially in the cases when the medium contains excessive amounts of biotin, the addition of antibiotics, surfactants, and antioxidants to the culture medium increases the yield of L-aspartic acid.

Cultivation is carried out aerobically at 24 to 37°C for 2 to 7 days. During the cultivation, the pH of the medium is adjusted to 5 to 9 with an organic or inorganic acid or alkali, or with urea, calcium carbonate, or gaseous ammonia. The aspartic acid which accumulates in the culture broth can be recovered by any conventional manner, for example, using ion exchange chromatography.

Fermentation of Hydrocarbons

V. Zangrandi and P. Peri; U.S. Patent 4,013,508; March 22, 1977; assigned to Liquichimica SpA, Italy describe a process which permits the obtaining of L-aspartic acid with a high yield by fermentation of hydrocarbons, with the twofold enormous advantage of applicability of the process on an industrial scale, joined to the availability and low cost of the raw material.

The fermentation is of the associate type and the microorganisms are of two kinds: the first presents morphologic and physiologic characteristics not dissimilar to those described for the species *Candida hydrocarbofumarica* (ATCC 20473); the second belongs to the genus *Bacillus* (ATCC 31177).

In the process, fermentation occurs at a temperature ranging from 25° to 40°C and the pH of the reaction is neutral. The hydrocarbons that may be generally used for the production of fumaric acid are specifically n-paraffins containing from 10 to 20 carbon atoms, preferably from 13 to 18. The concentration used is 5 to 10% w/v. The fumaric acid thus obtained constitutes the substratum for the second microorganism in order to obtain, after about 3 days, the transformation into L-aspartic acid.

According to the process, the culture medium, containing paraffin, mineral salts, and sources of assimilable nitrogen, is inoculated with an inoculum mass of the strain which metabolizes paraffin into fumaric acid and the culture is aerobically led under strong stirring to the almost complete transformation of the paraffin into dicarboxylic acid. At this point, the direct recovery of fumaric acid or the direct inoculation of the inoculum mass of the strain which catalyzes the transformation into L-aspartic acid may be performed.

The aeration of the culture medium obtained in the flasks merely by shaking with with a vibrator, and in the fermentation containers by stirring and direct aeration, is necessary to ensure a suitable concentration of oxygen dissolved into the culture broth, which is indispensable to the oxidizing metabolism of the n-paraffins and also to obtain a fine dispersion of the n-paraffin itself, which is more rapidly attackable by the microorganisms.

The fermentation of the two strains may occur at the same time or by an independent sequence. The fermentation is considered to be totally completed when all fumaric acid has been transformed into L-aspartic acid.

In order to effect the recovery of L-aspartic acid, the broth culture, which contains ammonium aspartate in the amount of 5 to 10% w/v, expressed as L-aspartic acid, is acidified with H_2SO_4 until it reaches a pH of 4 to 5. The acidified solution is purified by filtration through diatomaceous earth after a previous heating. The filtrate cooled to 10° to 18°C is brought to a pH of 2 to 3 with H_2SO_4 under stirring. Under these conditions L-aspartic acid is insoluble; therefore it crystallizes and precipitates. The recovery of the crystals is performed by centrifugation. The degree of purity reaches 90 to 95% and in order to obtain a higher degree, it is necessary to resort to a subsequent crystallization.

Example: The strain of *Candida hydrocarbofumarica* type, which metabolizes n-paraffins to fumaric acid, has been cultivated on an agar culture medium containing paraffins at a temperature of 30°C for 24 hours and then inoculated into a 3 liter flask, containing 500 ml of broth having the following composition:

	Grams per Liter
n-Paraffins	30
NH_4Cl	5
KH_2PO_4	1
$MgSO_4 \cdot 7H_2O$	0.5
$FeSO_4 \cdot 7H_2O$	0.03
$MnSO_4 \cdot 4H_2O$	0.03
Yeast extract	0.5
Meat flour	1
Polyoxyethylene 20 sorbitan monooleate (Tween 80)	0.5
H_2O (qs, 1 liter)	

The broth in the flask has been sterilized for 20 minutes in autoclave at 115°C, before inoculation. The development of the inoculum culture occurred at 30°C on a reciprocating vibrator at 100 strokes per minute. The period of incubation necessary to obtain a good cellular growth was 36 hours.

The entire broth culture of the inoculum flask has been transferred into a jar containing 20 liters of fermentation broth, having the composition reported hereinafter and maintained in incubation at a constant temperature of 30°C with a stirring speed of 800 rpm and an aeration of 1.5 v/v/min. The pH of the fermenting culture has been adjusted to 6.5 by addition of ammonia. The composition of the fermentation broth is as follows:

	Grams per Liter
n-Paraffins (C_{13-18})	80
NH_4Cl	5
KH_2PO_4	1
$MgSO_4 \cdot 7H_2O$	0.5
$FeSO_4 \cdot 7H_2O$	0.03
$MnSO_4 \cdot 4H_2O$	0.03
$ZnSO_4 \cdot 7H_2O$	0.05
Yeast extract (paste)	0.5
Meat flour	2
Polyoxyethylene 20 sorbitan monooleate (Tween 80)	0.5
Water (qs, 1 liter)	

The concentration of the fumaric acid after 72 hours of fermentation was 46 g/l. The yeast cells have been separated by centrifugation and the clear broth recovered has been inoculated into the same jar with 500 ml of an inoculum culture previously prepared of the strain of *Bacillus* species type which transforms fumaric acid into L-aspartic acid. The composition of the broth for the preparation of the second inoculum is the following:

	Grams per Liter
Ammonium fumarate	60
K_2HPO_4	2
$MgSO_4 \cdot 7H_2O$	0.5
Yeast extract (paste)	2
H_2O (qs, 1 liter)	

The broth, after adjustment of the pH to 7, has been distributed into 3-liter flasks in the ratio of 500 ml each. The flasks have been sterilized at 115°C in autoclave for 20 minutes. Each flask has been inoculated with a cellular suspension of the strain of *Bacillus* species type obtained from a slant of the strain cultivated on agar also containing fumaric acid, K_2HPO_4, $MgSO_4$ and yeast extract.

The development of the inoculum culture has been performed at 30°C on a reciprocating vibrator at 100 strokes per minute. The period of incubation was 18 hours. The strain of the *Bacillus* species inoculated into the broth containing fumaric acid has rapidly grown and has reached the stationary stage of growth in 18 to 20 hours. The concentration of L-aspartic acid in the broth culture at this point is only 0.5% w/v. From this moment on, the transformation of the fumaric acid continues rapidly and is completed in 56 to 60 hours. The final concentration of L-aspartic acid was 41 g/l. Therefore, the conversion yield in the paraffins minus L-aspartic acid fermentation is 51%.

CYSTEINE AND CYSTINE

Use of Cysteine Desulfhydrase

H. Yamada, H. Kumagai and H. Ohkishi; U.S. Patent 3,974,031; August 10, 1976; assigned to Mitsubishi Chemical Industries Ltd., Japan describe a process for producing L-cysteine or a derivative thereof in the presence of cysteine desulfhydrase which is derived mainly from microorganisms. L-cysteine is a quasi-essential amino acid, useful as a food additive and a pharmaceutical, e.g., as an antidote. In addition, L-cysteine is useful as an intermediate in the preparation of other pharmaceutical compounds. The derivatives of L-cysteine, for example, the sulfoxides of S-methyl-L-cysteine and S-allyl-L-cysteine, are known to suppress increasing levels of cholesterol in blood and liver.

The cysteine desulfhydrase employed in the process is an enzyme which is known as a catalyst for the reaction wherein L-cysteine decomposes into pyruvic acid, ammonia and hydrogen sulfide [(*Biochem. Biophys. Res. Commun.* 59 789 (1974)]. This enzyme is readily produced by a variety of microorganisms. Such microorganisms which produce the enzyme include, e.g., the microorganisms belonging to the following species: *Brevibacterium, Sarcina, Corynebacterium, Arthrobacter, Pseudomonas, Proteus, Micrococcus, Escherichia, Serratia, Alcaligenes, Bacillus, Agrobacterium, Enterobacter (Aerobacter), Citrobacter, Klebsiella, and Salmonella.* However, the microorganisms suitable for producing the enzyme are not limited to the aforementioned types. Any microorganism which produces the abovementioned enzyme may be employed.

The nutrients necessary for the incubation of these microorganisms are a carbon source, a nitrogen source and an inorganic salt source. Suitable carbon sources include glucose, sucrose, fructose, mannose, mannitol, xylose, glycerol, sorbitol, molasses, starch hydrolysate and the like; organic acids such as acetic acid, fumaric acid and the like; and n-paraffin and the like. Suitable nitrogen sources include ammonia; ammonium salts of organic acids and inorganic acids such as ammonium chloride, ammonium sulfate, ammonium carbonate, ammonium acetate and the like; nitrates such as sodium nitrate, potassium nitrate, ammonium nitrate and the like; corn steep liquor; yeast extract; meat extract; yeast powder; cotton seed powder; soybean powder; soybean hydrolysate; peptone; polypeptone and the like. Suitable inorganic salts include potassium phosphate, sodium phosphate, magnesium sulfate and the like.

The incubation temperature can range from 20° to 80°C, preferably from 25° to 50°C. The aerobic incubation is carried out for a period of 10 to 72 hours. It is preferred to maintain the pH of the medium in the range of 7 to 11 during the incubation. The presence of 0.1 to 1 wt % of at least one amino acid selected from the group consisting of L-cysteine, L-cystine, S-methyl-L-cysteine, S-ethyl-L-cysteine, L-serine and O-methyl-L-serine further enhances the yield of the enzyme. The cysteine desulfhydrase obtained by the above process is mainly present intracellularly. The usual methods, such as ultrasonic treatment, fractionation with ammonium sulfate and ion exchange chromatography, are applicable to the separation and purification of cysteine desulfhydrase. The molecular weight of the obtained enzyme is 150,000 to 500,000. Pyridoxal phosphate is usually obtained as a coenzyme.

According to the process, β-substituted L-alanine reacts with a thiol or a compound selected from the group consisting of hydrogen sulfide, ammonium hydrosulfide, a metal hydrosulfide, ammonium sulfide and a metal sulfide which is capable of producing such a metal hydrosulfide and/or hydrogen sulfide in the aqueous medium employed, using a pH normally in the range of 6 to 12, more preferably in the range of 7 to 11, in the presence of the cysteine desulfhydrase derived .mainly from microorganisms as described above. It is not required that the enzyme to be used be purified and crystallized. Any microorganism broth of living cells, dried cells, ground cells and cellular extracts which possesses the enzyme may be used. The amount of the enzyme to be used, defined as the weight of dried cells, is normally about 0.1 to 20 g/l, more preferably 1 to 5 g/l. The reaction temperature ranges generally from 20° to 80°C, more preferably from 30° to 50°C.

The reaction period varies from 1 to 100 hours dependent upon the activity of the enzyme, the concentration and the species of the substrate, and the reaction temperature. The time of reaction will normally be between 2 and 48 hours. The concentration of the β-substituted L-alanine and the thiol or the compound selected from the group consisting of hydrogen sulfide, ammonium hydrosulfide, a metal hydrosulfide, ammonium sulfide and a metal sulfide, are normally each 1 to 40 wt %, more preferably 3 to 20 wt %.

The addition of a carbonyl compound containing 1 to 20 carbon atoms to the reaction medium results in an increase in the yield of L-cysteine or derivative thereof. Representative of such compounds are α-keto carboxylic acids containing 3 to 20 carbon atoms, such as pyruvic acid, 2-oxobutyric acid, 2-oxoglutaric acid and the like; aldehydes containing 1 to 20 carbon atoms, such as acetaldehyde, propionaldehyde, isobutyraldehyde, benzaldehyde and the like; and ketones containing 3 to 20 carbon atoms, such as acetone, methyl ethyl ketone, diethyl ketone, methyl butyl ketone, cyclohexanone, benzophenone, acetophenone, benzil and the like. The carbonyl compound can be added such that the concentration is from about 0.01 to 20 M to effect an increase in yields, with from about 0.05 to 10 M being preferred.

The presence of a small amount of pyridoxal phosphate in the reaction medium enhances the activity of the enzyme. Pyridoxal phosphate should be added such that its concentration is up to 0.01 mM. The addition of pyridoxal phosphate to a concentration over 0.01 mM has no effect on the yield of L-cysteine or derivative thereof. Upon completion of the reaction, L-cysteine or derivative thereof can be separated employing the usual procedures, for example, treatment with an ion exchange resin.

Production from 2-Aminothiazoline-4-Carboxylic Acid

K. Sano, K. Matsuda, K. Mitsugi, K. Yamada, F. Tamura, N. Yasuda and I. Noda; U.S. Patent 4,006,057; February 1, 1977; assigned to Ajinomoto Co., Inc., Japan, have found that 2-aminothiazoline-4-carboxylic acid (ATC), which is available in a large quantity and at a lower cost than hair, is enzymatically converted in a very high yield to L-cysteine and L-cystine. It has further been found that the enzyme having the activity of converting ATC to L-cysteine and L-cystine (ATC-hydrolyzing enzyme) is produced by various microorganisms, especially bacteria. The microorganism which produces ATC-hydrolyzing enzyme can be grown in a medium containing ATC as a nitrogen source.

The microorganisms are easily obtained by the following method from natural sources: samples from a natural source containing microorganisms are inoculated on the screening medium which contains (per deciliter) 0.3 g DL-ATC·3H$_2$O, 1.0 g glycerol, 0.01 g yeast extract, 0.1 g KH$_2$PO$_4$, 0.05 g MgSO$_4$·7H$_2$O, and 2.0 g agar; and of 7.0 pH. The inoculated medium is incubated at 30°C for 1 to 5 days. Almost all the microorganisms which grow on the medium mentioned above produce the ATC-hydrolyzing enzyme.

Specimen cultures capable of producing the enzyme are as follows:

Sarcina lutea	AJ 1217	ATCC 272
Achromobacter delmarvae	AJ 1983	FERM-P 21
Alcaligenes denitrificans	AJ 2553	ATCC 15173
Bacillus brevis	AJ 1282	ATCC 8185
Brevibacterium flavum	AJ 1516	ATCC 13826
Enterobacter aerogenes	AJ 2643	FERM-P 2764
Erwinia carotovora	AJ 2753	FERM-P 2766
Escherichia coli	AJ 2592	FERM-P 2763
Micrococcus sodonensis	AJ 1753	ATCC 11880
Mycoplana dimorpha	AJ 2809	ATCC 4279
Serratia marcescens	AJ 2698	FERM-P 2765
Flavobacterium acidoficum	AJ 2494	ATCC 8366
Pseudomonas ovalis	AJ 2236	FERM-P 2762
Pseudomonas thiazolinophilum	AJ 3854	FERM-P 2810
Pseudomonas ovalis	AJ 3863	FERM-P 2811
Pseudomonas desmolytica	AJ 3868	FERM-P 2816
Pseudomonas desmolytica	AJ 3869	FERM-P 2817
Pseudomonas cohaerens	AJ 3874	FERM-P 2831
Pseudomonas ovalis	AJ 3864	FERM-P 2812
Pseudomonas ovalis	AJ 3865	FERM-P 2813
Pseudomonas ovalis	AJ 3866	FERM-P 2814
Pseudomonas ovalis	AJ 3867	FERM-P 2815
Pseudomonas desmolytica	AJ 3870	FERM-P 2818
Pseudomonas desmolytica	AJ 3871	FERM-P 2819
Pseudomonas desmolytica	AJ 3872	FERM-P 2820
Pseudomonas desmolytica	AJ 3873	FERM-P 2821

In order to produce ATC-hydrolyzing enzyme, the microorganisms mentioned above are cultured in conventional medium of a 6 to 9 pH, which contains carbon sources, nitrogen sources, inorganic ions, and when required, minor organic nutrients. Cultivation is carried out under aerobic conditions at 20° to 40°C for 1 to 3 days.

High enzyme activity is possessed by the resulting culture broth and especially in microbial cells. As the enzyme source, culture broth, intact cells, homogenate of cells, sonicate of cells, freeze-dried cells, cells dried with solvent and so on, are preferably used. Protein fractions separated from, for example, the homogenate of the cells or sonicate of the cells by conventional methods, such as gel-filtration or salting-out method, are also used as a preferable enzyme source. Especially, cells which have been contacted with an organic solvent or a surfactant, homogenate of cells, sonicate of cells, and cells treated with lytic enzymes of microbial cells, are preferably used.

The aqueous reaction mixture contains the enzyme or the enzyme source as mentioned above, ATC, and when required, pyridoxal phosphate and/or metal ions.

The amounts of ATC in the reaction mixture are preferably 0.1 to 30% and more preferably 0.5 to 10%. During the reaction, the pH of the reaction mixture is maintained at 5 to 11 and preferably 7 to 9.5. The reaction temperature is preferably maintained at 15° to 60°C and more preferably at 30° to 50°C.

Usually both L-cysteine and L-cystine are accumulated in the reaction mixture. However, it is possible to produce exclusively L-cystine by carrying out the reaction under oxidative conditions; on the other hand, higher amounts of L-cysteine are produced under reducing conditions.

Example: Soil was spread on the following screening medium, and the medium was incubated at 30°C for 4 days.

	Grams per Deciliter
DL-ATC·3H$_2$O	0.3
Glycerol	1.0
Yeast extract	0.01
KH$_2$PO$_4$	0.1
MgSO$_4$·7H$_2$O	0.05
Agar	2.0
pH 7 (NaOH)	

Strains grown on the screening medium were separated and inoculated in the following culture medium, and cultured at 30°C for 24 hours.

	Grams per Deciliter
Glycerol	1.0
Yeast extract	0.5
Peptone	0.5
Bouillon	0.5
NaCl	0.5
DL-ATC·3H$_2$O	0.2
pH 7.0 (KOH)	

Cells grown on the culture medium were collected and suspended in an aqueous reaction mixture containing (per deciliter) 1.0 g DL-ATC·3H$_2$O, and 1.0 g KH$_2$PO$_4$ (8 pH), and the suspension was held at 30°C for 24 hours. L-cysteine and L-cystine were produced in the culture medium.

GLUTAMIC ACID

Production from *Micrococcus glutamicus*

N.I. Zhdanova, L.M. Evstjugov-Babaev, R.M. Balitskaya, A.F. Sholin, T.B. Kasatkina and N.N. Kuznetsova; U.S. Patent 4,054,489; October 18, 1977 describe a method for preparing L-glutamic acid and its sodium salt by cultivating *Micrococcus glutamicus* with aeration on a culture medium containing sources of carbon, nitrogen, and mineral salts; treating the culture fluid to precipitate the propagated mass; removing colored admixtures; evaporating the obtained solution; and acidifying this solution to 3.0 to 3.2 pH with subsequent isolation of the end product.

Biotin concentration in liquid mineral medium, ensuring the maximum yield of glutamic acid, is 20 μg/l; the maximum cell concentration in the fermented broth is 3×10^9 cells per milliliter (medium contains 20% by weight of molasses). The production of L-glutamic acid is about 40 g/l. To precipitate the biomass, the culture fluid should be treated with orthophosphoric acid in the quantity 0.4 to 2% by weight with respect to the weight of the culture fluid (to 5 to 6 pH) at a temperature of 60° to 100°C.

This method employs the strain *Micrococcus glutamicus* VNIIGenetika 3144, (All Union Institute of Genetics and Selection of Industrial Microorganisms) differing from the known strain by some valuable properties that produce a significant effect on the fermentation process and isolation of L-glutamic acid, namely:

(a) The resistance of the synthesis of L-glutamic acid to the inhibiting action of biotin increases three times;

(b) The concentration of cells in a unit volume of the culture fluid decreases three times;

(c) The precipitate of the biomass can be easily separated by filtration;

(d) The productivity of a cell increases six times;

(e) L-glutamic acid is intensively synthesized at a lower (three times less) concentration of dissolved oxygen in the nutrient process; and

(f) The fermentation process is continued for a period of time, 30 to 35 hours less than in the known process.

Example: One day old culture of *Micrococcus glutamicus* VNIIGenetika 3144, grown on Hottinger slant agar, is used to inoculate the seeding culture medium in 750 ml flasks. The quantity of the nutrient medium is 75 ml, and it has the following composition:

	Weight Percent
Molasses	8
NH_4Cl	0.5
Corn steep liquor	0.3
K_2HPO_4	0.05
$MgSO_4 \cdot 7H_2O$	0.03
$CaCO_3$	1.0
Water to make 100%	

The pH of the medium is adjusted with a 60% solution of NaOH to 7.2. The medium is sterilized for 30 minutes at 0.8 atm. The seeding material is grown for 15 to 17 hours on a reciprocating shaker (220 to 240 rpm). The seeding material (75 ml) is transferred into a 5 liter fermentation vessel containing 3 liters of nutrient medium of the following composition:

	Weight Percent
Molasses	20
KH_2PO_4	0.5
$MgSO_4 \cdot 7H_2O$	0.3
Urea	0.8
$CaCO_3$	1.0
Water to make 100%	

The pH of the medium is 6.8 to 7.0. The fermentation medium (without urea) is sterilized in the fermentation vessel at a temperature of 124° to 126°C for an hour, and then cooled to 30°C. Urea (40% solution in water) is sterilized separately in an autoclave at 0.5 atm for 30 minutes and added to the medium immediately before inoculation.

The fermentation process is effected at a temperature of 30°C with continuous stirring (700 rpm) and aeration (0.8 volume of air per volume of medium per minute). The sulfite value is 2.4 g O_2/l/hr. As the pH of the medium falls below 7.0, a 40% solution of urea is automatically added. In 35 hour fermentation, 37 g/l of L-glutamic acid are accumulated in the culture fluid.

The specific gravity of the culture fluid is 1.05; the refractive index is 1.3600; the pH is 6.8. Then 1 liter, containing 37 g of L-glutamic acid, is heated to 50°C and calcium oxide is added in the quantity of 5 g/l. The fluid is stirred for 10 minutes and then conc. (73%) orthophosphoric acid is added gradually with stirring to adjust the pH to 5.5. The solution is heated to 80°C for 10 minutes, then cooled to 70°C, and filtered under pressure through calico and belting fabric. The precipitate is washed with 100 ml of water and the washings are added to the main filtrate. The optical density of the filtrate is 2, λ 10 mm, 535 nm.

The filtrate is passed through an ion exchange column packed with 200 ml of macroporous condensation type resin on the basis of metaphenylene diamine, formaldehyde, and phenol (resorcinol) at a rate of 200 ml/hr. The first 70 ml portion of the solution that emerges from the column is discarded. The next portions of the solution containing L-glutamic acid are collected together with the residual portion of the starting solution that is displaced from the column with 300 ml of water acidified to pH 5 with hydrochloric acid. The clarified solution is evaporated in vacuum at a temperature of 55°C to the residual volume of 200 ml.

Concentrated hydrochloric acid is added with stirring to the evaporated solution to adjust the pH to 3.2. The solution is cooled in a crystallization vessel to 10°C and kept for 20 hours with slow stirring. The precipitated crystals are separated on a filter, washed with water acidified to pH 3.2 with hydrochloric acid, and dried. The yield of L-glutamic acid is 31.45 g (85%). The assay is 98%.

Addition of Surfactants

K. Takinami, T. Tanaka, M. Chiba and Y. Hirose; U.S. Patent 3,971,701; July 27, 1976; assigned to Ajinomoto Co., Inc., Japan have found that from a glutamic-acid-producing microorganism of *Brevibacterium* there have been induced mutants sensitive to N-palmitoyl glutamic acid, which, when cultured in a medium containing an excessive amount of biotin, permit the amount of surfactant necessary for suppressing the excessive biotin to be greatly reduced without decreasing the yield of glutamic acid. The mutants of this process are more sensitive to N-palmitoyl glutamic acid than the parent strain; they include: *Brevibacterium flavum* AJ 3612 (FERM-P 2308) and *Brevibacterium lactofermentum* AJ 3611 (FERM-P 2307). The mutants were induced from *Brevibacterium flavum* ATCC 14067 and *Brevibacterium lactofermentum* ATCC 13869, respectively. The following glutamic-acid-producing microorganisms can be used also as the parent strains: *Brevibacterium divaricatum* NRRL B-2311, *Brevibacterium saccharoliticum* ATCC 14066, and *Brevibacterium roseum* ATCC 13825.

The mutant exhibits its superiority when it is cultured in aqueous culture media which contain too much biotin for the production of glutamic acid, such as those which contain beet molasses, cane molasses, fruit juice, raw sugar and starch hydrolyzate as a carbon source. The media of this process, therefore, contain too much biotin for glutamic acid production in the absence of surfactants, and other conventional ingredients such as a carbon source, a nitrogen source, inorganic ions, and minor organic nutrients.

The surfactants added to the media to suppress excessive activity of biotin are the same as those to which the mutants are sensitive, and are fatty acids having straight chains or compounds containing moieties of the fatty acids. The following compounds are suitable surfactants: lauric acid, myristic acid, palmitic acid, polyethylene glycol monopalmitate, polyoxyethylene sorbitan monolaurate, N-palmitoyl glycine and N-palmitoyl alanine.

The amount of surfactant added to the medium is usually less than 0.2 g/dl, and is one-half or one-quarter of the amount of surfactant employed in the known method. The surfactant is added to the medium prior to starting the cultivation or during the logarithmic phase of growth. Cultivation is aerobically carried out preferably maintaining pH of medium at 6 to 9, and the temperature at 30° to 40°C.

The yield of glutamic acid is superior to the known process and the amount of defoaming agent necessary in the process is smaller than in the known process.

Production Using Methanol

K. Nakayama, M. Kobata, Y. Tanaka, T. Nomura and R. Katsumata; U.S. Patent 3,939,042; February 17, 1976; assigned to Kyowa Hakko Kogyo Co., Ltd., Japan describe a process for producing L-glutamic acid whereby mutants derived from methanol-utilizing, L-glutamic acid-producing microorganisms belonging to the genus *Pseudomonas* or *Protaminobacter,* having at least one property selected from (a) a requirement for L-methionine; (b) a requirement for L-isoleucine; (c) a requirement for L-phenylalanine; and (d) a resistance to DL-lysine hydroxamate, are employed.

Particularly preferred mutants are *Pseudomonas insueta* K-015, ATCC 21966 (requiring L-methionine), *Pseudomonas insueta* K-038, ATCC 21967 (requiring L-isoleucine, leaky type), *Pseudomonas methanolica* LHX-8, ATCC 21968 (resistant to DL-lysine hydroxamate) and *Protaminobacter thiaminophagus* K-244, ATCC 21969 (requiring L-phenylalanine). These mutants have been deposited with the ATCC and are freely available to the public.

The culture medium used for culturing the microorganisms contains methanol (preferably maintaining a concentration of below 3% v/v) as a carbon source, a nitrogen source, inorganic materials and other growth-promoting factors which may be required by the specific strain employed. The total amount of methanol consumed may vary depending upon the specific microorganism and culturing conditions, particularly culturing period. Up to 40% v/v of methanol, based on the volume of the medium, may be used when a prolonged culturing period is employed.

Culturing is carried out under aerobic conditions at 20° to 40°C for 2 to 6 days.

In order to obtain a high yield of the product, it is desirable that the pH of the medium be maintained at about 4 to 9, preferably at around neutral during culturing. The pH may preferably be adjusted with ammonia.

After completion of culturing, the microbial cells are removed from the culture liquor by, for example, filtration. The L-glutamic acid accumulated in the culture liquor is isolated and purified by any of the methods well-known in the art, such as an ion exchange resin treatment, crystallization by concentration, or the like.

Example: In this example, *Pseudomonas insueta* K-038, ATCC 21967, is inoculated into 20 ml of a first seed medium having the following composition in a 250 ml Erlenmeyer flask.

Methanol	20 ml
$(NH_4)_2SO_4$	5 g
KH_2PO_4	1 g
K_2HPO_4	2.5 g
$MgSO_4 \cdot 7H_2O$	0.5 g
$FeSO_4 \cdot 7H_2O$	25 mg
$MnSO_4 \cdot 4H_2O$	8 mg
Thiamine hydrochloride	2 mg
Biotin	10 μg
Polypeptone	10 g
Water, to a volume of 1 liter	
pH	7.2

Culturing is carried out with shaking at 30°C for 27 hours. The first seed culture is then transferred to 300 ml of a second seed medium in a 2 liter Erlenmeyer flask and cultured with shaking at 30°C for 28 hours. The second seed culture is then transferred to 2.7 liters of a third seed medium in a 5 liter jar fermenter and cultured with aeration of 3 l/min and stirring at 600 rpm at 30°C for 24 hours, while maintaining the pH at 6.8 by the addition of concentrated aqueous ammonia. After 20 hours of culturing, 1% v/v of methanol is fed to the medium; 300 ml portions of the thus obtained third seed medium are then inoculated into 2.7 liters of a main fermentation medium in 5 liter jar fermenters. The second and third seed media and the main fermentation medium have the same composition as that of the first seed medium.

Culturing is carried out with aeration of 3 l/min and stirring at 600 rpm at 30°C for 58 hours while maintaining the pH at 6.8 by the addition of concentrated aqueous ammonia. After 10 hours of fermentation, 0.40 to 0.55% v/v per hour of methanol is continuously fed to the medium for a period of 36 hours. As a result, 32.8 mg/ml of L-glutamic acid is produced in the culture liquor while a total of 21.7% v/v is used for the fermentation. 3 liters of the culture liquor is then subjected to centrifugation to remove the microbial cells and the filtrate is concentrated and adjusted to pH 3.2 with hydrochloric acid. After crystallization, 75.3 g of L-glutamic acid is obtained.

Use of Ammonium Acetate

G.M. Miescher; U.S. Patent 3,929,575; December 30, 1975; assigned to Commercial Solvents Corporation describes a process for producing glutamic acid whereby a conventional nutrient fermentation medium consisting of a carbohydrate source (preferably glucose), a nitrogen source, a phosphorus source, a growth factor, and trace minerals (e.g., magnesium, iron and manganese) is employed as the base. When glucose is used as the carbohydrate, a growth starter (carmelized glucose as described by Miescher in U.S. Patent 3,156,627) is used. However, any carbohydrate source known in the art can be used in place of glucose. One suitable source is corn sugar molasses.

The organisms of the genera *Bacillus, Micrococcus, Brevibacterium, Microbacterium, Corynebacterium, Arthrobacter* and the like may be used as the glutamic acid-producing microorganism.

The fermentation is conducted in the presence of ammonium acetate at a concentration of about 3 to 7 g/l (equivalent to 2.34 to 5.45 g/l of acetic acid), preferably 5 g/l (equivalent to 3.9 g/l of acetic acid). The ammonium ion from the ammonium acetate, along with the ammonia used for pH adjustment, is used as the nitrogen source. The acetate ion is usable as a carbon source, but the amount present is insignificant compared to the glucose content. The growth factor employed can be biotin as known or preferably oleic acid. The amount of oleic acid used is in the range of 1 to 10 ml per 10 liters, but preferably 5 to 6 ml per 10 liters.

The fermentation generally is conducted in accordance with the prior art. The temperature is held at about 32°C for the first 14 hours and is then raised to about 38°C. A high degree of agitation is used to insure thorough aeration of the medium during the fermentation, and the CO_2 content of the exhaust gas is maintained at below 6.5%, preferably below 4.5%.

Analyses are made periodically for pH, carbohydrate content, and glutamic acid content for the remainder of the fermentation. The carbohydrate content is maintained at a concentration of from about 1 to 3% by incremental addition of a glucose solution. Typically about 2,250 ml of a solution containing about 1,600 g of glucose is required to complete the fermentation, so that typically about 2,810 g of carbohydrate is utilized. The pH of the fermentor contents is maintained at about 7.8 by the automatic addition of anhydrous ammonia. Hourly analyses indicate when glutamic acid production reaches a maximum. At this point, fermentation is complete and the glutamic acid is ready to be harvested.

Example: *Brevibacterium divaricatum* NRRL B 2311 was cultivated for 16 hours at 35°C on a rotary shaker at 385 rpm in a seed culture medium of the following composition:

Glucose	40 g
K_2HPO_4	1 g
$MgSO_4 \cdot 7H_2O$	0.5 g
Yeast extract (BYF-100)	1 g
Urea	8 g
Tap water	1,000 ml

A fermentation medium was prepared with the following ingredients:

Glucose	1,210 g
KH_2PO_4	12 g
K_2SO_4	12 g

(continued)

MgSO$_4$ (anhydrous)	6 g
FeSO$_4$·7H$_2$O	6 ppm
MnSO$_4$·H$_2$O	6 ppm
Antifoam agent*	1 ml
Tap water, to make 4,800 ml	

*Propylene glycol type compound (Hodag K-67)

This medium was sterilized by heating for 30 minutes at 15 psig of steam. To it was added: 50 g ammonium acetate and 5,000 ml tap water. This portion was sterilized as described above and to it was added the following growth starter, which had been sterilized by heating for 20 minutes at 15 psig of steam: 60 g corn sugar molasses, 60 ml water; this is adjusted with NH$_3$ to 8.3 pH.

The mixture prepared as above was cooled to 32 to 33°C and transferred to a fermentor equipped with an agitator, a temperature controller, an automatic pH-controlled system and a CO$_2$ detector to monitor the CO$_2$ content of the exhaust gas. Aeration was provided by introducing compressed, sterile air at the bottom of the fermentor through a sparger.

Agitation was begun and 600 ml of the seed culture prepared as described above was added to the fermentation medium. The pH was adjusted to 8.5 with ammonia and 6.5 ml oleic acid was added. The automatic pH control system was then set to maintain a pH of about 7.8. The CO$_2$ content of the exhaust gas was held below 6.5% by volume at all times and usually not above 4.5% by adjusting the air flow. After 14 hours the temperature was raised to 38°C and additional glucose feeding was commenced when the glucose content in the medium fell below 0.5 to 2% by incrementally adding a sterilized solution of 1,600 g of glucose dissolved in sufficient tap water to make 2,250 ml. The glucose was introduced at the rate of 5 ml at every 2 minute interval to the end of the fermentation. After 28.5 hours, 100 g/l glutamic acid was produced.

The foregoing experiment was repeated in all essential details, except that a solution of urea (20 g in 5,000 ml of water) was substituted for ammonium acetate and 5.5 ml of oleic acid was used. The CO$_2$ content was not monitored and the air flow remained constant. After 29 hours of fermentation, glutamic acid production was 75 g/l and at 36 hours it was only 83 g/l. Accordingly, the fermentation was judged to be complete.

Protein Hydrolysate Containing Glutamic Acid

T. Yokotsuka, T. Iwaasa and M. Fujii; U.S. Patent 3,912,822; October 14, 1975; assigned to Kikkoman Shoyu Co., Ltd., Japan describe a process for producing a protein hydrolysate having a high glutamic acid content by hydrolyzing a protein-containing raw material, which was subjected to a denaturing treatment such as heat treatment, in the presence of a proteolytic enzyme and glutaminase by means of a one step process, characterized in that the proteolytic enzyme to be used is in a pure form, and hydrolysis of the protein contained in the raw material is carried out in a closed system without contamination with other infectious microbes.

Any proteolytic enzyme of animal or vegetable origin or microorganism origin can be used in the process. Further, a culture of the microorganism or a crude enzyme or purified enzyme obtained from a culture of these microorganisms can also be used.

It is well-known that there are the proteolytic enzymes in almost all microorganisms. For example, *Aspergillus sojae* (ATCC 20235), *Aspergillus saitoi* (ATCC 14332), *Bacillus badus* (FERM-P 543), *Thermopolyspora polyspora* (ATCC 21451), etc. are preferable strains. Glutaminase of animal or vegetable origin can be used, but the utilization of a microorganism is industrially advantageous. Any microorganism capable of producing glutaminase can be used, for example, bacteria, ray fungus, Fungi Imperfecti, Ascomycetes, algae-like microorganisms, Basidiomycetes, etc., which are capable of producing glutaminase.

It is preferable to use a salt-resistant glutaminase when a soy sauce, an amino acid solution or an amino acid-containing seasoning is prepared by hydrolyzing protein by a proteolytic enzyme in the presence of 10% or more sodium chloride. It is generally preferable to effect hydrolysis at 50°C or higher to prevent contamination by foreign microorganisms when a protein hydrolysate is produced by enzymatic hydrolysis of the protein. In such a case, use of a heat-resistant glutaminase is preferable.

A strain capable of producing a glutaminase having a good salt resistance and a heat resistance at the same time, such as *Achromobacter liquefaciens* TR-9, *Bacterium succinicum* IAM 1017, *Bacillus megaterium* NRRL B-939 and *Cryptococcus albidus* ATCC 20294, is effectively utilized in the process, depending upon the desired object.

When the glutaminase produced by these microorganisms is used, a liquid medium is usually used and culturing is carried out according to the conventional method. The culturing can be carried out by state culture or by submerged culture. The ordinary substrates which the relevant microorganism can utilize are used in the medium. That is to say, as a source of carbon, for example, bran, glucose, maltose, sucrose, dextrin, starch, etc. are used. As a source of nitrogen, defatted soybean, soybean powders, gluten, yeast extract, peptone, meat extract, corn steep liquor, ammonium salts, nitrates, etc. are used alone or in a proper combination thereof. In addition, a very small amount of such salts of magnesium, calcium, potassium, sodium, phosphoric acid, iron, manganese, etc. is used if required. Furthermore, a substance capable of serving as a substrate of glutaminase, such as L-glutamine, can be added to a medium containing such nutritional sources to cultivate the microorganism.

It is preferable to use the culture in a state of suspension obtained by applying such a treatment as trituration, etc. to the culture and then adding water or a solution of various salts thereof, or in a state of a filtrate of the culture or in a state of crude or purified extract enzyme. The cells obtained from the culture can be used as they are or after the treatment with acetone or alcohol, or after the drying of the cells. For the extraction of the enzyme from the cells, the ordinary extraction methods can be widely utilized. For example, a crude enzyme solution obtained by applying an ultrasonic treatment, trituration, autolysis, etc. to the culture and adding an aqueous solution of salts such as a buffer solution to the thus treated culture to extract the enzyme can also be used.

Further, a purified enzyme obtained by applying to the crude enzyme a precipitation method by such a hydrophilic organic solvent as ethyl alcohol, acetone, etc.; salting out by ammonium sulfate, etc.; or dialysis; or adsorption and elution method by a cellulose ion exchanger; or ion exchange resin in a suitable combination can also be used.

The glutaminase thus obtained is added, together with a proteolytic enzyme preparation, to a raw material which may be denatured properly if necessary, in a proportion of 100 to 1,000 units of protease and 50 to 1,000 units of glutaminase per gram of the raw material, and the glutaminase and protease thus added are allowed to react upon the raw material at the same time. In another way, a microorganism such as *Aspergillus oryzae* or the like is inoculated into a raw material which may be denatured properly if necessary. The culture thus obtained is added to a solid or liquid medium to make the glutaminase and protease act upon the medium at the same time.

In that case, the pH of the reaction system is 2.5 to 9.0, particularly 5.0 to 8.0. The reaction temperature is selected in such a range that the enzyme undergoes no deterioration or the enzymatic activity is not sluggish; such a reaction temperature of 30° to 70°C is particularly preferable. The reaction time depends upon the raw material, the amount of enzyme, the reaction temperature, etc., but usually a reaction time of 10 hours to 120 days is sufficient.

Example: To 10 g of defatted soybean was added 13 ml of potable water, and after the heat treatment, 500 mg of proteolytic enzyme preparation, which enzyme preparation is a commercially available product of enzyme produced by *Aspergillus oryzae,* partially purified (Sigma Co.). 1 g of a glutaminase preparation (acetone powders of mycelia obtained by culturing *Aspergillus sojae* IAM 2665 in a Czapek's medium containing 1% L-glutamine and 0.5% glucose as a source of carbon at 30°C by shaking) and 50 ml of potable water were added thereto. Further, a small amount of toluene was added thereto, and the hydrolysis was then carried out at 37°C for 7 days with shaking.

As a control, the hydrolysis was carried out in the same manner as above, using only the proteolytic enzyme without the addition of the glutaminase. The ratio of free glutamic acid to soluble total nitrogen of the hydrolysate was 1.23 for the case that the glutaminase was added, and 0.69 for the case that no glutaminase was added. The amino acid solution thus obtained could be used as a seasoning or a raw material for others.

HISTIDINE

Production from *Serratia marcescens* Mutant

I. Chibata, M. Kisumi, M. Sugiura and N. Nakanishi; U.S. Patent 3,902,966; September 2, 1975; assigned to Tanabe Seiyaku Co., Ltd., Japan, have found that L-histidine can be prepared by cultivating an L-histidine ammonia-lyase-lacking and 2-methylhistidine-resistant mutant of *Serratia marcescens* in an aqueous nutrient medium. The mutants may be obtained by treating a wild type of *Serratia marcescens* with a mutagen and further treating the L-histidine ammonia-lyase-lacking mutant or the 2-methylhistidine-resistant mutant with a mutagen or ultraviolet rays. A viable culture of the mutant has been deposited with the ATCC under No. 31026.

The fermentation of an L-histidine ammonia-lyase-lacking and 2-methylhistidine-resistant mutant of *Serratia marcescens* may be accomplished by either shaking cultivation or submerged fermentation under aerobic conditions. The fermentation may be preferably carried out at a pH of 6 to 9. Calcium carbonate and ammonia may be employed for adjustment of the pH of the medium. The preferred temperature range for the fermentation is 25° to 37°C. The fermentation medium contains a source of carbon, a source of nitrogen and other elements. Suitable sources of carbon for the fermentation include saccharides; e.g., glucose, starch hydrolysate; organic acids, e.g., fumaric acid, citric acid; polyalcohols, e.g., glycerol; and hydrocarbons. Suitable sources of nitrogen include urea; organic ammonium salts, e.g., ammonium acetate; and inorganic ammonium salts, e.g., ammonium sulfate, ammonium nitrate. The preferred amount of the source of carbon and the source of nitrogen in the medium are within the range of 2 to 15 and 0.5 to 3% w/v, respectively.

Furthermore, organic nutrients, e.g., corn steep liquor, peptone, yeast extracts, and/or inorganics, e.g., potassium phosphate, magnesium sulfate, may be added to the medium. The fermentation can be accomplished in about 24 to 96 hours. L-histidine is accumulated in the fermentation broth.

After the fermentation is completed, cells and other solid culture compositions are removed from the fermentation broth by conventional procedures such as heating, followed by filtration or centrifugation. Known procedures may be employed in the recovery and/or purification of L-histidine from the filtrate or the supernatant solution.

Example: An aqueous nutrient medium (7.0 pH) comprising the following ingredients is prepared:

	Percent (w/v)
Glucose	3
Dextrin	10
Urea	2
Ammonium phosphate	1
Dibasic potassium phosphate	0.1
Magnesium sulfate 7 hydrate	0.05
Calcium carbonate	2

15 ml of the medium are charged into a 500 ml shaking flask and its contents are sterilized by autoclaving. Glucose and dextrin are separately sterilized and added to the medium aseptically. A loopful of the L-histidine ammonia-lyase-lacking and 2-methylhistidine-resistant mutant (Hd-MHr) ATCC No. 31026 of *Serratia marcescens* is inoculated into the medium. The medium is then cultivated for 72 hours at 30°C under shaking (140 rpm, 8 cm stroke). The fermentation medium thus obtained contains 5 mg/ml of L-histidine; 100 ml of the fermentation medium are heated at 100°C for 20 minutes and then filtered. The filtrate is passed through a column (1 x 10 cm) of strong cation exchange resin (H-form) (Amberlite IR-120). After washing with water, the column is eluted with 5% aqueous ammonia.

The fractions containing L-histidine are collected and concentrated under reduced pressure, and aqueous methanol is added to the concentrated solution. The crystalline precipitate is collected by filtration and recrystallized from aqueous methanol. 350 mg of L-histidine are obtained $[\alpha]_D^{25} = -37.38°$ (c = 2, in H_2O).

2-Thiazolealanine-Resistant Bacterial Strain

It has previously been disclosed in U.S. Patent 3,716,453 that histidine can be produced by fermentation by microorganisms of the genera *Brevibacterium*, *Corynebacterium* and *Arthrobacter*, which are mutants of parent strains incapable of producing histidine under the same conditions, the mutants being resistant to 2-thiazolealanine concentrations of more than 1 mg/ml of culture medium.

K. Kubota, H. Kamijo, O. Mihara, S. Okumura and H. Okada; U.S. Patent 3,875,001; April 1, 1975; assigned to Ajinomoto Co., Inc., Japan have found that other mutants belonging to the genera *Brevibacterium* and *Corynebacterium*, which combine resistance to 2-thiazolealanine with specific nutrient requirements, are substantially superior to the microorganisms of the earlier process in their ability of producing L-histidine by fermentation. The required nutrients are arginine, methionine, tryptophan, leucine, phenylalanine, lysine, threonine, xanthine, uracil, and/or shikimic acid.

The mutants are produced from the respective parent strains by conventional methods partly described in the prior process, as by exposure of the parent strain to ionizing radiation (ultraviolet rays, x-rays, gamma rays) or to chemical mutagenic agents (sodium nitrite, nitrosoguanidine, diethyl sulfate). Typically exposure to 250 μg/ml nitrosoguanidine at 30°C for 30 minutes will have the desired effect.

The mutants produced are inoculated on an otherwise conventional culture medium containing enough 2-thiazolealanine for suppressing the growth of the parent strain. This requires generally more than 1 mg/ml 2-thiazolealanine, and the best histidine-producing strains can grow on media containing 5 mg 2-thiazolealanine per milliliter. The resistant strains are collected and may be subjected to yet another treatment with mutagenic agents. The strains requiring the specific nutrients listed above are then selected from the first or second generation of mutants in a conventional manner.

They all produce histidine by aerobic fermentation of conventional aqueous culture media containing assimilable sources of carbon and nitrogen, and minor amounts of inorganic salts and organic nutrients including the specific required substance or substances.

The best histidine-producing strains available are listed below with their accession numbers and their specific nutrients in parentheses. They were derived from *Brevibacterium flavum* ATCC 14067, *Brevibacterium lactofermentum* ATCC 13869, and *Corynebacterium acetoacidophilum* ATCC 13870.

Brevibacterium flavum
   FERM-P 1561 (arginine)
   FERM-P 1562 (methionine)
   FERM-P 1563 (tryptophan)
   FERM-P 1564 (uracil)
   FERM-P 2168 (shikimic acid)
   FERM-P 2169 (xanthine)
   FERM-P 2170 (threonine)
Brevibacterium lactofermentum
   FERM-P 1565 (leucine)
   FERM-P 1566 (leucine and tryptophan)
   FERM-P 1567 (leucine and phenylalanine)
   FERM-P 1568 (phenylalanine)
Corynebacterium acetoacidophilum
   FERM-P 1569 (lysine)

For a good yield of histidine, the fermentation is carried out aerobically with aeration and agitation. Best yields require a pH controlled within the range from 5 to 9; the desired pH may be maintained by means of gaseous or aqueous ammonia, calcium carbonate, alkali metal hydroxide, urea, organic or inorganic acids, and some of these addition agents may also provide the necessary nitrogen supply. When the fermentation is carried out at 25° to 37°C, the histidine concentration in the broth reaches a maximum within 2 to 7 days. The histidine may be recovered from the culture medium by ion exchange resin treatment.

Example: *Brevibacterium flavum* FERM-P 1561 was cultured on bouillon agar slants. An aqueous fermentation medium containing, per deciliter, 10 g glucose, 5 g $(NH_4)_2SO_4$, 0.1 g KH_2PO_4, 0.04 g $MgSO_4 \cdot 7H_2O$, 0.2 mg Fe^{++}, 0.2 mg Mn^{++}, 50 μg biotin, 5.0 μg thiamine·HCl, 0.3 ml soy protein hydrolysate, 50 mg arginine and 5 g $CaCO_3$, was adjusted to pH 7.5; 20 ml batches of the medium were transferred to respective 500 ml shaking flasks, sterilized with steam, and thereafter inoculated with the microorganisms. Each culture medium was shaken for 72 hours at 31°C and the combined broths were then found to contain 1.11 g/dl histidine.

The microbial cells were removed from 1 liter of the broth by centrifuging, and the supernatant was stripped of histidine by passage over a column of the strongly acidic ion exchange resin, Amberlite IR-120 (H-type). The column was washed, and the histidine thereafter eluted with 3% ammonium hydroxide solution. The eluate was partly evaporated in a vacuum, and histidine crystallized from the concentrate was recovered in an amount of 7.3 g.

LEUCINE

Bacterial Strains Requiring Isoleucine, Threonine or Methionine

T. Tsuchida, H. Momose and Y. Hirose; U.S. Patent 3,970,519; July 20, 1976; assigned to Ajinomoto Co., Inc., Japan describe a process for producing L-leucine which comprises culturing an L-leucine-producing microorganism of the genus *Brevibacterium* or *Corynebacterium* under aerobic conditions in an aqueous medium containing a source of assimilable carbon and nitrogen, inorganic salts and nutriments at a pH of 5 to 9 until L-leucine accumulates in the medium, and recovering the accumulated L-leucine. The microorganism is characterized by: (a) requiring at least one of isoleucine, threonine or methionine as a growth nutriment, and (b) being resistant to feedback inhibition by leucine and analogs of leucine.

Particularly suitable mutant strains for use in the process are *Brevibacterium lactofermentum* AJ-3718 (FERM-P 2516), which is resistant to 2-thiazolealanine and β-hydroxyleucine and requires isoleucine and methionine; *Brevibacterium lactofermentum* AJ-3452 (FERM-P 1965), which is resistant to 2-thiazolealanine and β-hydroxyleucine and requires threonine; and *Brevibacterium lactofermentum* AJ-3719 (FERM-P 2517), which is resistant to 2-thiazolealanine and β-hydroxyleucine and requires isoleucine and methionine. These three mutants were all derived from the parent strain *Brevibacterium lactofermentum* ATCC 13869.

The following microorganisms: *Corynebacterium glutamicum* AJ-3453 (FERM-P 1966), which is resistant to 2-thiazolealanine and β-hydroxyleucine and requires isoleucine; *Corynebacterium glutamicum* AJ-3455 (FERM-P 1968, which is resistant to 2-thiazolealanine and β-hydroxyleucine and requires threonine; and *Coryne-bacterium glutamicum* AJ-3720 (FERM-P 2518), which is resistant to 2-thiazole-alanine and β-hydroxyleucine and requires isoleucine (all of which were derived from *Corynebacterium glutamicum* ATCC 13032) are also particularly useful in this process.

Conventional culture mediums may be used to produce L-leucine in accordance with the process. These normally include sources of assimilable carbon and ni-trogen and the usual minor constituents such as inorganic salts and organic nu-trients. Examples of the carbon source are carbohydrates such as glucose; fruc-tose; maltose; starch hydrolysate; cellulose hydrolysate or molasses; organic acids such as acetic, propionic or succinic; alcohols such as glycerol, methanol or etha-nol; any hydrocarbons such as n-paraffin. Useful nitrogen sources include ammo-nium sulfate, urea, ammonium nitrate, ammonium chloride or gaseous ammonia. Inorganic salts such as phosphates, magnesium, calcium, ferrous, manganese and other minor metallic salts are generally present. Amino acids, vitamins, soybean protein hydrolysate (Aji-Eki), yeast extracts, peptone and casamino acid are preferably employed for good bacterial growth. Of course, isoleucine, threonine or methioninc will be utilized.

The fermentation is performed at a pH of from 5 to 9, at a temperature of 24° to 37°C, under aerobic conditions for 2 to 7 days or until sufficient L-leucine accumulates. The pH of the culture medium can be adjusted by adding sterile calcium carbonate, urea, aqueous or gaseous ammonia, mineral acid or organic acid during the fermentation.

The L-leucine is recovered from the cultured broth by conventional methods. It may be identified by its R_f values in paper chromatography, its ninhydrin reaction, and by bioassay with *Leuconostoc mesenteroides* ATCC 8042.

Leucine-Antagonist-Resistant Bacterial Strains

S. Okumura, F. Yoshinaga, K. Kubota and H. Kamijo; U.S. Patent 3,865,690; February 11, 1975; assigned to Ajinomoto Co., Inc., Japan have found that some bacteria belonging to the genera *Brevibacterium* and *Corynebacterium*, and resis-tant to a leucine antagonist, produce a large amount of leucine when cultured in a nutrient medium. The L-leucine-producing mutants may be isolated from nat-ural sources, or can be derived by conventional mutant-inducing procedures such as x-ray radiation, ultraviolet light irradiation, or treatment with nitrosoguani-dines, diethyl sulfate or nitrite.

L-leucine-producing bacteria include *Brevibacterium flavum* AJ 3226 (FERM-P 420), which has been derived from *Brevibacterium flavum* ATCC 14067; *Brevi-bacterium lactofermentum* AJ 3427 (FERM-P 1769), which has been derived from *Brevibacterium lactofermentum* ATCC 13869; *Corynebacterium acetoacido-philum* AJ 3228 (FERM-P 421), which has been derived from *Corynebacterium acetoacidophilum* ATCC 13870; and *Corynebacterium glutamicum* AJ 3426 (FERM-P 1768), which has been derived from *Micrococcus glutamicus* ATCC 13032 (*Micrococcus glutamicus* is referred to as *Corynebacterium glutamicum*).

Leucine antagonists include 2-thiazolealanine; 4-azaleucine; 5,5,5-trifluoro-leucine; D-leucine, α-amino-isoamylsulfonic acid; norvaline; norleucine; methallyl-glycine; α-amino-β-chlorobutyric acid; δ-chloroleucine; β-hydroxynorleucine;

β-hydroxyleucine, cyclopentanealanine, 3-cyclopentene-1-alanine and 2-amino-4-methyl-hexanoic acid.

The culture medium used to produce L-leucine in the process may be entirely conventional. It includes an assimilable carbon source, an assimilable nitrogen source, and the usual minor nutrients. The fermentation is performed at a pH between 5 and 9, at a temperature of 24° to 37°C, under aerobic conditions for 2 to 7 days. The pH of the culture medium can be adjusted by adding sterile calcium carbonate, aqueous or gaseous ammonia, mineral acid or organic acid during the fermentation. The L-leucine is recovered from the cultured broth by conventional methods.

Example: A culture medium containing 10 g/dl glucose, 0.1 g/dl KH_2PO_4, 0.04 g/dl $MgSO_4 \cdot 7H_2O$, 4 g/dl $(NH_4)_2SO_4$, 100 γ/l biotin, 200 γ/l vitamin B_1 hydrochloride, 2 ppm Fe and Mn ions, 1 ml/dl soybean protein hydrolysate (Aji-Eki), and 5 g/dl $CaCO_3$ was prepared (pH 7.0). 300 ml batches of the medium were each placed in a small glass jar-fermentor, and sterilized.

The medium was inoculated with *Brevibacterium flavum* AJ 3226 (FERM-P 420) which had been previously cultured on a bouillon slant at 30°C for 24 hours and at 31°C for 48 hours, with stirring and aerating. The culture broth was found to contain 1.11 g/dl of L-leucine. 1 liter of broth was centrifuged to remove bacterial cells; the supernatant was passed through a column packed with a cation exchange resin (Duolite C x 20); and after washing with water, L-leucine was eluted with 1 N–NH_4OH solution. The elutate was concentrated to remove ammonia, treated with active charcoal, and pure crystalline L-leucine was obtained in an amount of 5.2 g.

LYSINE

Production Using Antibiotics, Surface-Active Agents and Antioxidants

L-lysine may be formed from fermentable carbohydrates by microorganisms as is well-known. Known L-lysine-producing microorganisms include a homoserine- or methionine- and threonine-requiring mutant of *Micrococcus glutamicus* disclosed in U.S. Patent 2,979,439; the threonine- or methionine-sensitive mutants and threonine-sensitive and threonine-requiring mutants of *Brevibacterium flavum* disclosed in U.S. Patent 3,616,218; mutants of *Brevibacterium flavum, Corynebacterium acetoglutamicum, Microbacterium ammoniaphilum,* or *Micrococcus glutamicus* resistant to the lysine analogue S-(2-amino-ethyl)-L-cysteine (AEC) as disclosed in U.S. Patent 3,707,441; and a mutant of *Arthrobacter* combining resistance to AEC with one or more nutrient requirements which is disclosed in Belgian Patent 798,890.

K. Kubota, Y. Yoshihara and Y. Hirose; U.S. Patent 3,929,571; December 30, 1975; assigned to Ajinomoto Co., Inc., Japan have found that the metabolism of mutant strains normally used for producing L-lysine can be influenced by antibiotics, surface-active agents and/or antioxidants in the culture media, and that L-lysine can be produced in high yields and high concentrations in such modified culture media.

The microorganisms used in the method may be of any L-lysine-producing strain such as those mentioned above, and may be derived by means of conventional mutagenic agents from parent strains of the genera *Brevibacterium, Corynebacterium, Microbacterium,* and *Arthrobacter,* and by screening of the mutants so produced for the necessary resistance to AEC, and/or nutrient requirement.

The addition agents suitable for the method include the following substances:

(a) Antibiotics: chloramphenicol, erythromycin, Oleandomycin, kanamycin, streptomycin, kasugamycin, tetracycline, oxytetracycline, mytomycin C, actinomycin D, cycloserine, members of the penicillin and cephalosporin groups, polymyxin and azaserine. Chloramphenicol, erythromycin, Oleandomycin, streptomycin and kasugamycin, which are thought to react with the protein synthesizing part of the microorganism, were found to be especially effective.

(b) Surface active agents: anionic surface active agents such as higher alcohol sulfates, alkylbenzene sulfonates, alkyl phosphates and dialkyl sulfo-succinates; cationic surface active agents such as alkylamine salts and quaternary ammonium salts; nonionic surface active agents such as polyoxyethylene alkyl ether, polyoxyethylene sorbitan monoalkyl ether and sorbitan monolaurate; amphoteric surface active agents such as imidazoline and betadine;

(c) Antioxidants: 4,4'-dihydroxy-3,3'-dimethyldiphenyl; 2,6-di-3-butylphenol; catechol; butylcatechol; protocatechuic acid; α-tocopherol; pyrogallol; gallic acid; esters of gallic acid; naphthol; phenolic compounds such as aminophenol; amines such as naphthylamine, diphenylamine, di-2-butyl-p-phenylenediamine, 6-ethoxy-2,3,4-trimethyl-1,2-dihydroquinoline, semicarbazide, phenothiazine and tetraphenylhydrazine; and sulfur-containing compounds such as thiodipropionic acid and thio-diglycolic acid.

The additives are added to the culture medium in an amount to inhibit the growth of the microorganism to some extent. They may be added in one batch at the beginning of the cultivation or during the cultivation, or added intermittently during the cultivation.

It is important that the pH of the fermentation medium be controlled for optimum production of L-lysine. It is preferred to maintain the pH between 5.0 and 9.0. This is conveniently accomplished by addition of urea, ammonia, calcium carbonate, organic or inorganic acids as may be necessary during the fermentation. The fermentation is carried out at temperatures of about 24° to 37°C, but the optimum temperature will vary depending upon the microorganism employed. The fermentation is conducted under aerobic conditions for about 2 to 7 days. The L-lysine thus produced may be recovered from the broth by methods known in the art. These include adsorption on and elution from suitable ion exchange resins, removal of the cells and concentration of the filtrate containing L-lysine, and the like.

Use of Methanol

K. Nakayama, M. Kohata, Y. Tanaka, T. Nomura and R. Katsumata; U.S. Patent 3,907,637; September 23, 1975; assigned to Kyowa Hakko Kogyo, Co., Ltd., Japan have found that mutant strains of the genus *Protaminobacter*, which can utilize methanol and have a resistance to both S-2-aminoethyl-L-cysteine and

L-threonine produces L-lysine in a high yield in a medium containing methanol as a main source of carbon. A particularly preferred mutant is *Protaminobacter thiaminophagus* SLR-77 (ATCC 21926), which is derived from a methanol-utilizing strain of *Protaminobacter thiaminophagus* (ATCC 21371) disclosed in U.S. Patent 3,663,370. The mutant is deposited with the ATCC and is freely available to the public.

Although the preferred mutant is derived from *Protaminobacter thiaminophagus*, any microorganism of the genus *Protaminobacter* which is capable of utilizing methanol may be mutated to possess a resistance to S-2-aminoethyl-L-cysteine and L-threonine. In obtaining mutants suitable for the process any of the conventional methods for inducing mutation to obtain a strain having a resistance may be employed.

Methanol to be used as the carbon source sometimes causes the growth inhibition of microorganisms when present at a high concentration in the medium. Generally, it is desirable that the concentration of methanol in the medium be maintained below about 3% v/v. Good results can be obtained when a medium initially having a low concentration, for example, 0.5 to 3% v/v of methanol, is used and culturing is carried out while feeding methanol to the medium continuously in an amount of 0.3 to 0.6% by volume based on the volume of the medium per hour, or intermittently in an amount of 0.5 to 2% by volume based on the volume of the medium at each feeding, as methanol is consumed by the microorganism.

As the nitrogen source, ammonium salts such as ammonium chloride, ammonium sulfate, ammonium phosphate and ammonium nitrate, and urea may be used. Casamino acid, peptone and yeast extract may also be used as the nitrogen source. Additionally, as inorganic materials, potassium phosphates, magnesium sulfate, iron and manganese salts may be used.

Where the strain used has a nutritional requirement, an appropriate source of the nutrient must, of course, be supplemented to the medium. *Protaminobacter thiaminophagus*, for example, requires thiamine for growth. Therefore, when the mutant of *Protaminobacter thiaminophagus* is used, it is necessary to add pure thiamine or a natural substance containing thiamine to the medium. Culturing is carried out under aerobic conditions at 20° to 40°C for 2 to 5 days. In order to obtain a high yield of the product, it is desirable that the pH of the medium be maintained at 4 to 9, preferably at around neutral, during culturing. The pH may be adjusted with calcium carbonate, various buffer solutions or alkaline solutions.

After completion of culturing, the microbial cells are removed from the culture liquor by, for example, filtration. The L-lysine thus obtained in the culture liquor is isolated and purified by any of the methods well-known in the art, such as an ion exchange resin treatment, crystallization by concentration, etc.

Example: In this example, *Protaminobacter thiaminophagus* SLR-77 (FERM-P 2020) (ATCC 21926) is used. This strain is inoculated in 10 ml of a seed medium having the following composition in a test tube and cultured at 30°C for 20 hours with shaking.

Methanol	20 ml
$(NH_4)_2SO_4$	10 g
KH_2PO_4	2 g
K_2HPO_4	7 g
$MgSO_4 \cdot 7H_2O$	0.5 g
$FeSO_4 \cdot 7H_2O$	10 mg
$MnSO_4 \cdot 4H_2O$	8 mg
Thiamine hydrochloride	1 mg
Biotin	10 μg
Yeast extract	0.1 g
$CaCO_3$	20 g
Water to make up 1 liter	
pH	7.2

The resulting seed culture is inoculated into 10 ml portions of a fermentation medium having the same composition as the seed medium in test tubes in a ratio of 1 ml. Fermentation is carried out with shaking at 30°C for 72 hours, while feeding 1% of methanol after 16 hours of culturing and 2% of methanol after 25 and 40 hours of culturing, respectively. After the completion of fermentation 1.5 mg/ml (as hydrochloride) of L-lysine is produced in the culture liquor.

The culture liquors in the test tubes are combined to make the total volume 2 liters, and the microbial cells are removed from the culture liquor. The cell-free culture liquor is subjected to adsorption on a strongly acidic cation exchange resin [Amberlite IR-120 B (Rohm & Haas Co.)]. Elution is carried out with an aqueous ammonia. The eluate is concentrated and neutralized with hydrochloric acid. After crystallization, 2.4 g of crystals of L-lysine hydrochloride is obtained.

In a similar process, *K. Nakayama, M. Kohata, Y. Tanaka, T. Nomura and R. Katsumata; U.S. Patent 3,907,641; September 23, 1975; assigned to Kyowa Hakko Kogyo Co., Ltd., Japan* describe a process for producing L-lysine, L-aspartic acid, L-alanine, L-valine, L-leucine and L-arginine, which comprises culturing *Protaminobacter thiaminophagus* ATCC 21927 in a nutrient medium containing methanol as the carbon source under aerobic conditions.

L-Lysine Analogues as Metabolic Antagonists

Y. Kurimura, Y. Furutani, N. Makiguchi and K. Souda; U.S. Patent 3,905,867; September 16, 1975; assigned to Mitsui Toatsu Chemicals, Incorporated, Japan have found that L-lysine analogues have a physiological activity as L-lysine metabolic antagonists, and that when a mutant which has a resistance to one of these metabolic antagonists is cultured in a suitable culturing medium, L-lysine is produced and accumulated in a large amount. The L-lysine produced is easily recovered from the medium.

The microorganism used in this process is a strain capable of producing L-lysine selected from mutants resistant to one of the L-lysine metabolic antagonists, and is easily obtained by mutant-inducing methods well-known in the art from parental strains selected from L-glutamic acid-producing microorganisms belonging to the genera *Arthrobacter, Corynebacterium* and *Brevibacterium.*

The culturing medium employed in this process must contain an assimilable carbon source, an assimilable nitrogen source and the usual minor nutrients.

The carbon sources are carbohydrates such as glucose, molasses, starch hydroly-sates; organic acids such as acetic acid; alcohols such as methyl alcohol; and hydrocarbons such as n-paraffins. The nitrogen sources are ammonium salts of inorganic acids such as ammonium sulfate and ammonium chloride; nitrate salts such as potassium nitrate and sodium nitrate; ammonia in an aqueous solution or in the gaseous state; and urea. Structures of L-lysine analogs follow:

Compound (a)	Compound (b)	Compound (c)
CH_2-NH_2	CH_2-NH_2	CH_2-NH_2
CH_2	CH_2	CH_2
CH_2	S	$S=O$
CH_2	CH_2	CH_2
$CH-NH_2$	$CH-NH_2$	$CH-NH_2$
$CONHOH$	$CONHOH$	$COOH$

Compound (d)	Compound (e)	Compound (f)
CH_2-NH_2	$CH_2-\!\!\langle\text{pyridine}\rangle$	$CH_2-\!\!\langle\text{pyridine}\rangle$
CH_2	CH_2	CH_2
$S-CH_3$	S	S
CH_2	CH_2	CH_2
$CH-NH_2$	$CH-NH_2$	$CH-NH_2$
$COOH$	$COOH$	$COOH$

Compound (a) was synthesized from L-lysine ethyl ester and hydroxylamine;

Compound (b) was synthesized from HCl, hydroxylamine and the ethyl ester of S-(β-amino ethyl)-L-cysteine·HCl;

Compound (c) was synthesized: S-(β-amino ethyl)-L-cysteine·HCl was dissolved in acidic solution, and then oxidized with hydrogen peroxide or potassium periodate or iodine, and compound (c) was obtained as hydrogen chloride salt;

Hydrogen chloride salts of compound (d) were synthesized from S-(β-amino ethyl)-L-cysteine·HCl and methyl iodate;

Compound (e) was synthesized from L-cysteine and 2-vinyl pyridine;

Compound (f) was synthesized from L-cysteine and 4-vinyl pyridine.

The L-lysine-producing mutants are selected from the group consisting of:

Arthrobacter B-1 1772-193, FERM-P 1295, ATCC 21868
Arthrobacter B-1 55-8, FERM-P 1296, ATCC 21867
Brevibacterium B-4 1506-13, FERM-P 1297, ATCC 21866
Brevibacterium B-4 1899-31, FERM-P 1298, ATCC 28165
Brevibacterium B-4 1433-54, FERM-P 1299, ATCC 21864
Corynebacterium 18S 728-26, FERM-P 1300, ATCC 21863
Corynebacterium 18S 351-23, FERM-P 1301, ATCC 21862
Arthrobacter B-1 874-75, FERM-P 1302, ATCC 21861
Brevibacterium B-4, 1304-31, FERM-P 1303, ATCC 28160
Arthrobacter B-1 1162-166, FERM-P 1304, ATCC 28159
Arthrobacter B-1 269-51, FERM-P 1305, ATCC 28158
Corynebacterium 18S 1202-138, FERM-P 1306, ATCC 21857.

Production Using *Brevibacterium* or *Corynebacterium* Mutants

O. Tosaka, H. Morioka, H. Hirakawa, K. Ishii, K. Kubota and Y. Hirose; U.S. Patent 4,066,501; January 3, 1978; assigned to Ajinomoto Co., Inc., Japan, have found that a mutant of the genus *Brevibacterium* or *Corynebacterium*, which is resistant to one of α-aminolauryllactam (ALL), γ-methyl-lysine (ML) and N^ω-carbobenzoxylysine (CBL) can produce L-lysine in higher yield than the known lysine-producing mutants. Specimens of the mutants used in this process are:

Brevibacterium lactofermentum AJ 3985 (FERM-P 3382) [CBL$^\gamma$ (resistant to CBL)]
Brevibacterium lactofermentum AJ 3986 (FERM-P 3383) (ML$^\gamma$)
Brevibacterium lactofermentum AJ 3987 (FERM-P 3384) (CBL$^\gamma$, AEC$^\gamma$)
Brevibacterium lactofermentum AJ 3988 (FERM-P 3385) (ALL$^\gamma$, AEC^{65})
Brevibacterium lactofermentum AJ 3989 (FERM-P 3386) [CBL$^\gamma$, AEC$^\gamma$, Ala⁻ (requiring alanine for growth)]
Brevibacterium lactofermentum AJ 3990 (FERM-P 3387) (ML^{65}, AEC$^\gamma$, Ala⁻) (ATCC 31269)
Corynebacterium acetoglutamicum AJ 3983 (FERM-P 3380) (CBL$^\gamma$, AEC$^\gamma$) (ATCC 31270)
Corynebacterium acetoglutamicum AJ 3984 (FERM-P 3381) (ALL$^\gamma$, AEC$^\gamma$)
Corynebacterium acetoglutamicum AJ 3991 (FERM-P 3414) (ML$^\gamma$) (AEC: S-(2-aminoethyl)L-cysteine)

The methods for producing L-lysine using the microorganisms mentioned above are conventional, and the microorganisms are cultured in a conventional medium containing carbon sources, nitrogen sources, inorganic salts, nutrients required for growth, and other minor nutrients.

As the carbon source, carbohydrates such as glucose, sucrose, molasses, or starch hydrolysate; organic acids such as acetic acid, propionic acid, or benzoic acid; alcohols such as ethanol, or propanol; and, for certain strains, hydrocarbons can be used. As the nitrogen source, ammonia, ammonium sulfate, ammonium nitrate, ammonium phosphate, urea, etc., can be used.

Nutrients required for growth can be used as purified ones or as natural substances containing the nutrients such as soybean hydrolysate, corn steep liquor, yeast extract or peptone.

Cultivation is carried out under aerobic conditions at a temperature of from 24° to 37°C for 2 to 7 days. During the cultivation the pH of the medium is adjusted to 5 to 9 by alkali or acid, or calcium carbonate, urea or gaseous ammonia.

L-lysine in the culture broth thus obtained can be separated by known methods such as by using ion exchange resins, or by directly crystallizing L-lysine from the culture broths.

Example: The culture medium mentioned below (20 ml) was placed in 500 ml shaking flasks and heated at 110°C for 5 minutes.

Glucose	10 g/dl
Ammonium sulfate	5 g/dl
KH_2PO_4	0.1 g/dl
$MgSO_4 \cdot 7H_2O$	0.04 g/dl
$FeSO_4 \cdot 7H_2O$	1.0 mg/dl

(continued)

$MnSO_4 \cdot 4H_2O$	1.0 mg/dl
Biotin	5.0 μg/dl
Thiamine·HCl	20.0 μg/dl
Soybean hydrolysate (7% total nitrogen)	1.5 ml/dl
Calcium carbonate (separately sterilized)	5 %
pH	7.0

Each of the microorganisms shown above was inoculated in the culture medium and the culture medium was held at 31°C for 72 hours with shaking. After the cultivation, the amount of L-lysine in the resultant culture medium was determined, and is shown below as the amount of L-lysine hydrochloride.

	L-Lysine Accumulated (g/l)
AJ 3985	2.0
AJ 3986	1.5
AJ 3987	27
AJ 3988	28
AJ 3989	37
AJ 3990	36
AJ 3445	18

AJ 3445 (AEC$^\gamma$) is an L-lysine-producing mutant of *Brevibacterium lactofermentum*, which does not have the resistance to CBL, ML or ALL.

Production from *Corynebacterium glutamicum*

K. Nakayama, K. Araki and Y. Tanaka; U.S. Patent 4,169,763; October 2, 1979; assigned to Kyowa Hakko Kogyo Co., Ltd., Japan, describe a process for producing L-lysine by fermentation, characterized by culturing a strain belonging to the genus *Corynebacterium* and having both an ability to produce L-lysine and a resistance to at least one member selected from the group consisting of aspartic acid analogs and sulfa drugs in a nutrient medium, and recovering L-lysine formed and accumulated from the culture liquor.

Examples of aspartic acid analogs are aspartic acid hydroxamate, α-methylaspartic acid, β-methylaspartic acid, cysteinesulfinic acid, difluorosuccinic acid, hadacidin, etc.; examples of sulfa drugs are sulfaguanidine, sulfadiazine, sulfamethazine, sulfamerazine, sulfamethizole, sulfamethomidine, sulfamethoxypyridazine, sulfathiazole, homosulfamine, sulfadimethoxine, sulfamethoxazole, sulfisoxazole, etc.

Among the strains to be used in the process, *Corynebacterium glutamicum* FERM-P 3633 (NRRL B-8182) may be mentioned as one example of the strain having a resistance to aspartic acid analog, and *Corynebacterium glutamicum* FERM-P 3634 (NRRL B-8183) as one example of the strain having a resistance to sulfa drugs.

Any of synthetic medium and natural medium may be used as the medium for the process, so long as it properly contains a carbon source, nitrogen source, inorganic materials, and other necessary nutrients.

The productivity of L-lysine by the microorganism can be further enhanced by addition of a leucine fermentation liquor to the medium. In this case, it is preferable to add a leucine fermentation liquor in an amount ranging from 0.2 to 15% by volume of the medium. Further, productivity of L-lysine by the microorganism can be also enhanced by adding other various additives, for example, various antibiotics, α-aminobutyric acid, cysteine, norleucine, leucine, aspartic acid, glutamic acid, etc., to the medium.

Culturing is carried out under aerobic conditions, for example, by shaking culture, agitating submerged culture, etc. The temperature for culturing is generally 20° to 40°C, and the pH of the medium is in a range of 3 to 9, and is preferably maintained at around neutral, but culturing can be carried out under conditions which are out of this range so long as the microorganism used can grow. The pH of the medium is adjusted with calcium carbonate, organic or inorganic acid or ammonia, alkali hydroxide, pH buffering agent, etc. Usually after culturing for 1 to 7 days, L-lysine is formed and accumulated in the resulting culture liquor.

After the completion of culturing, precipitates such as cells, etc., are removed from the culture liquor, and L-lysine can be recovered by use of the conventional methods such as ion exchange resin treatment, concentration, adsorption, salting-out, etc.

Example: *Corynebacterium glutamicum* FERM-P 3633, NRRL B-8182, (having a resistance to aspartic acid hydroxamate) is used as a seed strain. The seed strain is inoculated in a large test tube of 50 ml (190 x 20 mm) containing 7 ml of seed medium (7.2 pH) comprising 4 g/dl of glucose, 0.3 g/dl of urea, 0.15 g/dl of KH_2PO_4, 0.05 g/dl of K_2HPO_4, 0.05 g/dl of $MgSO_4 \cdot 7H_2O$, 50 μg of biotin, 2 g/dl of peptone, and 0.5 g/dl of yeast extract; and cultured at 30°C for 24 hours. 2 ml of the resulting seed culture is inoculated in an Erlenmeyer flask of 300 ml containing 20 ml of a fermentation medium (7.2 pH) comprising 8.5 g/dl of blackstrap molasses (as glucose), 2 g/dl of soybean cake acid hydrolysate (as soybean cake), 0.5 g/dl of ammonium sulfate, 0.3 g/dl of urea, 0.05 g/dl of $MgSO_4 \cdot 7H_2O$, 0.07 g/dl of KH_2PO_4 and 3 g/dl of calcium carbonate; and cultured with shaking at 30°C for 3 days.

As a result, 35 mg/ml of L-lysine (as monohydrochloride) is formed and accumulated in the culture medium. The amount of L-lysine by parent strain ATCC 21543, cultured at the same time under the same conditions as a control, is 25 mg/ml.

After completion of culturing, 1 liter of the culture liquor of the present strain is centrifuged to remove the cells and other precipitates. Supernatant is passed through a column of strongly acidic ion exchange resin (H^+) [Diaion SK-1 (Mitsubishi Chemical Industries Ltd.)] to adsorb L-lysine.

After washing the column with water, the column is eluted with dilute aqueous ammonia and then fractions containing L-lysine are collected and concentrated. After pH of the concentrate is adjusted to 2 by hydrochloric acid, the concentrate is cooled; by addition of ethanol thereto, L-lysine is crystallized. As a result, 26.5 g of crystals of L-lysine hydrochloride are obtained.

Production from *Nocardia alkanoglutinousa*

T. Tanaka, Y. Nakamura, K. Asahi, T. Shiraishi and K. Takahara; U.S. Patent 4,123,329; October 31, 1978; assigned to Kanegafuchi Kagaku Kogyo KK, Japan have found that mutants of the genus *Nocardia* produce unexpectedly large amounts of L-lysine when cultivated in a culture medium having an assimilable carbon source and an assimilable nitrogen source. These are *Nocardia alkano-glutinousa* 223-59 (ATCC 31220) and *Nocardia alkanoglutinousa* 223-15 (ATCC 31221).

Useful carbon sources which may be used in the culture media include, for example, carbohydrates such as glucose or fructose; a hydrolysis product of molasses or starch; organic acids such as acetic acid or citric acid; alcohols such as ethanol; animal oils and fats; vegetable oils and fats; fatty acids; n-alkanes containing 10 to 30 carbons; and hydrocarbons such as kerosene or crude oils. Examples of a nitrogen source which may be added to the culture media are ammonium acetate, ammonium sulfate, ammonium chloride, ammonium nitrate, ammonia water, amino acids, amino acid mixtures, yeast extract, peptone and meat extract. As required, inorganic salts such as phosphates, magnesium salts, calcium salts, potassium salts, sodium salts, iron salts, manganese salts and zinc salts, and traces of other metals may be added to the culture medium.

These aforementioned components of the fermentation medium may be added to the medium before sterilization or may be added to the medium in divided portions during the fermentation.

Cultivation is carried out with shaking or with aeration and stirring at a pH of from 6 to 9, preferably 6 to 8, and at a temperature of $20°$ to $40°C$, preferably $27°$ to $37°C$. After the completion of the cultivation, the culture broth has high viscosity and is sometimes difficult to filter by ordinary filtering methods. Its filterability can be improved by adding a mineral acid to adjust the pH to 1 to 3 or by adding an alkali to adjust the pH to at least 9, and then heating the culture broth to $80°C$ or more for at least 5 minutes.

L-lysine is recovered from the filtrate in a customary manner. Specifically, the filtrate is passed through an ion exchange resin such as IR-120 or IRC-84 for adsorption of the L-lysine, then washed with water, and eluted with an ammonia solution. The eluate is concentrated, neutralized with concentrated hydrochloric acid and dried to easily obtain L-lysine hydrochloride of high purity. The amount of L-lysine produced was measured by a bioassay method using a mutant strain of *Escherichia coli* or by using an amino acid analyzer.

Example: One loopful of cells was taken from a bouillon slant culture of each of *Nocardia alkanoglutinousa* 223-59 (ATTC 31220) and *Nocardia alkanoglutinousa* 223-15 (ATCC 31221) cultivated at $33°C$ for 1 day and inoculated into a large-sized test tube containing 10 ml of a lysine-producing culture medium having the composition set forth below:

n-Alkane (C_{14-18})	50.0 g
Ammonium sulfate	40.0 g
$CaCO_3$	30.0 g
K_2HPO_4	0.5 g
KH_2PO_4	0.5 g
$MgSO_4 \cdot 7H_2O$	0.5 g
NaCl	1.0 g
$FeSO_4 \cdot 7H_2O$	20.0 mg
$ZnSO_4 \cdot 7H_2O$	10.0 mg
$MnSO_4 \cdot 4H_2O$	10.0 mg
Tap water	940.0 ml
pH	7.0

The culture medium was steamed for 15 minutes at 120°C. It was then culti-vated with shaking at 33°C for 1 day. 1 ml of the resulting culture broth was inoculated into a 500 ml Sakaguchi flask containing 30 ml of the same lysine-producing culture medium and cultivated with shaking at 33°C for 5 days. The amounts (g/l) of L-lysine produced in the culture media (as hydrochlorides) were as follows: 19.5 g/l from *Nocardia alkanoglutinousa* 223-59 (ATCC 31220) and 28.7 g/l from *Nocardia alkanoglutinousa* 223-15 (ATCC 31221).

Production Using an *Acinetobacter* Strain

T. Tanaka, T. Hirakawa and K. Takahara; U.S. Patent 3,920,520; November 18, 1975; assigned to Kanegafuchi Chemical Industries Co., Ltd., Japan describe a fermentation process for producing L-lysine from hydrocarbons as a carbon source, which comprises cultivating an L-lysine-producing mutant strain of a microorganism belonging to the genus *Acinetobacter.*

A typical example of the microorganism belonging to the genus *Acinetobacter* is species No. 38 (ATCC 31023). This strain was found to produce only a small amount of L-lysine in a culture medium containing n-paraffins, but it was found that various mutant strains having an improved L-lysine-producing ability can be obtained from the above species when this species is subjected to a muta-tional treatment such as ultraviolet ray radiation or a treatment with mutagenic agents such as nitrosoguanidine, nitrogen mustard and the like according to the well-established mutant-producing technique. These mutant strains were desig-nated as No. 38-15 (ATCC 31024), No. 38-19 (ATCC 31025) and No. 38-20 (ATCC 31038). The process is based on the above finding and production of L-lysine can be advantageously conducted on an industrial scale using one of the above mutant strains.

The mutant strains having an excellent ability to produce L-lysine can be derived from the parent species by a mutagenic treatment or an adaptation using at least one of amino acids or amino acid analogues. More specifically, a culture of *Acinetobacter* No. 38 can be treated with a mutagenic agent such as nitroso-guanidine (N-methyl-N'-nitro-N-nitrosoguanidine) or nitrogen mustard or treated with ultraviolet irradiation, or can be subjected to an adaptation. The adaptation can be conducted by culturing the parent species in a culture medium containing a specific amino acid or amino acid analogue or a combination thereof, for ex-ample, amino acids such as threonine, valine, methionine, lysine, serine and the like; amino acid analogues such as norvaline, β-hydroxy-norvaline, α-amino-β-chlorobutyric acid, S-aminoethylcysteine, lysine hydroxide, serine hydroxide, methionine hydroxide, valine hydroxide and the like, advantageously at a con-centration higher than about 0.5 g of the amino acids or amino acid analogues or a combination thereof per 1 liter of the culture medium.

The mutant strain can then be obtained by diluting the resulting culture appro-priately and spreading the diluted culture on a minimum agar plate containing the same amino acids or acid analogues or the same combination thereof as used above and culturing the strain for 3 to 5 days to harvest the resulting colonies. Examples of the carbon source which can be used for the minimum agar medium are n-paraffins, organic acids, ethanol and the like.

The strain thus obtained is repeatedly subjected to the above treatment using a different amino acid or amino acid analogue or a combination thereof to obtain a strain which is resistant to various amino acids or amino acid analogues or a combination thereof. The resulting resistant strain has an excellent ability to produce L-lysine in the aerobic cultivation using n-paraffins.

The addition of surface-active agents often advantageously affects the fermentation. For example, polyoxyethylene sorbitan trioleate [Tween 85 (Atlas Powder Co.)] can be effectively used in the culture medium in an amount of approximately 0.01 to 0.5% by weight based on the total amount of the culture medium.

The pH of the culture medium is maintained in the range of about 5.5 to 9.0, preferably about 6.0 to 8.0 during the entire cultivation period. The pH can preferably be maintained at the above value by adding ammonium ions. The cultivation is carried out at a temperature of about 25° to 40°C, preferably from 30° to 37°C. It is necessary that the cultivation be conducted under aerobic conditions, for example, by stirring with aeration or by shake-culturing. Upon completion of the cultivation, the resulting microbial cells can be removed from the culture broth using well-known techniques such as filtration or centrifuging. In a preferred embodiment, the desired L-lysine can be separated from the thus obtained filtrate or the supernatant in accordance with the well-known procedures using an ion exchange resin such as Amberlite IRC-84, IRC-120, IRC-50 and the like.

The ion exchange resin is then eluted with aqueous ammonia and the resulting eluate is concentrated and neutralized with concentrated hydrochloric acid. The resulting L-lysine hydrochloride can then be dried to give a product having a purity higher than 98%.

Example: A platinum loopful amount of a bouillon slant culture of *Acinetobacter* No. 38-15 which had been cultured at a temperature of 33°C for 2 days was used to inoculate 20 ml of an L-lysine-producing culture medium contained in a 500 ml Sakaguchi flask. The inoculated strain was then shake-cultured at a temperature of 33°C for 4 days. The medium used in this example contained ethanol as a carbon source and had the following composition on a percentage basis.

Ethanol (fed incrementally)	5
K_2HPO_4	0.1
KH_2PO_4	0.1
$MgSO_4 \cdot 7H_2O$	0.05
$FeSO_4 \cdot 7H_2O$	0.001
$ZnSO_4 \cdot 7H_2O$	0.001
$MnSO_4 \cdot 4H_2O$	0.001
Polyoxyethylene sorbitan trioleate*	0.1
$CaCO_3$	1
$(NH_4)_2SO_4$	1

*Tween 85

The medium was sterilized for 15 minutes at 120°C. Upon completion of the cultivation, the resulting culture broth was found to contain 4.1 g L-lysine per 1 liter of the culture broth.

AEC-Resistant *Corynebacterium* Strain

K. Kubota, Y. Yoshihara and H. Okada; U.S. Patent 3,871,960; March 18, 1975;

assigned to Ajinomoto Co., Inc., Japan, have found that mutants of *Corynebacterium* combining resistance to feedback inhibition by lysine and its analogues such as AEC [S(2-aminoethyl)-L-cysteine] in concentrations of 1 mg/ml or more with methionine sensitivity or methionine sensitivity and a nutrient requirement for proline and/or arginine produce large amounts of L-lysine in a culture medium. A methionine-sensitive microorganism is defined as a strain whose growth during 24 hours' cultivation on a minimal medium supplemented with 0.1 mmol of L-methionine is inhibited severely and whose growth inhibition by L-methionine is not overcome by addition of L-threonine.

Three suitable strains were deposited as *Corynebacterium glutamicum* AJ 3400 (FERM-P 1639), AJ 3609 (FERM-P 2278) and AJ 3610 (FERM-P 2279). A synthetic culture medium or a natural nutrient medium is suitable for cultivation of the strains employed in the process as long as it contains the essential nutrients for the growth of the strain employed and which include a carbon source, a nitrogen source and inorganic compounds in appropriate amounts.

In order to obtain a good yield of lysine, the fermentation is preferably carried out aerobically with aeration and agitation. Best yields require pH control within the range of 5 to 9. The desired pH may be maintained by means of gaseous or aqueous ammonia, calcium carbonate, alkali metal hydroxides, urea, or organic or inorganic acids. When the fermentation is carried out at 24° to 37°C; the maximum concentration of L-lysine in the broth is usually reached within 2 to 7 days. The lysine accumulated in the fermentation broth can be recovered by conventional methods, as by adsorption on an ion exchange resin and precipitation from the eluate.

Example: 20 ml batches of a medium containing 10% glucose, 4.5% $(NH_4)_2SO_4$, 0.1% KH_2PO_4, 0.04% $MgSO_4 \cdot 7H_2O$, 2 ppm Mn^{++}, 2 ppm Fe^{++}, 50 μg/l biotin, 200 μg/l thiamine hydrochloride and 5% calcium carbonate (sterilized separately), with a pH of 7.0, were placed in separate 500 ml shaking flasks and sterilized.

Corynebacterium glutamicum AJ 3400 (FERM-P 1638), previously cultured on bouillon agar slants, was introduced into the flasks and cultured at 31°C with aeration and agitation for 72 hours. The cultured broth was found to contain 3.1 g/dl lysine and centrifuged. 1 liter of the supernatant liquid was passed through a column packed with an ion exchange resin, Amberlite IR-120 (H⁺ type), and lysine was eluted with 3% aqueous ammonia. The eluate was concentrated under reduced pressure. Hydrochloric acid was added to the concentrated solution which was then cooled in an ice box to precipitate L-lysine; 20.7 g of crude L-lysine hydrochloride dihydrate was obtained.

Production from L-Valine-Resistant Microorganism

K. Watanabe, T. Tanaka, T. Hirakawa, H. Kinoshita, M. Sasaki and K. Obayashi; U.S. Patent 3,905,866; September 16, 1975; assigned to Kanegafuchi Chemical Industries, Japan describe a process for producing L-lysine which comprises aerobically culturing an L-lysine-producing microorganism having a high resistance to L-valine, L-threonine or amino acid analogues of L-valine or L-threonine selected from the mutants belonging to the genus *Pseudomonas* and the genus *Achromobacter* in a culture medium containing a hydrocarbon as a main carbon source until a substantial amount of L-lysine is accumulated in the culture medium and recovering the thus accumulated L-lysine from the culture broth.

The microorganisms used in the process are designated as *Pseudomonas brevis* 22 (ATCC 21940) and *Achromobacter coagulans* 42 (ATCC 21934).

According to the process, L-lysine can be produced advantageously by isolating the amino acid analogue-resistant or amino acid-resistant strain after mutational treatment of microorganism of the genus *Pseudomonas* or the genus *Achromobacter* and culturing the thus isolated strain in a culture medium containing a hydrocarbon as a carbon source. Two or more amino acids or amino acid analogues may be used to produce a strain which is resistant to the amino compound used. In addition, it is also possible to effect the mutational treatments repeatedly using two or more amino acids or analogues thereof alternately.

The mutational treatment may be carried out by the conventional chemical or physical treatments commonly employed in producing mutants, for example, by using the well-known mutagenic agents such as N-methyl-N'-nitro-nitrosoguanidine, or the physical mutational treatment such as irradiation with ultraviolet rays, radioactive rays or other procedures.

In carrying out the L-lysine production in accordance with the process, an L-valine-resistant strain is aerobically cultured by a well-known culturing technique in the culture medium containing a hydrocarbon as a main carbon source, a nitrogen source, inorganic salts and other additives. Hydrocarbons which can preferably be used in the culture medium are n-paraffins containing 10 to 20 carbon atoms, preferably 13 to 18 carbon atoms, or kerosene.

Nitrogen sources which can be used in the culture medium include organic and inorganic ammonium salts such as ammonium sulfate, ammonium nitrate, ammonium chloride, ammonium acetate, ammonium citrate, ammonium succinate and the like; urea and ammonia. Inorganic salts include potassium phosphate, magnesium sulfate, manganese sulfate, zinc sulfate, copper sulfate, ferrous sulfate, calcium carbonate and the like, and are added at a concentration commonly employed.

In addition, surface-active agents, for example, polyoxyethylene sorbitan mono- or trioleate (Tween 80 or 85) may be used effectively in the culture medium in an amount of approximately 0.02 to 0.5% by weight based on the total amount of the culture medium.

The pH value of the culture medium is preferably maintained in the range of from about 6 to 9, preferably 6.5 to 8.0, during the whole period of cultivation. The cultivation is usually carried out at a temperature of from about 25° to 40°C, preferably 30° to 35°C.

It is necessary to conduct the cultivation under an aerobic condition, for example, by stirring with aeration and/or shake-culturing. Upon completion of the cultivation, the resulting microbial cells are removed from the culture broth by a well-known procedure such as filtration or centrifugation. The removal of the microbial cells can easily be conducted by heating the culture broth, for example, at a temperature of from 70° to 95°C for a period of from 10 to 30 minutes and removing the microbial cells, although heating is not essential. The desired L-lysine can be obtained from the filtrate or the supernatant in the form of L-lysine hydrochloride in accordance with the well-known procedure using an ion exchange resin such as Amberlite IRC-50, Amberlite IR-120 and the like.

Alternatively, a powder enriched in L-lysine can be obtained by extracting the culture broth or a filtrate (or supernatant) obtained from the culture broth with a solvent such as n-hexane to extract any residual hydrocarbon used as a carbon source, concentrating the resulting broth under reduced pressure with or without removal of the microbial cells by filtration and finally drying the resulting concentrate by drum-drying, spray-drying and the like.

Example: Valine-resistant strain, *Pseudomonas brevis* 56 (ATCC 21941), which had been obtained from *Pseudomonas brevis* 22, was cultivated on a bouillon agar slant medium at a temperature of 33°C overnight and then inoculated in a 2 liter shake-flask containing 600 ml of the previously sterilized seed culture medium having the following composition followed by being shake-cultured at a temperature of 33°C for 24 hours.

Phosphoric acid (75%)	12 ml
$(NH_4)_2SO_4$	6 g
NaCl	1 g
$MgSO_4 \cdot 7H_2O$	0.2 g
$CaCl_2 \cdot 2H_2O$	0.1 g
$FeSO_4 \cdot 7H_2O$	0.1 g
$ZnSO_4 \cdot 7H_2O$	0.03 g
$MnSO_4 \cdot 7H_2O$	0.002 g
KOH	14 g
Tap water	1,000 ml
n-Paraffin (C_{11-18})	7 ml

600 ml portion of the above seed culture was inoculated in a jar fermentor containing 20 liters of sterilized fermentation medium having the following composition and the inoculated culture was cultivated at a temperature of 35°C by agitating at a rate of 800 rpm with aeration at a rate of 27 l/min on a weight per volume percent basis. The following amounts are:

n-Paraffin (C_{11-18})	10
K_2HPO_4	0.1
KH_2PO_4	0.1
$MgSO_4 \cdot 7H_2O$	0.1
$FeSO_4 \cdot 7H_2O$	0.002
$ZnSO_4 \cdot 7H_2O$	0.001
$MnSO_4 \cdot 4H_2O$	0.005
Polyoxyethylene sorbitan trioleate*	0.05
$(NH_4)_2SO_4$	1.0
$CaCO_3$	1.0
pH	7.0

*Tween 85

During the cultivation, the medium was adjusted to a pH range of from 6 to 8 with an aqueous ammonia solution. The amount of L-lysine hydrochloride produced in the medium after 90 hours' cultivation was found to be 40.5 g/l, the yield being about 40%, based on the n-paraffin. The microbial cells were removed by centrifugation, and the L-lysine produced was adsorbed onto an ion exchange resin in a usual manner. After elution by ammonia, the eluate was concentrated, and hydrochloric acid and then an alcohol were added thereto to obtain crystals weighing 35.3 g containing L-lysine hydrochloride having a purity more than 98% from 1 liter of the culture broth.

Production from Mixed Microorganisms

K. Watanabe, T. Hirakawa, K. Takahara, Y. Nakamura, S. Iwasaki and T. Tanaka; U.S. Patent 3,888,737; June 10, 1975; assigned to Kanegafuchi Chemical Industries Co., Ltd., Japan describe a method for producing L-lysine with the process comprising aerobically culturing a mixture of (a) a hydrocarbon assimilating and L-lysine-producing bacterium and (b) at least one hydrocarbon nonassimilating microorganism in a nutrient medium in which a hydrocarbon or a mixture of hydrocarbons is the predominant carbon source, and recovering the lysine from the medium.

A wide range of microorganisms which have a hydrocarbon-assimilating and L-lysine-producing ability directly from hydrocarbons can be used. The L-lysine-producing strains can be obtained from the hydrocarbon-assimilating microorganisms using artificial mutation techniques known in the art, e.g., as disclosed in U.S. Patent 3,759,789 and British Patent 1,304,067. Among such strains are included the L-lysine-producing strains derived from bacteria belonging to the genera of *Pseudomonas, Arthrobacter, Achromobacter, Micrococcus, Corynebacterium, Brevibacterium, Acinetobacter, Alkaligenes* and *Nocardia*, which can assimilate hydrocarbons.

Examples of preferred mutants are *Pseudomonas brevis*, ATCC 21941 (valine-resistant strain), *Achromobacter coagulans*, ATCC 21936 (α-aminobutyrate- and norvaline-resistant strain) and *Arthrobacter alkanicus* M-1554 (threonine-resistant strain), which were derived using an artificial mutation technique and have a high productivity for L-lysine. The details of the mutation technique are described in U.S. Patent 3,222,258. These microorganism strains can be used alone or in combination as desired.

The strains which can be used for fermentation as the hydrocarbon nonassimilating microorganisms are as follows: bacteria belonging to the genera *Bacillus, Pseudomonas, Corynebacterium, Microbacterium, Micrococcus, Brevibacterium, Aerobacter, Aeromonas, Escherichia coli, Proteus, Flavobacterium* and the like.

As representative examples of bacterial strains *Bacillus subtilis* (IAM-1130), *Bacillus megatherium* (IAM-1145), *Bacillus polymyxa* (IAM-1189), *Pseudomonas xanthe* (IAM-1310), *Microbacterium flavum* (ATCC 10340), *Micrococcus candidus* (ATCC 14852), *Micrococcus lysodeikticus* (IAM 1313), *Corynebacterium fascians* (IAM 1079), *Brevibacterium ammoniagenes* (IAM 1641), *Acromonas formicans* (ATCC 13137), *Escherichia coli* K-13 and *Proteus vulgaris* (IAM 1025) can be used. Additional representative examples of microorganisms are as follows:

> Fungi belonging to the genera *Penicillium* and *Aspergillus*, with representative strains being *Aspergillus niger* (IAM 3008) and *Penicillium citrinum* (ATCC 9849);

> Actinomycetes belonging to the genera *Streptomyces* and *Actinomyces* with representative strains being *Streptomyces mitakaensis* (ATCC 15297), *Streptomyces hygroscopicus* (ATCC 10976), *Streptomyces griseus* (ATCC 23345), *Streptomyces virginiae* (ATCC 19817), and *Streptomyces albus* (ATCC 3381); and

> Yeast belonging to the genera *Rhodotorula, Zygosaccharomyces, Debaryomyces* and *Trichosporon*.

These strains can be used alone or in a combination of more than two strains. The mutant strains such as auxotrophs, analogue-resistant or temperature-sensitive mutants derived from abovementioned strains can also be used. These mutants can be obtained using techniques well-known in the art.

A seed culture medium is inoculated with the L-lysine-producing bacteria and the hydrocarbon nonassimilating strain or strains; the inoculated seed culture medium is cultured aerobically and a fermentation medium is then inoculated with the seed which contains the mixture of the two kinds of microorganisms, i.e., the L-lysine-producing microorganisms and the hydrocarbon nonassimilating microorganisms. Alternatively, the mixture of the two kinds of microorganisms can be cultured individually in seed media and the seed media thus obtained are mixed before the fermentation.

The hydrocarbon nonassimilating microorganisms can be mixed intermittently during the fermentation using an L-lysine-producing bacterium. The mixing ratio and the time at which the mixed strains are added to the fermentation medium can be appropriately selected in each case. A ratio (weight of cells) of about 1 to 100 of the mixed microorganisms to 100 of the L-lysine-producing strain is often preferable.

Either a synthetic culture medium containing hydrocarbon or a natural nutrient medium is suitable for the seed culture. For example, a synthetic medium containing n-paraffin(s), potassium phosphate, magnesium sulfate, ammonium sulfate and peptone, or a nutrient bouillon medium can be used as a seed culture medium. A synthetic medium containing 0.5 to 2.0% n-paraffin is often preferable.

In carrying out the L-lysine production, the two kinds of strains described above, i.e., the hydrocarbon-assimilating and hydrocarbon-nonassimilating, are aerobically cultured in the culture medium containing hydrocarbons as a main carbon source, nitrogen sources, inorganic salts and other additives. Hydrocarbons which can preferably be used in the culture medium in the process for the production of L-lysine are n-paraffins containing 10 to 20 carbon atoms with n-paraffins containing 13 to 18 carbon atoms being particularly preferred. The concentration of hydrocarbons in the medium for the production of L-lysine is usually 5 to 20% weight per volume.

The pH of the culture medium is maintained in the range of about 5.5 to 9.0, preferably about 6 to 8.5 during the entire cultivation period. The cultivation is carried out at a temperature of about 25° to 45°C, preferably between 28° and 37°C. It is necessary that the cultivation be conducted under aerobic conditions, for example, by stirring with aeration in a fermentor or by shake-culturing in a flask. Upon completion of the cultivation, the resulting microbial cells can be removed from the culture broth using well-known techniques such as filtration or centrifuging. The removal of the microbial cells can also be easily carried out by heating the resultant cultural broth at about 70° to 100°C.

The desired L-lysine can be obtained from the filtrate or the supernatant in the form of L-lysine hydrochloride in accordance with well-known procedures using an ion exchange resin such as Amberlite IR-120 (H$^+$ type), a strong cationic exchange resin, or Amberlite IRC-50 (H$^+$ type), a weak cationic exchange resin.

Use of L-Leucine-Producing Microorganism

K. Inuzuka and S. Hamada; U.S. Patent 3,959,075; May 25, 1976; assigned to Kyowa Hakko Kogyo, Co., Ltd., Japan have found that the yields of L-lysine are greatly improved by culturing L-lysine-producing mutants in a medium supplemented by the culture liquor of L-leucine-producing mutants. Preferred L-lysine-producing mutants include *Micrococcus glutamicus* ATCC 13286, ATCC 13287; *Brevibacterium flavum* ATCC 21475, ATCC 21127, ATCC 21128, ATCC 21517, ATCC 21518, ATCC 21528, ATCC 21529; and *Corynebacterium glutamicum* ATCC 21299, ATCC 21300, ATCC 21513, ATCC 21514, ATCC 21515, ATCC 21516, ATCC 21527, ATCC 21544, KY 10403, KY 10031.

Preferred L-leucine-producing mutants include *Brevibacterium flavum* (FERM-P 1838) ATCC 21889; *Brevibacterium lactofermentum* (FERM-P 1837) ATCC 21888; and *Corynebacterium glutamicum* ATCC 21301 through 21308, ATCC 21885, ATCC 21886 (FERM-P 1835); and *Corynebacterium acetoacidophilum* (FERM-P 1836) ATCC 21887.

Either a synthetic medium or natural medium may be used for the culturing of the L-lysine-producing mutant so long as it properly contains an assimilable carbon source, a nitrogen source, inorganic materials and other growth-promoting factors which may be required by the specific strains used. Molasses, blackstrap molasses, acetic acid and glucose are preferred as the carbon source.

According to the process, the addition of the culture liquor obtained by culturing the L-leucine-producing mutant to the L-lysine fermentation medium is effective to enhance the yields of L-lysine. Such culture liquor to be added to the L-lysine fermentation medium may be employed either as is or after removal of the microbial cells. In either case, it is, of course, necessary to sterilize the culture or culture filtrate prior to use. It is preferred that the L-lysine fermentation medium contains the culture liquor at a concentration of 2 to 150 ml/l based on the volume of the nutrient medium for L-lysine fermentation, the optimum amount being readily determined for each particular application.

Fermentation of the L-lysine-producing mutant is carried out under the conditions normally used in L-lysine fermentation, that is, under aerobic conditions; for example, with aeration and stirring or with shaking at a temperature of 25° to 40°C and a pH of 6 to 8.5. Usually, after 30 to 150 hours of culturing, a considerable amount of L-lysine is accumulated in the culture liquor. After completion of culturing, L-lysine is isolated and purified by any of the methods well-known in the art, such as an ion exchange resin treatment, crystallization by concentration, and the like.

Culturing of the L-leucine-producing mutant is carried out under the conventional culturing conditions. Generally, culturing is carried out under aerobic conditions, for example, with aeration and stirring or with shaking at a temperature of 25° to 40°C and a pH of 6 to 9 for 48 to 120 hours.

Optical Resolution of DL-Lysine

H. Hirohara, S. Nabeshima and T. Nagase; U.S. Patent 4,108,723; August 22, 1978; assigned to Sumitomo Chemical Company, Ltd., Japan describe a method of producing L-lysine or acid-addition salts thereof from DL-lysine alkyl ester or acid-addition salts thereof through enzymatic optical resolution, which comprises contacting an aqueous solution of the DL-lysine alkyl ester or acid-addition salts thereof with a nonspecific protease having ability to resolve the DL-lysine alkyl ester or acid-addition salts thereof produced by culturing a microorganism of *Streptomyces sp.* under a weakly acidic condition at a temperature lower than 60°C, and recovering L-lysine or acid-addition salts thereof thus produced.

The amount of the nonspecific protease to be used is variable with the degree of purification and the activity involved, but this enzyme need not be used in a large amount. The amount is sufficient in the order of 0.3 to 3%, particularly 0.1 to 1.5%, by weight based on the amount of the substrate. The nonspecific protease may be used in the immobilized form on an insoluble carrier. Any of the bacteria of genus *Streptomyces* that produce nonspecific proteases can be effectively used insofar as it is able to produce an enzymatic component which exhibits activity to the lysine alkyl ester. Particularly, *Streptomyces griseus, Streptomyces fradiae, Streptomyces erythreus, Streptomyces rimosus* and *Streptomyces flavovirens* are advantageously used since they produce highly active lysine alkyl ester hydrolases.

Example: In 20 ml of 0.004 M monopotassium phosphate solution, 3.0 g of DL-lysine methyl ester dihydrochloride were dissolved. The resultant solution was adjusted to pH 6.0 by addition of 1 N sodium hydroxide solution. To the solution, 1 ml of an enzyme solution containing 36 mg of Pronase (nonspecific protease) obtained from *Streptomyces griseus* were added to start an enzymic reaction. The hydrolysis was allowed to proceed at 30°C, with the pH value of the reaction solution kept at 6.0 by addition of a 1 N sodium hydroxide solution. After a lapse of a period of 25 minutes, which a calculation performed on the basis of the amount of sodium hydroxide added to the reaction system showed as necessary for the reaction to proceed to the extent of consuming about 39% of the DL-lysine methyl ester, the reaction was stopped by adding methanol in an amount of twice the reaction solution to the reaction system.

The solvent was expelled through evaporation from the reaction solution by heating the solution at temperatures not exceeding 50°C, and the residue obtained from the evaporation was dissolved in a small amount of methanol to permit separation of insolubles by filtration. The insolubles were dissolved in water, and an amount of trichloroacetic acid corresponding to 5% by weight was added thereto. Subsequently, the insoluble enzyme was removed by centrifugal separation. Further, the aqueous solution was evaporated to a final volume of about 1 ml. The evaporation residue was mixed with about 15 ml of methanol. The precipitate consequently formed in the mixture was filtered off, washed and dried to afford 1.20 g of purified crystals of L-lysine dihydrochloride.

This product had a specific optical rotation $[\alpha]^{20}_{546}$ + 20.0°(c = 2, 6 N HCl) as compared with the standard pure L-form substance, which has specific optical rotation $[\alpha]^{20}_{546}$ + 20.4°(c = 2, 6 N HCl). This means that the selectivity for the purified lysine was 99%.

In the meantime, the ester which had as its main component the D-lysine methyl ester dihydrochloride in the initial filtrate was withdrawn by adding ether to the filtrate therby causing precipitation of the dihydrochloride and crystallizing out the precipitate. Specific optical rotation, $[\alpha]^{20}_{546}$, of the crystals was found to be –12.4°(c = 2, 6 N HCl), indicating that the selectivity was 98.5%.

METHIONINE

Production from DL-N-Carbamoylmethionine

H. Yamada; S. Takahashi, K. Sumino, H. Fukumitsu and K. Yoneda; U.S. Patent 4,148,688; April 10, 1979; assigned to Kanegafuchi Kagaku Kogyo KK, Japan, describe a method whereby optically active L-methionine can be effectively prepared from DL- or L-N-carbamoylmethionine by the catalytic action of enzymes of microorganisms.

In general, it is preferable to use DL-N-carbamoylmethionine which may be prepared by reacting DL-methionine with potassium cyanate according to a known method or may be derived from DL-5-(2-methylthioethyl)hydantoin obtained as an intermediate in the course of the synthesis of DL-methionine. The microorganisms employed are those having a capability of asymmetrically hydrolyzing the carbamoyl group of N-carbamoylmethionine to produce L-methionine, and are selected from bacteria, actinomycetes, molds, yeasts and deuteromycetes.

It is preferable to add a small amount of DL- or L-N-carbamoylmethionine to the reaction substrate, or DL-5-(2-methylthioethyl)hydantoin to the culture medium in order to adaptively enhance the activity of the desired enzyme. The culture conditions are selected from the temperature range of 20° to 85°C and pH range of 4 to 11 in accordance with the optimum growth conditions of the employed strain, and usually microorganisms are cultured at a temperature of 20° to 40°C at a pH of 5 to 9 for 10 to 75 hours. During the culture, the growth of microorganisms may be accelerated by aeration and agitation.

The reaction substrate is usually admixed with the cultured broth, cells or treated cells in an aqueous medium to make the enzymes of microorganisms act catalytically on the substrate. The reaction may also be effected by adding the reaction substrate to a liquid culture medium during the culture of microorganisms.

The concentration of the reaction substrate, DL- or L-N-carbamoylmethionine, is selected from 0.1 to 50% by weight. The substrate, usually present in a form of the solution, may be present in a form of the suspension. The pH of the aqueous medium for the hydrolysis reaction is selected from 6 to 11, and preferably the reaction is carried out at a pH of 7 to 10. The hydrolysis reaction is carried out at a temperature suitable for the enzyme of the employed microorganism, and usually at a temperature of 20° to 85°C. The reaction time varies depending on the activity of the employed microorganism and the reaction temperature, and is usually selected from 10 to 100 hours. The produced L-methionine is isolated from the reaction mixture in a conventional manner.

According to the process, only L-form of N-carbamoylmethionine is converted to L-methionine and the D-form remains in the reaction mixture without being converted.

Example: A liquid culture medium of pH 7.0 containing the following components was prepared, and 9 ml portions thereof were poured into 70 ml test tubes and steam-sterilized at a temperature of 120°C for 10 minutes.

Glucose	0.5 %
Yeast extract	0.1 %

(continued)

Peptone	0.5 %
KH_2PO_4	0.5 %
$MgSO_4 \cdot 7H_2O$	0.02 %
$FeSO_4 \cdot 7H_2O$	10 ppm
$MnSO_4 \cdot 4H_2O$	10 ppm

1 ml of a 3.3% aqueous solution of DL-N-carbamoylmethionine sterilized under sterile conditions (pH 7.0) was added to each test tube. Each microorganism shown in the table below was inoculated into the medium with a platinum loop, and then cultured at a temperature of 33°C for 22 hours with shaking. Cells were separated from each cultured broth by centrifugation and washed with 5 ml of 0.9% saline water. The thus obtained cells from each cultured broth were suspended into a mixture having the following composition: (a) 5.0 ml of an aqueous solution (7.6 pH) containing 2.0% of DL-N-carbamoylmethionine (substrate content: 100 mg); and (b) 5.0 ml of 0.2 M phosphate buffer solution (7.6 pH).

Into the mixture of the above (a) and (b), the cells were suspended and the reaction was carried out on standing at 33°C for 40 hours. After completion of the reaction, the reaction mixture was centrifuged and the amount of methionine in the resulting supernatant liquid was determined by a bioassay using a strain requiring L-methionine. Further, the product was subjected to a silica-gel thin-layer chromatography (solvent: n-butanol/acetic acid/water = 4/1/1) to separate the methionine spot, and the amount of methionine was colorimetrically determined by developing color with ninhydrin. The results are shown in the table.

Strain	L-Methionine (mg/ml)	Conversion (% by mol)
Achromobacter polymorph	0.35	4.5
Achromobacter superficialis	0.30	3.9
Aerobacter cloacae IAM 1221	0.70	9.0
Aeromonas punctata IAM 1646	0.35	4.5
Agrobacterium rhizogenes IFO 13259	1.45	18.7
Alcaligenes faecalis IAM 1015	0.30	3.9
Arthrobacter ureafaciens IFO 12140	0.10	1.3
Bacillus circulans IFO 3329	0.25	3.2
Bacillus licheniformis IFO 12195	0.10	1.3
Bacillus megaterium IFO 3003	0.40	5.2
Bacillus pumilis IFO 3028	0.40	5.2
Brevibacterium flavum ATCC 21129	0.10	1.3
Cornynebacterium sepedonicum IFO 3306	3.70	47.7
Enterobacter cloacae IFO 13535	0.05	0.6
Erwinia aroideae IFO 12380	0.10	1.3
Escherichia coli ATCC 8739	0.45	5.8
Klebsiella pneumoniae IFO 3319	0.20	2.6
Microbacterium flavum ATCC 10340	0.17	2.2
Micrococcus roseus IFO 3764	0.28	3.6
Mycobacterium smegmatis ATCC 607	2.60	33.5
Nocardia corallina IFO 3338	0.20	2.6
Protaminobacter ruber IFO 3708	0.40	5.2
Proteus mirabilis IFO 3849	0.10	1.3
Pseudomonas aeruginosa IFO 3445	0.55	7.1
Pseudomonas solanacearum IFO 12510	0.10	1.3
Sarcina lutea IFO 3232	0.55	7.1
Sarcina variabilis IFO 3067	0.40	5.2
Staphylococcus aureus IFO 12732	0.10	1.3
Xanthomonas campestris IAM 1671	0.45	5.8

N-Acyl-L-Methionine

C.E. Stauffer; U.S. Patent 3,963,573; June 15, 1976; assigned to The Procter & Gamble Company describes a process for producing optically pure N-acyl-L-methionine comprising (1) subjecting N-acyl-D,L-methionine ester to the action of a proteolytic enzyme selected from the group consisting of (a) sulfhydryl proteinases and (b) microbially-derived serine proteinases; and (2) separating the resulting N-acyl-L-methionine.

A variety of specific N-acyl-D,L-methionine ester compounds can be employed. Preferably, the acyl group is derived from fatty acids containing from 1 to 9 carbon atoms. The ester group can be derived from a variety of alcohols containing from 1 to 10, preferably 1 to 6, carbon atoms. Especially suitable examples of ester groups are methyl, ethyl, n-propyl, isopropyl, n-butyl and isobutyl. Especially suitable examples of acyl groups are formyl, acetyl and propionyl. Racemic N-acetyl-D,L-methionine methyl ester is most preferred for use in the process. Suitable sulfhydryl proteinases are, for example, papain, ficin and bromelain.

Serine proteinases suitable for use are derived from microorganisms such as bacteria, fungi and mold. Microbially-derived serine proteinases are preferred. These proteinases are relatively inexpensive and commercially available. An example of preferred serine proteinases are those derived from the bacterial organism *Bacillus subtilis* and termed subtilisins. A preferred subtilisin is the *Bacillus subtilis*-derived Carlsberg strain. The Carlsberg strain is a known subtilisin strain, the amino acid sequence of which is described in Smith et al, "The Complete Amino Acid Sequence of Two Types of Subtilisin, BPN' and Carlsberg", *J. of Biol. Chem.* vol. 241, p. 5974 (Dec. 25, 1966).

An x-ray-mutated *Bacillus subtilis*-derived subtilisin constitutes another preferred subtilisin of the process. This mutation can be effected in accordance with U.S. Patent 3,031,380 issued April 24, 1972 to Minagawa et al, by irradiation of a *Bacillus subtilis* organism with x-rays. Subsequent treatment in a conventional manner can be employed to result in the preparation of an enzymatic composition.

Other examples of suitable serine proteinases for use herein include the following: serine proteinases derived from *Aspergillus oryzae, Streptomyces griseus* (ATCC 3463), and *Aspergillus sydowi.* Other suitable examples of microbially-derived serine proteinases are *Aspergillus* alkaline proteinase (EC 3.4.21.15), *Alternaria endopeptidase* (EC 3.4.21.16) and *Arthrobacter* serine proteinase (EC 3.4.21.17). These particular enzymes have been identified according to a systematic nomenclature involving an EC number. [See *Enzyme Nomenclature,* Commission of Biochemical Nomenclature, Elsevier Publishing Company (1973), U.S. Library of Congress Card No. 73-78247].

The action of the proteolytic enzyme on the N-acyl-D,L-methionine ester is suitably conducted in an aqueous medium maintained at a pH of from about 5 to 10, preferably about 7 to 8, and at a temperature of from about $10°$ to $60°C$. Preferably, the temperature is maintained in the range of from about $20°$ to $40°C$.

Because of the high selective esterase activity of the particular proteolytic enzymes employed toward N-acyl-L-methionine ester in N-acyl-D,L-methionine ester mixtures, very small amounts of the proteolytic enzyme are required in order to rapidly produce N-acyl-L-methionine. For example, aqueous solutions, preferably containing 0.005 to 0.5% by weight of enzymes, are employed. (Enzymes herein refer to pure crystalline enzymes.) The amount of N-acyl-D,L-methionine employed will generally preferably amount to and exceed the maximum solubility of the N-acyl-D,L-methionine ester in the aqueous medium.

Example: *(A) Preparation of N-Acetyl-D,L-Methionine* — 100 g of D,L-methionine were mixed with 500 ml of methanol containing 30 g of NaOH to form a stirrable slurry. To this mixture there was slowly added, with stirring, 1.5 mol equivalents (based on D,L-methionine) of acetic anhydride. This mixture was continuously stirred until it cleared. The resulting clear solution contained N-acetyl-D,L-methionine dissolved in methanol.

To this solution was added with stirring sufficient anhydrous H_2SO_4 to neutralize the NaOH and render the solution slightly acidic. Under these conditions the N-acetyl-D,L-methionine was converted to N-acetyl-D,L-methionine methyl ester. The remaining methanol and methyl acetate (formed from the reaction of residual acetic acid and acetic anhydride with NaOH) was removed by warming the mixture under vacuum. The remaining mixture was slurried with chloroform and filtered to remove sodium sulfate salts. The ester-chloroform solution was then water-washed to remove any remaining salts. The resulting ester-chloroform solution was then warmed under vacuum to remove the chloroform. The resulting material was N-acetyl-D,L-methionine methyl ester.

(B) Preparation of N-Acetyl-L-Methionine — An aqueous solution containing 20% by weight N-acetyl-D,L-methionine methyl ester was formed. The pH was adjusted to about 7.5 with 1 N NaOH. To this solution there was slowly added 0.04% by weight of the aqueous solution of crystalline subtilisin Carlsberg, a serine proteinase. During this addition, the solution was continuously stirred, and the pH was kept constant by the automatic addition of 1 N NaOH from an automatic titrator. When no further NaOH is needed to maintain the pH, the reaction is complete (in this case about 30 minutes).

The resulting mixture of N-acetyl-D-methionine methyl ester and N-acetyl-L-methionine was separated in the following manner. The aqueous mixture was raised to a pH of 10 to 11 by the addition of NaOH. To the mixture was added an equal volume of chloroform. The resulting mixture was agitated briefly. The water-immiscible chloroform was then allowed to separate and removed. The chloroform extraction procedure was repeated twice. The separated chloroform layers were combined; the chloroform was removed by warming under vacuum. The resulting N-acetyl-D-methionine methyl ester can be added to water or methanol containing NaOH and heated for a time sufficient to racemize the ester material. The resulting D,L-material can then by recycled in the process.

The water layer, containing N-acetyl-L-methionine, was acidified by the addition of H_2SO_4 to lower the pH to 2 or less. This acidified water was then extracted with an equal volume of chloroform with a brief agitation. The chloroform was allowed to separate from the water and removed. This chloroform extraction procedure was repeated twice. The separated chloroform layers were combined. The chloroform solution was then warmed gently under vacuum to remove the chloroform. The product was N-acetyl-L-methionine having an optical purity of better than 95%.

PHENYLALANINE

Tyrosine-Requiring Strain of Brevibacterium

T. Tsuchida, H. Matsui, H. Enei and F. Yoshinaga; U.S. Patent 3,909,353; September 30, 1975; assigned to Ajinomoto Co., Inc., Japan have found that L-phenylalanine is accumulated in high concentrations in cultures of certain tyrosine-requiring, artificially induced mutant strains of *Brevibacterium* whose growth is not inhibited by tryptophan analogs nor by phenylalanine analogs in concentrations sufficient to inhibit the growth of the respective parent strains showing neither the tyrosine requirement nor the resistance to tryptophan or phenylalanine analogs.

The mutants are derived from the parent strains by mutagenic doses of ionizing radiation (ultraviolet, x-rays, gamma rays) or of chemical agents (sodium nitrite, N-methyl-N'-nitro-N-nitrosoguanidine, diethyl sulfate), and by screening the treated parent strains for mutants having the desired properties. Tyrosine-requiring mutants are isolated by the replication method, and the tyrosine-requiring strains resistant to phenylalanine analogs and tryptophan analogs are identified by their ability of growing vigorously on otherwise conventional media containing enough of the analogs to suppress or virtually suppress growth of the parent strains.

Comparison between the parent and the mutants is made by determining the relative growth for each tested strain in the presence of the analog, relative growth being the percentage of the growth observed in the absence of the analog.

Commonly known phenylalanine analogs include β-amino-β-phenylpropionic acid, o-fluorophenylalanine, m-fluorophenylalanine, p-fluorophenylalanine, β-2-thienylalanine, β-3-thienylalanine, β-2-furylalanine, β-3-furylalanine, o-aminophenylalanine, p-aminophenylalanine, m-aminophenylalanine, etc.

Known tryptophan analogs include 6-methyltryptophan, 4-methyltryptophan, naphthylalanines, indoleacrylic acid, naphthylacrylic acid, β-(2-benzothienyl)-alanine, styrylacetic acid, α-amino-β-3-indazole-3-yl-propionic acid, 5-fluorotryptophan, 6-fluorotryptophan, 4-fluorotryptophan, and 7-azatryptophan.

The aqueous fermentation media in which the mutants produce L-phenylalanine are largely conventional. They must contain sources of assimilable carbon and nitrogen and tyrosine, and should further contain inorganic ions and minor organic nutrients.

Aerobic conditions are maintained by aeration and/or agitation, and pH is held between 5 and 9 for good yields. When ammonia is used for pH control, it may also serve as a nitrogen source. The phenylalanine concentration in the broth reaches its maximum within 2 to 7 days if the fermentation is carried out at 24° to 37°C.

Example: *Brevibacterium lactofermentum* No. 2256 (ATCC 13869) was exposed to 250 γ/ml N-methyl-N'-nitro-N-nitrosoguanidine and 5 tyrosine-requiring, phenylalanine-producing mutant strains were isolated. Strain AJ 3435 (FERM P-1912) did not show more resistance to phenylalanine or tryptophan analogs than the parent. Strain AJ 3436 (FERM P-1913) resisted p-fluorophenylalanine.

Strain AJ 3432 (FERM P-1844) resisted 5-methyltryptophan. The strains AJ 3437 (FERM P-1914) and AJ 3699 (FERM P-2476) resisted both p-fluorophenyl-alanine and 5-methyl-tryptophan. Strain AJ 3699, in addition to tyrosine, required tryptophan and methionine for its growth.

Brevibacterium lactofermentum AJ 3436 and AJ 3437 were cultured separately with shaking at 31.5°C for 16 hours in an aqueous culture medium containing, per deciliter, 3 g glucose, 0.1 g KH_2PO_4, 0.04 g $MgSO_4 \cdot 7H_2O$, 1 mg $FeSO_4 \cdot 7H_2O$, 1 mg $MnSO_4 \cdot 4H_2O$, 40 mg L-tyrosine, 40 mg DL-methionine, 5 ml soy protein acid hydrolyzate, 10 γ biotin, 30 γ thiamine·HCl, and 0.2 g urea.

An aqueous fermentation medium was prepared to contain, per deciliter, 0.4 g ammonium acetate, 0.4 g sodium acetate, 0.1 g KH_2PO_4, 0.04 g $MgSO_4 \cdot 7H_2O$, 1 mg $FeSO_4 \cdot 7H_2O$, 1 mg $MnSO_4 \cdot 4H_2O$, 40 mg L-tyrosine, 40 mg DL-methionine, 5 ml soy protein acid hydrolyzate, 10 γ biotin, 30 γ thiamine·HCl, and 0.2 g urea, and adjusted to pH 7.0. 300 ml batches of the fermentation medium in 1-liter fermentors were sterilized with steam and inoculated each with 15 ml of the previously prepared seed cultures of strains AJ 3436 and 3437 respectively.

The fermentation was carried out at 31.5°C with agitation at 1,200 rpm and aeration of 300 ml/min. Starting 3 hours after inoculation, 70% acetic acid and gaseous ammonia were fed to each culture to maintain a pH of 7.0 to 7.7. After 48 hours cultivation, strain AJ 3437 had consumed 57 g acetic acid and produced 2.01 g/dl L-phenylalanine. Strain AJ 3436, in the same period, consumed 30 g acetic acid and produced 0.75 g/dl phenylalanine.

Production from *Microbacterium ammoniaphilum*

K. Nakayama and H. Hagino; U.S. Patent 3,917,511; November 4, 1975; assigned to Kyowa Hakko Kogyo KK, Japan have found that strains of *Microbacterium ammoniaphilum* which exhibit resistance to at least one compound selected from the group consisting of tyrosine, phenylalanine, and analog thereof (such as, e.g., 2-fluorophenylalanine, 3-fluorophenylalanine, 4-fluorophenylalanine, 2-methylphenylalanine, 3-methylphenylalanine, 4-methylphenylalanine, 2-hydroxy-phenylalanine, 3-hydroxyphenylalanine, 2-aminophenylalanine, 3-aminophenyl-alanine, 4-aminophenylalanine, 2-nitrophenylalanine, 4-nitrophenylalanine, β-2-thienylalanine, β-3-thienylalanine, 2-indole alanine, 1-naphthyl alanine, 2-naphthyl alanine, 2-pyridyl alanine, 2-thiazole alanine, 3-thiazole alanine, 3-aminotyrosine, 3-fluorotyrosine, 3-hydroxytyrosine, 3-nitrotyrosine, 5-hydroxy-2-pyridyl alanine, phenyl alanine hydroxamate, etc.) are capable of producing and accumulating a large amount of L-phenylalanine.

Specimen cultures of the *Microbacterium ammoniaphilum* microorganism identified by ATCC No. 21645 have been deposited with the American Type Culture Collection located at Rockville, Maryland and specimens thereof are freely available to qualified persons.

The strains which may be used for the process can be obtained by subjecting certain strains belonging to the bacteria to mutation such as, e.g., irradiation by ultraviolet light, x-ray, cobalt 60, etc. and chemical treatment, etc. and then screening the strains for isolation from colonies capable of growing on agar plate media containing tyrosine, phenylalanine, or analogs thereof.

When a parent strain is subjected to a series of mutation treatments, it is possible to obtain the mutant strain having resistance to two or more of the compounds.

Any of synthetic or natural media may be used if they contain suitable amounts of nitrogen sources, carbon sources, inorganic materials as well as trace amounts of nutrients required for growth of the used strain, as shown in the example.

The strain is cultured under aerobic conditions, for example, by culturing with shaking, aeration-agitation, etc. The preferable culture temperature is, in general, from 20° to 40°C. The pH of the medium is preferably maintained at from 4 to 8 during the cultivation. However, other temperatures and pH conditions may also be used, if desired. The time for culture is usually from 2 to 5 days to accumulate a substantial amount of L-phenylalanine in the medium.

After the completion of the cultivation, microbial cells are first removed from the fermented broth, and then L-phenylalanine is recovered therefrom in a suitable manner such as by using activated carbon treatment, ion exchange resin treatment, etc., as shown in the example.

Example: An L-phenylalanine-producing strain of *Microbacterium ammoniaphilum* (ATCC 21645) previously screened as being resistant to an analog of phenylalanine (4-fluorophenylalanine) was cultured at 30°C for 24 hours in a seed medium containing glucose (2%), peptone (1%), yeast extract (1%) and NaCl (0.3%). The resultant seed (1 ml) was inoculated into a fermentation medium (10 ml) in a 250 ml Erlenmeyer flask and cultured with shaking at 30°C for 4 days to produce 4.7 mg/ml of L-phenylalanine from the culture broth.

The composition of the fermentation medium used herein was glucose (10%), K_2HPO_4 (0.05%), KH_2PO_4 (0.05%), $MgSO_4 \cdot 7H_2O$ (0.025%), ammonium sulfate (2%), NZ-amine (0.5%), biotin (30 mg/l) and $CaCO_3$ (2%) (pH: 7.2).

After the completion of the fermentation, 1 liter of the cultured broth was centrifuged to remove microbial cells and $CaCO_3$. The resultant liquor was mixed with active charcoal which had previously been treated with acetic acid and was thoroughly mixed to adsorb phenylalanine thereon. The charcoal was separated by filtration and was washed with water. The charcoal was eluted with 20% (v/v) acetic acid containing phenol (5% v/v) and the eluate was treated with ether. After the removal of phenol, the eluate was concentrated and was passed through a column of a strongly acidic cation exchange resin (Diaion SK-1A) (H^+ form) to adsorb phenylalanine which was then eluted with 0.3% aqueous ammonia. The phenylalanine fraction of the eluate was concentrated and precipitated with alcohol to recover 1.6 g of L-phenylalanine.

SERINE

Production Using *Pseudomonas* sp. DSM 672

F. Wagner, H. Sahm and W.H. Keune; U.S. Patent 4,060,455; November 29, 1977; assigned to Gesellschaft fur Biotechnologische Forschung mbH, (GBF), Germany describe a process for the production of L-serine in which methanol and glycine are mixed with a submerse culture containing inorganic nutrients and the bacterium *Pseudomonas* sp. DSM 672 and/or mutants of this bacterium in a reactor supplied with a gas which is air or oxygen-enriched air at a pH in the range 8.0 to 9.0 and at a temperature in the range 20° to 40°C, the mixture is allowed to react, the resulting cell mass is separated from the culture filtrate and the L-serine is isolated from the culture filtrate.

The submerse culture is preferably cultivated at a pH in the range 6.5 to 8.0 with a methanol concentration of from 0.1 to 1.0% by volume as carbon and energy source to facilitate multiplication of the bacterium. The cultivation may be effected continuously in a first reactor and a portion of the culture transferred to a second reactor for the production of L-serine of the cultivation or L-serine production may be effected in a single reactor.

The glycine is preferably added to the submersed culture in an amount to give an initial concentration of 0.5 to 2.0 wt % and methanol is added to maintain a constant value in the range 0.5 to 2.0 wt %. The temperature of the reactor is preferably maintained at 28° to 33°C and the contents may be mixed with a mechanical stirrer and/or by aeration with the gas.

After the formation of L-serine the cell mass is separated from the culture filtrate and the L-serine isolated. The L-serine-free culture filtrate may be wholly or partially returned to the reactor or discarded.

The bacterium *Pseudomonas* sp. DSM 672 has been filed in the German Collection of Microorganisms in Gottingen under the number DSM 672 and the mutant under the number DSM 673.

Example: An agar slanting tube culture of *Pseudomonas* sp. DSM 672 on a defined medium was introduced into 100 ml of a liquid medium in a flask, which was sterilized beforehand for 20 minutes at 120°C and to which then 1% of methanol was added under aseptic conditions.

The medium contained the following nutrients: 0.10 g of KH_2PO_4, 0.45 g of Na_2HPO_4, 0.20 g of $(NH_4)_2SO_4$, 0.04 g of $MgSO_4 \cdot 7H_2O$, 2.0 mg of $CaCl_2 \cdot 2H_2O$, 1.0 mg of $FeSO_4 \cdot 7H_2O$, 0.5 mg of $MnSO_4 \cdot H_2O$, 0.1 mg of $(NH_4)_6Mo_7O_{24} \cdot 4H_2O$ and 1 ml of methanol to 100 ml of distilled water. The culture was aerobically incubated for 40 hours at 30°C on a rotary shaker machine. 1% of this grown culture served as inoculum for a liquid culture having the same composition as the medium and the same incubation conditions, apart from the incubation time. During the incubation, altogether another 1% of methanol was twice added, at intervals from 15 to 20 hours. Before each addition, the pH value of the submerse culture was adjusted with 1 N NaOH to 7.0. After a multiplication phase of 45 to 60 hours, the pH value was adjusted to 8.5 with 1 N NaOH, and again 2% (v/v) of methanol and simultaneously 2% (w/v) of glycine were added. After another 30 to 50 hours, 4 g/l of L-serine were detected in the nutrient medium.

Production from Mutant Strains of *Corynebacterium glycinophilum*

It is known from U.S. Patent 3,623,952 that L-serine may be produced by fermentation of glycine by certain microorganisms which include *Corynebacterium* sp. ATCC 21341. *K. Kageyama, I. Maeyashiki, K. Kubota, S. Konishi and S. Okumura; U.S. Patent 3,880,741; April 29, 1975; assigned to Ajinomoto Co., Inc., Japan* have discovered that artificially induced mutants of this species of

Corynebacterium produce serine in significantly greater yields than could be obtained by the earlier method. The mutants are characterized by specific nutrient requirements, but may differ from each other in the required nutrient which may be leucine, isoleucine, methionine, tryptophan, serine, or glycine, or more than one of these amino acids.

The mutants are derived from *C. glycinophilum* by means of any conventional mutagenic agent, as by treating cells of the parent strain with 0.1 M diethyl sulfate solution at 30°C for 20 hours, screening the treated cells by the replication method, and examining them for their ability of producing L-serine.

The five best strains of *C. glycinophilum* obtained so far have been deposited with the Fermentation Research Institute, Agency of Industrial Science and Technology, the Ministry of Industrial Trade and Industry, in Chiba-shi, Chiba-ken, Japan, and are freely available from the Institute. Their accession numbers and the specific nutrients required by them are listed below:

> FERM-P 1685 (leucine)
> FERM-P 1686 (tryptophan)
> FERM-P 1687 (leucine + methionine)
> FERM-P 1688 (leucine + isoleucine)
> FERM-P 1689 (leucine + serine or glycine)

The culture media on which the mutant strains are cultured may be entirely conventional aside from the required specific nutrients. Thus, carbohydrates (glucose, sucrose, maltose, fructose, starch hydrolyzate, molasses), organic acids (acetic, propionic, citric, benzoic acid), or alcohols (ethanol) may provide carbon, and ammonia, ammonium salts (sulfate, chloride) and organic nitrogen bearing compounds (amino acids, urea) are suitable sources of assimilable nitrogen. The usual inorganic ions (phosphate, calcium, magnesium, iron, manganese) are to be supplied, and the yield and rate of production of L-serine from glycine in the culture medium are improved by known growth promoting agents, such as vitamins, soybean protein hydrolyzate, yeast extract, corn steep liquor, peptone, and casein hydrolyzate. The growth promoting agents may contain enough of the specifically required nutrients.

The glycine concentration in the medium is preferably 0.5 to 4 g/dl, and glycine may be added to the medium before inoculation, early during culturing, or gradually.

Microbial fermentation of glycine is preferably carried out under aerobic conditions maintained by shaking, stirring, or by aeration. The pH should be held between 5.0 and 9.0 by adding acids, alkali hydroxide or carbonates of alkali or alkaline earth metals in a known manner. The preferred cultivation temperature is 24° to 37°C. Maximum serine concentration is reached within 2 to 5 days depending on process variables.

The cells may also be harvested from the broth and dispersed as an enzyme source in an aqueous glycine solution at pH 5.0 to 9.0 for 5 to 48 hours at 25° to 40°C to produce L-serine by enzyme action from the glycine and a source of carbon assimilable by the microorganisms.

L-serine is practically the only amino acid produced by the cultures or during enzymatic reaction so that L-serine may be recovered from a cell-free filtrate by means of a cation exchange resin of the H type, eluted, and crystallized from the eluate after partial evaporation of the water present.

Serine and Serinol Derivatives

Aromatic L-serine-derivatives and aromatic L-serinol-derivatives are important intermediates for synthesizing epinephrine and norepinephrine. Higher aliphatic β-hydroxy-L-amino acids are intermediates for synthesis of surface active agents. p-Nitrophenyl-L-serine and p-nitrophenyl-L-serinol may be prepared by the process as intermediates for synthesis of chloramphenicol.

H. Nakazawa, H. Enei, K. Kubota and S. Okumura; U.S. Patent 3,871,958; March 18, 1975; assigned to Ajinomoto Co., Inc., Japan have found that an aldehyde having at least three carbons reacts with glycine or ethanolamine in the presence of an enzyme source prepared by cultivation of microorganisms and serine-derivatives of formula (1) and serinol-derivatives of formula (2) can be produced.

(1)
$$\underset{\underset{OH}{|}}{\underset{|}{RCHCHCOOH}}\overset{\overset{NH_2}{|}}{}$$

(2)
$$\underset{\underset{OH}{|}}{\underset{|}{RCHCHCH_2OH}}\overset{\overset{NH_2}{|}}{}$$

R is the organic residue having at least two carbons derived from the aldehyde employed. The enzyme source may be a broth in which a microorganism is cultivated, or a cell-free filtrate of the broth, intact cells of the microorganism recovered from the broth, an aqueous suspension of ground cells or a filtrate of the suspension. It is believed that the enzyme catalyzing the reaction between the aldehyde and glycine may be a threonine aldolase.

The microorganisms capable of producing the enzyme are distributed very widely, representative microorganisms belonging to the genera *Escherichia, Citrobacter, Klebsiella, Aerobacter, Serratia, Proteus, Bacillus, Staphylococcus, Arthrobacter, Bacterium, Xanthomonas, Candida, Debaryomyces, Corynebacterium* and *Brevibacterium*.

The reaction can be carried out by culturing the microorganisms mentioned above in a conventional nutrient medium containing aldehyde and glycine and/or ethanolamine under aerobic condition as usual, or by adding the enzyme source to a reaction mixture containing an aldehyde and glycine and/or ethanolamine.

The media employed for culturing the microorganisms may be conventional, containing sources of assimilable carbon and nitrogen and the usual minor nutrients.

In certain cases the activity of the enzyme sources is improved by addition of threonine to the culture medium.

The fermentation is carried out under aerobic conditions, at 25° to 40°C. It is preferred to adjust the pH of the culture medium to 5.5 to 8.5 during cultivation. Culturing is generally carried out for 10 to 72 hours.

The broth may be used as an enzyme source as is without removing the bacterial cells. A cell-free filtrate of the culture broth and a suspension or extract of crushed bacterial cells prepared by trituration, autolysis or ultrasonic oscillation, are also used as the enzyme source of the process. Crude enzyme and pure enzyme recovered by centrifuging, salting out and solvent precipitation can also be used.

The amount of aldehyde in the reaction mixture should be controlled so that the activity of the enzyme is not inhibited, and preferably its range is from 0.1 to 10 wt % of the reaction mixture.

The amounts of glycine and ethanolamine in the reaction system should be equimolar with the aldehyde or in excess for good yield.

The reaction is normally carried out at pH 5 to 10, preferably at pH 7.5 to 8.5, at a temperature of 5° to 60°C, preferably at 10° to 37°C, with stirring or without stirring. In order to maintain the pH value of the reaction system during the reaction, buffer solutions, for example, phosphate buffer, tris buffer and ammonium chloride-ammonia buffer, are employed.

The reaction of the process is promoted by nonionic, anionic, cationic, and amphoteric surface active agents, and preferably by nonionic surface active agents. The surface active agents are added to the reaction system in an amount preferably from 0.1 to 0.5 wt %.

Serine-derivatives having a substituent in the β-position and obtained from an aldehyde and glycine include β-ethylserine, β-isopropylserine, β-n-propylserine, β-butylserine, β-isobutylserine, β-neobutylserine, β-amylserine, β-caprylserine, β-palmitylserine, β-vinylideneserine, β-phenylserine, β-α-furfurylserine, β-p-chloro-phenylserine, etc. Similarly, serinol-derivatives having substituents in the β-position are obtained from an aldehyde and ethanolamine.

The reaction products may be recovered by means of ion exchange resin, adsorption on and desorption from activated charcoal, solvent extraction and so on.

Example: An aqueous culture medium was prepared to contain 2.0 g/dl L-threonine, 0.2 g/dl KH_2PO_4, 0.1 g/dl $MgSO_4$ and 0.4 g/dl $(NH_4)_2SO_4$, and was adjusted to pH 5.0. 50 ml batches of the medium were placed in 500 ml shaking flasks and inoculated with the microorganisms shown in Table 2 which were previously cultured on bouillon agar slants at 31°C for 20 hours, and cultivation was carried out at 31°C for 24 hours with shaking.

The microbial cells in 100 ml of each cultured broth were harvested by centrifuging, washed and suspended in 50 ml of reaction solutions (A), (B), (C) and (D) whose compositions are shown in Table 1 at pH 8.5 and at 31°C for 20 hours.

Table 1

Ingredients	. . . . . . Reaction Solution (g/dl)			
	(A)	(B)	(C)	(D)
p-Nitrobenzaldehyde	2.0	2.0	–	–
Isobutyraldehyde	–	–	2.0	2.0
Glycine	2.0	–	2.0	–
Ethanolamine	–	2.0	–	2.0
Sorpol W-200*	0.2	0.2	0.2	0.2

*Surface active agent (polyoxyethylene alkyl phenol ether).

β-p-nitrophenyl-L-serine, β-p-nitrophenyl-L-serinol, β-isopropyl-L-serine and β-isopropyl-L-serinol were produced in solutions (A), (B), (C) and (D) respectively in the yields shown in Table 2.

Table 2

Microorganisms Employed	. . . . Reaction Solution Yield (g/dl)			
	(A)	(B)	(C)	(D)
Escherichia coli ATCC 15289	0.2	0.1	0.2	0.2
Citrobacter freundii ATCC 6750	0.8	0.7	0.5	0.6
Klebsiella pneumoniae ATCC 10031	0.5	0.4	0.3	0.3
Aerobacter aerogenes IFO 3319	0.1	0.1	0.2	0.1
Aerobacter cloacae IAM 1020	0.2	0.2	0.1	0.2
Serratia marcescens ATCC 14223	0.9	0.8	0.7	0.6
Proteus vulgaris ATCC 881	0.3	0.1	0.2	0.1
Bacillus mesentericus IAM 1026	0.2	0.1	0.3	0.2
Bacillus subtilis ATCC 13953	0.4	0.3	0.4	0.3
Staphylococcus aureus ATCC 4776	0.6	0.5	0.5	0.4
Arthrobacter simplex ATCC 13260	0.5	0.3	0.4	0.4
Arthrobacter globiformis ATCC 4336	0.6	0.6	0.2	0.5
Arthrobacter pascens ATCC 13346	0.6	0.6	0.5	0.4
Arthrobacter ureafaciens ATCC 7562	0.1	0.1	0.1	0.2
Arthrobacter aurescens ATCC 13344	0.7	0.6	0.6	0.6
Bacterium cadaveris ATCC 9760	0.8	0.7	0.8	0.6
Bacterium succinicum IAM 12090	0.4	0.3	0.4	0.3
Xanthomonas campestris IAM 1671	0.7	0.6	0.7	0.6
Candida rugosa ATCC 10571	0.6	0.5	0.5	0.4
Candida humicola ATCC 14438	1.0	0.8	0.9	0.9
Candida utilis ATCC 9950	0.7	0.6	0.5	0.4
Candida guilliermondii ATCC 9058	0.5	0.4	0.3	0.2
Candida parapsilosis ATCC 7330	0.3	0.2	0.2	0.1
Debaryomyces hansenii ATCC 18110	0.1	0.1	0.2	0.1

TRYPTOPHAN

Production from Indole and β-Chloroalanine

It is known that tryptophan and hydroxytryptophan are produced from indole and 5-hydroxy-indole, respectively, reacting with an α-amino acid such as serine or cysteine (Japanese Patent Publication No. 46348/1972) by the action of an enzyme produced by a microorganism of genus Proteus, Erwinia, Escherichia, Pseudomonas or Aerobacter.

S. Konosuke and K. Mitsugi; U.S. Patent 3,929,573; December 30, 1975; assigned to Ajinomoto Co., Inc., Japan have found that the same enzyme catalyzes the formation of tryptophan from indole and β-halogenoalanine, and hydroxytryptophan from 5-hydroxyindole and β-halogenoalanine.

According to this process, the yield of tryptophan or hydroxytryptophan is higher than in the known methods, and moreover, the reaction rate is very high as compared with the known methods.

The microorganisms are cultured in entirely conventional media containing a carbon source, nitrogen source, inorganic ions, and preferably a minor organic nutrient.

Preferably, 0.01 to 0.5 g/dl tryptophan or hydroxytryptophan are added to the medium to increase the enzyme activity. When the microorganisms are cultured in the medium mentioned above for 2 to 7 days at pH 4 to 9, and at 27° to 40°C, the enzyme capable of producing tryptophan and hydroxytryptophan is accumulated in the cells or in the broth. The cells or the broths themselves are used as an enzyme source.

The isolation and purification of the enzyme and enzymological properties of the enzyme are disclosed in FEBS' 48 (1) 56 (1974), H. Yoshida et al. The enzyme is a tryptophanase since it catalyzes degradation of tryptophan to indole, pyruvate and ammonia.

The reaction mixture contains the enzyme or a source of the enzyme, β-chloroalanine, and indole or 5-hydroxyindole. Pyridoxal-5-phosphate is necessary as the coenzyme when purified apoenzyme is used. A reducing agent such as thiosulfate is preferably added to the reaction mixture to increase the yield of tryptophan or hydroxytryptophan.

The reaction mixture is held at 5° to 60°C and at pH 5 to 10 until indole or 5-hydroxyindole is substantially converted to tryptophan or hydroxytryptophan. The tryptophan or hydroxytryptophan formed in the reaction mixture can be isolated and purified by conventional methods such as ion exchange chromatography, or precipitation at the isoelectric point.

Example: *Proteus morganii* IFO 3848 was cultured in 60 ml batches of the following medium held in 500 ml flasks with shaking at 30°C for 20 hours.

Medium

L-tryptophan, g	2
KH_2PO_4, g	0.5
$MgSO_4 \cdot 7H_2O$, g	0.5
$FeSO_4 \cdot 7H_2O$, mg	10
$MnSO_4 \cdot 4H_2O$, mg	7
Corn steep liquor, ml	20
Yeast extract, g	2
Casein hydrolyzate, g	5
Tap water, l	1
pH (KOH)	7.0

Cells were collected by centrifuging from one liter combined culture broth thus obtained, and added to one liter of 0.1% NH_4Cl–NH_4OH buffer solution (pH 8.0) containing 20 g indole and 20 g β-chloro-L-alanine. The reaction mixture was held at 37°C for 20 hours with stirring. 24 g of L-tryptophan was produced (by microbiological assay method).

After dissolving of the precipitated tryptophan with NaOH, the reaction mixture was filtered to remove cells, adjusted to pH 4 with HCl, and passed through a column packed with active carbon. L-tryptophan was eluted with 2 N NH_4OH, the eluate was concentrated, and 20.2 g L-tryptophan crystals were obtained.

Production from Indole and Methanol

S.V. Gatenbeck and P.O. Hedman; U.S. Patent 3,963,572; June 15, 1976; assigned to AB Bofors, Sweden describe a method for fermentative preparation of L-tryptophan and derivatives of the general formula

where R is hydrogen, hydroxyl, alkyl or alkoxy groups from indole or indole derivatives of the general formula

where R has the same meaning as above, characterized in that the fermentation is carried out in a substrate consisting essentially of methanol as the main source for carbon and is carried out using methanol-using bacteria of a type that assimilates methanol in such a way that the methanol after oxidation reacts with intracellular glycine, whereby serine is formed.

A preferable procedure according to the process implies that under aerobic conditions, submerged in the fermentor, one grows such isolates of the families *Pseudomonas* and *Methylomonas* that are able to utilize methanol as the carbon source via serine. This process can be accomplished either by batch culture or by continuous culture. A mineral salt solution is used as the substrate with a composition as described in the following example. No organic substrate component needs to be included with the exception of methanol and indole. The growth is suitably accomplished within a temperature interval of between 18° and 45°C, therewith preferably a temperature of between 25° and 30°C should be employed. The pH must be carefully regulated continuously during the cultivation, and ought to be held within an interval of 5.5 to 8.5 and preferably between 6.5 to 7.5.

The supply of methanol should be carried out continuously and this regardless of whether it is batch or continuous culture. The methanol concentration should not exceed the marginal toxic value for the cultured isolate, but should lie about the optimal value for tryptophan production. The marginal toxic value for these microorganisms, which are used in this connection, lies around 5 volume percent methanol. The suitable concentration of methanol for optimal production of tryptophan lies between 0.1 and 1.0% and preferably around 0.5 volume percent methanol.

The concentration of free indole in the culture medium must be kept below a value of 500 mg/l. The period of growth ought to lie between 30 and 90 hours with batch culture. After the cells have been separated an ion exchanger is used for the isolation and purification of produced tryptophan.

Example: *Pseudomonas* AM 1 is grown in 500 ml flasks on a mechanical shaker, at a temperature of 25°C and containing 100 ml of a substrate of the following composition:

KH_2PO_4, g/l	1.4
Na_2HPO_4, g/l	5.4
$(NH_4)_2SO_4$, g/l	0.5
$MgSO_4 \cdot 7H_2O$, g/l	0.2
$CaCl_2$, mg/l	7.4
$FeSO_4 \cdot 7H_2O$, mg/l	5.0
$Na_2MoO_4 \cdot 2H_2O$, mg/l	2.5
$MnSO_4 \cdot H_2O$, mg/l	1.8
Methanol, g/l	5

After the microorganism has grown out 11.2 mg of indole is added, and after 48 hours of growth 3.3 mg of tryptophan is obtained which corresponds to an exchange of 17% calculated on added indole.

Tryptophan Derivatives from *Candida humicola*

L.-E. Paulsson, L.I. Wiberger and A.W. Ohlsson; U.S. Patent 3,915,796; Oct. 28, 1975; assigned to Aktiebolaget Bofors, Sweden describe derivatives of tryptophan, which advantageously may be prepared by means of *Candida humicola* and which have the following general formula:

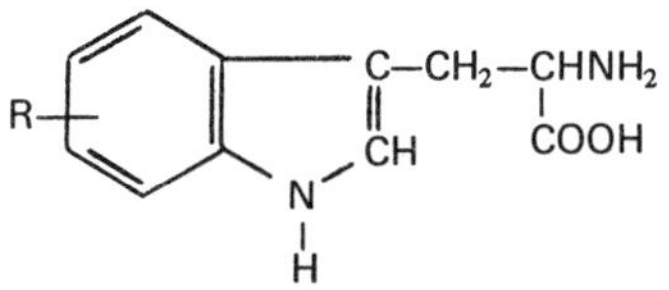

wherein R may be selected from hydrogen, hydroxy, halogen, lower alkyl and lower alkoxy. Many of these derivatives are of great interest to the pharmaceutical industry and are extremely expensive.

The process is substantially characterized by culturing the microorganism *Candida humicola* under aerobic conditions in a continuously supplied nutrient substrate with a separate continuous supply of the corresponding indole derivative. The indole derivative has the formula shown on the following page wherein R is hydrogen or hydroxy or halogen or alkyl or alkoxy. The micro-organism as used is characterized by a very high productivity of L-tryptophan from indole. As the microorganism is sensitive to a too high concentration of indole in the culture liquor, a continuous supply of the precursor must be made. The microorganism may be cultured on a completely defined substrate and besides carbohydrates and thiamine it does not require any other organic substances for its growth.

In order to stimulate the formation of mutants which give no tryptophan degradation, niacin may be added to the substrate in a concentration of about 0.05 to 0.5 g/l of the substrate.

The feature that appears to have the greatest influence on the conversion velocity of indole to tryptophan and on the degradation of tryptophan via kynurenine, is the concentration of biologically available iron, e.g., in the form of ferrous citrate. The concentration of iron (Fe^{2+}) in the culture liquor must be below the content that will give degradation of tryptophan but be above the content at which the growth of the culture will be severely disturbed and continuous culture thus be impossible.

At a low supply of oxygen gas (aeration) it is possible to maintain a somewhat higher Fe^{2+}-content in the culture liquor while the Fe^{2+}-content should be low at a higher supply of oxygen gas. These observations are valid for nonmutated *Candida humicola*.

In order to control the process so that the indole supply will not give high values of residual indole in the fermentor contents, which will cause the ceasing of growth, a number of analyses are carried out. The titration frequency and the level of oxygen saturation will increase strongly at an increased content of residual indole due to the fact that the aerobic metabolism is changed to an anaerobic one. These values are followed up and may be used to control the supply of ferrous citrate or the aeration.

The analyses on residual indole are carried out with unseparated fermentor contents on a gas chromatograph with o-nitrophenyl ethanol dissolved in methanol as an internal standard. A greater supply velocity of indole may be maintained with a glycine addition of about 1 to 10 g/l.

In order to insure that the tryptophan production is not disturbed by too high iron supplies, continuous analyses are carried out on a dextran gel column (Sephadex) with UV source and thereafter separate UV analysis of the tryptophan fractions. These analyses give a good indication of metabolites of tryptophan principally kynurenic acid. UV source is a photometer for UV-light (LKB-Produkter, Sweden).

For the isolation and purification of the produced tryptophan an ion exchange method was used on comparatively cell-free fermentor liquor. The ion exchange matrix had a low degree of crosslinking or was of a so-called macro-reticular type. The flow used was 0.1 bed volumes (BV) per hour. Washing was carried out with distilled water and elution with 50% ethanol and 50% 1 N NH$_4$OH by back wash. The tryptophan solution thus obtained was neutralized with HCl and tryptophan crystals were obtained. The yield was about 95%.

The product as obtained had 100% biologic activity according to a test with *S. fecalis*. No contaminations being positive in the Ehrlich and ninhydrin tests were present.

L-5-Methoxytryptophan

N. Kitajima, S. Watanabe and I. Takeda; U.S. Patent 3,915,795; October 28, 1975; assigned to Asahi Kasei Kogyo KK, Japan describe a process of producing L-5-methoxytryptophan comprising contacting 5-methoxyindole in a culture medium at a pH of from 4 to 10 and at a temperature of from 20° to 40°C with a microorganism selected from the group consisting of:

> *Brevibacterium ammoniagenes* ATCC 6872
> *Corynebacterium* sp. No. 14001 NRRL B-5395
> *Corynebacterium melassecola* ATCC 17965
> *Arthrobacter globiformis* ATCC 8010
> *Microbacterium ammoniaphilum* ATCC 15354
> *Micrococcus luteus* ATCC 21102
> *Sarcina lutea* ATCC 15176
> *Staphylococcus citreus* ATCC 4012
> *Bacillus subtilis* ATCC 14593
> *Serratia marcescens* IFO 3046
> *Escherichia coli* ATCC 9637
> *Aerobacter aerogenes* IFO 3317
> *Proteus mirabilis* IFO 3849
> *Erwinia carotovora* IFO 3057
> *Pseudomonas fragi* IFO 3458
> *Xanthomonas pruni* IFO 3511
> *Protaminobacter alboflavus* IFO 3707
> *Flavobacterium arborescens* ATCC 4358
> *Agrobacterium tumefaciens* IFO 3058
> *Saccharomyces cerevisiae* ATCC 7754
> *Candida petrophilum* ATCC 20226
> *Torulopsis petrophilum* ATCC 20225
> *Brettanomyces petrophilum* ATCC 20224
> *Hansenula anomala* IFO 0118
> *Nocardia asteroides* IFO 3384
> *Streptomyces aureus* ATCC 3309
> *Aspergillus oryzae* IFO 4075
> *Penicillium chrysogenum* ATCC 15241

An L-5-lower alkoxy tryptophan is an intermediate for use in the production of L-5-hydroxytryptophan (hereinafter to as "L-5-HTP"), which is useful as an antidepressant and has various hormonal actions against Down's syndrome (Mongolism) and the like diseases. Further, L-5-HTP is an extremely important metabolic intermediate for synthesizing in the metabolic system of a living body such intracerebral amine type hormones as L-5-hydroxytryptamine (serotonin), L-5-methoxytryptamine and N-acetyl-L-5-methoxytryptamine (melatonin) from an L-tryptophan.

Cultivation: The medium used contains 5-MI (5-methoxyindole), together with or without serine, in addition to a carbon source, a nitrogen source, inorganic salts and natural organic nutrients which are suitable for the good growth of the microorganism used and for the smooth production of L-5-MTP (L-5-methoxy-tryptophan).

The manner of addition of 5-MI, together with or without serine, is as follows: In a process in which the growth of microorganism is conducted simultaneously with the production of L-5-MTP, 5-MI, together with or without serine, may be added directly to the culture liquor. In a process in which is used a microorganism, which has been cultured and has once been taken out of the culture liquor, 5-MI together with or without serine is added to a suspension of the microorganism.

The 5-MI and serine may be used at various concentrations, but are desirably used individually at a concentration of about 0.2 to 20 g/l. These may be added either at one time or divisionally.

The cultivation may be effected either under aerobic conditions according to shaking culture or aerobic stirring culture, or under anaerobic conditions according to stationary culture. The type of cultivation is decided according to the kind of microorganism used. The pH of the medium is preferably from 4 to 9. The cultivation temperature is preferably from 25° to 37°C. In case a microorganism isolated from the culture liquor is used, the reaction is preferably carried out at a pH from 7 to 10 and a temperature from 25° to 40°C.

For the control of pH, there may be used a known neutralizing agent such as ammonia, sodium hydroxide, potassium hydroxide, calcium carbonate or hydrochloric acid. The cultivation time or the reaction time varies depending on the manner of addition and the concentration of starting material and on the kind of strain used. Ordinarily, however, L-5-MTP is produced and accumulated in a period of 1 to 5 days. From 0.1 to 50 g/l of L-5-MTP can be accumulated in the culture liquor (or the reaction liquid).

The thus accumulated L-5-MTP can be isolated according to a process known per se. That is, the L-5-MTP is adsorbed on active carbon or such ion exchange resin as Dowex-1 (CH_3COO^- type) or the like, eluted with hot ethanol or like organic solvent or with acetic acid, concentrated, and then recovered by addition of alcohol.

Production of 5-HTP from 5-MTP: The production of 5-HTP from 5-MTP can be easily accomplished according to demethylation or the like reaction using hydriodic acid. The demethylation reaction is described in further detail below.

The reaction liquid used in the reaction comprises L-5-MTP, hydriodic acid (specific gravity 1.7), acetic anhydride, etc. The reaction is sometimes promoted by use of a catalyst such as red phosphorus. The injection of nitrogen gas or carbon dioxide into the reaction liquid is effective for inhibiting the reaction product from decomposition. The L-5-MTP may be used at various concentrations, but is desirably used at a concentration of about 10 to 150 g/l. It may be added either at one time or incrementally. The reaction temperature is in the range from 20° to 90°C, preferably from 40° to 80°C. The reaction time varies depending on the concentration of L-5-MTP, the presence or absence of catalyst and the reaction temperature. Ordinarily, however, the reaction is complete within several minutes to several hours (about 5 to 6 hours). According to the process, 5 to 50 g/l of L-5-HTP can be produced in the reaction liquid. The thus produced L-5-HTP can be isolated according to a process known per se. That is, the L-5-HTP is adsorbed on active carbon or such ion exchange resin as Dowex-1 or the like, desorbed, concentrated, and then recovered by addition of alcohol or the like.

Example: A microorganism *Corynebacterium* sp. No. 14001, NRRL B-5395 was inoculated in a shaking flask containing 50 ml of a medium (pH 7.0) containing 1.0% of peptone, 1.0% of meat extract, 0.4% of sodium chloride and 0.1% of yeast extract, and was cultured while shaking the flask reciprocally at 30°C for 24 hours. The thus cultured microorganism was collected and suspended in a phosphate buffer of pH 8.5. To the resulting suspension were added 20 mg of 5-MI, 20 mg of L-serine and 120 mg of glucose, and the total volume of the suspension was made 20 ml. Subsequently, the suspension was allowed to react under stirring at 30°C for 16 hours, while maintaining the suspension at a pH of 8.5. As a result, 1.35 mg/ml of L-5-MTP was accumulated in the reaction liquid.

Subsequently, the same reaction as above was effected to obtain a reaction liquid. This reaction liquid was mixed with the aforesaid reaction liquid to obtain 950 ml of a mixed liquid. The mixed liquid was subjected to active carbon treatment to isolate 470 mg of crude crystals of L-5-MTP. 400 mg of the crude crystals were allowed to react, together with 4 ml of hydriodic acid and 4 ml of acetic anhydride, at 50°C for 90 minutes while introducing nitrogen, to produce 15.7 mg/ml of L-5-HTP.

OTHER AMINO ACIDS

Citrulline

I. Chibata, T. Tosa, T. Sato and K. Yamamoto; U.S. Patent 3,886,040; May 27, 1975; assigned to Tanabe Seiyaku Co., Ltd., Japan describe a process whereby L-citrulline or a mixture of L-citrulline and D-arginine can be prepared by polymerizing at least one acrylic monomer in an aqueous suspension containing an L-arginine deiminase-producing microorganism, and subjecting the resultant immobilized L-arginine deiminase-producing microorganism to enzymic reaction with L-arginine, DL-arginine or their inorganic acid addition salt.

The polymerization reaction can be preferably carried out in the presence of a polymerization initiator and a polymerization accelerator. Typical of suitable polymerization initiators are potassium persulfate, ammonium persulfate, vitamin B_2 and Methylene Blue. β-(dimethylamino)-propionitrile and N,N,N',N'-tetramethylethylenediamine are among polymerization accelerators which can be employed. It is preferred to carry out the reaction at 0° to 50°C, especially at 20° to 40°C. The reaction may be completed within 10 to 60 minutes.

The acrylic monomers employed in the process are, among others, acrylamide, N,N'-lower alkylene-bis-acrylamide and bis-(acrylamidomethyl)ether. N,N'-methylene-bis-acrylamide and N,N'-propylene-bis-acrylamide are suitable as the N,N'-lower alkylene-bis-acrylamide. Moreover, preferred examples of L-arginine deiminase-producing microorganisms which are employed in the process include

Pseudomonas putida ATCC 4359, *Pseudomonas fluorescens* IFO (Institute for Fermentation, Osaka, Japan) 3081, *Pediococcus cerevisiae* P-60 ATCC 8042, *Sarcina lutea* IAM (Institute of Applied Microbiology, Tokyo University, Japan) 1099, *Mycobacterium avium* IFO 3154, *Streptomyces griseus* IFO 3122 and *Streptococcus faecalis* ATCC 11420. All of these microorganisms are publicly available from the abovementioned depositories.

L-citrulline can be prepared by enzyme reaction of the resultant immobilized microorganism with L-arginine or an inorganic acid addition salt thereof. Alternatively, L-citrulline and D-arginine can be prepared by using DL-arginine or an inorganic acid addition salt thereof instead of L-arginine. Suitable examples of the inorganic acid addition salts of L- or DL-arginine include hydrochloride, sulfate and nitrate. The enzymic reaction can be preferably accelerated by carrying it out in the presence of a surfactant. Any one of a cationic surfactant (e.g., cetyltrimethyl ammonium bromide), an anionic surfactant (e.g., triethanolamine laurylsulfate) and a nonionic surfactant (e.g., glyceryl monoalkylate) is employed for this purpose. Preferred concentration of the surfactant in the reaction solution is about 0.005 to 1.0 w/v percent.

The concentration of a substrate employed is not critical. For example, L- or DL-arginine is dissolved in water at any concentration. The aforementioned immobilized microorganism is suspended in the solution of L- or DL-arginine, and the mixture is stirred at a temperature of 15° to 65°C. After the reaction is completed, the mixture is filtered or centrifuged to recover the immobilized microorganism for subsequent use. L-citrulline or a mixture of L-citrulline and D-arginine is recovered from the filtrate or supernatant solution.

Example: An aqueous nutrient medium containing the following ingredients is prepared:

Glucose, w/v %	1
Yeast extract, w/v %	1
Polypeptone, w/v %	0.5
L-arginine hydrochloride, w/v %	0.5
Ammonium chloride, w/v %	0.1
Dipotassium phosphate, w/v %	0.1
Magnesium sulfate heptahydrate, w/v %	0.05
Manganese sulfate tetrahydrate, w/v %	0.1
Ferrous sulfate heptahydrate, w/v %	0.0005
Sodium chloride, w/v %	0.2

The aqueous nutrient medium is adjusted to pH 6.2. *Pseudomonas putida* ATCC 4359 is inoculated into 300 ml of the medium, and the medium is cultivated at 30°C for 40 hours under shaking. The medium is then filtered by centrifugation. The microbial cells thus collected are suspended in 12 ml of a physiological saline solution. 2.25 g of acrylamide, 0.12 g of N,N'-methylene-bis-acrylamide, 1.5 ml of 5% β-(dimethylamino)-propionitrile and 1.5 ml of 2.5% potassium persulfate are added to the suspension. The suspension is then allowed to stand at 25°C for 10 minutes. The insoluble product is ground and washed with a physiological saline solution. 25 g of an immobilized preparation of *Pseudomonas putida* ATCC 4359 are obtained.

2.25 g of the immobilized preparation of *Pseudomonas putida* ATCC 4359 are charged into a 1.6 cm x 15.5 cm column, and one liter of an aqueous 10% L-arginine hydrochloride solution is passed through the column at 37°C at a flow rate of 15 ml/hr. One liter of the effluent is concentrated at 60°C to bring the total volume to 250 ml. Then, the concentrated solution is allowed to stand at 5°C overnight. The crystalline precipitate is collected by filtration and washed with methanol. 74.0 g of L-citrulline are obtained. $(\alpha)_D^{25} = +21.3°$ (c = 10, 1 N HCl).

Glutamine

F. Yoshinaga, T. Tsuchida, K. Kikuchi and S. Okumura; U.S. Patent 3,886,039; May 27, 1975; assigned to Ajinomoto Co., Inc., Japan have found that L-glutamine is produced at higher concentrations than were available heretofore under comparable conditions by mutant strains of microorganisms which have resistance to at least one sulfa drug when these mutants are cultured on otherwise conventional media including sources of assimilable carbon and nitrogen, inorganic ions, and unspecific organic growth promoting agents.

The mutants are derived by means of conventional mutagenic agents, such as ultraviolet light, x-rays, or gamma rays in mutagenic doses, and sodium nitrite, nitrosoguanidine or diethyl sulfate solution, from glutamic-acid-producing microorganisms of genera *Brevibacterium, Corynebacterium* and *Microbacterium*, and by screening of the mutants so produced for the necessary resistance to growth inhibition by sulfa drugs. Such a mutant may also be obtained without using mutagenic agents, by screening strains having resistance to sulfa drugs from colonies of the parent strain cultured on a medium containing the sulfa drugs.

The most effective glutamine-producing mutants found so far are *Brevibacterium flavum* FERM P-1684, *Brevibacterium flavum* FERM P-2371, *Corynebacterium glutamicum* FERM P-2372, *Corynebacterium glutamicum* FERM P-2373, and *Microbacterium flavum* FERM P-2374.

For a good yield of glutamine, the fermentation should be carried out aerobically with aeration and/or agitation. To produce an optimum yield, the pH should be controlled within the range from 5.5 to 8.5. The desired pH may be maintained by adding gaseous or aqueous ammonia, calcium carbonate, alkali metal hydroxide, urea or organic or inorganic acids to the medium from time to time; certain of these additives also supply assimilable nitrogen. When the fermentation is carried out at a temperature in the range from 24° to 37°C, the concentration of glutamine in the broth usually reaches its maximum within from 2 to 5 days.

Conventional methods may be used for recovering the accumulated glutamine from the culture broth, preferably after removal of the microbial cells by filtration or centrifuging. Ion exchange resins conveniently absorb glutamine from the cell-free broth. Crystalline L-glutamine is obtained from the eluate by conventional methods.

Example: A fermentation medium was prepared with the following composition:

Glucose, %	10
Ammonium sulfate, %	1
KH_2PO_4, %	0.25
$MgSO_4 \cdot 7H_2O$, %	0.04
$FeSO_4 \cdot 7H_2O$, mg %	1

(continued)

Thiamine hydrochloride, γ/l	350
Biotin, γ/l	5
Soybean protein hydrolyzate, ml/dl	0.5
pH	7

300 ml batches of the medium were introduced in glass jar fermentors, and were sterilized by steam. The media were inoculated with *Brevibacterium flavum* FERM P-1684 (mutant strain having resistance to sulfaguanidine), which had previously been cultured on bouillon slants for 24 hours at 31.5°C. The fermentation was carried out at 1,200 rpm at 31.5°C for 30 hours while aerating with one-quarter volume of air per minute and volume of broth. The pH of each medium was held at about 6.5 by addition of gaseous ammonia. 3.9 g/dl of L-glutamine were found in the broth after 30 hours cultivation.

The cultured broth was centrifuged to remove the cells, and one liter of the supernatant was stripped of L-glutamine by passage over a column of an ion exchange resin. 15.5 g of crystalline L-glutamine was recovered from the broth by a conventional method.

On the other hand, only 0.85 g/dl of L-glutamine was found in the broth in which *Brevibacterium flavum* ATCC 14067 was cultured in the same way.

Valine

T. Tsuchida and F. Yoshinaga; U.S. Patent 3,893,888; July 8, 1975; assigned to Ajinomoto Co., Inc., Japan describe a method of producing L-valine by bacterial fermentation.

It was found that some bacteria having a resistance to 2-thiazolalanine (hereinafter abbreviated as 2-TA) produce a large amount of L-valine when cultured in a nutrient medium. It was also found that a mutant strain having a resistance to 2-TA together with a certain nutrient requirement has a better ability to produce L-valine than a strain having only resistance to 2-TA or having only a certain nutrient requirement.

A microorganism employed in the process is a strain belonging to genera *Brevibacterium* and *Corynebacterium* which resists growth inhibition by 2-TA. The microorganism may also have a nutrient requirement and/or resistance to other reagents in addition to the resistance to 2-TA. The required nutrient may be leucine, isoleucine, threonine, or isoleucine and methionine, and the other reagents α-amino-β-hydroxyvaleric acid and ethionine.

Representative strains useful for the process are as follows:

(1)   2-TA resistant mutants such as *Corynebacterium glutamicum* FERM P-1768 whose parent strain is *Corynebacterium glutamicum (Micrococcus glutamicus)* ATCC 13032, *Corynebacterium acetoacidophilum* FERM P-1314 whose parent strain is *Corynebacterium acetoacidophilum* ATCC 13870 and *Brevibacterium lactofermentum* FERM P-1945 whose parent strain is *Brevibacterium lactofermentum* ATCC 13869.

(2)   Mutants having nutrient requirements in addition to 2-TA resistance such as *Corynebacterium glutamicum* FERM P-1967

(leucine-requiring mutant), FERM P-1968 (threonine-requiring mutant) and FERM P-1966 (isoleucine-requiring mutant) which were obtained by a conventional artificial mutant-inducing method from *Corynebacterium glutanicum* FERM P-1768 (2-TA resistant mutant), *Brevibacterium lactofermentum* FERM P-1964 (leucine-requiring mutant), FERM P-1963 (isoleucine-requiring mutant) and FERM P-1965 (threonine-requiring mutant) which were obtained by a conventional artificial mutant-inducing method from *Brevibacterium lactofermentum* FERM P-1945 (2-TA resistant mutant), and *Brevibacterium lactofermentum* FERM P-1845 (isoleucine plus methionine-requiring mutant having resistance to 2-TA) which were induced from *Brevibacterium lactofermentum* FERM P-1858 (isoleucine plus methionine-requiring mutant).

The culture medium used to produce L-valine in the process may be entirely conventional. It should include an assimilable carbon source, an assimilable nitrogen source, and the usual minor nutrients.

The conditions for cultivation are also quite conventional. The fermentation is performed at a pH between 5 and 9, at a temperature of 20° to 40°C under aerobic conditions for 1 to 4 days. The pH of the culture medium can be adjusted by adding sterile calcium carbonate, aqueous or gaseous ammonia, mineral acid or organic acid during the fermentation. The L-valine is recovered from the cultured broth by conventional methods.

Example: 20-ml batches of the fermentation medium shown hereunder were each placed in a 500-ml shaking flask, and sterilized. The medium was inoculated with *Brevibacterium lactofermentum* FERM P-1845 (2-TA, isoleu⁻ plus met⁻) which had previously been cultured on a bouillon slant at 30°C for 24 hours, and cultured at 30°C for 72 hours with stirring and aerating. The cultured broth was found to contain 2.38 g/dl of L-valine. As a control, *Brevibacterium lactofermentum* FERM P-1858 (isoleu⁻ plus met⁻) having no resistance to 2-TA was cultured by the same way mentioned above, and 0.9 g/dl of L-valine was found in the broth after 72 hours cultivation, the mutant strain FERM P-1845 having produced more than twice as much valine. Composition of the medium is as follows:

Glucose, %	8
$(NH_4)_2SO_4$, %	4
KH_2PO_4, %	0.1
$MgSO_4 \cdot 7H_2O$, %	0.04
Biotin, γ/l	50
Thiamine HCl, γ/l	300
Fe^{++}, ppm	2
Mn^{++}, ppm	2
L-isoleucine, mg/dl	20
DL-methionine, mg/dl	40
$CaCO_3$ (sterilized separately), %	5
pH	7.0

MISCELLANEOUS PROCESSES

Amino Acid Production from Hydrogen-Oxidizing Bacteria

Y. Morinaga, A. Ishizaki and S.-i. Otsuka; U.S. Patent 3,943,038; March 9, 1976; assigned to Ajinomoto Co., Inc., Japan have found that certain gram positive hydrogen-oxidizing bacteria, especially bacteria belonging to *Arthrobacter, Brevibacterium and Mycobacterium,* can produce a number of amino acids from carbon dioxide as the carbon source and hydrogen as the energy source. It is further found that amino acid production is remarkably improved by adding penicillin to the culture medium. Amino acids produced in accordance with this process include lysine, histidine, arginine, aspartic acid, threonine, serine, glutamic acid, proline, glycine, alanine, cysteine, valine, isoleucine, leucine, tyrosine, phenylalanine and methionine.

The bacteria useful in the process are: *Brevibacterium* sp. AJ 3588 (FERM P-2234), *Mycobacterium* sp. AJ 3589 (FERM P-2235) and *Arthrobacter* sp. AJ 3786 (FERM P-2638).

Aqueous culture media for culturing the microorganisms of this process contain nitrogen sources and inorganic salts. Suitable nitrogen sources include ammonium salts, ammonia, and nitrate salts. Inorganic salts useful in the media are the conventional salts of potassium or sodium, the phosphate salts, the calcium salts, the sulfate salts, the ferrous salts, the manganese salts, the zinc salts and others.

Sometimes microbial growth is promoted by the addition of small amounts of organic nutrients such as amino acids or vitamins, and small amounts of materials containing those organic nutrients, such as yeast extract, bouillon, corn steep liquor, casein hydrolyzate, malt extract, soy protein or its hydrolyzate.

The addition of penicillin, such as benzylpenicillin, phenoxymethylpenicillin, or ampicillin, to the culture medium usually improves the production of amino acids. Preferably 5 to 100 I.U./ml penicillin are added during 50 to 80 hours cultivation.

During the cultivation, hydrogen, oxygen and carbon dioxide are maintained in or supplied to the fermentation vessel containing the culture medium. The gases are introduced into the fermentation vessel preferably in the ratio of 10 to 80% by volume hydrogen, 5 to 25% by volume oxygen and 3 to 20% by volume carbon dioxide.

Culturing is carried out desirably at 20° to 35°C and pH 4 to 9 for 1 to 7 days. pH is maintained by supplying gaseous ammonia which also suitably serves as a nitrogen source.

Amino acids accumulated in the culture broth are recovered by any known method. For example, after removing the cells by centrifuging and evaporating the supernatant, amino acids are precipitated by adjusting the pH to isoelectric point, or the amino acids may be separated by using ion exchange resin.

Amino Acids from Aldehyde-Containing Hydrocarbons

D.O. Hitzman; U.S. Patent 3,965,985; June 29, 1976; assigned to Phillips Petroleum Company describes a process whereby hydrocarbons are oxidized and contacted with an aldehyde reactive nitrogen-containing compound, and the water-soluble mixture formed thereby is fed to a fermentor for microbial fermentation resulting in an uncontaminated microbial production product suitable as a protein food source.

Examples of those products which can be employed as microbial feedstocks include the water-soluble aliphatic alcohols, ketones, aldehydes, carboxylic acids, ethers, and polyols, preferably containing as many as 10 carbon atoms.

The microbial feed is contacted with nitrogen-containing compounds that are reactive with aldehydes. The aldehydes modified by or in the presence of the nitrogen-containing compound become effective nutrients and the resultant mixture can be fed directly into a fermentor as a carbon and hydrogen nutrient feedstock for cellular production by the microorganisms under conditions suitable for fermentation. It is preferred that only the water-soluble products of the resultant mixture be fed to the fermentor.

Efficient utilization of almost any hydrocarbon or carbonaceous raw material such as natural gas, petroleum, naphtha, coal, peat, asphalt and the like; and conversion thereof via oxidation and subsequent microbial conversion to microbial production products essentially free of oil and other hydrocarbon contaminants is obtained with an excellent fermentation productivity rate according to this process.

It is believed probable that the conversion of biodeleterious aldehydes, such as formaldehyde, to products such as hexamethylenetetramine and the like is largely responsible for the microbial suitability of these materials as feedstocks.

It is critical that the nitrogen-containing compounds that are reactive with aldehydes be admixed with the oxygenated hydrocarbon feedstock containing aldehydes before introduction of the resultant mixture to the fermentor.

Illustrative examples of suitable nitrogen-containing compounds which can be employed include ammonia, ammonium hydroxide, ammonium sulfate, ammonium nitrate, ammonium phosphate, acetonitrile, urea, guanidine, uric acid, and the like. Ammonia or ammonium compounds are preferred.

Sufficient amounts of the nitrogen-containing compound should be added to render innocuous a substantial amount of the deleterious material in the feedstock. Normally, from about 0.01 to 10 mol equivalents of the nitrogen-containing compound should be provided for each mol of aldehyde.

The fermentation process is carried out according to the conditions generally known in the art to support microbial fermentation. Generally, temperatures in the range of about 15° to about 60°C and pressures in the range of about 0.1 to 100 atmospheres are employed. Normally pressures in the range of about 1 to 30 atmospheres are used.

One of the most important limitations to increased cell production is the dissolved oxygen level in the fermentor. The dissolved oxygen can be increased by running the fermentor under increased pressures. Pressures of about 0.1 to 50 atmospheres gage are usually employed.

The particular microorganism employed in this process is not critical and many are suitable for employment. Exemplary of the microorganisms are *Pseudomonas methanica* (NRRL B-3449), *Pseudomonas fluorescens* (NRRL B-3452), *Methanomonas methanica* (NRRL B-3450), *Methanomonas methanooxidans* (NRRL B-3451), *Arthrobacter parafficum* (NRRL B-3452), and *Corynebacterium simplex* (NRRL B-3454). *Bacillus, Mycobacterium, Actinomyces* and *Nocardia* genera are other illustrative examples of bacteria which have been tested and found to be suitable. Other examples of bacteria include the genera: *Micrococcus; Rhodobacillus; Chromatium; Nitrosomonas; Serratia; Nitrobacter; Rhizobium; Azotobacter; Aerobacter; Escherichia; Streptococcus; Bactridium; Clostridium;* and *Corynebacterium.* Other suitable classes of microorganisms include the yeasts, molds, fungi, and the like. Combinations of microorganisms can also be employed.

Suitable minerals, growth factors, vitamins, and the like are generally added in amounts sufficient to provide for the particular needs of the microorganisms utilized.

Upon completion of the desired degree of fermentation, the microbial fermentation products can be separated by any means known to the art such as centrifugation, filtration, solvent extraction, stripping of volatiles, heating, and the like.

Pseudomonas methanica was particularly high in the production of tryptophane, lysine, and threonine; *Pseudomonas fluorescens* in the production of lysine, threonine, leucine, tryptophane and valine; and *Corynebacterium simplex* in the production of leucine, lysine, threonine, tryptophane and alanine.

Sulfur-Containing L-Amino Acids

K. Soda, H. Tanaka, H. Nakazawa and K. Mitsugi; U.S. Patent 4,071,405; January 31, 1978; assigned to Ajinomoto Co., Inc., Japan describe the production of L-amino acids such as L-cysteine, L-homocysteine and their S-substituted derivatives.

The process is carried out by contacting β- or γ-substituted L-amino acid in an aqueous medium containing methioninase in the presence of the precursor of the substituent. A wide variety of sulfur-containing chemical compounds can be employed as substituent precursors in the process. These may be represented by the generic formula: Y-H wherein Y is a sulfur-containing, nucleophilic moiety in which there is at least one unshared electron pair associated with the sulfur atom.

Specific compounds which may be employed include cysteine thiol, cysteine, homocysteine, cysteine sulfenic acid, homocysteine sulfenic acid, homocysteine sulfinic acid, propyl mercaptan, phenyl mercaptan, o-methyl phenyl mercaptan, benzyl mercaptan, α-naphthyl mercaptan, hexyl mercaptan, hexenyl mercaptan, phenyl sulfenic acid, p-ethyl phenyl sulfinic acid and α-thio-β-methyl naphthalene.

The amino acid starting compounds which may be used in the process may be represented by the generic formula (1): $X(CH_2)_nCH(NH_2)COOH$ wherein n is 1 or 2. They are therefore β-substituted alanines or γ-substituted-α-amino acids.

Typically useful substituents which may be represented by X include those represented by halogen, and by R_1O-, R_2S-, R_2SO-, R_2SO_2-, and HSO_3- wherein R_1 is hydrogen, acyl, alkyl, aryl and aralkyl, and R_2 is hydrogen, alkyl, aryl, aralkyl, $HOOCCH(NH_2)CH_2S-$, or $HOOCCH(NH_2)CH_2CH_2S-$. In these definitions, alkyl radicals may contain up to 6 carbon atoms, and aryl and aralkyl radicals may contain up to 12 carbon atoms, and may be substituted with reaction inert substituents; and acyl refers to substituents, including formyl, containing up to about 5 carbon atoms.

Methioninase (methionine-α-deamino-γ-mercaptomethanelyase) is a known enzyme which decomposes L-methionine to methylmercaptan, ammonium ion and α-keto butyric acid. The enzyme is known to be produced by various microorganisms such as *Clostridium sporogenes* (Cancer Research 33, 1862-1865), *Escherichia coli (Medical Journal of Osaka University* 2, 111-117), *Pseudomonas ovalis, Pseudomonas taetrolens, Pseudomonas striata* and *Pseudomonas desmolytica* (Summary of Report for the *Annual Meeting of Agricultural Chemical Society* of Japan 96, 1974). It can be isolated from these sources by known methods.

Methioninase can be used in various forms including purified or crude methioninase preparations, intact cells of the microorganism which contain the methioninase activity, freeze-dried cells of the microorganisms, cells of the microorganism which have been dehydrated with acetone, homogenates of cells of the microorganism, or sonicates of the cells of the microorganism.

The aqueous reaction medium contains the enzyme or the enzyme source, the amino acid of formula (1) and Y—H, and, typically, pyridoxal phosphate and/or inorganic ions.

The preferred concentration of the amino acid of formula (1) in the reaction mixture is from 0.1 to 20%, and the concentration of Y—H is usually less than 10%. The substituent precursor may be utilized in the form of a salt, preferably an alkali metal salt such as potassium or sodium.

The reaction temperature is preferably maintained at from 1° to 70°C. During the reaction, the pH of the reaction mixture is maintained at from 5 to 12. When the reaction is carried out for 1 to 40 hours, high yields of sulfur-containing L-amino acid are accumulated in the reaction mixture.

The sulfur-containing L-amino acid in the reaction medium can be recovered by any known method such as precipitating at a pH of isoelectric point.

Example: An aqueous culture medium was prepared containing, per deciliter, 0.25 g L-methionine, 0.1 g peptone, 0.2 g glycerin, 0.1 g KH_2PO_4, 0.1 g K_2HPO_4, 0.01 g $MgSO_4 \cdot 7H_2O$, and 0.025 g yeast extract, adjusted to pH 8.0 with NaOH, of which 1,000 ml was placed in a 2,000 ml shaking flask, and heated by steam.

Each of the microorganisms shown in the following table was inoculated into a shaking flask, and cultured at 27°C for 24 hours. Cells in the resultant culture broth were collected by centrifuging, and freeze-dried.

Aqueous reaction media were prepared containing 32 mmol of the amino acid shown in the following table, 50 mmol n-propyl mercaptan, 0.01 mmol pyridoxal phosphate and 0.5 g/dl of the freeze-dried cells, and 10 ml portions of each mixture were placed in 500 ml shaking flasks, and shaken at 37°C for 16 hours.

Molar yields of propylthio-substituted L-amino acid accumulated in the resultant reaction media were determined and are shown in the table.

Amino Acid	Microorganism	Molar Yield (%)
L-homocysteine	*Pseudomonas taetrolens* IFO 3460	20
	Pseudomonas ovalis IFO 3738	75
	Pseudomonas striata IFO 12996	60
	Pseudomonas desmolitica IFO 12570	51
L-γ-chloro-α-amino-	IFO 3460	28
butyric acid	IFO 3738	85
	IFO 12996	60
	IFO 12570	50
L-methionine	IFO 3460	25
	IFO 3738	80
	IFO 12996	63
	IFO 12570	57

Optically Active Amino Acids from Hydantoin Compounds

K. Mitsugi, K. Sano, K. Yokozeki, K. Yamada, I. Noda, T. Kagawa, C. Eguchi, N. Yasuda, F. Tamura and K. Togo; U.S. Patent 4,016,037; April 5, 1977; assigned to Ajinomoto Co., Inc., Japan have found that *Flavobacterium aminogenes* AJ 3912 (FERM P-3133) produces an enzyme having remarkably high activity for converting hydantoin compounds of formula (1)

$$R-CH_2-CH \underset{\underset{CO}{NH}}{\overset{}{\rule{0pt}{0pt}}} \overset{}{\underset{NH}{CO}}$$

(R is phenyl or substituted phenyl radicals or indolyl or substituted indolyl radicals) to L-amino acids of formula (2)

$$R-CH_2-CH-COOH \atop NH_2$$

(R is the same as in formula 1).

When *Flavobacterium aminogenes* is cultured in conventional media, the enzyme for converting the hydantoin compounds to L-amino acid is produced mainly in the cells and slightly in the culture broth. The culture media used contain a carbon source, nitrogen source, inorganic ions, and where required organic nutrients. When culture media contain the hydantoin compound of formula (1) such as 5-indolylmethyl-hydantoin, the enzyme activity is generally much higher.

Cultivation is carried out at pH 6 to 9 and preferably 20° to 40°C under aerobic conditions. After 1 to 5 days cultivation, the enzyme is produced in the cells.

Hydantoin compounds are, for example:

> 5-benzyl-hydantoin;
> 5-(4'-hydroxybenzyl)-hydantoin;
> 5-(3',4'-dihydroxybenzyl)-hydantoin;
> 5-(2',4'-dihydroxybenzyl)-hydantoin;
> 5-(3',4'-methylenedioxybenzyl)-hydantoin;
> 5-(3',4'-dimethoxybenzyl)-hydantoin;
> 5-[3'(4')-methoxy-4'-(3')-hydroxybenzyl]-hydantoin;
> 5-(3',4'-isopropylidenedioxybenzyl)-hydantoin;
> 5-(3',4'-cyclohexylidenedioxybenzyl)-hydantoin;
> 5-indolylmethyl-hydantoin;
> 5-(5'-hydroxy-indolylmethyl)-hydantoin;
> 5-(5'-methyl-indolylmethyl)-hydantoin; and
> 5-(3',4',5'-trihydroxybenzyl)-hydantoin

The following L-amino acids are produced from the corresponding hydantoin compounds:

> phenylalanine;
> tyrosine;
> 3,4-dihydroxyphenylalanine;
> 2,4-dihydroxyphenylalanine;
> 3,4-methylenedioxyphenylalanine;
> 3,4-dimethoxyphenylalanine;
> 3(4)-methoxy-4(3)-hydroxyphenylalanine;
> 3,4-isopropylidenedioxyphenylalanine;
> 3,4-cyclohexylidenedioxyphenylalanine;
> tryptophane;
> 5-hydroxytryptophane;
> 5-methyltryptophane; and
> 3,4,5-trihydroxyphenylalanine

The reaction mixtures contain the hydantoin compound and the enzyme or the enzyme source of *Flavobacterium aminogenes*. The amounts of hydantoin compounds in the reaction mixture are usually less than 50 g/dl, and preferably less than 10 g/dl. Usually water is used as the solvent. Hydrophilic or hydrophobic organic solvents are used with water in order to increase the solubility of the hydantoin compounds.

In the reaction mixture containing 5-dihydroxybenzyl-hydantoin, 5-trihydroxybenzyl-hydantoin, and 5-(5'-hydroxyindolylmethyl)-hydantoin, reducing agents such as sodium sulfite are added to prevent the oxidation of the hydantoin compounds.

The reaction mixture is maintained at pH 5 to 11 and preferably 7 to 10, and at 15° to 70°C and preferably 30° to 50°C.

After 5 to 150 hours, L-amino acid is accumulated in the reaction mixture. The accumulated L-amino acid can be recovered by conventional means such as ion chromatography.

Example: An aqueous culture medium was prepared to contain 0.5 g/dl glucose, 1.0 g/dl yeast extract, 1.0 g/dl peptone, 0.5 g/dl NaCl, and 0.2 g/dl DL-5-indolylmethyl hydantoin. 50 ml batches of the aqueous culture medium were placed in 500 ml shaking flasks, heated with steam, and inoculated with *Flavobacterium* *aminogenes* AJ 3912. Each flask was shaken at 30°C for 16 hours. Cells were separated from 500 ml of culture broth by centrifuging, washed with physiological saline, and suspended in 500 ml of 0.1 M phosphate-buffer (pH 8.0).

Each 5 ml of the suspension was added with 50 mg of the hydantoin compounds listed in the following table and 15 mg of sodium sulfite in the cases of DL-5-(3',4'-dihydroxybenzyl)-hydantoin, and D-5-(5'-hydroxyindolylmethyl)-hydantoin, and held at 37°C for 24 hours. The amount of L-amino acid shown in the table was accumulated in the reaction mixture.

Hydantoin Compound	L-Amino Acid Accumulated (mg/ml)	
DL-5-benzylhydantoin	L-phenylalanine	8.7
DL-5-(3',4'-dihydroxybenzyl)hydantoin	L-3,4-dihydroxyphenylalanine	8.9
DL-5-(3',4'-methylenedioxybenzyl)-hydantoin	L-3,4-methylenedioxyphenyl-alanine	9.0
DL-5-(3',4'-dimethoxybenzyl)hydantoin	L-3,4-dimethoxyphenylalanine	0.5
DL-5-(3'-methoxy-4'-hydroxybenzyl)-hydantoin	L-3-methoxy-4-hydroxyphenyl-alanine	8.7
DL-5-indolylmethylhydantoin	L-tryptophane	2.6
DL-5-(5'-hydroxyindolylmethyl)hy-dantoin	L-5-hydroxytryptophane	0.72
D-5-(5'-hydroxyindolylmethyl)hy-dantoin	L-5-hydroxytryptophane	0.73

In a similar process, *L. Degen, A. Viglia, E. Fascetti and E. Perricone; U.S. Patent 4,111,749; September 5, 1978; assigned to Snamprogetti, SpA, Italy* describe a method of converting racemic hydantoins into optically active amino acids which comprises the steps of producing the specific enzyme which is formed during the growth of microorganisms of the *Pseudomonas* class and hydrolyzing hydantoins into derivatives of D,L-amino acids with this enzyme, wherein the enzyme can be used directly as a bacterial suspension or culture medium, or it can be extracted from the cells and the culture medium.

The bacterial strains listed in the following table grow well in all the common laboratory media. They are all capable of using hydantoin as a single nitrogen source. The growth takes place at a temperature ranging from 5° to 40°C, the optimum being between 25° and 30°C.

Strain	N-carbamylphenylglycine as a % of the Theoretical Yield
Pseudomonas sp.	
940	41
941	50
942	64
943	50
944	50
945	57
946	50
947	15
948	15
949	20
950	20
951	20
952	20
953	20

(continued)

Strain	N-carbamylphenylglycine as a % of the Theoretical Yield
Pseudomonas sp.	
954	20
955	20
Pseudomonas fluorescens 956	10
Pseudomonas sp. 957	41
Pseudomonas ATCC 11299	9
Pseudomonas oleovorans CL 59	17
Pseudomonas desmolyticum NCIB 8859	25
Pseudomonas fluorescens ATCC 11250	25
Pseudomonas putida ATCC 12633	14

The temperature of the enzymatic reaction can be maintained between 10° and 60°C. For practical reasons, however, temperatures between 25° and 40°C are preferred.

Example 1:

Meat peptone, g/l	10
Yeast extract, g/l	10
Glucose, g/l	5
NaCl, g/l	3
pH	7.2

To 100 ml of a broth culture in the above reported medium of the strain *Pseudomonas* sp. 942 in a 500-ml flask there were added at the 24th hour of incubation (orbital stirring) at 30°C, 100 ml of a phosphate buffer (0.14 M) pH = 8.5, which contained 30 μmol/ml of 5-(D,L)-phenylhydantoin.

After 5 additional hours of incubation, under the same conditions, the as-formed N-carbamylphenylglycine was determined.

From 530 mg of 5-(D,L)-phenylhydantoin, there were formed 525 mg of N-carbamylphenylglycine, which correspond to a yield of about 90%.

Example 2: A culture broth was prepared, having the above reported composition and which contained 1 g/l of 5-(D,L)-methylhydantoin.

The pH was adjusted to 7.2 with soda and the medium was distributed into 50-ml portions in 250-ml flasks. Upon sterilization at 110°C during 30 minutes, the flasks were inoculated with a culture of *Pseudomonas* sp. 942 from a slant which contained the same medium with the 2% of agar (Difco) and incubated at 30°C during 20 hours with orbital stirring (220 rpm).

With this preculture (OD, at 550 nm: 0.400; dilution 1:10) there were inoculated 5 ml in 500-ml flasks which contained 100 ml of the same medium and the culture was incubated at 30°C with orbital stirring (220 rpm) during 18 hours (last phase of the exponential growth). The cells were then collected by centrifugation (5,000 *g*, 20 minutes) and washed three times in a buffered physiological salt solution and finally suspended in a phosphate salt buffer pH = 8.5, 0.07 M to prepare resting cells.

For the enzymatic hydrolysis there were incubated at 30°C with orbital stirring (220 rpm) in 250-ml flasks, 64 ml of the reaction mixture composed of 260 mg of bacteria (dry weight) and 20 μmols/ml of D,L-phenylhydantoin (3.52 mg/ml). At various time intervals, the product of the hydrolysis, that is, D-carbamyl-phenylglycine, was determined with a colorimetric method [*J. Biol. Chem.* 238, 3325 (1963)] at 438 nm. Figure 2.1 shows the results, in terms of percent of the theoretical yield of the as-formed carbamyl derivative: on the abscissae there are reported the times (hours) and on the ordinates the percentages of D(-)-N-carbamylphenylglycine which had been formed.

Figure 2.1: Production of D(-)-N-Carbamylphenylglycine

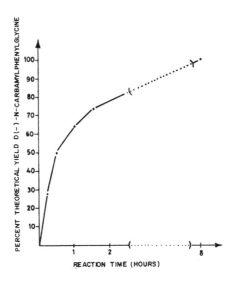

Source: U.S. Patent 4,111,749

Optical Resolution of N-Acylamino Acids

T. Uzuki, M. Takahashi, M. Noda, Y. Komachiya and H. Wakamatsu; U.S. Patent 3,907,638; September 23, 1975; assigned to Ajinomoto Co., Inc., Japan describe a method of selectively deacylating N-acyl-DL-amino acids in an aqueous medium in such a manner that the reaction products and the enzyme are recovered separately with minimal losses, a method which is simple, inexpensive, and does not require the introduction of significant amounts of alien chemicals into the process.

In the process, water, N-acyl-DL-amino acid, and acylase are stirred or otherwise held in contact with the organic solvent at a pH below 7 and at a temperature suitable for acylase activity until the desired, selective deacylation has taken place. It is then found that the acylase is entirely in the aqueous liquid, the optically active N-acylamino acid and the carboxylic acid formed by deacylation are in the organic solvent phase, and the desired, deacylated amino acid is in the aqueous phase as a solute or precipitates in the solid state.

The process can be made continuous by continuously or periodically withdrawing a portion of the organic solvent from the reaction zone and replacing it with solvent containing less optically active N-acylamino acid and carboxylic acid. The withdrawn solvent may be purified and returned to the process, and the optically active N-acylamino acid may be racemized and recycled. The optically active amino acid may be similarly recovered from the aqueous phase continuously or periodically, and such recovery is particularly simple when the amino acid crystallizes from the aqueous liquor. It is normally more convenient to feed N-acyl-DL-amino acid to the continuous system with the organic solvent.

The solvent may be chosen conveniently to permit the optically active N-acylamino acid to be racemized by heating while dissolved in the solvent so that the resulting solution or mixture of solvent and racemic starting material may be fed to the reaction zone without isolating the N-acetyl-DL-amino acid. The acid formed by the enzyme must be stripped at least partly from the recycled solvent since it must not accumulate in the reaction system.

The solvents employed may be chosen widely among organic liquids not miscible with water and capable of dissolving the N-acylamino acid and the carboxylic acid originating in the acyl groups removed by the acylase. Obviously, the solvent must be inert to the enzyme and to the amino acid and its N-acyl derivative. When the solvent is intended to provide a medium for the thermal racemization of the optically active N-acylamino acid, its boiling point is to be chosen above the racemization temperature which typically is 90° to 200°C.

Solvents meeting the more restrictive conditions include the triesters of phosphoric acid having at least five, and preferably at least seven, carbon atoms in the combined alcohol moieties, such as triethyl, tripropyl, tributyl, triisobutyl, trichloroethyl, methyldibutyl, and methylpropylbutyl phosphate. Also suitable are the lower alkyl esters of the lower alkanoic acids, the term "lower alkyl" and its analogs relating to carbon chains of up to four members. Examples of such lower alkyl esters are methyl acetate, ethyl acetate, butyl acetate and ethyl propionate, these ester solvents being chosen mainly for their ready commercial availability and low cost. Other suitable solvents include the dialkyl ketones having at least four carbon atoms, such as methyl ethyl ketone, diethyl ketone, methyl isobutyl ketone, and alcohols having at least four carbon atoms, such as n-butanol and amyl alcohol (1-pentanol).

The optimum ratio of organic solvent and aqueous medium in the reaction zone must be determined experimentally for a given set of other conditions since it is affected by the nature of the solvent, the kind of base employed for setting the pH of the aqueous phase, the nature and concentration of the N-acylamino acid in the water phase, and the like. Generally, good results are obtained with a volume of solvent which is at least 70% of the volume of the aqueous phase, and equal volumes of organic and aqueous phase may be employed initially in determining the best solvent:water ratio. This ratio, however, is not critical.

The most advantageous concentration of N-acylamino acid or its salts in the aqueous phase varies somewhat with the nature of the compound, of the base employed for pH adjustment, and of the acylase. A range from 5 to 60 g/dl is usually effective, and most advantageous results are often achieved at concentrations of 10 to 40 g/dl.

The base employed for setting a desired pH in the aqueous phase, while affecting other process variables to a minor extent, is not a partner in the reaction and thus not critical. The alkali metal hydroxides, such as sodium and potassium hydroxide, calcium hydroxide, ammonium hydroxide, and organic amines are all effective, and the choice will normally be dictated by cost and availability.

All known acylases have been found to be operative if they are capable of selectively hydrolyzing either the L- or the D-form of the N-acylamino acid. Acylases found effective were products of fungi, such as *Aspergillus* and *Penicillium,* of bacteria such as *Achromobacter, Pseudomonas, Micrococcus,* and *Alcaligenes,* and of the ray fungus *Streptomyces.* The amount of acylase needed is small, but depends on the enzyme activity of the product in an obvious manner. The acylase may amount to as little as 0.1 wt % of the aqueous phase or to as much as 3%, but acylase of reasonable purity is usually employed in amounts of 0.5 to 1%.

The reaction temperature is chosen to suit the specific acylase employed, a temperature between 20° to 60°C being usually best, and a temperature near 35° or 40°C being beneficial to many acylases.

Example: 90 g N-acetyl-DL-phenylalanine was dissolved in 250 ml water and enough ammonium hydroxide solution to make the solution neutral. Its ultimate volume was 300 ml, and it was transferred to a two-liter reactor together with 900 ml tributyl phosphate containing 8.73 g L-phenylalanine, and 30.6 g N-acetyl-DL-phenylalanine. When equilibrium was established by stirring, the aqueous phase had a pH of 6.03. At this stage, 71.4 mg $CoCl_2 \cdot H_2O$ and 6.03 g acylase (19,000 U/g) were added and the contents of the reactor were stirred while a constant temperature of 37°C was maintained.

As the enzymatic deacylation of the N-acetyl-L-phenylalanine proceeded, crystals of L-phenylalanine precipitated. After 48 hours, the organic solvent phase was drawn off, and the L-phenylalanine crystals were filtered from the aqueous phase which was saturated with L-phenylalanine. The crystals were washed with a small amount of ice-cold water and dried. 40.43 g of L-phenylalanine of 100% optical purity was recovered, corresponding to an 84.1% yield based on the entire N-acetyl-DL-phenylalanine initially present.

The organic solvent phase was found to contain 5.21 g acetic acid and 25.0 g N-acetylphenylalanine which was D-enantiomorph of 72.6% optical purity. The acetic acid and a portion of the solvent were distilled off in a vacuum, and the residue was held at about 180°C until the N-acetyl-D-phenylalanine was racemized. The resulting suspension of N-acetyl-DL-phenylalanine was recycled.

The aqueous mother liquor obtained from the crystallization of the L-enantiomorph was returned to the reactor together with 900 ml tributyl phosphate and 29.1 g N-acetyl-DL-phenylalanine. The aqueous layer had a pH of 5.97. The two-phase mixture was stirred at 37°C for 12 hours, whereupon the two liquids were separated, and 11.6 g crystalline L-phenylalanine was recovered from the aqueous phase. The procedure outlined above was repeated four times, and a total of 44.5 g crystalline L-phenylalanine was harvested.

Optically Active α-Amino Acids from Racemic α-Amino Acid Amides

W.H.J. Boesten and L.R.M. Meyer-Hoffman; U.S. Patent 4,080,259; March 21, 1978; assigned to Novo Industri A/S, Denmark describe a process of preparing

L-α-amino acid and D-α-amino acid amide from DL-α-amino acid amide by contacting the DL-α-amino acid amide with a preparation having L-α-amino acyl amidase activity.

This process is characterized in using a preparation obtained by cultivation of a microorganism selected from the group consisting of *Pseudomonas putida, Pseudomonas reptilivora,* and *Pseudomonas arvilla* in the presence of a nutrient medium containing assimilable sources of carbon, nitrogen, and phosphorus.

Preferred strains are *Pseudomonas putida* ATCC 12633, ATCC 25571, ATCC 17390, ATCC 17426 and ATCC 17484, *Pseudomonas reptilivora* ATCC 14039, and *Pseudomonas arvilla* ATCC 23974. *Pseudomonas putida*, and, in particular, the strain ATCC 12633 is a particularly preferred microorganism.

The microorganisms are preferably cultivated at a temperature within the range of from 30° to 35°C under aerobic conditions.

In most cases, the addition of growth factors or inductors is unnecessary. The addition of yeast extract appears to have a favorable influence on the production of the enzyme. After an incubation period of between about 2 and about 30 hours, the cells may be harvested, preferably during the period of exponential growth.

A preparation having α-amino acyl amidase activity may be obtained by precipitating the cells, optionally by using a flocculating agent. The cells may also be crosslinked or bonded to or absorbed on a carrier. In some cases, it may be desirable to modify the cell walls, e.g., by a heat treatment, to render the enzyme more accessible. A crude preparation may also be obtained by destroying the cells and recovering the enzyme by extraction, filtration and optionally spray-drying.

A preparation consisting of pure enzyme may be recovered in a conventional manner from the crude product described above. Pure enzyme or enzyme preparations may also be obtained from the culture medium by well-known techniques.

The preparation having L-α-amino acyl amidase activity is preferably contacted with the DL-α-amino acid amide in an aqueous medium at a temperature of between 0° and 60°C, and most preferred at a temperature of between 20° and 40°C, and at a pH value of between 6 and 10.5, and more preferred of between 7.5 and 9.5.

Outside these ranges, the activity and/or the stability of the enzyme is generally insufficient for practical use. The enzyme may be activated in a well-known manner, e.g., by the addition of a metal compound such as a magnesium, manganese, or zinc compound.

The weight ratio of the (unpurified) enzyme to the substrate may vary within wide ranges, e.g., between 1:25 and 1:750. If a pure enzyme is used, a higher ratio may be utilized.

When the hydrolysis of the L-α-amino acid amide has been completed, the free acid may be separated from the remaining D-α-amino acid amide, and the latter compound may then be hydrolyzed so as to form D-α-amino acid.

The process is suitable for isolating optically active natural or synthetic α-amino acids such as the D- and/or L-form of phenylalanine, 3,4-dihydroxyphenylalanine, tyrosine, methionine, leucine, alanine, phenylglycine, 4-hydroxyphenylglycine, 4-alkoxyphenylglycine and other substituted phenylglycines.

Example 1: *Pseudomonas putida* ATCC 12633 was incubated at 28° to 30°C in a flask placed on a rotating shaker. The growth was measured with a spectrophotometer at λ = 680 nm. A nutrient medium of pH 6.85 was prepared by mixing 1 liter of distilled water, 8.95 g secondary sodium phosphate dodecahydrate, 3.4 g primary potassium phosphate, 1.0 g ammonium sulfate, 200 mg of nitrilotriacetic acid, 580 mg magnesium sulfate heptahydrate, 67 mg calcium chloride dihydrate, 2.0 mg ferrous sulfate heptahydrate, 0.2 mg ammonium p-molybdate, and 1 ml "Hutner's metals 44" (a diluted solution of zinc, iron, manganese and copper sulfate, cobalt nitrate, sodium perborate and EDTA). The mixture was then sterilized and subsequently cooled, and finally 2 g asparagine and 2 g DL-mandelic acid were added.

The cells were harvested during the exponential growth phase by centrifugation (30 minutes at 10,000 rpm with cooling). The solid thus-obtained was washed with 0.1 M phosphate buffer at a pH of 6.8 and once again centrifuged (20 minutes at 10,000 rpm). The solid was suspended in the phosphate buffer (40 g wet cells in 100 ml buffer), whereafter the cell walls were destroyed with an ultrasonic cell disintegrator (20 kc/s for 20 minutes at 0°C). A crude extract was obtained by removing the solid particles by centrifugation (30 minutes, 10,000 rpm at 4°C). The yield of cell extract, calculated as dry substance, amounted to 0.8 g/l of culture liquid.

Example 2: The procedure of Example 1 was repeated, except that a culture medium containing 10 g of yeast extract was used instead of asparagine and mandelic acid. The yield of cell extract amounted to 1.25 g of dry substance per liter of culture liquid.

Example 3: In a flask provided with a stirrer, 1.5 ml of 0.125 M $MgCl_2$, 0.5 ml of 0.025 M $MnCl_2$, and 0.1 ml crude cell extract (dry weight 7 mg) prepared as described in Example 2 were added to a solution of 2.0 g (13.3 mmol) of L-phenylglycineamide in 48 ml of water, with stirring, at 25°C. During the reaction the pH of the reaction mixture rose from 9.6 to 9.7.

After 20 hours, the reaction mixture was acidulated with 4 N hydrochloric acid to a pH of 6.5. L-phenylglycine which then crystallized out was removed by filtration on a glass filter and washed on the filter with 2 portions of 10 ml of water and subsequently with 2 portions of 10 ml acetone. After drying, 1.6 g of L-phenylglycine (yield: 80%) were obtained.

The specific rotation of the L-phenylglycine was $[\alpha]_D^{20}$ = 157.7° (c = 1.6; 2.6 wt % of HCl). From literature is known that $[\alpha]_D^{20}$ = 157.5° (c = 1.6; 2.6 wt % of HCl).

Decarboxylation of Amino Acids

Arylethylamines such as tryptamine, serotonin and 3,4-hydroxyphenylethyl-amine are used in medicine. It is known that animal tissues contain a small amount of decarboxylase active for aryl amino acids. Decarboxylase separated from animal tissues is expensive.

K. Sano, K. Matsuda, H. Nakazawa and K. Mitsugi; U.S. Patent 3,930,948; January 6, 1976; assigned to Ajinomoto Co., Inc., Japan have found that certain microorganisms of genus *Micrococcus* produce large amounts of decarboxylase active for aryl amino acids. Various arylethylamines can be produced at a lower cost using this bacterial decarboxylase.

Microorganisms which produce aryl amino acid decarboxylase are, for example, *Micrococcus percitreus* AJ 1065 (FERM P-2200), and *Micrococcus conglomeratus* AJ 1015 (FERM P-2199).

The aqueous medium employed for culturing the microorganisms is entirely conventional, and contains assimilable sources of carbon and nitrogen, inorganic salts and minor organic nutrients.

Preferably tryptophan or phenylalanine is added to the aqueous medium to promote formation of the decarboxylase. Cultivation is carried out for 1 to 5 days, preferably at pH 4 to 10, and at 25° to 40°C under aerobic conditions.

The broth may be used as an enzyme source as it is, without removing the bacterial cells. Washed cells, acetone-treated cells, freeze-dried cells, cell homogenate, or sonicate can also be used as the enzyme source.

The reaction mixture contains an aryl amino acid and the decarboxylase of this process, and a minor amount of vitamin B_6. The mixture is desirably maintained at a temperature of 10° to 50°C and at pH 5 to 12.

Amino acids decarboxylated by the enzyme of this process are phenylalanine, tryptophan, and tyrosine, and derivatives thereof. The amount of the amino acids contained in the reaction mixture is usually 0.1 to 10 g/dl. The arylethylamines produced in the reaction mixture are recovered by conventional methods such as extracting with organic solvents.

Example: An aqueous culture medium was prepared to contain, per deciliter, 1 g yeast extract, 1 g peptone, 0.5 g KCl and 0.2 g L-tryptophan, and was adjusted to pH at 7.0 with KOH. 0.50 ml batches of the medium were placed in 500 ml shaking flasks, and sterilized with steam. *Micrococcus percitreus* AJ 1065 was inoculated in the medium, and cultured at 30°C for 24 hours. Cells were separated from the cultivation broth by centrifuging, and freeze-dried.

A reaction mixture containing, per deciliter, 1 g L-tryptophan, 1 g KH_2PO_4, 0.01 g pyridoxal phosphate and 1 g freeze-dried cells and having a pH of 9.5 was maintained at 30°C for 40 hours. Thereafter cells were removed by centrifuging, and the supernatant was adjusted to pH 13.5. Tryptamine was extracted with 800 ml ethyl alcohol. 3.1 g of tryptamine hydrochloride was obtained from 400 ml of reaction mixture.

Tryptamine was identified by paper chromatography, paper electrophoresis, NMR spectrum, infrared spectrum, and ultraviolet spectrum.

PEPTIDES

Dipeptide Derivative and Amino Acid Derivative Addition Compounds

Y. Isowa, M. Ohmori, K. Mori, T. Ichikawa, Y. Nonaka, K. Kihara, K. Oyama, H. Satoh and S. Nishimura; U.S. Patent 4,165,311; August 21, 1979; assigned to Toyo Soda Manufacturing Co. Ltd. and (Zaidanhojin) Sagami Chemical Research Center, Japan describe addition compounds having the formula

(1)
$$R_3-\overset{O}{\underset{R_2}{\overset{\|}{C}}-CH-NH_2}\cdot HO\overset{O}{\overset{\|}{C}}-(CH_2)_n-\overset{NH}{\overset{R_1}{\overset{|}{C}}}H-\overset{O}{\overset{\|}{C}}-NH-\overset{R_2}{\underset{|}{C}}H-\overset{O}{\overset{\|}{C}}-R_3$$

wherein R_1 represents an aliphatic oxycarbonyl group, benzyloxycarbonyl group which can have nuclear substituents, or benzoyl, aromatic sulfonyl or aromatic sulfinyl group; R_2 represents methyl, isopropyl, isobutyl, isoamyl or benzyl group; R_3 represents a lower alkoxyl, benzyloxy, benzhydryloxy group and n represents 1 or 2.

The addition compounds are produced by reacting an N-substituted monoamino-dicarboxylic acid having the formula

(2)
$$HO\overset{O}{\overset{\|}{C}}-(CH_2)_n-\overset{NH}{\overset{R_1}{\overset{|}{C}}}H-\overset{O}{\overset{\|}{C}}OH$$

with an amino carboxylic acid ester having the formula

(3)
$$H_2N-\overset{R_2}{\underset{|}{C}}H-\overset{O}{\overset{\|}{C}}-R_3$$

wherein R_1, R_2 and R_3 have the same meanings as in formula (1), in an aqueous medium in the presence of a protease and reacting the resulting dipeptide ester with the amino carboxylic acid and separating the addition compound. The N-substituted monoaminodicarboxylic acid having the formula (2) and the amino acid ester having the formula (3) are in L-form or DL-form.

The process produces α-L-aspartyl-L-phenylalanine alkyl esters which have as sweet a taste as sugar. α-L-aspartyl-L-phenylalanine methyl ester has a sweetness of about 200 times that of sugar.

The starting compounds of N-substituted monoaminodicarboxylic acids are N-substituted aspartic acid in the case of n = 1 and N-substituted glutamic acid in the case of n = 2.

The N-substituted monoaminodicarboxylic acids can be easily obtained by intro-ducing the protective group of R_1 to the monoaminodicarboxylic acid by the conventional processes.

The amino carboxylic acid esters used as the other starting compounds are amino acid esters having a hydrophobic group at the side chain and they are alanine esters in the case of R_2 = methyl group; valine esters in the case of R_2 = isopropyl group; leucine esters in the case of R_2 = isobutyl group; isoleucine esters in the case of R_2 = isoamyl group and phenylalanine esters in the case of R_2 = benzyl group. Of these R_2 groups, the benzyl group which gives phenyl-alanine esters as the amino acid ester is the most typical case.

The proteases used in the process are preferably metalloproteases which have a metal ion in the active center. Suitable metalloproteases are enzymes originat-ing from microorganisms, such as neutral proteases from ray fungus, prolisin, thermolysin, collagenase, Crotulus atrox protease, etc.

In the syntheses of the process, the peptide linkage formation reaction is per-formed in an aqueous medium, preferably aqueous solutions, under the pH con-dition wherein the protease exerts the enzymic activity.

The reaction for forming the addition compound of peptide and amino carboxylic acid ester is also pH dependent. It is preferable to perform the reactions in a pH range of about 4 to 9 especially about 5 to 8. Accordingly, the starting com-pounds of the N-substituted monoaminodicarboxylic acids and the amino car-boxylic acid esters can be in a free form or a salt form, however, when both of them are dissolved in the aqueous medium, it is necessary to adjust pH in the range given above.

The reaction is carried out in a temperature range of about 10° to 90°C, prefer-ably 20° to 50°C from the viewpoint of maintaining enzymatic activity. The re-action is usually completed in about 30 minutes to 24 hours though it is not critical.

The ratio of the used starting compounds is not critical. However, the reaction is to bond one molecule of the N-substituted monoaminodicarboxylic acid to two molecules of the amino carboxylic acid ester whereby the stoichiometric molar ratio of the starting compounds is 1:2, and actually used ratio is usually in a range of 100:1 to 1:100 preferably 5:1 to 1:5 especially 2:1 to 1:3.

The amount of the protease used in the process is not critical. When the con-centration of the enzyme is high, the reaction can be completed in a short time. When the concentration of the enzyme is low, the reaction time is prolonged. Thus, it is usually in a range of about 2 to 400 mg (5×10^{-5} to 1×10^{-2} mmol) per 1 mmol of the starting compounds, preferably about 5 to 100 mg (1×10^{-4} to 3×10^{-3} mmol) per 1 mmol of the starting compounds.

Example: 1,335 mg (5 mmol) of N-benzyloxycarbonyl-L-aspartic acid and 1,078 mg (5 mmol) of L-phenylalanine methyl ester hydrochloride were charged in a 30 ml flask and 10 ml of water was added to dissolve them and pH was adjusted to 6 with 7% ammonia water.

The resulting solution was admixed with 50 mg of thermolysin and shaken at 38° to 40°C over one night. The precipitate was collected and separated from the solution and dried to obtain 1,504 mg of an addition compound of N-ben-zyloxycarbonyl-L-aspartylphenylalanine methyl ester and L-phenylalanine methyl ester (1:1) (yield: 99.1% based on L-phenylalanine methyl ester hydrochloride) (melting point: 104° to 113°C).

Hydrolysis of Soy Protein

J.L. Adler-Nissen; U.S. Patent 4,100,024; July 11, 1978; assigned to Novo Industri A/S, Denmark describes a process which prepares polypeptides soluble in aqueous media of pHs in the range of from 2 to 7 from soy protein by hydrolyzing soy protein with microbial, alkaline proteinase in a concentration which corresponds to a proteolytic activity in the range of from 4 to 25 Anson units per kg soy protein at a substrate concentration of between 5 and 20% w/w soy protein, preferably 8 to 15% w/w soy protein, at a pH in the range of from 7 to 10.

Desirably the temperature is between about 15°C below the temperature optimum and the temperature optimum for the enzyme. Hydrolysis is carried out until a degree of hydrolysis (DH) in the range of from about 8 to about 15%, preferably from 9 to 12%, and more preferably 9.5 to 10.5% is attained, whereupon the enzyme is inactivated by reduction of pH through addition of a food grade acid preferably to a pH of 4.2 or lower, whereafter the supernatant is separated from the precipitate and treated with activated carbon.

The peptides are suitable for use as an additive to low protein acid food products, including soft drinks, for example, carbonated soft drinks, marmalade and jams.

The degree of hydrolysis (DH) is defined by the equation

$$DH = \frac{\text{Number of peptide bonds cleaved}}{\text{Total number of peptide bonds}} \times 100\%$$

Reference is made to J. Adler-Nissen, *J. Agr. Food Chem.* 24, Nov.-Dec. 1976 for a more detailed discussion of the degree of hydrolysis (DH).

The DH plays an important part in the process. The hydrolysis is controlled by means of the DH. When DH reaches a critical value, hydrolysis is interrupted. The DH is both the monitoring parameter of the hydrolysis, and a measure of product suitability, i.e., DH of the product is 8 to 15%, more preferably 9.5 to 10.5%.

A preferred embodiment of the process comprises the use of a microbial alkaline proteinase, originating from *Bacillus licheniformis*. By using this embodiment, a fast reaction is attained, and in addition this enzyme is acceptable from a toxicological point of view. A suitable enzyme is Alcalase (Novo Industri A/S). Another alkaline microbial enzyme is Esperase (produced by submerged fermention of an alcalophilic bacillus).

Example: 4,000 ml of a suspension of soy protein isolate, which suspension contains approximately 8% protein (N x 6.25), was hydrolyzed with 0.2% of

Alcalase S 6.0 (calculated with respect to the weight of the protein as corresponding to a proteolytic activity of 12 AU/kg) at pH 8.0 and a temperature of 50°C. Alcalase S 6.0 has a proteolytic activity of 6.0 Anson units/g. During hydrolysis, which was followed by means of the pH-Stat (Radiometer), the pH was kept constant by addition of 4 N NaOH. After 2 hours' hydrolysis time, a degree of hydrolysis of 10% corresponding to 0.8 meq/g was obtained [complete hydrolysis (DH = 100) of soy protein corresponds to 7.9 meq of base consumption per gram of soy protein], and 4 M citric acid was then added until the pH reached 3.5.

The hydrolysis mixture was then allowed to stand for 30 minutes before the supernatant was decanted through a paper filter using diatomaceous earth as filter aid. The supernatant was treated twice with 0.01% w/v activated carbon powder for 30 minutes at a temperature of 30°C. The activated carbon was Coporafin B.G.N. (Lurgi Apparate-Technik GmbH). This produced a protein hydrolyzate in 68% yield and with a nonbitter, pleasant taste.

A beverage with 3% of protein hydrolyzate, 10% of sucrose and 0.005% of Firmenich Tetrarome Lemon P 05.51 was produced and found to be organoleptically acceptable. The beverage was pasteurized in closed vessels and stored in a refrigerator for several weeks. No precipitation or growth of microorganisms occurred, and only a slight discoloration because of Maillard reactions was observed.

Use of Serine or Thiol Proteinase

Y. Isowa, M. Ohmori, H. Kurita, T. Ichikawa, M. Sato and K. Mori; U.S. Patent 4,086,136; April 25, 1978; assigned to (Zaidanhojin) Sagami Chemical Research Center, Japan describe a process for producing a peptide having the formula X–A–B–Y wherein A and B are the same or different and represent an amino acid residue or a peptide residue, X represents an amino protective group, Y represents a carboxyl protective group selected from the group consisting of substituted or unsubstituted tertiary alkoxy, and benzyloxy, benzylamino and benzhydrylamino, by reacting an amino acid or peptide having an N-terminal protective group or a salt thereof of the formula X–Z–OH with an amino acid or peptide having a C-terminal protective group or a salt thereof of the formula H–B–Y in the presence of a thiol proteinase or serine proteinase enzyme in an aqueous solution having a pH sufficient to maintain the enzyme activity of the thiol proteinase or serine proteinase.

Typical thiol proteinase enzymes include papain, stembromelein, ficin, cathepsin B, chymopapain, streptococcal proteinase, asclepain, *Clostridium histolyticum* proteinase B and yeast proteinase B.

Typical serine proteinases include subtilisin, *Aspergillus* alkaline proteinase, elastase, α-lytic proteinase, chymotrypsin, Metridium proteinase A, trypsin, thrombin, plasmin, kininogenin, enteropeptidase, acrosin, *Phaseolus* proteinase, Altemaria endopeptidase, *Arthrobacter* serine proteinase and *Tenebrio* α-proteinase.

The starting materials are usually used in a ratio of 0.8 to 2 mols, preferably 1 to 1.5 mols of the acid component per 1 mol of the amine component. If the starting materials are not too soluble in the aqueous medium, it is possible to improve the solubility of the reactants by adding a solvent such as an alcohol, e.g., methanol, or ethanol; dimethylformamide; dioxane; tetrahydrofuran; dimethylsulfoxide, or the like, to the aqueous solution. The amount of the added solvent should be limited so as not to inhibit the activity of the enzyme in the reaction. If a solvent is employed, it is usually used in an amount of less than one part by weight of water. The reaction is performed in an aqueous medium, and it is necessary to decrease the relative solubility of the reaction product preferably to a sparingly soluble or insoluble state in the system.

The amount of thiol proteinase or serine proteinase enzyme employed is in a range of 10 to 500 mg, preferably 10 to 400 mg, especially 50 to 300 mg/mmol of the amine component. An enzyme activator such as cysteine or a salt thereof or 2-mercaptoethanol or a salt thereof can also be added to the solution. The reaction temperature employed is usually in a range of 20° to 55°C, preferably 30° to 40°C which is sufficient to maintain enzyme activity. The reaction proceeds smoothly under these conditions for 1 to 24 hours. The reaction product precipitates from the reaction system and the reaction product can be easily isolated.

Use of Metalloproteinase

Y. Isowa, M. Ohmori, H. Kurita, T. Ichikawa, M. Sato, K. Mori and K. Oyama; U.S. Patent 4,116,768; September 26, 1978; assigned to (Zaidanhojin) Sagami Chemical Research Center, Japan and Y. Isowa, M. Ohmori, H. Kurita, T. Ichikawa, M. Sato and K. Mori; U.S. Patent 4,119,493; October 10, 1978; assigned to (Zaidanhojin) Sagami Chemical Research Center, Japan describe a similar process using metalloproteinase as the enzyme.

These enzymes are produced from microorganisms such as *Bacillus subtilis, Bacillus thermoproteoliticus, Streptomyces caespitosus, Bacillus megaterium, Bacillus polymyxa, Streptomyces griseus, Streptomyces naraensis, Streptomyces fradiae, Pseudomonas aeruginosa, Aspergillus oryzae, Clostridium histolyticum, Proteus aeruginosa* and the like (prolisin, thermolysin, collagenase, thermoase, tacynase N, and pronase).

A catalytic amount of an enzyme is employed, preferably 10 to 500 mg of enzyme per 1 mmol of the amine component. When a purified enzyme is used, the amount of the enzyme can be 5 to 30 mg. The reaction temperature is usually in a range of 20° to 80°C, preferably 20° to 50°C in order to maintain enzyme activity. The reaction proceeds smoothly under these conditions for 1 to 24 hours. The reaction product precipitates from the reaction system and the reaction product can be easily isolated.

N-Acylpentapeptides

N. Kuwana, K. Suzuki and T. Sugitani; U.S. Patent 3,929,572; December 30, 1975; assigned to Eisai Co., Ltd., Japan describe a method for the production of N-acylpentapeptides in a high yield by cultivating *Streptomyces naniwaensis* in a nutrient medium containing a precursor or precursors of the N-acylpentapeptides.

N-acylpentapeptides obtained in accordance with the process are represented by the general formula shown on the following page.

```
                      CH₃   CH₃                      CH₃   CH₃
                        \   /                          \   /
  CH₃   CH₃ CH₃   CH₃    CH                             CH
    \   /     \   /      |                    CH₃       |
     CH        CH        CH₂ OH                |        CH₂ OH
     |         |         |   |                 |        |   |
R-NH-CH-CO-NH-CH-CO-NH-CH-CH-CH₂-CO-NH-CH-CO-NH-CH-CH-CH₂-COOH
```

wherein R is acyl radical.

The term "precursor" herein used means the substances which will provide a moiety to result in a corresponding N-acylpentapeptide including the substances, which may hereinafter be called "N-acyl radical donor," exemplified by 4-amino-3-hydroxy-6-methylheptanoic acid (AHMHA) per se or in a form of its derivatives, for example.

As for the N-acyl radical donors, there may be mentioned aliphatic acids in a form of their inorganic salts, esters and amides; such as sodium, potassium and ammonium salts of acetic, propionic and butyric acids; such as the esters derived from a higher alcohol, polyalcohol and other alcohols and an aliphatic acid, for example, stearyl and oleyl acetates, propionates and butyrates; and such as the amides derived from amino radical-containing compounds, for example, an amino acid and amino sugars, especially, N-acetylvaline and N-acetylglucoamine. In this connection, it has been found that use of a coenzyme-type acetyl radical donor such as thiol esters of the fatty acids exemplified by S-acetylglutathion is particularly preferable for the predominant production of Pepsidine C.

The amount of the precursor(s) to be used is not critical. It is, however, usually preferable to use in an amount of 0.05 to 2.0% by w/v of a single precursor or an admixture thereof on the basis of the employed nutrient medium in order to obtain a desired result.

The precursor or precursors may be incorporated into a nutrient medium in a form of its aqueous solution. They may thus be added at once or intermittently to the medium.

The compositions of the nutrient medium for the concomitant use with the precursor(s) and the conditions employed for the cultivation of *Actinomyces naniwaensis* EF 44-20 (ATCC 21689) strain in the medium are not critical. Usually, the cultivation may be carried out under shaking and/or aerating at a temperature of about 20° to about 24°C and for 10 to 120 hours.

According to the process, N-acylpentapeptides are obtained in higher yields than those obtained by the known arts. Furthermore, it is particularly notable that a marked enhancement in the yield of an intended specific N-pentapeptide can be attained by a selected use of the precursor.

Modification of Porcine or Bovine Insulin

It is well known that administration of insulin is of immense value in the management of diabetes in humans. For this purpose, the insulin normally used is obtained from pigs, so-called "porcine insulin" or from cows, so-called "bovine insulin." It is also known that administration of either porcine insulin or bovine insulin, as normally available, often causes an antigenic response. In humans, this antigenic response can lead, over a prolonged period of dosing, to an increase in the daily dose required by the diabetic patient to control the disease.

H. Gregory and P.L. Walton; U.S. Patent 3,903,069; September 2, 1975; assigned to Imperial Chemical Industries Limited, England have found that if porcine or bovine insulin, as normally available, is treated with a lysine specific aminoendopeptidase, that is, a peptidase which cleaves a polypeptide at the peptide link attached to the amino group of lysine and does not require the lysine residue to be at the C-terminus of the peptide chain, then the products obtained are as potent as porcine or bovine insulin in causing a hypoglycemic response, and furthermore, each of them causes a significantly smaller production of antibodies which bind insulinlike materials on subsequent administration than does the initial porcine or bovine insulin respectively.

According to the process there is provided a polypeptide of the formula:

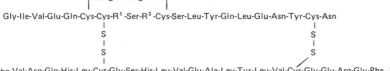

wherein either R^1 is threonine and R^2 is isoleucine, or R^1 is alanine and R^2 is valine.

The polypeptide wherein R^1 is threonine and R^2 is isoleucine differs from porcine insulin only in that the B chain, i.e., the longer chain, lacks the C-terminal lysyl and alanyl residues which form the 29th and 30th residues of the B chain of porcine insulin. This polypeptide can therefore be named des-Lys^{29}-Ala^{30}-porcine insulin, and this name will be used hereafter.

The polypeptide wherein R^1 is alanine and R^2 is valine differs from bovine insulin only in that the B chain lacks the C-terminal lysyl and alanyl residues. It can therefore be named as des-Lys^{29}-Ala^{30}-bovine insulin, and this name will be used hereafter.

As examples of suitable lysine specific aminoendopeptidases there may be mentioned the enzyme AM protease and the enzyme myxobacter AL-1 protease II.

The enzyme AM protease may be obtained from the mature fruiting bodies of the fungus *Armillaria mellea* as fully described and claimed in U.K. Patent specification No. 1,263,956.

The enzyme myxobacter Al-1 protease II may be obtained from myxobacter strain AL-1 as described by Wingard, Matsueda and Wolfe in the *Journal of Bacteriology*, Vol. 112, pages 940-949.

The exposure of the porcine or bovine insulin to the lysine specific aminoendopeptidase may be carried out with the enzyme simply dissolved in the aqueous medium, or it may be carried out with the enzyme bound to a support which may or may not be soluble in the aqueous medium.

When the enzyme AM protease is used, the reaction will proceed at a wide variety of enzyme/substrate ratios, and indeed, as little as 1 part of enzyme to 10,000 parts of substrate can be used. The reaction will also proceed at a pH of the incubation medium from 3 to 10, and at a temperature from 10° to 60°C. Preferred reaction conditions to obtain essentially complete cleavage of lysyl-alanine in a relatively short period, for example, 3 to 12 hours, are the use of a concentration of insulin near to its solubility in the incubation medium, a pH of from 7 to 9, a temperature of from 30° to 50°C and as low an enzyme-substrate ratio as possible, for example, one one-thousandth or less. It is also convenient to include metal ions in the incubation medium, for example, calcium or magnesium in the form of their chlorides at a concentration 10^{-2} to 10^{-4} molar.

When the enzyme myxobacter AL-1 protease II is used, the reaction conditions are very similar to those which may be used with AM protease; an enzyme/substrate ratio as low as one ten-thousandth may be used at a pH from 4 to 10 at a temperature from 25° to 75°C. Preferred conditions are the use of a concentration of insulin near to its solubility in the incubation medium, a pH of from 6 to 9, a temperature from 30° to 50°C, and as low an enzyme/substrate ratio as possible, for example, one one-thousandth or less. Metal ions, for example, calcium or magnesium, may be included in the incubation medium.

As indicated above, the reaction may be carried out using the enzyme bound, preferably covalently, to a support. A variety of supports and methods of attachment are possible, and for details, a review of E. Katchalski in "Biochemical Aspects of Reactions on Solid Supports," ed. G.R. Stark, Academic Press 1971, should be consulted. One convenient method of attaching AM protease to a support is to activate agarose gel beads with cyanogen bromide, with or without spacer groups derived, for example, from hexylamine or hexanoic acid, and then allow the activated agarose to react with AM protease. Alternatively, the agarose gel beads may be activated using 2,4-dichloro-6-carboxymethylamino-s-triazine as coupling agent. Another useful alternative is the use of carboxymethylcellulose hydrazide.

When the exposure to enzyme is carried out with the enzyme simply dissolved in the aqueous medium, the required des-Lys^{29}-Ala^{30}-porcine (or bovine) insulin may be separated from the other components of the incubation mixture by any conventional technique for the separation of polypeptides, but the use of a molecular sieve technique is particularly convenient. Thus, the aqueous medium remaining after all the starting insulin has been degraded may be filtered through a column of crosslinked dextran gel or a polyacrylamide gel suitable for fractionating compounds of molecular weight 5,000 to 10,000 at any convenient pH, preferably removed from, and preferably substantially below, the isoelectric point of the polypeptides, and the column eluted with a buffer composed of components which are volatile on freeze drying under high vacuum, for example, aqueous acetic acid, ammonium carbonate or ammonium acetate, to give the desired product which is isolated by freeze drying the eluate.

The polypeptides of the process are used for the management of diabetes in essentially the same way as porcine or bovine insulin. Thus, they are administered parenterally, usually subcutaneously, either as a solution or as depot formulations having differing durations of action. Such formulations allow for a duration of action for up to 24 hours. Doses are selected for individual patients as required for the control of the diabetes and may be as high as 200 units daily.

PROTEINS

Extraction of Microbial Cellular Proteins

Microbial cells such as bacteria, yeasts, fungi, algae and protozoa are rich in proteins and are more similar to animal proteins, which are nutrient, than to vegetable proteins. Recently, the microbial cellular proteins have been noticed as an important protein source in view of both quantitative and qualitative significance, and there is now considered an opportunity for utilizing certain microbial cells, in particular, petroleum assimilating microbial cells, as a feed stuff. Mass production of the above microbial cells now becomes promising.

Indirect use of such microbial cellular proteins as feed stuff for cattle and fish, however, is not considered efficient utilization of these proteins, and so direct use of microbial cellular proteins is considered desirable. Since microbial cells are covered with tough cell walls resistant to digestion, accompanied with peculiar odor and coloration, these disadvantages make direct adoption of cells as edible foodstuff for humans difficult.

M. Ishida, Y. Oguri, T. Muroi and N. Shimizu; U.S. Patent 3,862,112; January 21, 1975; assigned to Hitachi, Ltd., Japan describe an alkali extraction of microbial cellular proteins which is carried out in the presence of 0.1 to 1 wt % of a strong alkali substance at a temperature of 100° to 200°C.

The process may be applicable to alkali extraction of microbial cells including yeasts, fungi, algae, bacteria, protozoa and mixtures thereof. Preferable yeasts, which assimilate natural gases, paraffins, aromatic hydrocarbons and other hydrocarbons, are those belonging to *Candida* and genus *Torulopsis*. Also bacteria belonging to genus *Pseudomonas*, genus *Methanomonas* and genus *Micrococcus* may be used. Other usable yeasts are such as genus *Saccharomyces* which assimilate sugars, organic acids, alcohols or other nonhydrocarbon compounds. Bacteria such as genus *Bacillus* and genus *Escherichia* are also usable. Algae such as chlorella and protozoa such as paramecium are usable. Either wet or dry cells can likewise be employed. The microbial cells which assimilate hydrocarbons are preferred.

A suitable amount of the microbial cells is dispersed in water or other aqueous media. In view of extraction efficiency, 6 to 15 wt % of microbial cells are preferably dispersed in an aqueous medium. To the dispersion thus prepared is added about 0.1 to about 1 wt % of a strong alkali substance to adjust the pH value of the aqueous dispersion to about 9.5 to about 13.5. It has been found that among the strong alkali substances sodium hydroxide and potassium hydroxide are suitable for obtaining a high yield. In particular, sodium hydroxide is the most useful extracting agent because of low cost.

Example: 1 kg of dried *Candida lipolytica*, which is a yeast capable of assimilating petroleum, was dispersed in 9.95 liters of an alkali solution in which 44 g of sodium hydroxide (0.4 wt %) was dissolved, whereby a pH value of the dispersion was adjusted to 12.0.

The dispersion was charged in an autoclave and heated at 120°C for 10 minutes with stirring so as to carry out alkali extraction. During the extraction, the pressure of the autoclave was kept at 4 kg/cm^2.

After cooling the dispersion, 16 ml of concentrated hydrochloric acid was added to the dispersion to neutralize the alkali (NaOH), and an extract was separated by centrifugation therefrom. The extraction rate was 85%. The extract was then subjected to an isoelectric point precipitation after concentrated hydrochloric acid was added to the extract to adjust the pH value to the isoelectric point 4.2 so that desired proteins were precipitated. The white precipitate collected was washed once with water and thereafter dried to obtain 391 g (protein yield 75%) of purified microbial cellular proteins of 91% purity, white and colorless.

Protein from *Cellulomonas* Strain

V.R. Srinivasan and Y.-C. Choi; U.S. Patent 4,104,124; August 1, 1978; assigned to Louisiana State University Foundation describe a process for the preparation of comestible, digestible protein by the cultivation of a mutant strain of *Cellulomonas* (ATCC-21399) which has the ability to excrete L-glutamic acid or L-lysine, or both, preferably *Cellulomonas* sp. 21399 strain LC-10 (ATCC-31230), *Cellulomonas* sp. 21399 strain A^r-1 (ATCC-31231), or *Cellulomonas* sp. 21399 strain A^r-156 (ATCC-31232), using an aqueous nutrient medium, or culture broth, which contains assimilable sources of carbon, nitrogen and mineral nutrients.

The mutants can be cultivated on various energy or carbon substrates, alone or in compatible association with other microorganisms. Carbohydrates are also suitable, e.g., the monosaccharides such as glucose, and the like, the disaccharides such as sucrose, lactose, maltose, and the like, and the polysaccharides such as the starches, celluloses, hemicelluloses, and the like. Cellulose is a particularly preferred carbon source. Chemical celluloses, e.g., nitrocellulose, ethylcellulose, cellulose acetate, methylcellulose, carboxymethylcellulose, and regenerated cellulose products, e.g., viscose, rayon, cupraammonium rayon, cellophane, and the like, and particularly delignified cellulose from any source, e.g., bagasse, waste paper, etc., can be used as the energy or carbon substrate.

In the preferred embodiment, cellulose from any suitable source is first delignified, preferably by treatment with an alkali. Preferably, the cellulose is treated with an aqueous alkali metal hydroxide solution, wherein the alkali metal hydroxide is present in concentration ranging from about 0.5 to about 75%, preferably from about 2 to about 20%, based on the weight of the total solution. Suitably, the physical size of the cellulose, if large, is reduced by cutting, sawing, grinding or crushing to assure intimate contact with the alkali. Generally, the cellulose is simply immersed in a solution of the alkali. Generally, the alkali treatment is conducted for periods ranging from about 0.1 hour to about 20 hours. When using strong alkalis, e.g., alkali metal hydroxides such as sodium hydroxide, the alkali cellulose treatment is conducted at temperatures ranging from about –10° to about 70°C, preferably 10° to about 30°C, from about 1 to about 24 hours.

After contact between the cellulose and alkali, the treated cellulose is separated from the alkali by standard procedures known in the art, e.g., filtration, decantation, centrifugation, screening, passage between squeeze rolls, and the like. The alkali-treated cellulose, generally after washing, is then placed in an oxidation, or circulating air oven and heated, generally at temperatures ranging from about –10° to about 150°C, and preferably at temperatures ranging from about 25° to about 100°C. The time period that the treated cellulose remains in the oven is not critical, periods generally ranging from about 0.1 to about 20 hours. Air is blown on the cellulose during this period to accelerate the rate of oxidation and hydrolysis of the cellulose.

To initiate fermentation, the oxidized, alkali-treated cellulose is next removed from the oven, treated ex situ to a pH ranging between 5 and 9, e.g., a pH of 7, and then charged into a fermentor (or charged into the fermentor and then treated in situ to a pH ranging between 5 and 9). Adjustment of the pH is generally accomplished by addition of an aqueous acid solution, or proton donor, e.g., hydrochloric acid, with agitation sufficient to thoroughly contact the cellulose with the acid and establish equilibrium conditions.

The temperature of the broth is then adjusted to that desired for the most rapid growth of the microorganism, suitably from about 20° to about 40°C. The fermentation medium is then inoculated with the mutant microorganism, or microorganisms, to be cultivated, and harvested. Within the fermentor, comprising a closed vessel, a draft tube and air lift are provided to maintain vigorous agitation and suitable aerobic conditions. The pH of the medium is maintained while providing optimum growth temperatures for the microorganisms to be harvested.

The presence of oxygen is essential in the broth or cultivation medium. Conveniently, the oxygen is supplied as an oxygen-containing gas, e.g., air, which contains from 19 to 22 wt % oxygen. While it is preferable to employ air, oxygen or oxygen enriched air, e.g., in excess of 22 wt % oxygen, can be used.

The presence of nitrogen is also essential to biosynthesis. A convenient and satisfactory method of supplying nitrogen is to employ ammonium phosphate or ammonium acid phosphate, which can be added as the salt, per se, or can be produced in situ in the aqueous fermentation media by bubbling ammonia through the broth to which phosphoric acid was previously added, thereby forming ammonium acid phosphate.

The following chemically defined medium has been found capable of supporting a satisfactory rate of growth of *Cellulomonas* sp. ATCC-21399 strain LC-10. This simple medium, it will be observed, includes glucose as a carbon source, a mixture of salts, inclusive of trace salts, referred to hereinafter as a basic salt mixture (or basic salts), and specific vitamins in distilled water, to wit:

Glucose, g	5
$(NH_4)_2SO_4$, g	0.5
K_2HPO_4, g	1.5
NaH_2PO_4, g	0.5
$MgCl_2 \cdot 6H_2O$, g	0.1
$FeSO_4 \cdot 7H_2O$, g	0.005
$CaCl_2 \cdot 2H_2O$, g	0.05
Sodium citrate, g	0.3
Trace salts solution, ml	1
Biotin, μg	5
Thiamine, μg	10
Distilled water, ml	1,000
Trace salts solution:	
$\quad ZnSO_4 \cdot 5H_2O$, g	0.1
$\quad CoCl_2 \cdot 6H_2O$, g	0.05

(continued)

CuSO$_4$·5H$_2$O, g	0.005
MnCl$_2$·6H$_2$O, g	0.005
Distilled water, ml	1,000

In the example, the parent microorganism, *Cellulomonas* sp. ATCC-21399, was grown in such solution necessarily inclusive of an added yeast extract, and after separation of the mutant strain therefrom, it was found that the mutant microorganism, unlike the parent, could be grown in a medium which did not include the yeast extract.

Example: *Isolation and Growth of Cellulomonas sp. ATCC-21399 Strain LC-10 — Cellulomonas* sp. ATCC-21399 was grown overnight in a medium containing glucose, yeast extract and basic salt mixture. About 10^8 organisms were spread on a number of plates containing 1.5% agar, 0.5% glucose, basic salts and vitamins. After 1 hour crystals of groups of amino acids were placed on the inoculated agar plates and incubated. After 3 days of incubation at 30°C a few colonies grew on one single group of amino acids, consisting of sienna, leucine and cystine. One such colony was isolated and grown in a liquid medium containing glucose (0.5%), vitamins (50 μg/ml), basic salts and the amino acids serine, leucine, and cystine (100 μg/ml each).

About 10^8 organisms from such a culture were again spread on a plate containing 1.5% agar, glucose, vitamins and salts and incubated at 30°C for 3 days. Several colonies grew on such a plate. One such colony was isolated and purified by repeated streaking on a minimal agar plate and isolating a single colony. This strain of the *Cellulomonas* sp. 21399 is termed LC-10. This strain was grown in different carbon sources in the medium outlined previously and the yield of cells is given below:

Carbon Source	Volume of Fermentation (l)	Time of Fermentation (hr)	Biomass Dry Weight (g/l)
Glucose	5	10	29
Solka floc	5	24	18-20
Bagasse	5	36	18

Single-Cell Protein from Methanol

F. Wagner and H. Sahm; U.S. Patent 4,048,013; September 13, 1977; assigned to Gesellschaft fur Molekularbiologische Forschung mbH, Germany describe a process for production of single-cell protein which comprises inoculating a sterile liquid medium with a culture of the methanol-assimilating bacterium *Methanomonas* sp. DSM 580 containing assimilable sources of nitrogen, methanol as the sole source of carbon and energy, essential mineral salts and, if necessary, growth promoting agents.

The fermentation is carried out under aerobic conditions providing the system with air or oxygen enriched air, the cultivating temperature is from 20° to 45°C. After culturing, the microbial cells are removed out of the three phase system and dried, presenting a biomass with a crude protein content from 60 to 70% (w/w), a nucleic acid content from 1 to 17% (w/w), an ash content of 3 to 6% (w/w) and a lipid content from 3 to 8% (w/w). The culture fluid removed in the separation step may be recycled back in the process. Throughout the fermentation, a constant pH was maintained in the range of from 4.5 to 9.0 by adding alkali or acids.

Example: An 80-liter capacity fermenter is charged with 50 liters mineral salt medium [composition: $(NH_4)_2SO_4$, 200 g; KH_2PO_4, 150 g; Na_2HPO_4, 125 g; $MgSO_4 \cdot 7H_2O$, 25 g; $Ca(NO_3)_2 \cdot 4H_2O$, 1.25 g; $FeSO_4 \cdot 7H_2O$, 0.25 g; $ZnSO_4 \cdot H_2O$, 0.25 g; KCl, 0.25 g in 50 liters distilled water] and sterilized at 121°C for 10 minutes, cooled to 15°C, aseptically mixed with 1,000 ml methanol, inoculated with 500 ml of a culture of *Methanomonas* sp DSM 580 and incubated at 35°C for 28 hours under stirring with a turbostirrer with a stirring rate of 300 rpm and a aeration rate of 0.7 v/v/min. Throughout the fermentation the pH is automatically maintained at 6.4 by the addition of a 6% ammonium hydroxide solution..

After 22 hours of fermentation, the medium is cooled to 15°C and adjusted to pH 3 by sulfuric acid and the precipitated biomass is recovered by filtration, washed with water and then dried (324 g cell dry weight). The cell composition of the dried biomass is 71% crude protein, 9% nucleic acids, 4% ash and 7% lipid.

Glycoproteins

J.-C. Drouet, M.-O. Martin, D. Biard and R. Zalisz; U.S. Patent 4,154,821; May 15, 1979; assigned to Roussel Uclaf, France describe the preparation of water-soluble glycoproteins which comprises cultivating a Hafnia microbial strain in a solid or liquid nutritive media until complete development of the microbial bodies, collecting the bodies, subjecting the latter to lysis, treating the resulting lysate with at least one organic solvent, dissolving the resulting product in water and subjecting the aqueous solution to diafiltration with a porous membrane calibrated to have a threshold for retaining substances with a molecular weight of at least 300,000 and subjecting the resulting solution to lyophilization to obtain the water-soluble glycoprotein.

Preferably, the Hafnia strain is cultivated in a stirred liquid media under aerobic conditions. The culture medium used is a conventional media containing, for example, meat extracts, casein peptone, soybean papainic peptone, yeast autolysates, sugars, mineral elements and distilled water.

The lysis of the microbial bodies may be physical, chemical or enzymatic. Physical lysis is preferably effected with ultrasonic means or by heating or by penetrating radiation. Chemical lysis is preferably effected with adjunction of a surface active agent such as polyethyleneglycol sorbate or with an antiseptic organomercurial agent such as sodium mercurothiolate and even with a mineral or organic acid such as trichloroacetic acid. Enzymatic lysis is preferably effected with an enzyme such as lysozyme, trypsin, pronase, papain or α-chymotrypsin. The lysis, whether physical, chemical or enzymatic, is preferably 7 to 60 days long. The lysate may be lyophilized.

The resulting lysate is treated with one or more organic solvents and preferably with at least two solvents in separate steps such as first acetone and then methanol. The treatment is intended to eliminate lipids and pigments and in practice the mixtures of lysate and solvent are vigorously stirred for several hours.

The porous membranes calibrated to retain substances with a molecular weight equal to or greater than 300,000 are preferably commercial membranes (XM 300, Romicon and Amicon). These membranes can occur in the form of hollow fibers and commercial hollow fibers of this type are HIP 100 (Amicon).

The method for relieving inflammation and inducing immunostimulating activity in warm-blooded animals, including humans, comprises administering to warm-blooded animals an effective amount of glycoprotein of the process. The product may be administered locally, orally, rectally or parenterally and the usual effective dose is 0.001 to 1 mg/kg depending on the method of administration.

Example: *Step (A) Culture* — A culture medium was prepared by successively mixing into about 20 liters of distilled water 690 g of meat extract, 690 g of sodium chloride, 690 g of casein peptone, 690 g of yeast autolysate, 483 g of dipotassium phosphate, 207 g of monopotassium phosphate and the pH of the resulting solution was adjusted to about 7. The medium was sterilized at 120°C for 40 minutes and then a solution of 4,104 g of glucose and 2,760 g of soybean papainic peptone was added to the culture medium at the moment of seeding after having first been sterilized. The Hafnia strain (Pasteur Institute No. 5731) to be cultivated in the gelose media was added to 50 ml of the culture media and this solution acting as the inoculum was added to the rest of the culture bouillon and the total volume of the culture medium was adjusted to 138 liters by the addition of sterile distilled water.

The culture medium was held at 37°C and the pH was automatically adjusted to 7 by addition of hydrochloric acid solution or ammonium hydroxide solution. The increase in germs was determined photometrically to calculate the number of germs as the function of the optical density determined by comparison with a standard curve. After complete development or after about 7 hours, the medium contained about 1,000,000,000 germs per ml.

Step (B) Lysis — An aqueous solution of lysozyme hydrochloride (sterilized by filtration through an 0.22 micromillipore membrane) was added to 138 liters of the culture of Step (A) to obtain a final concentration of 160 γ of lysozyme hydrochloride per ml of culture and the mixture remained in contact at 56°C in the presence of 0.25 g of EDTA, 862.5 mg of sodium mercurothiolate and 80 g of polysorbate (Tween 80) per liter of culture bouillon. The lysis was continued for 7 days at 37°C under sterile conditions and the resulting lysate was homogenized by stirring and was lyophilized to obtain 7,900 g of a brown powder.

Step (C) Treatment — The brown powder from Step (B) was suspended in 138 liters of cold acetone and the suspension was vigorously stirred for 3 hours at 3,500 rpm and was then filtered through a glass frit. The mixture was rapidly vacuum filtered to obtain 7,295 g of a yellow powder which was suspended in 138 liters of cold methanol. The mixture was vigorously stirred for 3 hours at 1,500 rpm and the mixture was then decanted. The major part of the supernatant was drawn off with a siphon and the rest was filtered through a glass frit. The powder was vacuum filtered and dried at room temperature under reduced pressure for 24 hours to obtain 3,988 g of a clear beige powder.

Step (D) Diafiltration — 3,600 g of the powder of Step (C) was dissolved in 60 liters of distilled water containing 1 g/l of Merthiolate and the solution was stirred at 4°C for 24 hours and was then centrifuged for 2 hours at 4,000 rpm and then was added to a continuous centrifuge at 90,000 g at a range of 6 liters per hour and the recovered solution was adjusted to a volume of 10 liters with distilled water filtered through a 0.22 μ millipore membrane. The solution was introduced into diafilter apparatus equipped with porous membranes with a retention threshold of 300,000 MW and an apparent diameter of the pores being

approximately 2Å (commercial membranes XM 300, Amicon). 50 volumes of
distilled water (500 liters) were circulated through the apparatus for about 48
hours and the diafiltration solution was then placed in a continuous centrifuge
at 90,000 g at a rate of 6 l/hr. 9.5 liters of solution was obtained which was
then lyophilized to obtain 105 g of glycoproteins in the form of a cottony
white-beige powder which was very hygroscopic.

Analysis — C, 37.9%; H, 6%; N, 7.5%. Content: 9% water; 0.01% phosphorus;
0% chlorine. Proteins: 51% biuret. Sugars: 29% orcinol (neutral hexoses) and
1.5% carbazole (uronic acids). UV Spectrum: max at 216 mμ and 258 mμ.
IR Spectrum: confirmed the nature of peptidic part of the products.

Glucoproteins

H.-H. Nagel; U.S. Patent 4,153,686; May 8, 1979 has found that under the influence of yeast ferments, monosaccharides can be directly reacted with albumin
to give soluble glucoproteins.

In the process, it is possible to use an aqueous suspension of yeast, e.g., *Saccharomyces cerevisiae* or *Torula utilis*. The reaction is advantageously performed in
the neutral range, that is to say, at pH values of approximately 6 to 7.5 and at
ambient temperature, that is to say, approximately 18° to 22°C.

As starting substances lactalbumin and gelatin have proved suitable as albumin
components and lactose and glucose as monosaccharide components.

The working up of the reaction product for separating the glucoproteins obtained
can take place by methods known per se from albumin chemistry. The glucoprotein is preferably separated from the impurities by dialysis and is subsequently
dried under optimum gentle conditions, for example, by centrifuging, solvent extraction or preferably lyophilization (freeze-drying). The glucoproteins obtained
are water-soluble and can be redissolved from water. Their average molecular
weight can vary within wide limits and in particularly preferred manner is in
the range 55,000 to 65,000.

The pure glucoproteins prepared according to the process are not stable, making
stabilization necessary prior to their further use. It has been found that both
inorganic electrolytes and saccharide/albumin mixtures are suitable for stabilization purposes. The first group in particular includes salts of metals of the first,
second and eighth Groups of the Periodic system, for example, chlorides, carbonates and phosphates of potassium, sodium, magnesium, calcium and iron.
The second group preferably includes mixtures of lactalbumin, lactose and D-
glucose, although it is also possible to use other albumin substances and other
aldehyde sugars. According to a particularly preferred embodiment, stabilization
is performed by adding the inorganic salts and an albumin/sugar suspension to
the aqueous glucoprotein solution, followed by drying under gentle conditions
in the manner described hereinbefore.

The albumin/sugar mixture is used in a large weight excess based on the glucoprotein, preferably in a ratio of 10^3 to 10^6:1, and more particularly at approximately 10^5:1. The thus obtained glucoprotein is stable and suitable for further
use. It has been found that glucoproteins increase the action of therapeutics,
hormones and chemotherapeutics and in part widen their action spectrum. Al-

though no complete explanation can be given for this observation, it can be assumed that transfer improvement effects are involved because glucoproteins are constituents of the cell membranes. For example, the increased action in the case of antibiotics could be based on an increased inhibition of specific transfer peptidases necessary in bacteria cell wall synthesis.

The potentiating action is observed even with very low glucoprotein concentrations so that they are preferably used in a ratio of 1 part by weight glucoprotein to 10^2 to 10^6 parts by weight, and preferably approximately 10^4 parts by weight of active medicament substance. According to the observations, the action is optimum if the average molecular weight of the glucoproteins is in the range of approximately 55,000 to 65,000.

Example: 10 g of lactalbumin, 2 g of lactose and 1 g of yeast were suspended in 250 ml water and dialyzed for 24 hours in a dialyzer at 20°C relative to distilled water. After only 8 hours the passage of albumin was detected with the aid of the ninhydrin reaction. After 24 hours dialysis dialysate was carefully concentrated in vacuo and the solid residue was redissolved from a little water, whereupon the crystals were dried in a desiccator. A glucoprotein with an average molecular weight of 60,000 was obtained.

For stabilization purposes in each case 1 mg of glucoprotein was added to a suspension of 65 g of albumin/sugar (50% lactalbumin, 25% lactose, 25% D-glucose) which also contained traces of sodium and potassium chloride, calcium and iron carbonate and magnesium phosphate. After thorough mixing the suspension was carefully concentrated at ambient temperature under reduced pressure, whereupon the concentrate was dehydrated by freeze-drying. Glucoprotein is present in stabilized form in the thus-obtained powder so that its potentiating action remains unchanged.

Metal Proteinates

N.L. Jensen; U.S. Patent 3,969,540; July 13, 1976; assigned to Albion Laboratories, Inc. describes enzymatically prepared metal proteinates wherein the proteinate is a chelate of an enzymatically digested protein with bivalent essential metal wherein the naturally occurring vitamins and hormones in the protein are not destroyed.

The enzymatic hydrolysis is brought about by placing a comminuted protein source in an aqueous solution. In general, the solution will contain about 20% by weight/volume of proteinaceous material to water. Preferred sources of protein include naturally occurring tissues such as muscle, heart, liver, brain, pancreas, spleen, kidneys, duodenum, thymus, and orchic. These tissues can be used as carrier materials for minerals without destroying the valuable vitamins and hormones which may be destroyed by treatment with acids, bases or heat. Other protein sources such as casein, gelatin, collagen or albumin may be used.

Any protease may be utilized as the enzyme. Typical of such proteases are pepsin, pancreatin, trypsin, papain, bacterial protease and fungal protease. The enzyme is added in an amount of about 1 to 10 wt % based on the protein.

The hydrolysis is carried out at between about 25° and 70°C so that none of the amino acids in the polypeptides are destroyed in the hydrolysis process. The hydrolysis period may last anywhere from a matter of hours to a matter of weeks depending upon the amount of hydrolysate to be formed and the ease of hydrolysis. In general, the hydrolysis will be carried out over a period of about 2 hours to about 5 days. Preferably, the hydrolysis is carried out under neutral conditions and it may be necessary to adjust the pH of the protein-enzyme digestion solution by the use of an acid or base. Small amounts of acids or bases such as hydrochloric acid and sodium hydroxide may be utilized to neutralize the enzymatic hydrolysis mixture, but not in amounts sufficient to cause acid or base hydrolysis. If desired, a small amount of preservative such as toluene may also be present during the digestion or hydrolysis process.

As the proteinaceous material is hydrolyzed into polypeptides it will be brought into solution. The hydrolysis is brought about by constantly stirring the mixture and maintaining the temperature relatively constant.

At the end of the digestion period the mixture may, if desired, be brought to a boil for a short period of time to kill the enzyme so that further hydrolysis will not take place. However, this tends to destroy natural materials such as vitamins and hormones. The solubilized hydrolysate may then be directly treated with a soluble metal salt to form a metal proteinate or may first be filtered to separate the hydrolysate from undigested tissue.

Preferably the pH will be adjusted to a point that is on the alkaline side of the isoelectric point. In general, pHs of about 7.5 to 10 are preferred. Since the hydrolysate has previously been formed, the pH adjustment does not materially affect the stability of the hormones and vitamins in the protein hydrolysate. The appropriate amount of metal salt is added to an aqueous solution of the protein hydrolysate to form a precipitate which is filtered and then washed. The amount of metal salt that is added is adjusted such that there are at least 2 mols of protein hydrolysate or polypeptide per mol of metal salt. Otherwise, a true chelate will not be formed. The metal chelates readily precipitate, and unlike the protein hydrolysates, do not have a net charge. In other words, there are no free ions associated with the metal proteinates.

The product thus obtained can be dried and mixed in appropriate amounts as a dietary supplement in the form of a powder, tablet, liquid or in any other form desired, and then administered to an animal having a need for the particular metal proteinate thus formed, or if desired, the metal proteinates may be mixed in an appropriate base for topical or cosmetic application.

Example: Into a 300-gallon jacketed tank equipped with a stirrer and a temperature controller was placed 1,000 lb of water. While stirring, 200 lb of casein was added. 6 lb of sodium hydroxide was added to neutralize and solubilize the acid casein. To this mixture was added a mixture of enzymes as follows: 3 lb of papain, 3 lb of bacterial protease, and 3 lb of fungal protease. 20 lb of toluene was added as a preservative. The mixture was covered and stirred for a period of 4 days at a temperature of 50°C. The pH was then adjusted to 8.5 with sodium hydroxide and the mixture was brought to a boil for about 15 minutes to kill the enzymatic action. The hydrolysate thus obtained was filtered while hot through muslin. To the filtrate thus obtained was added 33 lb of zinc chloride whereupon a precipitate was formed, which, when washed and dried, consisted of about 200 lb of zinc proteinate containing 8 wt % zinc.

Ammonia as By-Product

F.T. Sherk and D.O. Hitzman; U.S. Patent 3,897,303; July 29, 1975; assigned to Phillips Petroleum Company describe a process for the production of cellular protein material in combination with the synthesis of ammonia. The process is illustrated in the schematic process flow diagram of Figure 2.2.

Figure 2.2: Integrated Process of Ammonia Production and Biosynthesis

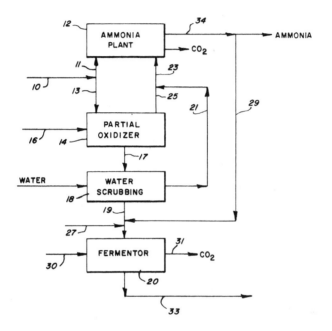

Source: U.S. Patent 3,897,303

Referring to the drawing, air is passed through line **10** with a portion thereof being directed through line **13** into partial oxidizer **14** and the remainder being directed through line **11** into ammonia plant **12**. The ratio of feed directed through line **11** and through line **13** is generally about 4:1 on a mol basis.

In a typical ammonia plant utilizing a synthesis gas feed stream, methane, air and steam are mixed and allowed to react, initially to form carbon monoxide, carbon dioxide and hydrogen and further, utilizing the steam water gas reaction, the carbon monoxide and steam react to form more hydrogen and carbon dioxide. A gas stream is formed which contains the nitrogen originally present in the air feed and hydrogen in the proper proportion to form ammonia, i.e., 3 mols hydrogen to 1 mol nitrogen. These are allowed to react at high temperature and pressure over a suitable catalyst such as metal oxide catalyst to form ammonia.

In this process, the methane feed stream is first introduced into the partial oxidizer **14** through line **16**, admixed with air from stream **13** allocated from line

10 and is partially oxidized to methanol, formaldehyde, carbon monoxide and carbon dioxide; some hydrogen is also produced here. The gases, i.e., the residual methane, nitrogen, carbon monoxide, carbon dioxide and traces of hydrogen, constitute the feed stream to the ammonia plant after removal of the water-soluble oxidation products in scrubber **18**.

The partial oxidizer **14** can be operated over a range of temperatures and pressures. The temperature may vary from 400° to 450°C, preferably about 430°C, and the pressure may vary from 300 to 1,500 psia, preferably about 900 to 1,200 psia. The feed comprising fresh feed and optionally recycle gases from scrubber **18**, contain methane, nitrogen, oxygen and traces of carbon monoxide, carbon dioxide and hydrogen. The mol ratio of methane to oxygen is in the range of 10 to 50:1, preferably about 20 to 30:1. At 430°C approximately 51% of the methane oxidized is converted to methyl alcohol and approximately 4.1% to formaldehyde.

The effluent from partial oxidizer **14** is passed via line **17** to water scrubber **18**. As a result of water scrubbing at **18**, the methyl alcohol and formaldehyde are dissolved and the aqueous effluent is then passed via line **19** to fermentor **20**. The undissolved gases from scrubber **18** consisting mainly of methane, nitrogen, carbon dioxide, carbon monoxide and a little hydrogen are sent via line **21** from which they can be diverted either through line **23** into ammonia plant **12** or through line **25** into oxidizer **14**. The gas stream from **21** may be fed to the ammonia plant **12** in its entirety via line **23**, or may be divided into a recycle stream via line **25** to the partial oxidizer and the ammonia plant feed stream via line **23** in some suitable ratio of recycle stream to feed stream up to about 4:1, preferably about 2:1.

In fermentor **20** the aqueous effluent from water scrubber **18** containing methanol and a minor amount of formaldehyde is subjected to microbial fermentation by methods known to the art. The microbial fermentation of water-soluble alcohols and aldehydes, and particularly methanol and formaldehydes, is disclosed in U.S. Patent 3,642,578. Among the microorganisms suitable for this type of fermentation are bacteria, yeasts and fungi exemplified in a nonlimiting manner.

Bacteria:	*Brevibacterium, Nocardia, Corynebacterium, Micrococcus, Arthrobacter, Mycobacterium, Streptomyces, Pseudomonas, Bacillus, Actinomyces*
Yeasts:	*Candida, Saccharomyces, Torulopsis, Rhodotorula, Hansenula, Brettanomyces, Pichia, Debaryomyces*
Fungi:	*Aspergillus, Penicillium, Monilia, Fusarium, Rhizopus, Mucor, Altenaria, Fungi imperfecti*

A suitable liquid growth medium, growth factors, minerals, etc., as disclosed in the referenced patent, for example, are introduced through line **27**. Inoculation of the fermentor with the desired microorganism may also be accomplished through line **27**. Ammonia from ammonia plant **12** is introduced via line **29** and admixed with the aqueous solution of methanol and formaldehyde in line **19** prior to its introduction into fermentor **20**. This provides the nitrogen required for the fermentation while air or oxygen for the fermentation is supplied through separate stream **30**. Carbon dioxide produced in fermentor **20** is vented through line **31**. The proteinaceous cellular material is removed via line **33** and can be subjected to usual recovery procedures such as coagulation/centrifugation or coagulation/filtration.

The advantage of this combined integrated process lies in the fact that the common feed stock, methane, can be used to provide the carbon source to the fermentation and the hydrogen to the ammonia plant. The air used to oxidize the methane will be stripped of its oxygen, normal for an ammonia plant, to provide the required nitrogen, but at least a part of this oxidation is utilized to produce the carbon source for the fermentation. Less than one-half of 1% of the ammonia per day is required to furnish the fixed nitrogen requirement for the fermentation, a hardly noticeable amount even though **30,000** lb/day of cells may be produced. The plant effluents are largely innocuous or can be recycled so that the process is also ecologically sound. Both products are valuable and contribute to the food supply of the nation.

NUCLEIC ACIDS
AND THEIR SUBSTITUENTS

GUANOSINE

Mutants Resistant to Psicofuranine or Decoyinine

H. Enei, K. Sato and Y. Hirose; U.S. Patent 3,960,660; June 1, 1976; assigned to Ajinomoto Co., Inc., Japan have found that much greater yields of guanosine are produced, compared with known methods, by culturing, in a culture medium, a mutant of Bacillus which is resistant to psicofuranine or decoyinine, and which requires adenine for growth.

The preferred guanosine-producing mutants are:

Bacillus subtilis AJ 3727 (FERM-P 2540)(resistant to decoyinine),

Bacillus subtilis AJ 3728 (FERM-P 2541)(resistant to 8-azaguanine, decoyinine),

Bacillus subtilis AJ 3729 (FERM-P 2542)(resistant to psicofuranine),

Bacillus subtilis AJ 3730 (FERM-P 2543)(resistant to 8-azaguanine, psicofuranine).

The culture media in which the mutants produce guanosine are largely conventional. They must contain sources of assimilable carbon and nitrogen and adenine, and should further contain inorganic ions and minor organic nutrients. Suitable carbon sources include glucose, fructose, sucrose, starch hydrolyzate and molasses. Nitrogen may be derived from nitrates, ammonium salts, ammonium hydroxide, urea, and like inorganic and organic compounds.

Aerobic conditions are maintained by aeration and/or agitation, and pH is held between 5 and 9 for optimum yields. When ammonia is used for pH control, it may also serve as a nitrogen source. The guanosine concentration in the broth reaches its maximum within 2 to 7 days if the fermentation is carried out at 24° to 37°C.

The guanosine accumulated in the fermentation broth can be recovered by conventional methods, such as removing cells by filtration or centrifuging or passing the broth over an ion exchange resin.

Mutants Resistant to Sulfa Drugs

In a similar process *H. Enei, K. Sato, Y. Anzai and Y. Hirose; U.S. Patent 3,969,188; July 13, 1976; assigned to Ajinomoto Co., Inc., Japan* have found that a remarkably higher amount of guanosine is produced, when it is compared with the known method, by culturing in a culture medium a mutant of Bacillus which is resistant to at least one compound selected from sulfa drugs, and which requires adenine for growth.

The sulfa drugs used in this process contain the group:

and have antimicrobial action which is suppressed at least in part by p-aminobenzoic acid.

The known sulfa drugs useful in the process are sulfapyridine, sulfathiazole, sulfadiazine, sulfaguanidine, sulfamethazine, sulfamerazine, sulfadimethoxine, sulfamethomidine, sulfamethoxypyridazine, sulfisomidine, sulfisoxazole, acetosulfamine, sulfamethizole, sulfaethidole, sulfapyrazine, irgafen, Irgamide, sulfanylamide, sulfisomezole, and sulfaphenazole.

The most effective guanosine-producing mutants found so far are as follows:

Bacillus subtilis AJ 3617 (FERM-P 2313)(sulfaguanidine), and

Bacillus subtilis AJ 3618 (FERM-P 2314)(8-azaguanine, sulfamerazine).

Culture conditions are the same as in the previous process.

Mutants Resistant to Methionine or Azaserine

In another similar process, *H. Enei, Y. Anzai, K. Sato, H. Eguchi and Y. Hirose; U.S. Patent 3,912,587; October 14, 1975; assigned to Ajinomoto Co., Inc., Japan* have found that remarkably higher amounts of guanosine are produced by culturing in a culture medium a mutant of Bacillus which is resistant to at least one compound selected from methionine, methionine analogue, and azaserine, and which requires adenine for growth.

The most effective guanosine-producing mutants found so far are as follows:

Bacillus subtilis AJ 3473 (FERM-P 2107)(methylmethioninesulfonium chloride)

Bacillus subtilis AJ 3474 (FERM-P 2108)(azaserine)

Bacillus subtilis AJ 3475 (FERM-P 2109)(methioninesulfoxide)

Bacillus subtilis AJ 3476 (FERM-P 2110)(ethionine)

Bacillus subtilis AJ 3477 (FERM-P 2111)(methionine)

Bacillus subtilis AJ 3478 (FERM-P 2112)(8-azaguanine, methioninesulfoxide)

Bacillus subtilis AJ 3479 (FERM-P 2113)(8-azaguanine, azaserine),

Bacillus subtilis AJ 3480 (FERM-P 2114)(methionine + methioninesulfoxide)

Bacillus subtilis AJ 3481 (FERM-P 2115)(8-azaguanine, methionine + methioninesulfoxide), and

Bacillus pumilus AJ 3482 (FERM-P 2116)(methioninesulfoxide).

The + shows that mutant was screened in the medium containing both methionine and methioninesulfoxide.

(2'-Amino-2'-Deoxypentofuranosyl)Guanine

T. Suzuki and T. Nakanishi; U.S. Patent 4,019,957; April 26, 1977; assigned to Kyowa Hakko Kogyo Kabushiki Kaisha, Japan describe a process whereby a compound represented by the following general formula is obtained.

(2'-amino-2'-deoxypentofuranosyl)guanine (herein designated as 2'-APG) which is the guanosine analogue according to this process can easily be isolated in the form of white plate-like crystals having alkaline nature by concentrating, under reduced pressure, a solution of it in a suitable solvent such as, for example, water.

Any and all microorganisms which belong to the genus Aerobacter (Enterobacter) may be used for this process. The species of microorganism which has been found especially useful is *Aerobacter cloacae (Enterobacter cloacae)*(FERM-P 1893).

In carrying out the process, the culture medium containing a carbon source, nitrogen source, inorganic compounds such as phosphate, sulfate, hydrochloride, metallic salt such as iron, manganese, magnesium, potassium, sodium, etc. and growth-promoting factors is sterilized and then inoculated with the aforementioned microorganism for aerobic culturing. It is possible to improve the yield of the compound by adding purine nucleotides such as inosinic acid, xanthylic acid, guanylic acid, etc. or its precursor to the medium.

The cultivation may be effected at a temperature of 20° to 40°C and a pH of 5 to 8, preferably 6 to 7. In this case, the pH is adjusted by the addition of urea solution, aqueous ammonia or ammonium carbonate. The cultivation is completed when the amount of 2'-APG produced reaches a maximum. A period of from 1 to 7 days is usually sufficient for effecting cultivation.

When isolating the 2'-APG produced, cell bodies are removed from the culture medium by means of centrifugation or filtration. The 2'-APG is then recovered from the culture liquor by passing the liquor through an acidic ion exchange resin to selectively absorb the 2'-APG thereon. If necessary, absorbing agents may be used. The 2'-APG is eluted using, for example, water or aqueous ammonia and the eluate is concentrated under reduced pressure to produce purified crystalline 2'-APG.

Example: *Aerobacter cloacae* (*Enterobacter cloacae*) (FERM-P 1893) was inoculated as a seed in a seed medium composed of meat extract (1.0%), peptone (1.0%), yeast extract (0.3%), table salt (0.3%) and sorbitol (2.0%). The pH of the seed medium was adjusted to 7.3 before sterilization, and the medium cultured at a temperature of 30°C for 24 hr with shaking. The resultant seed was put into a 2-liter Florence flask containing a fermentation medium as follows:

Sucrose, %	10.0
Ammonium Sulfate, %	1.0
$MgSO_4$, %	0.1
$FeSO_4$, mg/l	40
Yeast extract, %	0.1
KCl, %	0.8
K_2HPO_4, %	0.02
$ZnSO_4$, mg/l	3

The pH of the fermentation medium was adjusted to 7.0 before sterilization and the cultivation was carried out at 30°C for 24 hr using a rotary shaking method to yield 205 mg of 2'-APG/l of the fermented liquor which was then centrifuged to obtain the supernatant solution. The supernatant (3 liters) was passed through a resin column packed with Amberlite IRC-50 (a weakly acidic ion exchange resin in H-form, Rohm & Haas) and eluted with 0.5 N aqueous ammonia. It was then passed through a resin column packed with Diaion SK 1B (a strongly acidic ion exchange resin in NH_4-form, Mitsubishi Kasei Kogyo Kabushiki Kaisha, Japan) and eluted with 10 mM aqueous ammonia. After concentration, the solution was made up with 10 mM aqueous ammonia and was passed through a column packed with Diaion HP (a synthetic absorbing agent, Mitsubishi Kasei Kogyo Kabushiki Kaisha, Japan).

The eluted fractions containing 2'-APG were collected, concentrated and then acetone added (10 times the quantity of the fractions). The supernatant was concentrated to obtain crystals of 2'-APG (440 mg) having a melting point of 252° to 254°C (decomposition).

VARIOUS NUCLEOSIDES

Inosine

It is known that mutant strains of *Bacillus subtilis* requiring adenine, or resistant to 8-azaguanine and requiring adenine produce inosine in culture medium.

H. Enei, K. Sato, M. Ishii and Y. Hirose; U.S. Patent 3,960,661; June 1, 1976; assigned to Ajinomoto Co., Inc., Japan have found that certain mutants of the known strains which are resistant to at least one sulfa drug and require adenine for growth, produce higher yields of inosine than the previously known strains.

The preferred inosine-producing strains are:

> *Bacillus subtilis* AJ 3721 (FERM-P 2534)(requiring adenine, lysine and thiamine, and resistant to sulfaguanidine),
>
> *Bacillus subtilis* AJ 3722 (FERM-P 2535)(requiring adenine, lysine and thiamine, and resistant to sulfadiazine),
>
> *Bacillus subtilis* AJ 3723 (FERM-P 2536)(requiring adenine, and resistant to sulfathiazole),
>
> *Bacillus subtilis* AJ 3771 (FERM-P 2554)(requiring adenine, resistant to sulfaguanidine and having weaker activity to nucleotidase),
>
> *Bacillus subtilis* AJ 3772 (FERM-P 2555)(requiring adenine, resistant to sulfaguanidine and having weaker activity to IMP-dehydrogenase),
>
> *Bacillus subtilis* AJ 3773 (FERM-P 2556)(requiring adenine, and resistant to methionine sulfoxide and sulfaguanidine), and
>
> *Bacillus subtilis* AJ 3774 (FERM-P 2557)(requiring adenine, and resistant to 8-azaguanine and sulfaguanidine).

The media for culturing the microorganisms are conventional except that they contain adenine which is required for growth. They will normally contain sources of carbon, nitrogen, inorganic and, where required, organic nutrients.

Carbohydrates (such as glucose, sucrose, molasses, starch, or starch hydrolyzate), organic acids (such as benzoic acid, acetic acid, propionic acid, higher fatty acids, or fumaric acid), and alcohols (such as ethanol, propanol, sorbitol or glycerine) are the normally preferred carbon sources.

Ammonium salts, nitrate salts, ammonia water, gaseous ammonia and urea can be used as nitrogen sources. Minor organic nutrients are, for example, amino acids and vitamins and materials containing organic nutrients such as corn steep liquor, soy protein hydrolyzate, beef extract, casein hydrolyzate, or yeast extract. Cultivation is carried out aerobically, preferably while maintaining a pH of 5 to 9 and a temperature of 25° to 40°C.

Double Salts of S-Adenosyl-L-Methionine

A. Fiecchi; U.S. Patent 3,954,726; May 4, 1976; assigned to Bioresearch Limited, Italy describes a process relating to extremely stable salts of S-adenosyl-L-methionine (SAM), to a process which enables it to be prepared simply and economically on an industrial scale and to pharmaceutical compositions which contain them as the active principal, for use in numerous fields of human therapy.

SAM is notably a product of natural origin, found in all living organisms from bacteria to plants, from single cell organisms to superior mammals including man, the structure of which has been known for some time and is identified by the following formula.

in which X is a generic anion.

In living organisms SAM is formed by the intervention of enzymes (S-adenosyl-methioninsynthetasis or S-adenosyltransferasis) in the cytoplasmatic ambit starting from methionine assumed with the nutriments or from ATP present as energy reserve in every living cell.

It has also been known for some time that SAM is a product of fundamental importance in a large number of biological reactions of enzymatic transmethylation, because of which it has always been considered a very important reagent in biochemistry. The big problem with this substance has, however, always been its extreme instability at ambient or above ambient temperatures, and its methods of production which are laborious and cannot easily be carried out on an industrial scale.

The salts according to this process are double salts of SAM with p-toluenesulfonic acid and sulfuric acid, corresponding to the formula:

$$SAM^+ \cdot HSO_4^- \cdot H_2SO_4 \cdot 2CH_3C_6H_4SO_3H \quad \text{and} \quad SAM^+ \cdot HSO_4^- \cdot H_2SO_4 \cdot CH_3C_6H_4SO_3H$$

respectively.

The process for preparing the salts comprises essentially the following stages.

    (a) Preparation of a solution rich in SAM either by extraction from natural substances which contain it or by enzymatic synthesis from adenosine triphosphate (ATP) and methionine.

    (b) Precipitation of the SAM present in the filtered aqueous solution by a saturated aqueous solution of picrolonic acid or by solutions of the same acid in organic solvents soluble in water such as methyl, ethyl, propyl, isopropyl, n-butyl, or isobutyl alcohols; or acetone, methylethylketone, methylisobutylketone, ethyl acetate, tetrahydrofuran, 2-methoxyethanol, 2-ethoxyethanol, dioxan or dimethylformamide.

    (c) Dissolving the filtered precipitate in a mixture consisting of equal parts by volume of a solvent partially miscible with water such as methylethylketone, methylisobutylketone, n-butanol or isobutanol and a solution of equal normality of p-toluenesulfonic acid and sulfuric acid.

    (d) Separation of the organic layer and addition to the aqueous solution of an organic ketone or alcohol solvent completely soluble in water.

(e)   Redissolving the precipitate in a 10 to 20% solution of n-tolu-
       enesulfonic acid in methanol and treating the solution with
       decoloring charcoal.

(f)   Addition to the concentrate of an organic solvent able to pre-
       cipitate the pure SAM in a well crystalline and easily filterable
       form.

Stage (a) of the process can be carried out in different ways which are equally
efficient for the purposes of obtaining a concentrated solution of SAM. According to one alternative, yeast (*Saccharomyces cerevisiae, Torulopsis utilis, Candida
utilis,* etc.), enriched in SAM by the addition of methionine under suitable conditions [Schlenk, *Enzymologia 29,* p 283 (1965)], is reacted with ethyl acetate
and then sulfuric acid of a normality between 0.1 and 0.5, preferably 0.35 N at
ambient temperature, so as to cause the lysis of the cells and the passage into
solution of practically 100% of the SAM present.

Preferably quantities of water and acetate between $\frac{1}{20}$ and $\frac{1}{5}$ of the weight of
the humid cells are used and the treatment is protracted for a time between 15
and 45 min, preferably for 30 min. Sulfuric acid is then added, and lysis is
carried out for a time between 1 to 2 hr, preferably 1½ hr.

The stage (b) of the process enables the SAM to be separated in a state of high
purity. In fact, in an acid environment the only compound precipitated by picrolonic acid is SAM, as is shown by thin layer chromatography in accordance with
Anal. Biochem. 4, pp 16–28 (1971). Picrolonic acid has thus an extremely and
surprisingly selective action.

The stage (c) is preferably carried out with solutions containing p-toluenesulfonic
acid and sulfuric acid in concentrations both between 0.05 and 0.2 N, preferably
0.1 N, and with an organic solvent particularly miscible with water such as
methylethylketone or n-butanol. The use of the organic solvent enables the
aqueous acid solutions to be very much reduced and practically eliminates all
the picrolonic acid.

The stage (d) of the process is carried out by preferably using between 4 and 8
volumes (with respect to the volume of the aqueous solution) of a solvent chosen
from the group comprising acetone, methyl alcohol, ethyl alcohol and propyl
alcohol.

It has also been found that if in stage (e) the minimum quantity of methanol
necessary to dissolve the precipitate originating from stage (d) is used, the double
salt $SAM^+ \cdot HSO_4^- \cdot H_2SO_4 \cdot 2CH_3C_6H_4SO_3H$ separates in the subsequent precipitation
stage (f).

If, however, a volume of methanol equal to at least double the necessary volume
is used in stage (e), the double salt $SAM^+ \cdot HSO_4^- \cdot H_2SO_4 \cdot CH_3C_6H_4SO_3H$ separates
in the subsequent precipitation stage (f).

The use of intermediate quantities of methanol leads to the formation of mixtures of the two salts.

The final precipitation of one or other of the salts according to the process
stage (f) requires the use of an organic solvent chosen from the group consisting

of methanol, ether, chloroform, n-propanol, isopropanol, n-butanol, isobutanol, secondary butyl alcohol, isoamyl alcohol and tetrahydrofuran. The double salts of SAM can be preserved indefinitely in the dry state, as stated, practically unaltered.

The fields of use already ascertained for the SAM salts are: treatment of hepatopathy, hyperdyslipidemias, generalized or local arteriosclerosis, psychiatric manifestations of depressive and neurological type, degenerative arthropathies, neurological algic manifestations and disturbances of the sleeping-waking rhythm, whereas many other fields of use still remain to be examined and ascertained. The SAM salts are preferably administered by intramuscular or intravenous injection, or in oral or sublingual tablets, or in capsules.

Example: To 90 kg of yeast enriched in SAM (6.88 g/kg) in accordance with Schlenk [*Enzymologia 29*, p 283 (1965)] are added 11 liters of ethyl acetate and 11 liters of water at ambient temperature. After energetic agitation for 30 min, 50 liters of 0.35 N sulfuric acid are added, continuing agitation for a further hour and a half. After filtering and washing with water, 140 liters of solution are obtained containing 4.40 g/l of SAM, equal to 99.5% of that present in the starting material. A solution of 2.3 kg of picrolonic acid in 24 liters of methylethylketone is added to the previous solution under agitation. After standing for one night, the precipitate is separated by centrifuging and washed with water.

The solid thus obtained is added under agitation to a mixture of 18 liters of a 0.1 N solution of sulfuric acid and p-toluenesulfonic acid, and 18 liters of methylethylketone. After standing, the organic phase separates and is fed to the picrolonic acid recovery system, the aqueous phase is treated with a little methylethylketone to eliminate residual traces of picrolonic acid, decoloring charcoal is added and it is then filtered.

This solution (16.5 liters) contains 33.8 g/l of SAM, equal to 90% of the compound present in the yeast. When analyzed by thin layer chromatography in accordance with *Anal. Biochem. 4*, pp 16-28 (1971) it is known to contain only SAM without traces of its decomposition products or other organic bases. The above solution is poured into 100 liters of acetone under agitation. After standing, it is decanted from the solvent and the solid is dissolved in 3.3 kg of a 15% solution of p-toluenesulfonic acid in methanol. After adding decoloring charcoal, the mixture is filtered and added to 25 liters of ethyl ether.

1,184 g of a well-crystallized salt precipitate results; it is easily filterable, not very hygroscopic, and very soluble in water (more than 20%) with formation of a colorless solution. The salt is only slightly soluble in methanol and ethanol, and insoluble in acetone, methylethylketone, chloroform, higher alcohols and benzene. From thin layer chromatography in accordance with *Anal. Biochem. 4,* pp 16-18 (1971) the product is shown to be free from any impurity.

Analysis—Calculated for $C_{29}H_{42}N_6O_{19}S_5$ (MW 938.98): C, 37.09%; H, 4.51%; S, 17.7%; N, 8.95%. Found: C, 36.39%; H, 4.6%; S, 16.7%; N, 8.8%. Additional Analysis—Calculated: H_2SO_4, 20.89%; p-toluene sulfonic acid, 36.67%; SAM 41.54%. Found: H_2SO_4, 20.5%; p-toluene sulfonic acid, 36.0%; SAM, 41.7%.

Humidity determined in accordance with K. Fischer: 1.7 to 2%. The UV spectrum of the compound shows an absorption maximum at 260 nm, $E_1^{1\%}$ em of 182. Such data agree with a compound of the following formula.

S-Adenosylmethionine and Methylthioadenosine

T. Tsuchida, F. Yoshinaga and S. Okumura; U.S. Patent 3,962,034; June 8, 1976; assigned to Ajinomoto Co., Inc., Japan describe a method for producing S-adenosylmethionine [S-(5'-desoxyadenosin-5'-yl)methionine] or methylthioadenosine (1-adenyl-5-methylthioribose). S-adenosylmethionine and methylthioadenosine will be referred to as SAM and MTA, respectively, herein.

It has been found that high amounts of SAM and/or MTA are accumulated in the culture broth and/or in cells of various yeasts, when the yeasts are cultured in otherwise conventional medium containing 0.5 to 2.0 g/dl L-methionine while maintaining a pH of the medium at 2 to 5 and preferably 3 to 5.

Yeasts which produce SAM or MTA, found so far, belong to genus *Candida, Pichia, Rhodotorula, Cryptococcus, Hansenula, Trichosporon, Kloechera, Torulopsis, Hanseniaspora, Sporoboromyces, Lipomyces* or *Debaryomyces*. Most of the yeasts produce SAM and MTA at the same time in the culture broth and in the cells.

The culture media are conventional except for containing 0.5 to 2.0 g/dl L-methionine. The media also contain a carbon source, a nitrogen source, and inorganic ions. For a certain yeast, minor organic nutrients such as vitamins or amino acids are necessary for or promote the growth.

Carbohydrates (such as glucose, sucrose, maltose, fructose, starch, or cellulose), alcohols (such as methanol, ethanol, propanol, glycerol, sorbitol, or xylulose), organic acids (such as acetic acid, propionic acid, butyric acid, glyceric acid, higher fatty acid, fumaric acid, or benzoic acid), esters, alkanes or aldehydes, or raw materials which contain those carbon sources (such as starch hydrolyzate, molasses, soy whey, fruit juice, waste water of fish processing, fermentation processing, or pulp processing water) are used as the carbon source. Suitable nitrogen sources are ammonium salts, ammonia water, gaseous ammonia, urea, nitrate salts, and amino acids. Inorganic salts are conventional such as phosphate, sulfate, and salts of potassium, magnesium, sodium, iron, manganese and calcium.

Cultivation is carried out maintaining a pH of 2 to 5, preferably 3 to 5 with alkali or acid, and temperature of 25° to 40°C. Normally, the fermentation period is from 2 to 10 days, although some variation is possible to accumulate SAM or MTA.

SAM accumulated in culture broth or cells can be recovered by using an ion exchange resin or other known methods. Also cells themselves which contain S-adenosylmethionine are recovered as the SAM. MTA accumulated in culture broth or cells can be recovered also by using an ion exchange resin or other known method. The MTA in the cells is recovered in the form contained in the cells, or from the homogenate of the cells by conventional methods. Cells containing SAM and/or MTA are especially useful as an additive of feed. It is also possible to use the culture broth, as it is or after processing as a feed additive.

Nucleoside Cyclic Phosphates from RNA

Y. Norimoto, S. Ohmura, Y. Shimizu and T. Tatano; U.S. Patent 3,920,519; November 18, 1975 describe the enzymatic hydrolysis of ribonucleic acid (herein referred to as RNA), into nucleoside 2',3'-cyclic phosphates with intact cells of microorganisms having a ribonuclease activity.

In accordance with this process, any microorganism that exhibits a ribonuclease activity may be used. Any of the methods well known in the art may be utilized for the determination of whether a specific microorganism exhibits a ribonuclease activity. Preferred microorganisms are Gram-negative bacteria, particularly those belonging to the genera *Escherichia, Salmonella, Aerobacter* or *Pseudomonas.*

Culturing may be carried out under the conditions suitable for the microorganisms to be employed. After the completion of culturing, the cells are collected, for example, by centrifugation.

In carrying out the hydrolysis of RNA, the cells are suspended in an aqueous medium at a concentration of 10 to 40 mg/ml, preferably about 20 mg/ml, determined as dry weight. To the suspension is added RNA, a chelating agent and a predetermined amount of phosphate ion. The RNA to be used may be obtained in any well-known manner, and is usually obtained from microorganisms. For hydrolysis, a concentration of 1 to 40 mg/ml, preferably 10 to 20 mg/ml of RNA is used.

As the chelating agent, any of those usually used for chelating metal ion may be used. Preferred are ethylenediaminetetraacetic acid (herein referred to as EDTA), cyclohexanediaminetetraacetic acid (herein referred to as Cy-DTA) and hydroxyethylenediaminetriacetic acid (herein referred to as EDTA-OH). The chelating agent is employed at a concentration of 5 to 30 mM, preferably 10 to 30 mM.

The formation ratio of nucleoside 2',3'-cyclic phosphates and nucleoside 3',5'-cyclic phosphates is controlled by the concentration of phosphate ion in the reaction medium.

When the concentration of phosphate ion in the medium is from 0 to 0.15 M, the reaction products are rich in nucleoside 2',3'-cyclic phosphates; and when the concentration is 0.17 M and higher, the reaction products are rich in nucleoside 3',5'-cyclic phosphates. Where the concentration is between 0.15 and 0.17 M, nucleoside 2',3'- and 3',5'-cyclic phosphates are produced almost equally.

As a source of phosphate ion, various phosphates that liberate phosphate ion in an aqueous medium such as phosphoric acid, sodium phosphate, potassium phosphate, ammonium phosphate, etc., may be employed.

Generally, the aqueous medium containing RNA, microbial cells, a chelating agent and various concentrations of phosphate ion is allowed to react at 35° to 45°C, preferably 38° to 42°C for 1 to 96 hr, preferably 24 to 48 hr. During the reaction, the pH is maintained at 4 to 9. Where the selective production of nucleoside 2',3'-cyclic phosphates is desired, the pH is preferably kept at about 5 and where the selective production of nucleoside 3',5'-cyclic phosphate is desired, the pH is preferably kept at about 8.

Suitable aqueous reaction media may be various buffer solutions such as an acetate buffer and a phosphate buffer. In view of the pH adjustment, the use of a phosphate buffer solution is most convenient. After completion of the reaction, the products are isolated and purified by well-known means.

Example 1: In this example, *Escherichia coli*, ATCC 10798, is cultured in 500 ml of a medium comprising 30 mg/ml glucose, 6 mg/ml yeast extract, 8.5 mg/ml potassium dihydrogen phosphate, 11 mg/ml dipotassium hydrogen phosphate and 0.25 mg/ml magnesium sulfate (pH 6.8) in a 3-liter Erlenmeyer flask at 38°C for 16 hr. Thereafter, the cells are collected from the culture liquor by centrifugation. Then, 4 g (in the examples, the weight of the cells is dry weight) of the cells are suspended in 200 ml of water. To the suspension are added 3 g of RNA and 876 mg of EDTA. The mixture is allowed to react at 40°C for 48 hr. After the reaction, 14 mg/ml of a mixture of nucleoside 2',3'-cyclic phosphates is formed in the reaction mixture.

The cells are removed from the resulting reaction mixture by filtration and the filtrate is passed through a column of Dowex 1X2 (Cl⁻). Elution is carried out with an aqueous solution of ammonium bicarbonate and the eluate is concentrated and freeze-dried. 1.8 g of a 90% pure powdery mixture of nucleoside 2',3'-cyclic phosphates comprising adenosin 2',3'-cyclic phosphate, guanosin 2',3'-cyclic phosphate, uridin 2',3'-cyclic phosphate and cytidin 2',3'-cyclic phosphate is obtained.

Example 2: In this example, 4.4 g of the cells of *Escherichia coli*, ATCC 10798, are suspended in 200 ml of 0.33 M phosphate buffer solution having a pH of 8.0. To the suspension are added 2 g of RNA and 876 mg of EDTA and the mixture is allowed to react at 40°C for 48 hr. As a result, 7.2 mg/ml of a mixture of nucleoside 3',5'-cyclic phosphates is formed in the reaction mixture.

The cells are then removed by filtration and the filtrate is subjected to adsorption of the resin. Elution is carried out with 50% acetone and the eluate is concentrated. The concentrate is subjected to adsorption on Dowex 1X2 (Cl⁻) and elution is carried out with a calcium chloride-hydrochloric acid buffer solution. The eluate is concentrated and freeze-dried. As a result, 0.6 g of a powder of a mixture of nucleoside 3',5'-cyclic phosphates of 86% purity is obtained.

1-β-D-Ribofuranosyl-1,2,4-Triazole-3-Carboxamide

During the past decade, many nucleoside analogs have been found to exhibit good antitumor and antiviral activities. Among the known synthetic nucleosidic antiviral agents, the more important generally are considered to be 5-iodo-2'-deoxyuridine (IDU), 9-β-D-arabinofuranosyladenine (ara-A), and 1-β-D-arabino-furanosylcytosine (ara-C). These compounds, however, are only active against a limited spectrum of viruses which does not include those causing respiratory diseases in man (influenza, common cold).

The only nucleosidic analog that is active against these respiratory disease viruses is 1-β-D-ribofuranosyl-1,2,4-triazole-3-carboxamide.

Certain derivatives of this latter compound have also been found to have significant activity against these viruses, and it has also been discovered that the triazole bases of certain of such compounds, 1,2,4-triazole-3-carboxamide and 1,2,4-triazole-3-thiocarboxamide, likewise have significant antiviral activity against these respiratory viruses. The chemical structure and synthesis of each compound have been previously reported [*Latvijas PSR Zinatnu Akad. Vestis*, Kim. Ser., (2) 204–208, see *Chem. Abst. 63*, 13243 (1965)].

J.T. Witkowski and R.K. Robins; U.S. Patent 3,976,545; August 24, 1976; assigned to ICN Pharmaceuticals, Inc. have found that 1,2,4-triazole-3-carboxamide may be caused to undergo enzymatic conversion to 1-β-D-ribofuranosyl-1,2,4-triazole-3-carboxamide, the aforenoted nucleoside which is a significantly effective antiviral compound. As will be shown herein, the triazole base may be reacted with the enzyme nucleoside phosphorylase to effect the indicated conversion.

Example 1: *1,2,4-Triazole-3-Carboxamide* – Methyl 1,2,4-traizole-3-carboxylate was heated with excess aqueous ammonia until the reaction was complete. The mixture was cooled and the product was collected. Recrystallization from water afforded a nearly quantitative yield of 1,2,4-triazole-3-carboxamide with a melting point of 313° to 315° dec. Analysis–Calculated for $C_3H_4N_4O$: C, 32.14%; H, 3.60%; N, 49.99%. Found: C, 32.37%; H, 3.73%; N, 50.09%.

1,2,4-Triazole-3-carboxamide may be converted to 1-β-D-ribofuranosyl-1,2,4-triazole-3-carboxamide by reaction with the enzyme nucleoside phosphorylase at a pH within 5 to 9, preferably 7 to 8, at an enzyme concentration of 0.015 to 0.75 mg/ml, preferably about 0.15 mg/ml, and a temperature of 0° to 50°C, with the preferred temperature of 25° to 35°C. Satisfactory results have been obtained when the triazole base is present in a concentration greater than 5×10^{-5} M and ribose-1-phosphate is present at a concentration greater than 2×10^{-5} M.

Generally, 0.1 to 2 hr, preferably 0.5 to 1 hr, are required for the reaction. The source of the enzyme may be animal, tissue, or bacteria. The principal bacterial sources are *E. coli* and yeast, while a variety of animal sources exist, including beef spleen, rat liver, calf liver, calf thymus, beef liver, monkey brain, horse liver, calf spleen, human erythrocytes, fish skin, and fish muscle. The synthesis will be better understood from the following example.

Example 2: *1-β-D-Ribofuranosyl-1,2,4-Triazole-3-Carboxamide* – Synthesis of 1-β-D-Ribofuranosyl-1,2,4-Triazole-3-Carboxamide from 1,2,4-Triazole-3-Carboxamide via Purified Calf Spleen Nucleoside Phosphorylase: The incubation samples contained, in a final volume of 0.135 ml, Tris HCl, pH 7.4, 50 μmol; ribose-1-phosphate, 0.25 μmol; 1,2,4-triazole-3-carboxamide (H³), 42 μCi/μmol, 0.05 μmol; and calf spleen nucleoside phosphorylase (Sigma Chemical Co.), 80 μg.

The samples were incubated at 25°C for 5 min and then frozen in Dry Ice/isopropanol to stop the reaction. Aliquots of the thawed samples were then spotted on silica gel together with standard solutions of 1,2,4-triazole-3-carboxamide.

and 1-β-D-ribofuranosyl-1,2,4-triazole-3-carboxamide and separated in isopropanol:NH_4OH:H_2O (7:1:2). Areas of the chromatograms coinciding with 1,2,4-triazole-3-carboxamide were removed and counted to determine the percent of conversion of 1,2,4-triazole-3-carboxamide to 1-β-D-ribofuranosyl-1,2,4-triazole-3-carboxamide.

Results of Nucleoside Phosphorylase Assay with 1,2,4-Triazole-3-Carboxamide

Sample	% Conversion
– Enzyme	0.8
– Ribose-1-P, + enzyme, 80 μg	1.1
+ Enzyme, 20 μg	37.9
+ Enzyme, 40 μg	48.4
+ Enzyme, 60 μg	54.1

It is apparent from the foregoing that the indicated conversion occurred, with greater conversion achieved with greater concentration of enzyme.

4-Carbamoyl-1-β-D-Ribofuranosylimidazolium-5-olate

K. Mizuno, T. Ando, M. Tsujino, M. Takada, M. Yoshizawa, T. Matsuda and M. Hayashi; U.S. Patent 3,888,843; June 10, 1975; assigned to Toyo Jozo Kabushiki Kaisha, Japan have found that an immunosuppressive agent, 4-carbamoyl-1-β-D-ribofuranosylimidazolium-5-olate, henceforth designated bredinin, having the formula:

can be produced using a microorganism belonging to *Eupenicillium.*

According to an embodiment of this process *Eupenicillium brefeldianum* NRRL 5734 (FERM-P 1104) is cultured in a conventional manner. It can be cultured as a solid culture or in a liquid culture medium; however, for industrial production submerged aeration culture may preferably be applied.

For the medium, there may be used a conventional nutrient medium for microorganism cultivation. The cultivating temperature will be selected so as to control growth of the microorganism and production of bredinin, and preferably is 26° to 30°C. Cultivation time may depend on the conditions and it may preferably be 40 to 70 hr. When the produced bredinin has its highest activity, the cultivation should be terminated.

The bredinin is recovered in a liquid filtrate and not in a mycelial cake. Bredinin is a water-soluble substance, difficultly soluble in most organic solvents and especially insoluble in water-immiscible organic solvents, so it is impossible to extract by solution in organic solvents. Therefore, it may preferably be isolated by weak acidification of bredinin or insolubility in an organic solvent. Since bredinin has a weak antimicrobial activity, it can be assayed by the conventional cup method using *Candida albicans* as a test organism.

Example 1: 100 ml of aqueous medium (pH 6.5) containing glucose 2%, potato extract (prepared from 300 g of potato slices and 1 liter water, boiled for 1 hr) 10%, cotton seed powder 0.5%, KH_2PO_4 0.5% and $MgSO_4 \cdot 7H_2O$ 0.25% were introduced in a 500 ml flask and sterilized at 120°C for 15 min. Into this medium, spores of *Eupenicillium brefeldianum* NRRL 5734 (FERM-P 1104) were inoculated and rotatory shake cultured at 300 rpm, at 26°C. After 48 hr, the cultured medium was transferred to 20 liters of the same medium as hereinabove in a 30 liter jar fermenter and cultured at 26°C, with agitation at 300 rpm and aeration of 20 l/min for 49 hr.

The thus-cultured medium was transferred into 200 liters of previously sterilized aqueous medium (pH 6.5) containing glucose 2%, peptone 1%, corn steep liquor 1%, KH_2PO_4 0.2%, $MgSO_4 \cdot 7H_2O$ 0.1% and antifoaming agent 0.1% in a 300 liter stainless steel fermentation tank and cultured at 26°C, with agitation at 350 rpm, and aeration of 200 l/min, for 55 hr to obtain the cultured broth (pH 5.9) containing 50 μg/ml of bredinin. The broth was adjusted to pH 9.0 by adding 50% aqueous sodium hydroxide and filtered to obtain 170 liters of clear filtrate. The filtrate was passed through a column (diameter 15 cm) of 20 liters of Amberlite IRA-411 (OH type) at a flow rate of 300 ml/min to absorb the material and washed with 50 liters of water.

Elution was carried out with 2% aqueous acetic acid and each 5 liters of eluate was fractionated. Active fractions were found in fractions No. 7 to 9 assayed by *Candida albicans* as a test organism. The active fractions were collected, adjusted to pH 9.6 with 50% aqueous sodium hydroxide, then passed through 4 liters of Amberlite IRA-411 (OH type) in a column (diameter 7.5 cm), washed with water, thereafter eluted with 2% aqueous acetic acid to obtain fractions of 500 ml each.

Active fractions were found at fractions No. 13 to 18, which were collected and concentrated in vacuo to obtain 200 ml of oily residue. The residue was well mixed with 400 ml of methanol and 200 ml of acetone and precipitated by centrifugation at 3,000 rpm for 10 min. The precipitate was washed with acetone and dried in vacuo to obtain 50 g of crude bredinin (purity 10%) as a greyish-white powder.

Example 2: Crude bredinin powder obtained in Example 1 is dissolved in a small amount of water and suspended in a solvent mixture defined hereinbelow and charged on 500 ml of silica gel (60 to 80 mesh) in a column (diameter 4.0 cm) packed with a solvent mixture of n-butanol:acetic acid:water (10:1:2), and thereafter developed with the same solvent mixture. Each 500 ml of the eluent was fractionated and active fractions were found in fractions No. 5 to 7 showing violet color. The active fractions were collected and dried in vacuo to obtain 14.3 g of dark blue violet powder (purity 26%). The powder was dissolved in 100 ml of water and saturated with hydrogen sulfide to liberate the chelated metal as a sulfide. The dried material was dissolved in 10 ml of 0.1 mol pyridine-acetic acid buffer (pH 6.0) and charged on 400 ml of DEAE-Sephadex A-25 in a column (diameter 2 cm) packed with the same buffer. Eluate was fractionated to fraction of 10 ml each and activity was found in fractions No. 70 to 135, which were collected and dried in vacuo to obtain 1.8 g of white powder (purity 90%). Recrystallization from hot methanol gave 1.1 g of the crystals of bredinin (purity 100%).

VARIOUS NUCLEOTIDES

Cyclic-3',5'-Cytidylic Acid

J. Ishiyama and T. Yokotsuka; U.S. Patent 3,926,725; December 16, 1975; assigned to Kikkoman Shoyu Co., Ltd., Japan describe a process for producing cyclic 3',5'-cytidylic acid by culturing a microorganism belonging to the genus *Corynebacterium, Arthrobacter* or *Microbacterium*.

Cyclic-3',5'-cytidylic acid (hereinafter referred to as CCMP) is a substance having the below-mentioned structural formula, and the importance thereof has recently been recognized in the field of biochemistry. CCMP is used as a reagent for hormone mediators and the like, and is quite expensive. The structural formula of CCMP is as follows.

Concrete examples of these microorganisms are *Corynebacterium murisepticum* No. 7 (ATCC 21374, FERM-P No. 206) and *Corynebacterium* No. MT-11 (ATCC 31019, FERM-P No. 2384) as the microorganisms belonging to the genus *Corynebacterium, Arthrobacter* 11 (ATCC 21375, FERM-P No. 207) and *Arthrobacter* No. MT-12 (ATCC 31020, FERM-P No. 2482) as the microorganisms belonging to the genus *Arthrobacter*, and *Microbacterium* No. 205 (ATCC 21376, FERM-P No. 106), *Microbacterium* No. 205CM7 (ATCC 21979, FERM-P No. 1557) and *Microbacterium* No. MT-3 (ATCC 21981, FERM-P No. 787) as the microorganisms belonging to the genus *Microbacterium*.

For the production of CCMP according to this process, any of the aforesaid strains is inoculated to a medium comprising natural substances or to a synthetic medium prepared by properly blending carbon sources, nitrogen sources and inorganic nutrients capable of being utilized by the strain, and if necessary, other inorganic salts and components and is cultured at a pH of 5 to 9 at a temperature of 20° to 40°C for a period until the production of CCMP becomes maximum, e.g., for about 10 to 160 hr.

The production of CCMP is further increased when the medium is incorporated with at least one member selected from the group consisting of cytosine, uracil, orotic acid, ribosides having the compounds as bases (i.e., cytidine, uridine and orotidine), and their ribotides [e.g., 2' (or 3' or 5')-cytidylic acid, cytidine-2' (or 3' or 5')-diphosphate, cytidine-2' (or 3' or 5')-triphosphate, 2' (or 3' or 5')-uridylic acid, uridine-2' (or 3' or 5')-di- or triphosphate orotidine-2' (or 3' or 5')-monophosphate, orotidine-2' (or 3' or 5')-diphosphate and orotidine-2' (or 3' or 5')-triphosphate]. Herein, these will be referred to as the cytosine and the like substances.

In case the cultivation is carried out in the presence of cytosine and the like substances, the microorganism is inoculated to, for example, such an ordinary medium as mentioned above and cultured at a pH of 5 to 9 at 20° to 40°C for a suitable period, e.g., 6 to 20 hr (this step is referred to as preculture). Subsequently, the culture liquor is incorporated with at least one member selected from the group consisting of the cytosine and the like substances, and the cultivation is further continued at a pH of 5 to 9 at a temperature of 20° to 40°C for a period until the amount of CCMP produced becomes maximum, e.g., for 4 to 140 hr. (This step is referred to as post culture.)

The amount of the cytosine and the like substances to be added to the medium is not particularly limited, but is preferably a concentration of 0.05 to 2% (w/v), in general.

When the microorganism is cultured in a medium in the presence of at least one member selected from the group consisting of fluorides, boric acid and borates, the production of CCMP is further increased. In this case, the amount of the fluorides, boric acid and borates to be added to the medium is preferably 0.1 to 500 mg/l. The fluorides, boric acid and borates may be previously added to the medium or may be added within about 30 hr after initiation of the cultivation.

In addition to the abovementioned nutrients, vitamins such as biotin, vitamin B_1 and B_2 and pantothenic acid or related compounds thereof are effectively added as a minor components.

The productivity of CCMP increases when caffeine, theophylline, theobromine or the like methylxanthine, or 2,3-, 2,4- or 2,5-pyridinedicarboxylic acid, dipicolinic acid, 8-hydroxyquinoline, polyphosphoric acid or pyrophosphoric acid, which is a growth inhibitor for cyclic 3',5'-nucleotide phosphodiesterase, is previously incorporated into the medium or added during the course of cultivation to a concentration of 500 mg/l.

The cultivation may be carried out according to a proper culture method such as shaking culture, stirring culture or aerobic culture.

The pressure of oxygen present in the culture liquor is controlled to 0.1 to 0.6 atm and that of carbon dioxide present therein to less than 0.08 atm, while at the CCMP production stage (16 to 60 hr after initiation of cultivation), the pressure of the oxygen is controlled to 0.2 to 0.8 atm and that of the carbon dioxide to less than 0.05 atm.

When the amount of CCMP produced has reached the maximum, the cultivation is discontinued, and the CCMP is recovered, if necessary, from the culture and is purified as necessary. The recovery of CCMP is carried out according to any adsorption method or precipitation method, which may be employed either singly or in combination. The adsorption method referred to herein is a method in which the culture liquor freed from the cells is contacted with such adsorbent as anion exchange resin, cation exchange resin, active carbon or alumina to adsorb the desired CCMP onto the adsorbent, thereby separating the CCMP from the culture liquor. The precipitation method is a method in which the culture liquor freed from the cells is charged with nonsolvent for CCMP such as a lower alcohol or acetone, thereby separating and recovering the CCMP from the culture liquor.

Inosinic Acid

H. Enei, H. Matsui and Y. Hirose; U.S. Patent 3,925,154; December 9, 1975; assigned to Ajinomoto Co., Inc., Japan describe the production of inosinic acid (inosine-5'-monophosphate), and more particularly a method of converting hypoxanthine or inosine to inosinic acid.

It is known from U.S. Patent 3,586,604 that *Corynebacterium sp.* ATCC 21251 (AJ 1562) converts inosine to inosinic acid in a culture medium containing phosphate ions. It has been found that certain discovered mutants of the known strain, which require at least hypoxanthine for their growth, produce higher yields of inosinic acid from inosine or hypoxanthine than the known parent strain.

The mutant strains were obtained by exposing cells of the parent strain to mutagenic agents in a conventional manner, for example, by exposing the cells of the parent strain to a solution of 250 γ/ml N-methyl-N'-nitro-N-nitrosoguanidine for 30 min at 30°C. The hypoxanthine-requiring mutants were then isolated by the replication method from the exposed culture of the parent strain. The two best inosinic-acid-producing strains found so far are *Corynebacterium sp.* AJ 3614 (FERM-P 2310) and *Corynebacterium sp.* AJ 3615 (FERM-P 2311) which are available from the Fermentation Research Institute, Japan, under the indicated accession numbers. *Corynebacterium sp.* AJ 3614, in addition to hypoxanthine, requires guanine for its growth. Strains producing high yields of inosinic acid are readily screened from others by their resistance to 8-azaguanine.

The aqueous culture media employed for producing inosinic acid, in addition to inosine or hypoxanthine and phosphate ions, contain assimilable sources of carbon and nitrogen, and the conventional inorganic ions and minor organic nutrients necessary for the growth of the microorganisms as described in more detail in the aforementioned patent which also provides detailed identifying characteristics of the microorganisms.

Hypoxanthine and inosine may be added to an otherwise conventional, inoculated culture medium either at the start of the culturing period or gradually, and it is immaterial whether these starting materials are brought into contact with the medium and the microorganisms in the form of crude crystals, in practically pure aqueous solutions, or as ingredients of a culture broth in which they were produced by other microorganisms.

Phosphoric acid and its water-soluble salts with cationic moieties nontoxic to the microorganisms may furnish the phosphate ions which should amount to 0.5 to 2.5 g/dl for best yields. The various mono-, di, and triphosphates of potassium, sodium, ammonium, and magnesium are merely typical sources of phosphate ions.

Glucose, fructose, mannose, sucrose, starch hydrolyzate, molasses, acetic acid, ethanol, fatty acids, and gluconic acid may furnish assimilable carbon, and ammonium salts, nitrates, ammonia, or urea are typical nitrogen sources. Ions of potassium, magnesium, iron, manganese and sulfate are normally needed, and minor organic nutrients include vitamins, amino acids, or may be ingredients of such ill-defined mixtures as yeast extract, bouillon, corn steep liquor, peptone, soy-protein hydrolyzate, casein hydrolyzate, and malt extract.

A pH of 5 to 8, and a temperature of 25° to 40°C should be maintained, and the highest inosinic acid concentration is reached within 2 to 5 days, depending on process variables. The inosinic acid is recovered by any one of the several known methods.

Example: *Corynebacterium sp.* AJ 3615 was cultured at 30°C for 24 hr in an aqueous seed culture medium containing 2 g/dl glucose, 1 g/dl yeast extract, 0.5 g/dl peptone, 0.5 g/dl NaCl, and 5 mg/dl hypoxanthine. 2 ml of the seed culture were inoculated on each of a series of 20 ml batches of culture medium containing 10 g/dl glucose, 2 g/dl KH_2PO_4, 1 g/dl $MgSO_4 \cdot 7H_2O$, 0.5 g/dl $(NH_4)_2PO_4$, 0.5 g/dl yeast extract, 3 ml/dl soy protein hydrolyzate, 20 γ/dl biotin, 1 mg/dl $FeSO_4 \cdot 7H_2O$, 1.0 g/dl inosine, and 5 g/dl $CaCO_3$ which has been adjusted to pH 6.5 with KOH. The 500 ml flasks containing respective batches of inoculated medium were shaken at 34°C for 72 hr when they were found to contain 1.45 g/dl inosine-5'-monophosphate.

The combined broths were filtered to remove cells, and 1 liter of the cell-free broth was adjusted to pH 1.2 with hydrochloric acid and passed over an ion exchange resin column (Diaion SK 1 in the H^+ form). The eluate was adjusted to pH 7.2 with sodium hydroxide solution, partly evaporated, and cooled to precipitate sodium inosine-5'-monophosphate in an amount of 6.8 g. The parent strain *Corynebacterium sp.* AJ 1562 produced a broth containing 0.95 g/dl inosine-5'-monophosphate under otherwise identical conditions.

5'-Guanylic Acid

In a similar process, *H. Enei, H. Matsui and Y. Hirose; U.S. Patent 3,922,193; November 25, 1975; assigned to Ajinomoto Co., Inc., Japan* have found that a mutant of the known *Corynebacterium* strain which requires hypoxanthine for its growth and which is defective in guanosine-5'-monophosphate reductase can produce 5'-guanylic acid (guanosine-5'-monophosphate) from hypoxanthine or inosine and phosphate ions in an otherwise conventional culture medium, and that the yield of 5'-guanylic acid can be very high. The best 5'-guanylic acid producers among the artificially induced mutants of *Corynebacterium sp.* AJ 1562 are resistant to 8-azaguanine, and the highest yields have been obtained from the strain AJ 3616 (FERM-P 2312). Culture conditions are the same as in the previous process.

3',5'-Cyclic Adenylic Acid

J. Ishiyama; U.S. Patent 3,904,477; September 9, 1975; assigned to Kikkoman Shoyu Co., Ltd., Japan describes a process for producing 3',5'-cyclic adenylic acid (CAMP) by culturing microorganisms.

It is well known that 3',5'-cyclic adenylic acid participates in various biochemical reactions in vivo and that it plays an active part as a mediator for various hormones. It has, therefore, been a highly valued biochemical reagent.

Typical of the microorganisms suitable for this process are *Microbacterium* MT-3 (ATCC 21981, FERM-P 787) and *Microbacterium* 205-M-32 (ATCC 31001, FERM-P 1559). These two strains are those resistant against Mn^{++}, or Fe^{+++}. The term resistant used in this process does not mean inhibition with metal ions for growing the microorganisms but for the productivity of CAMP. The *Microbacterium* MT-3 and *Microbacterium* 205-M-32 are artificial mutants obtained from *Microbacterium* 205 (ATCC 21376, FERM-P 106) as a parent strain.

CAMP can be produced by inoculating the strain capable of being employed in this process in a medium containing 0.02 mg/l or more of Mn^{++} in terms of $MnCl_2 \cdot 4H_2O$, 10 mg/l or more of Fe^{++} in terms of $FeCl_2 \cdot 7H_2O$ and/or 10 mg/l or more of Fe^{+++} in terms of $FeCl_3 \cdot 7H_2O$ as well as carbon sources, nitrogen sources assimilable to the strain, inorganic phosphates, inorganic salts other than phosphates if desired and other components in an appropriate amount and culturing until the accumulation of CAMP goes up to the maximum. It is preferred, for example, to effect the culture at a pH of 5 to 10 for 24 to 120 hr at a temperature of 20° to 40°C.

The accumulation of CAMP can be increased by adding in advance or in the course of culturing to the medium an inhibitor of cyclic-3',5'-nucleotide phosphodiesterase, such as for example, methylxanthines such as caffeine, theophylline, theobromine or the like, 2,3-, 2,4- or 2,5-pyridinedicarboxylic acid, dipicolinic acid, 8-hydroxyquinoline, polyphosphoric acid, pyrophosphoric acid and the like in the rate of 0.001 to 500 mg/l.

The culturing may be carried out by an appropriate method, e.g., with shaking, with agitation, with aeration or the like. In these cases, it is preferable to effect the stirring and the aeration under such conditions that oxygen transferring rate (Kd) and oxygen supplying rate (KGa) in the fermentor is set during a growing period (8 to 16 hr after the beginning of cultivation) to control a partial pressure of oxygen dissolved in the culture broth at 0.1 to 0.6 kg/cm^2, and a partial pressure of carbon dioxide at 0.08 kg/cm^2 or less, and during a CAMP producing period (16 to 60 hr after the beginning of cultivation), the former is controlled at 0.2 to 0.8 kg/cm^2 and the latter at 0.05 kg/cm^2 or less.

When the accumulation of CAMP attains its maximum, the culture is stopped, and then CAMP is isolated and purified. In the isolation and purification thereof, means such as treatment with active carbon, treatment with a cationic or anionic exchange resin, addition of CAMP-insoluble solvent may be properly employed in combination. For example, CAMP contained in the culture broth from which the fungal bodies have been removed is adsorbed on an active carbon, and the adsorbed CAMP is eluted with ammoniac aqueous alcohol solution, ammoniac aqueous acetone solution or the like. After the excess ammonia is removed by subjecting the eluate to concentration under reduced pressure or the like, CAMP is adsorbed on an anionic exchange resin, [e.g., Dowex I, chloride form (Dow Chemical Co.), Dowex I, formate form or the like] and then the adsorbed CAMP is eluted with an appropriate solvent [e.g., with dilute hydrochloric acid or calcium chloride + dilute hydrochloric acid system for Dowex I (chloride form) or with dilute formic acid or dilute formic acid + sodium formate system for Dowex I (formate form)]. CAMP in the eluate is again adsorbed on an active carbon and the adsorbed CAMP is eluted with ammoniac aqueous alcohol solution, ammoniac aqueous acetone solution or the like. Thereafter, the excess ammonia is removed by subjecting the eluate to concentration under a reduced pressure and CAMP is adsorbed on a cationic exchange resin [e.g., Dowex 50 (H^+ form)].

The adsorbed CAMP is eluted with dilute hydrochloric acid. CAMP can be separated in the form of crystal by concentrating the thus-obtained eluate under reduced pressure and leaving the resultant in a cold chamber or adding CAMP-insoluble solvent such as alcohol, acetone or the like to the eluate.

In a similar process, *J. Ishiyama, M. Kato, F. Yoshida and T. Yokotsuka; U.S. Patent 4,028,184; June 7, 1977; assigned to Kikkoman Shoyu Co., Ltd., Japan* describe a method of producing CAMP which is characterized by culturing in a medium containing nutrient sources such as carbon sources, nitrogen sources, inorganic salts and the like a microorganism belonging to the genus selected from the group consisting of *Corynebacterium, Arthrobacter* and *Microbacterium*.

Typical of these are, as a microorganism belonging to *Corynebacterium*, artificial mutants of *Corynebacterium murisepticum* No. 7 (ATCC 21374, FERM-P No. 206) such as for example, *Corynebacterium murisepticum* No. 7-10 (ATCC 21977, FERM-P No. 1555) or the like, as one belonging to *Arthrobacter*, artificial mutants of *Arthrobacter* 11 (ATCC 21375, FERM-P No. 207) such as, for example, *Arthrobacter* 11-211 (ATCC 21978, FERM-P No. 1556) or the like, as one belonging to *Microbacterium*, artificial mutants of *Microbacterium* No. 205 (ATCC 21376, FERM-P No. 106) such as, for example, *Microbacterium* No. 205-CM7 (ATCC 21979, FERM-P No. 1557), *Microbacterium* No. 205-CM-XA3 (ATCC 21980, FERM-P No. 1558), *Microbacterium* No. 205-MP-197 (ATCC 21976, FERM-P No. 2449) or the like.

According to the process, CAMP can be produced by inoculating the strain capable of being employed in this process in a medium containing carbon and nitrogen sources assimilable to the strain, inorganic phosphates, inorganic salts other than phosphates if necessary and other components in an appropriate amount and culturing until the accumulation of CAMP goes up to the maximum. It is preferred to culture at a pH of 5 to 9 for 24 to 80 hr at a temperature of 20° to 40°C.

Example: *Microbacterium* No. 205-CM7 (ATCC 21979, FERM-P No. 1557) was precultured in a slant culture medium composed of 0.5% of $(NH_4)_2SO_4$, 0.5% of KH_2PO_4, 0.05% of $MgSO_4 \cdot 7H_2O$, 1% of casamino acid, 0.3% of yeast extract, 1% of glucose, 2% of agar and of pH 7.0 (adjusted with 3 N KOH aqueous solution).

Separately, 30 ml each of a medium composed of 5% of glucose, 0.01% of $ZnSO_4 \cdot 7H_2O$, 0.5% of urea, 0.5% of $(NH_4)_2SO_4$, 1% of KH_2PO_4, 1% of K_2HPO_4, 0.5% of arginine, 30 γ/l of biotin, 1% of $MgSO_4 \cdot 7H_2O$ and of pH 7.5 (adjusted with 3 N KOH aqueous solution) was poured into a 500 ml flask for shaking culture and subjected to sterilization at 115°C for 10 min using an autoclave. The obtained seed culture was inoculated in the medium and cultured with shaking at 30°C for 48 hr. As a result, 1.0 mg/ml of CAMP was accumulated in the medium.

The cultured broth was centrifuged to remove the fungal bodies and the supernatant of the broth adjusted to pH 4 with 3 N HCl aqueous solution was adsorbed on an active carbon. The adsorbed CAMP was eluted with ethyl alcohol containing 0.7% of ammonia and the eluate was concentrated under reduced pressure to remove excess ammonia, and then adjusted to pH 8.0 with ammonia. The resultant eluate was passed through a column packed with Dowex I formate form having a mesh of 100 to 200 to adsorb CAMP. Subsequently, the column was washed with 0.02 N formic acid solution and the absorbed CAMP was eluted with 0.15 N formic acid solution.

The eluate was again adsorbed on an active carbon and the adsorbed CAMP was eluted with 0.7% ammonia-containing ethyl alcohol. The eluate was concentrated under reduced pressure and adjusted to pH 2.0 with hydrochloric acid. The resultant eluate was passed through a column packed with Dowex 50 hydrogen form having a mesh of 100 to 200 to adsorb CAMP and the adsorbed CAMP was eluted with 0.05 N HCl aqueous solution. The eluate was again concentrated under reduced pressure and left in a cold chamber (2° to 3°C) to obtain 600 mg of CAMP crystals out of 1,000 ml of the cultured broth.

The parent strain of the mutant, *Microbacterium* No. 205 (ATCC 21376, FERM-P No. 106) was cultured in the same manner as described above, but CAMP was hardly accumulated in the cultured broth.

NUCLEOTIDE BASES

4-Thiouracil

J.H. Coats, A. Dietz, L.A. Dolak, O.K. Sebek and W.T. Sokolski; U.S. Patent 4,010,075; March 1, 1977; assigned to The Upjohn Company describe a method of producing 4-thiouracil, which can be shown by the following structural formula.

This compound is active against a variety of bacteria, for example, *Staphylococcus aureus, Bacillus subtilis, Escherichia coli* and *Proteus vulgaris*. Accordingly, 4-thiouracil can be used in various environments to eradicate or control these microorganisms. For example, it can be used for treating breeding places of silkworms and to prevent or minimize infections caused by *Bacillus subtilis*. Since 4-thiouracil is active against *E. coli*, it can be used to reduce, arrest, and eradicate slime production in paper mill systems caused by this microorganism.

The microorganism used for the production of 4-thiouracil is *Streptomyces libani* subsp. *soldani*. A subculture of this microorganism can be obtained from the permanent collection of the Northern Regional Research Laboratory, U.S. Department of Agriculture, Peoria, Ill. Its accession number in this depository is NRRL 8173.

4-thiouracil is produced when the elaborating organism is grown in an aqueous nutrient medium under submerged aerobic conditions. It is to be understood, also, that for the preparation of limited amounts surface cultures and bottles can be employed. The organism is grown in a nutrient medium containing a carbon source, for example, an assimilable carbohydrate, and a nitrogen source, for example, an assimilable nitrogen compound or proteinaceous material. Preferred carbon sources include glucose, brown sugar, sucrose, glycerol, starch, cornstarch, lactose, dextrin, molasses, and the like.

Preferred nitrogen sources include cornsteep liquor, yeast, autolyzed brewer's yeast with milk solids, soybean meal, cottonseed meal, cornmeal, milk solids, pancreatic digest of casein, fish meal, distillers' soluble solids, animal peptone liquors, meat and bone scraps, and the like. Combinations of these carbon and nitrogen sources can be used advantageously. Trace metals, for example, zinc, magnesium, manganese, cobalt, iron, and the like, need not be added to the fermentation medium since tap water and unpurified ingredients are used as components of the medium prior to sterilization of the medium.

Production of 4-thiouracil can be effected at any temperature conducive to satisfactory growth of the microorganism, for example, between $18°$ and $40°C$, and preferably between $20°$ and $28°C$. Ordinarily, optimum production of the compound is obtained in 3 to 15 days. The medium normally remains acidic during the fermentation. The final pH is dependent, in part, on the buffers present, if any, and in part on the initial pH of the culture medium.

A variety of procedures can be employed in the isolation and purification of 4-thiouracil from fermentation beers, for example, solvent extraction, partition chromatography, silica gel chromatography, liquid-liquid distribution in a Craig apparatus, adsorption on resins, and crystallization from solvents.

Example: *Part (A) Fermentation* — One agar plug of *Streptomyces libani* subsp. *soldani,* NRRL 8173, stored above liquid nitrogen, is used to inoculate each of a series of 500 ml Erlenmyer flasks each containing 100 ml of sterile seed medium consisting of the following ingredients.

Glucose monohydrate	25 g/l
Pharmamedia*	25 g/l
Tap water qs 1 liter	

*Pharmamedia is an industrial grade of cottonseed
flour produced by Traders Oil Mill Company,
Fort Worth, Texas.

The seed medium presterilization pH is 7.2. The seed inoculum is grown for 3 days at $28°C$ on a Gump rotary shaker operating at 250 rpm and having a 2½" stroke. Seed inoculum (5%), prepared as described above, is used to inoculate a series of 500 ml fermentation flasks, each containing 100 ml of sterile fermentation medium consisting of the following ingredients.

Glycerol	10 g/l
Yeast extract	2.5 g/l
Brer Rabbit Molasses	10 g/l
Tryptone*	5 g/l
Tap water qs 1 liter	

*Difco Laboratories, Detroit, Michigan

The inoculated fermentation flasks are incubated at a temperature of 28°C for 5 days while being shaken on a Gump rotary shaker operating at 250 rpm and having a 2½" stroke. Foaming in the fermentation flasks is controlled by the synthetic defoamer Ucon (Union Carbide). A representative 5-day fermentation has the following titers of antibiotic in the fermentation broth.

Day	Assay, BU*/ml
2	6.7
3	15.0
4	10.0
5	16.0

*A biounit (BU) is defined as the concentration of the antibiotic which gives a 20 mm zone of inhibition under assay conditions.

Part (B) Recovery — Whole fermentation beer (800 ml) at pH 8.0 is filtered with the aid of diatomaceous earth as a filter aid after the adjustment to pH 7.0 with 3 N HCl. The resulting 750 ml of filtrate is extracted twice with 2 750 ml portions of ethyl acetate. The organic phases are combined, washed with a saturated sodium chloride solution, dried with magnesium sulfate, filtered and concentrated on a rotary evaporator; yield 166.0 mg of a preparation of 4-thiouracil.

Part (C) Purification — A preparation of 4-thiouracil (116 mg), prepared as described above, is dissolved in about 5 ml of ethyl acetate and the solution is injected via a syringe onto a size C Merck prepacked column of silica gel. The column is developed isocratically at 5 ml/min to tube 35 whereupon the solution is switched to 10:1 EtOAc:MeOH (v/v). The 25 ml fractions are assayed by dipping 12.7 mm pads and placing them onto agar trays seeded with *Salmonella schottmuelleri*. The trays are incubated for 18 hr at 32°C and the zones are read. Tubes 23 through 28 which give zones ranging from 24 to 38 mm are combined. Concentration of this pool gives 112.2 mg of essentially pure 4-thiouracil.

NUCLEIC ACIDS

Production by Fermentation

High-molecular weight nucleic acids have heretofore been obtained by extraction from microorganism cells as well as from higher animals or higher plants, which have been destroyed in various manners. However, the extraction and recovery procedures of these methods are more or less hazardous owing to the considerable loss of nucleic acids as well as to the presence of the residues of the destroyed cells. In contrast, according to the following process, it is not necessary to destroy cells. Furthermore, undesired high-molecular-weight compounds other than the high-molecular-weight nucleic acids can easily be removed.

*T. Suzuki, F. Tomita and S. Ito; U.S. Patent 3,880,738; April 29, 1975; assigned
to Kyowa Hakko Kogyo Kabushiki Kaisha, Japan* describe a process to produce
high-molecular-weight nucleic acids by fermentation, which can advantageously
be applied to commercial preparation because almost all of the thus-produced
high-molecular-weight nucleic acids are dissolved in aqueous solution.

All of the strains referred to in the specification by ATTC number are freely
available to the public, and are on deposit at the American Type Culture Collection. Microorganisms which may preferably be used in the process include,
for example, the following.

Arthrobacter paraffineus	ATCC 15590
Arthrobacter simplex	ATCC 15799
Brevibacterium ketoglutamicum	ATCC 15587
Micrococcus ureae	ATCC 21288
Arthrobacter hydrocarboglutamicus	ATCC 15583
Corynebacterium hydrocarboclastus	ATCC 15592
Pseudomonas fluorescens	ATCC 948
Pseudomonas aeruginosa	ATCC 15246
Aspergillus oryzae	ATCC 7252
Candida lipolitica	ATCC 8861

As apparent from the above, various microorganisms can be used regardless of
the genus and family in taxonomy. The group consisting of *Arthrobacter, Brevibacterium, Micrococcus, Corynebacterium, Pseudomonas, Aspergillus* and *Candida*
is, however, preferred.

Better results can be obtained by the use of n-paraffin and various hydrocarbon
fractions containing n-paraffin as a carbon source. Preferably such n-paraffins
contain C_{10} to C_{22}, especially C_{12} to C_{18} carbon atoms. It is possible, however,
to use various other carbon sources such as any kind of assimilable hydrocarbons
and carbohydrate, e.g., glucose, sorbitol, etc., organic acids or alcohols, preferably C_{12} to C_{20} acids or alcohols, etc., which are assimilable to the microorganism.
As the nitrogen source both inorganic and organic nitrogen sources can be used.

In carrying out the fermentation of the process, a medium containing hydrocarbon or carbohydrate as main carbon source along with the aforementioned nitrogen source is employed. Inorganic salts and required growth-promoting substances are added to the medium. The medium is inoculated and cultured under aerobic conditions, generally at 20° to 60°C and preferably at a temperature
of from 30° to 42°C. During the cultivation urea solution, ammonia water, ammonia or ammonium carbonate solution is added to give an adjusted pH of from
4 to 10, preferably 6 to 9.

The fermentation is generally completed in about 7 days, but it is preferred to
continue the fermentation until the determination of nucleic acids indicates a
maximum.

Example: *Arthrobacter simplex* (ATCC 15799) was cultured in a medium containing meat extract (1.0%), peptone (1.0%), NaCl (0.5%) and glucose (2%) and
having an adjusted pH of 7.2 (before sterilization). Culturing took place for 24
hr with shaking to obtain a seed culture. The seed culture was then used to inoculate a medium (3.0 liters) containing meat extract (1.0%), peptone (1.0%),
NaCl (0.5%) and glucose (10%) and having an adjusted pH of 7.2 (before sterilization). A 5-liter jar fermenter and a volume ratio of 10% was employed. The

cultivation was carried out at 30°C for 30 hr with agitation (400 rpm) and aeration (1 l/min). The medium was adjusted to a pH 6.5 to 7.5 with ammonia water. When the cultivation was completed, the glucose was almost consumed.

To 1 liter of the fermented liquor obtained by removing microbial cells from the fermentation broths was added about 500 ml of phenol which had been saturated by a solution containing NaCl (0.15 mol) and sodium citrate (0.01 mol). The treated liquor was subjected to shaking and centrifuged. After this, the resultant upper layer was collected.

Repeating the procedure twice, the upper layers were combined to obtain an aqueous solution, to which an equivalent of 95% ethanol (or methanol) was added. The resultant fibrous precipitates were recovered to give crude deoxyribonucleic acids. The crude deoxyribonucleic acid was purified by subjecting it twice to the ribonuclease treatment and a similar phenol treatment to give 0.7 g of purified deoxyribonucleic acid. It was found that the molecular weight of the preparation was more than 5×10^5 when determined by means of the gradient sucrose density centrifugation method using deoxyribonucleic acid of phage λ as standard [modified method of J. Marmur: *J. Mol. Biol.* 3, 208 (1961)].

After the removal of the fibrous precipitates, an equivalent of 95% ethanol was added to the fermentation liquor and the resultant precipitates were collected and combined to give crude ribonucleic acid. The crude ribonucleic acid was subjected to deoxyribonuclease treatment and phenol treatment to give 2.0 g of purified ribonucleic acid. In order to obtain soluble ribonucleic acid, NaCl (1 mol) was added to the purified ribonucleic acid and it was stirred. The supernatant was collected and twice its volume of 95% ethanol was added. The resultant precipitates were dissoved in NaCl (0.1 mol) and further precipitated with isopropanol to give 0.1 g of soluble ribonucleic acid.

ALCOHOLS

ETHANOL

Production from Cellulose in One-Step Process

Heretofore, production of alcohol (ethanol) has been attempted by a procedure comprising the steps of reacting a cellulase upon cellulose as the substrate to enzymatically saccharify the cellulose to glucose, and subsequently separately causing the resultant glucose to be reacted upon by an alcohol-producing microorganism to produce alcohol. According to this conventional method, the conversion of cellulose to glucose by a cellulase is low and, consequently, large amounts of unconverted cellulosic residue are obtained.

W.F. Gauss, S. Suzuki and M. Takagi; U.S. Patent 3,990,944; November 9, 1976; assigned to Bio Research Center Company Limited, Japan found that greater yields of alcohol can be obtained from a cellulosic material when there are simultaneously reacted under anaerobic conditions the cellulosic material, a cellulose, and an alcohol-producing microorganism.

The cellulosic substrates which are useful as starting materials include purified cellulose, agriculturally produced materials such as cotton, wood, rice straw, wheat straw, maize ears (corncobs) and other substances composed preponderantly of cellulose such as newspaper, corrugated paper, magazine paper and scrap paper.

For these substances to be used effectively as substrates for the saccharification reaction in the presence of cellulase, it is desirable to pulverize or disintegrate them. For the hydrolysis of these cellulosic substrates, use of a commercially available cellulase will suffice. An enzymatic preparation such as, for example, Cellulase Onotsuga may be used. A liquid containing a cellulase, namely a culture liquid from a cellulase-producing microorganism such as, for example, a culture liquid from *Trichoderma viride* may also be used.

As the alcohol-producing microorganism to be simultaneously used with the cellulase, there can be employed such microorganisms as, for example, *Saccharomyces cerevisiae* and *Rhizopus javanicus* which have heretofore been used for the conversion of glucose into ethanol.

In order for the cellulosic substrate to be simultaneously reacted upon by a cellulase and an alcohol-producing microorganism, an aqueous suspension containing from 1 to 30% by weight of cellulose or a substance composed predominantly of cellulose is prepared and thermally sterilized so as to serve as the substrate, a cellulase (or a cellulase-containing liquid) is added to the substrate and at the same time, an alcohol-producing microorganism cultured in advance is added thereto so that the reaction will proceed anaerobically at temperatures of from about 25° to 35°C.

When the reaction is carried out as described above, the production of alcohol in a yield approximately four times as high as the described conventional method becomes possible.

Example: A substrate was prepared by suspending 12.5 g of pulp obtained from wood (having 80% by weight of cellulose content) in 100 g of water containing in solution 0.25 g asparagine as a nitrogen source, 0.1 g potassium hydrogen phosphate (KH_2PO_4), 0.3 g magnesium sulfate ($MgSO_4 \cdot 7H_2O$) and 0.02 g yeast extract. The resulting mixture was adjusted to pH 4.0 by addition of an acetate buffer and then thermally sterilized. To the above mixture there was added 1 g of refined commercially available cellulase and two platinum loopfuls of *Saccharomyces cerevisiae* mycelium directly from an agar slant thereof, and the mixture was allowed to react at about 30°C for 96 hours. When the reaction mixture was analyzed at this point, formation of 2 g of alcohol in the substrate was confirmed. When the reaction was continued for a further 96 hours, there were eventually formed 4 g of alcohol.

In essentially the same process, *W.H. Hoge; U.S. Patent 4,009,075; February 22, 1977; assigned to Bio-Industries, Inc.* describes a method for producing ethanol which may be outlined as follows:

(1) Sterilization — Cellulose fiber-containing material, such as waste, is subjected to a steam-treatment sufficiently to eliminate unwanted bacterial strains which later may cause unwanted reactions. Thereafter, the resultant sterilized mass is adjusted to the proper solids content, temperature and pH for effective fermentation reactions;

(2) Concurrent digestion and fermentation — The cellulosic fiber mass of step (1) is reduced to fermentable sugars by inoculation and fermentation of the mass with yeast and enzymes, e.g., cellulase, to produce alcohol; and

(3) Alcohol removal — The alcohol formed by step (2) is removed as by stripping, e.g., vacuum distillation; the yeasts and enzymes are then recycled for use.

Improvement features of this process of making alcohol from cellulosic materials include, for example:

(I) The use of nonproteolytic yeasts to ferment sugars in conjunction with vacuum distillation removal of carbon dioxide and ethyl alcohol in the same reaction vat with the enzymatic hydrolysis of the cellulose to sugars, and thereby obtain the following desirable results:

(a)   Elimination of large volumes of cooling water, such
as normally required to control the temperature of
fermentation. In processes heretofore known, it is
typical to use 50 to 80 gallons of cooling water per
gallon of alcohol, whereas this process uses practically
none;

(b)   The removal of the reaction products may be carried
out in a continuous or intermittent manner in order
to force the reactions to completion. Sugar formed
by the enzymatic breakdown of the cellulose is immediately removed when it is fermented by the yeasts.
The alcohol formed by the fermentation of the sugars
is likewise continuously removed by the vacuum distillation.

(c)   The bottoms or solution which remains after the sugar
and alcohol production is completed will still retain
active cellulase enzyme and yeasts, and this inoculum
solution, therefore, may be recycled to dilute new quantities of incoming sterilized cellulosic fibers.

(II) In the embodiment wherein mineral acid is used in the sterilizer in combination with enzymatic hydrolysis in the later fermentation step, a reduction in the reaction time and the sizes of the tanks or vats is realized
over the nonacid hydrolysis embodiment.

(III) By combining the sterilization with partial acid hydrolysis the maximum
benefit with respect to the cost of this treatment step is obtained.

Example 1: Approximately 100 g (dry weight) of fibrous cellulosic material,
such as recovered from a municipal waste slurry after being dewatered in a filter
press, and which resultant mass comprises a total solids content of about 38%
and less than 20% of nonfibrous material, is subjected to a sterilizing treatment.
To accomplish this, the fibrous material is placed in a 1 liter stainless steel pressure vessel or container fitted with a pressure gauge and steam inlet line for the
introduction of high-pressure steam. Noncondensable gasses, e.g., air, etc., first
are removed from the fibrous material being treated by heating the container,
e.g., in an oil bath to a temperature of about 90°C for 10 to 15 minutes while
the steam inlet line remains open to permit the gasses to escape.

High-pressure saturated steam is then injected into the container until the pressure gauge indicates about 140 psi. This steam pressure is maintained for approximately 15 minutes after which the steam line is shut off and the container
is allowed to cool to room temperature. To shorten the cooling time, the container is immersed in a stream of cold water.

Following sterilization, the fibrous mass of material is transferred from the container to a jacketed vacuum vessel equipped with a 40 rpm stirrer, the vessel being connected to a pump to maintain a vacuum of about 26" of Hg in the vessel.
The sterilized fibrous mass is diluted to a total volume of approximately 2 liters
by the addition of 400 ml of inoculum solution, as hereinafter described, and
1.2 liters of sterile water. The inoculum solution consists of two primary components, although minor amounts of other compatible substances which do not
prevent microbial growth may be present. The primary component is cellulase
enzyme culture broth as prepared by the action of *Trichoderma viride* on cellulose as described, for example, in U.S. Patent 3,764,475. The second compo-

nent of the inoculum solution is a nonproteolytic yeast which converts simple sugars to ethanol. An example of such yeast is *Saccharomyces cerevisiae* var. *ellipsoideus.* The culture which contains 0.3 to 1.0 mg of protein/ml is adjusted to a pH of 4.5 by the addition of oleic acid and other nutrients commonly used in such yeast fermentations. (A pH range of 3.0 to 5.0 is permissible.)

The allowable temperature in the jacketed vessel which the cellulase can tolerate is 55°C without damage, whereas yeasts have maximum temperatures of 30° to 45°C. *Saccharomyces cerevisiae* can withstand only about 30°C. Accordingly, the inoculated fibrous cellulose material in the jacketed vessel is maintained at a temperature of 30°C under mild agitation. The reaction is permitted to proceed for about eight hours with periodic application of vacuum as needed to avoid the temperatures above about 30°C which might kill the yeast. After about 16 hours, the reaction is essentially completed as judged by the fact that it is no longer necessary to apply vacuum to prevent the temperature from rising above 30°C. During the reaction the vapor phase material is removed from the jacketed vessel and condensed to form a liquid condensate which is distilled to recover a yield of about 25 ml of 95% ethyl alcohol.

The removal of the alcohol at this relatively low temperature permits recycling and reuse of the nonvolatile portion of the reaction mixture to replace dilution water and to reduce the amount of inoculum needed for the new batch. This nonvolatile portion contains yeast, cellulase enzyme and unreacted materials. Where desired, the culture or nonvolatile portion of the reaction mixture may be concentrated by ultrafiltration as commonly practiced.

The process may be carried out as a continuous or intermittent procedure, products formed throughout the process being removed to force the reaction to completion. Likewise, the alcohol produced during the fermentation is removed, continuously or intermittently by the vacuum treatment.

Example 2: The process as described in Example 1 is repeated using a sterilizing cellulosic mixture consisting of 100 g of dry fiber and 163 g of water and 5 ml of 10% sulfuric acid. The resulting slurry mixture, after sterilization, is treated with slaked lime to raise the pH to 4.5. The resultant mixture forms a more fluid mass than the sterilized mixture of Example 1, and is diluted with 0.6 liter of water and 0.4 liter of inoculum solution as described in Example 1. This provides a total volume of 1 liter, which mixture is easily stirred with the 40 rpm agitator. The presence of the acid in the sterilization treatment serves to permit a higher solids content of the fibrous slurry mass and this increases the enzyme concentration during the fermentation step which results in increasing the productivity of the fermentation.

Thermophilic Mixed Culture

W.D. Bellamy; U.S. Patent 4,094,742; June 13, 1978; assigned to General Electric Company describes a process for producing ethanol which comprises admixing cellulose or a cellulosic material with nutrient mineral broth to form a suspension having a pH ranging from about 7 to 8, admixing with the suspension a mixed culture of thermophilic cellulolytic sporocytophaga and thermophilic ethanol-producing bacillus, and fermenting the resulting mixture at a pH ranging from about 7 to 8 and at a temperature ranging from about 50° to 65°C to produce ethanol in an amount of at least about 10% by weight of the cellulose present in the fermentor.

The source of cellulose, i.e., natural or waste, need only be treated, if necessary, to make its cellulose component available at least in a major amount, preferably higher than 60% by weight of the cellulose present, for contact with the mixed culture. Such treatment may be physical or chemical.

According to one method the lignocellulosic material is comminuted to a fine particle size, preferably less than 100 microns, to expose and make available at least a major portion of the cellulose component for contact with the present mixed culture during fermentation. Comminution can be carried out by conventional techniques such as by mechanical grinding or pulverizing the lignocellulosic material.

Alternatively, to free the cellulose component, the lignocellulosic material can be treated with acid or alkali metal hydroxide which penetrates the lignin and degrades or depolymerizes it sufficiently to make the cellulose available for contact with the present mixed culture.

Representative species which can be utilized include *Bacillus stereothermophilus* and *Bacillus coagulans.* Cultures of these bacilli are contained in the American Type Culture Collection in Washington, D.C. and in other repositories. It is believed that the present thermophilic cellulolytic gram-positive bacillus is a *Bacillus stereothermophilus* which grows slowly at 65°C.

The present mixed culture can be formed in a conventional manner, i.e., the thermophilic cellulolytic sporocytophaga and ethanol-producing bacillus can be grown together submerged in a standard nutrient solution at temperatures ranging from about 50° to about 65°C.

The present mixed culture comprised of thermophilic celluloytic Gram-negative sporocytophaga designated US and ethanol-producing Gram-positive thermophilic bacillus designated OK was deposited with the U.S. Department of Agriculture. This mixed culture is identified as NRRL B-11077.

Example: In this example the mixed culture assigned the number NRRL B-11077 and comprised of thermophilic cellulolytic sporocytophaga US and thermophilic cellulolytic ethanol-producing bacillus OK was used. The nutrient mineral broth used contained $(NH_4)_2SO_4$, 5.0 g; NaCl, 1 g; $MgSO_4 \cdot 7H_2O$, 0.2 g; $ZnSO_4 \cdot 7H_2O$, 0.008 g; $FeSO_4 \cdot 7H_2O$, 0.2 g; $MnSO_4 \cdot 4H_2O$, 0.02 g; $CaCl_2$, 0.02 g; Versenol, 0.2 g; and 0.067 M phosphate buffer per liter of distilled water and was adjusted to pH 7.5.

The apparatus used was comprised of a one liter fermentation vessel associated with a 500 ml water trap, plus necessary connectors, and condensers. The fermentor vessel had a pH probe and an inlet for either carbon dioxide gas or ammonia. The water trap had a vacuum gauge, and the condenser in the alcohol trap operated on ground water at 15° to 18°C. Both the fermentor and the water trap were maintained at desired temperature by thermostated Glascol heaters. During operation, the pH and the partial vacuum in the fermentor vessel were maintained at desired values by manual operations. The pH of the fermenting mixture was maintained at 7.5 by periodic addition of ammonia as ammonium hydroxide because organic acids were produced during growth. Carbon dioxide gas from an associated lecture cylinder was periodically bled into the atmosphere of the fermentor slowly to help sweep out the ethanol vapor from the fermentor atmosphere and to replace the carbon dioxide removed by the partial vacuum.

A suspension of 500 ml of the mineral broth plus 0.5 g yeast extract and 5 g of Whatman #1 filter paper which had a particle size ranging from about 100 to 1,000 microns was inoculated with the present mixed culture (NRRL B-11077) at room temperature. The filter paper consists of cellulose only. Although yeast extract was included in the mineral broth, it is not a required nutrient for growth of this mixed culture, and specifically it is not required for growth of the ethanol-producing bacillus in the present mixed culture since the sporocytophaga provides the required nutrients supplied by the yeast extract.

The resulting fermentation mass was maintained in the closed fermentation vessel at 55°C at a pH ranging from 7 to 8 at atmospheric pressure for 12 hours. At the end of that time, there was heavy growth of the mixed culture in the fermentation vessel. The atmospheric pressure was then gradually lowered to 153 mm Hg. The fermentation vessel was maintained at 55°C, the water trap at 40°C, and the ethanol trap was at room temperature which was about 25°C while the condenser water was 15° to 18°C. Under these conditions, 10 ml of 50% ethanol was collected in the trap during the first two hours, and approximately 2 ml of 50% ethanol per hour was collected for the next 7 hours. It was necessary to return the water from the water trap to the fermentation vessel to make up the water lost at the end of the two hours and five hours. The yield of ethanol based on the amount of cellulose was about 40% of theoretical.

OTHER ALCOHOLS

D-Mannitol from D-Glucose

M. Takemura, M. Iijima, Y. Tateno, Y. Osada, and H. Maruyama; U.S. Patent 4,083,881; April 11, 1978; assigned to Towa Kasei Kogyo Co., Ltd.; Japan and W.M. Kruse; U.S. Patent 4,173,514; November 6, 1979; assigned to ICI Americas Inc. describe a process for preparing D-mannitol, which comprises isomerization(s) of D-glucose to convert a portion of the D-glucose into D-mannose alone or into D-mannose and D-fructose, followed by catalytic hydrogenation of the resulting mixture of sugars under high hydrogen pressure.

According to the first step (i.e., epimerizing reaction of D-glucose) of the process, a mixture of D-glucose aqueous solution and molybdic acid compound is adjusted to pH 2.0 to 4.5, preferably to 3.0 to 4.0, and heated at a temperature of 110° to 160°C, thereby the reaction is performed within a short time, e.g., about 40 minutes, at high formation ratio of D-mannose based on the D-glucose material used. The reaction is further performed effectively even when a high concentration solution of D-glucose (e.g., 40 to 80%) and a small amount of molybdic acid compound are employed.

Molybdic acid compounds used for the epimerization reaction of this process include ammonium molybdate, sodium molybdate, potassium molybdate, molybdenum trioxide, molybdic acid, and the like. Favorable amounts of the molybdic acid compound are however, only 0.05 to 0.20% relative to the initial D-glucose which corresponds to about $\frac{1}{20}$ to $\frac{1}{5}$ the amount of molybdic acid used for the conventional process.

According to the second step of the process, the epimerized mixture or its refined solution is subjected to the catalytic hydrogenation reaction under high
pressure. For example, the refined solution is adjusted to have 50 to 70% dried
matter. Added thereto is a hydrogenating catalyst such as Raney nickel or a platinum group catalyst (e.g., platinum, ruthenium or palladium) and heated, while
agitated, at a temperature of 110° to 160°C for 1.0 to 2.5 hours under a hydrogen pressure of 100 to 200 kg/cm^2. Dried matter composition of the finished
solution of the hydrogenating reaction (or hydrogenated mixture) thus obtained
is, when crystalline D-glucose has been employed as the material for the epimerization reaction, for instance, 70 to 64% D-sorbitol and 30 to 36% D-mannitol.

The enzymatic isomerization reaction of the process is, for instance carried out
in such a manner that the epimerized mixture is adjusted to pH 6.5 to 8.5 and
to have 50 to 65% dried matter. Added thereto is a predetermined amount of
glucose-isomerase followed by warming the resultant aqueous mixture at a temperature of 65° to 75°C for 48 to 72 hours, while agitating gently and maintaining the pH at 6.5 to 8.5. The finished solution of the isomerization reaction
(or isomerized mixture) is, for example, filtered to separate the enzyme, treated
with active carbon and further with ion exchange resins to give a colorless,
transparent solution. Refined solution obtained exhibits pH 5.0 to 7.0 and its
dried matter composition is, when crystalline D-glucose has been employed as
the material for the initial epimerization reaction, for instance, 39 to 35% D-
glucose, 31 to 29% D-fructose and 30 to 36% D-mannose.

The isomerization reaction of the process may also be conducted by using
the so-called immobilized enzyme system. For instance, glucose-isomerase is adsorbed onto a suitable carried, with which is filled a column having a predetermined
length, and the epimerized mixture is made to flow continuously through the
column at a suitable rate, thereby the remaining D-glucose in the epimerized mixture is isomerized enzymatically while having been retained in the column.

As the glucose-isomerase to be used for the enzymatic isomerization reaction
of the process, are available all the enzyme preparations of this kind on the
market, which have been prepared from the cultured materials of the microorganisms belonging to genus *Streptomyces, Pseudomonas, Aerobacter,* and
the like. With an increase in the amount of glucose-isomerase used, the reaction velocity of the isomerization becomes greater and the formation ratio
of D-fructose is improved.

For the industrial application, however, it is preferable to use 0.5 to 2.0%, relative
to the dried matter of the D-glucose material for the epimerization reaction, of
an enzyme preparation having about 1,000 GIU/g of enzyme activity, if 1 GIU
is defined to represent such an enzyme activity that produces 1 mg of D-fructose
at a temperature of 70°C in 60 minutes.

The refined solution of the isomerized mixture can be subjected to the catalytic
hydrogenation in the similar manner to that applied to the refined solution of the
epimerized mixture described previously.

A flowsheet of the reaction steps of the process using crystalline D-glucose as the
material is illustrated in the figure on the following page.

Figure 4.1: Process Flowsheet

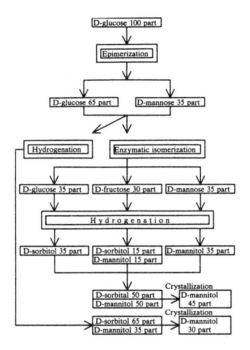

Source: U.S. Patent 4,083,881

4-Aminobenzyl Alcohol Derivatives

Well known as schistosomacidal agents are a wide variety of 4-[amino-(polycar-bon-lower alkyl)-amino] -2-halotoluenes which differ from 4-[amino-(polycarbon-lower alkyl)-amino] -2-halobenzyl alcohols or ethers in having methyl instead of hydroxymethyl or alkoxymethyl as a substituent on the benzene ring ortho to halo and para to the aminoalkylamino substituent.

All heretofore known compounds having high schistosomacidal activity required the presence of the methyl substituent on the benzene ring, as well as the halo and amino-(polycarbon-lower alkyl)-amino substituents at ortho and para posi-tions, respectively. Workers in the field have reported that the methyl group cannot be replaced by another radical without losing the schistosomacidal activ-ity.

It has been found that compounds of this type where the ring methyl substituent is replaced by hydroxymethyl, i.e., 4-[amino-(polycarbon-lower alkyl)-amino] -2-halobenzyl alcohols, or by lower alkoxymethyl not only have high schistoso-macidal activity but also are more active as schistosomacidal agents in hamsters and less toxic than the corresponding methyl compounds.

S. Archer and D. Rosi; U.S. Patent 3,917,703; November 4, 1975; assigned to Sterling Drug Inc. describe a process for producing a 4-[amino-(polycarbon-lower alkyl)-amino]-2-halobenzyl alcohol which comprises subjecting a 4-[amino-(polycarbon-lower alkyl)-amino]-2-halotoluene to the fermentative enzymatic action of an organism capable of effecting oxidation of the 1-methyl group to 1-hyhydroxymethyl. The organism is classified in the orders *Moniliales, Mucorales, Sphaeriaeles, Sphaeropsidales, Melanconiales* and *Actinomycetales.*

These compounds are those of the following general formula:

$$CH_2OR_6$$

R_5 ──X ──R_3 NB

where X is halo, i.e., chloro, bromo, iodo or fluoro; R_3, R_5 and R_6 are each hydrogen or lower alkyl; NB is $N(R)-Y-NR_1R_2$ or

$$N \overset{CH_2CH_2}{\underset{CH_2CH_2}{\diagup\diagdown}} N-R_4;$$

R is hydrogen or lower alkyl; Y is polycarbon-lower alkylene; R_1 and R_2 are each hydrogen or lower alkyl, lower alkenyl or lower hydroxyalkyl and can be the same or different; R_1 and R_2 taken with N also comprehend saturated N-heteromonocyclic radicals having from five to six ring atoms, illustrated by piperidino, pyrrolidino, morpholino, piperazino, hexamethyleneimino and lower alkylated derivatives thereof; and R_4 is hydrogen, lower alkyl, lower alkenyl, lower hydroxyalkyl, carbamyl, thiocarbamyl, lower alkanoyl, lower carbalkoxy, carboxy-(lower alkanoyl), carboxy-(lower alkenoyl), phenyl-O-(lower alkyl), or phenyl-(lower alkyl).

It will be understood the benzene ring of phenyl can bear any number and kind of substituents such as would occur to the person skillled in organic chemistry, e.g., such substituents, solely for illustration and without limitation, including lower alkyl, lower alkoxy, halo (chloro, bromo, iodo or fluoro), nitro, lower-alkylmercapto, lower alkanoylamino, lower alkanoyloxy, lower alkylamino, lower alkenyl, and the like.

Example: Fermentative enzymatic oxidation of 2-chloro-4-(1-piperazinyl)toluene to 2-chloro-4-(1-piperazinyl)-benzyl alcohol was accomplished as follows: Four 10 liter fermentations employing *Aspergillus sclerotiorum* (SWRI A_{24}, Sterling-Winthrop) in sterile soy-dextrose medium of the following composition were carried out.

Dextrose, g	1,000
Soybean meal, g	150
Yeast, g	50
NaCl, g	50
$MgSO_4 \cdot 7H_2O$, g	2.5
$NaH_2PO_4 \cdot H_2O$, g	13.8

(continued)

$Na_2HPO_4 \cdot 12H_2O$, g	301
Tap water, liters	9.5
pH adjusted to 7.3 with 10 N HCl; autoclave at 120°C and 15 psi for 15 minutes	

The stock culture was initially grown at 26°C on slants in a nutrient medium
(e.g., maltose, 40 g/l, and Proteose Peptone No. 3, 10 g/l) in 22 x 175 mm tubes
for 10 to 14 days. These slants were used to prepare seed for the 10 liter fermenters as follows: Sterile distilled water (10 ml) was added to a slant and the
spores and some vegetation growth were scraped with a sterile hooked needle.
The resulting suspension was added to a 2 liter flask containing 700 ml of sterile
soy-dextrose medium of the following composition:

Soybean meal, g	15
Dextrose, g	20
Yeast, g	5
NaCl, g	5
K_2HPO_4, g	5
Tap water, liter	1
pH adjusted to 6.4 with 10 N HCl prior to autoclaving at 121°C and 15 psi for 15 min- utes	

These cotton-plugged flasks were incubated at 26°C for approximately 70 hours
on a rotary shaker with a 1" throw at 240 rpm. Each flask provided seed for
one 10 liter fermenter. The fermentations were carried out at 28°C in a water
bath with an air flow of 5 liters per minute and were agitated at 450 rpm. After
an initial 24 hours' growth, the substrate, 2-chloro-4-(1-piperazinyl)toluene as
its hydrochloride salt, 212 g (53 g to each fermenter), was added in portions
(5 to 10 g) over a period of 5 days. The tanks were assayed for 2-chloro-4-(1-
piperazinyl)benzyl alcohol content periodically as follows.

Samples from each fermenter were made basic with 0.2 ml of 10 N sodium hydroxide solution and extracted with methylene dichloride. The extracts were
evaporated to dryness and the residues dissolved in 3 ml of methylene dichloride.
Aliquots (20 μl) were transferred to thin-layer silica gel plates impregnated with
1% of a phosphor (e.g., Radelin GS-115). The plates were developed in a solvent
containing 8:1:1 parts by volume of $CH_2Cl_2 : CH_3OH : N(C_2H_5)_3$. The components
were viewed as blue spots on a yellow fluoroescent background. Under these
conditions, the R_f value of the intermediate toluene derivative was 0.46 and
that of the corresponding benzyl alcohol was 0.36.

Isolation — The fermentations were terminated by the addition of 130 ml of
10 N NaOH and each tank was extracted with methylene dichloride (2 x 20 liters).
The extracts were reduced under vacuum and combined. Further reduction in
volume resulted in a yellow crystalline precipitate weighing 150 g. Thin layer
chromatography of this material showed traces of 2-chloro-4-(1-piperazinyl)tol-
uene with the major component chromatographically identical with a sample
of 2-chloro-4-(1-piperazinyl)benzyl alcohol prepared by reduction of the corresponding 2-chloro-4-(1-piperazinyl)benzaldehyde.

Purification — The material was recrystallized from ethyl acetate (about 800 ml).
An insoluble fraction (4.5 g) was filtered off; this material was probably a salt

of the product since in the above basic TLC system it behaved like 2-chloro-4-(1-piperazinyl)benzyl alcohol. On cooling the filtrate in an ice bath, 120 g of cream colored crystalline 2-chloro-4-(1-piperazinyl)benzyl alcohol, MP 120° to 122°C (corr.), was obtained. A second crop of 23.5 g was obtained from the mother liquor. Analysis calculated for $C_{11}H_{15}ClN_2O$: N, 12.36; Cl, 15.64. Found: N, 12.43; Cl, 15.44. Infrared, ultraviolet and nuclear magnetic resonance spectral data of the above product show it to be identical with the product obtained chemically by reduction of the corresponding 2-chloro-4-(1-piperazinyl)benzaldehyde.

STEROLS

Ergosterol and Its Esters

Y. Nakao, M. Kuno, and M. Suzuki; U.S. Patent 3,884,759; May 20, 1975; assigned to Takeda Chemical Industries, Ltd., Japan describe a method for the production of ergosterol and its esters by cultivating a microorganism capable of accumulating ergosterol or its esters extracellularly belonging to the genus *Trichoderma,* the genus *Fusarium* or the genus *Cephalosporium.* Ergosterol and its esters have been important as a starting compound for the synthesis of vitamin D and as a precursor for the preparation of synthetic steroid hormones.

The typical examples of the microorganisms used in this process are *Trichoderma sp.* IFO 6355, *Trichoderma viride* IFO 9066, *Fusarium sp.* S-19-5 IFO 8884 (ATCC 20192), *Fusarium oxysporum* IFO 4471, *Fusarium roseum* IFO 7189, *Cephalosporium coremioides* IFO 8579, and *Cephalosporium sclerotigenum* IFO 8385. Their mutants and variants, which have an ability of accumulating ergosterol or its esters are induced by treating the abovementioned microorganism with irradiation of ultraviolet rays, x-rays or γ-rays or with a chemical agent such as sodium nitrite, N-methyl-N'-nitro-N-nitrosoguanidine.

Various fatty acid esters of ergosterol may be produced solely or concomitantly with free ergosterol through the cultivation of the microorganism used. These fatty acid esters are exemplified by those with fatty acids having from 2 to 19 carbon atoms. The typical examples of those are lower fatty acids such as acetic acid and higher fatty acids having 14 to 18 carbon atoms such as palmitic acid, stearic acid and oleic acid. By the addition of the preferable fatty acid or its esters to the culture broth, various fatty acid esters of ergosterol can be prepared.

The culture medium contains assimilable carbon sources, digestible nitrogen sources, inorganic salts and growth factors.

The conditions of cultivation, i.e., pH of the medium, cultivation temperature and cultivation period, are adequately employed for the maximum accumulation of ergosterol or its esters by the microorganism. For example, in case of shaking or submerged culture, the cultivation is advantageously carried out at a temperature ranging from 15° to 45°C, preferably 20° to 38°C, at a pH of 2 to 9, preferably 4 to 8 and for 10 to 360 hours, preferably 72 to 192 hours.

The separation and purification of the thus-accumulated ergosterol and its esters are carried out in a per se conventional manner. For example, the culture broth is extracted with an organic solvent such as n-hexane, cyclohexane, benzene, ethyl acetate, butyl acetate or chloroform. The extraction may be carried out after eliminating the mycelia, if desired. The products are transferred to the organic solvent by the extraction procedure, and then the organic solvent layer is collected and concentrated under reduced pressure, whereby a residue of brown syrup is obtained. The residue is dissolved in chloroform and the solution is passed over a column filled with silica gel or alumina by using chloroform, petroleum ether or petroleum benzine as a developer to elute. Portions which show a strong absorption at 283 mμ are collected.

The thus-collected portions are dissolved in a solvent such as hexane or ethanol and kept standing in a cold room, whereby the desired product is obtained as crystals.

Example: A culture of *Fusarium sp.* S-19-5 (IFO 8884, ATCC 20192) is inoculated on a slant (pH 6.5) containing 1% of malt extract, 1% of glucose, 1% of Polypeptone and 2% of agar and cultivated for 168 hours.

50 parts by volume of a seed culture medium (pH 6.5) containing 2% of glucose, 2% of soybean powder, 0.1% of KH_2PO_4 and 0.5% of soybean oil is poured into a fermenter of 300 parts by volume in capacity. After sterilization at 120°C for 20 minutes, the medium is inoculated with mycelia growing as the slant and is then kept at 24°C for 48 hours to prepare a seed culture.

1,000 parts by volume of a fermentation medium (pH 6.5) containing 13% of n-paraffin (a mixture of decane, undecane, dodecane and tridecane), 7% of Polypeptone S (made by Daigo Nutritive Chemicals, Ltd.), 0.8% of KH_2PO_4, 0.1% of $FeSO_4 \cdot 7H_2O$, 0.05% of $MgSO_4 \cdot 7H_2O$, 0.01% of $CaCl_2 \cdot 2H_2O$, 0.5% of soybean oil, 0.5% of Tween 60 and 1% of $CaCO_3$ is poured into a fermenter of 3,000 parts by volume in capacity. After sterilization at 120°C for 20 minutes, the medium is inoculated with 50 parts by volume of the seed culture and is then cultivated at 24°C for 90 hours.

The thus-obtained culture broth is subjected to centrifugation to eliminate the mycelia. The culture liquid is extracted with one-half volume of cyclohexane three times. The cyclohexane layers are collected and dried by treating with sodium sulfate and then concentrated under reduced pressure. The thus-obtained concentrate is dissolved in chloroform. The chloroform solution is passed over a column packed with silica gel (Merck & Co., Inc.) for chromatography using petroleum ether as a developer. Among the petroleum ether fractions, those showing a strong absorption at 283 mμ, are collected and concentrated. The residue is dissolved in ethanol and kept standing in a cold room for 2 days. The precipitating crude colorless fine needles are recovered by filtration and recrystallized from ethanol, whereby 0.650 part by weight of ergosteryl palmitate is obtained as colorless plates.

The mother liquor obtained by eliminating the crystals of ergosteryl palmitate by filtration is concentrated and dissolved in chloroform. The chloroform solution is passed over a column packed with silica gel using chloroform to elute. The portions following the elution of ergosteryl palmitate are collected and concentrated, whereby 0.2 part by weight of ergosterol is obtained as yellow powder. Physiochemical properties of the product are virtually identical with those of an authentic sample of ergosterol.

Cholesterol

C.H. Krauch, F. Hill, R. Lehmann, H. Pfeiffer and J. Schindler; U.S. Patent 3,919,045; November 11, 1975; assigned to Henkel & Cie, GmbH, Germany describe a process for the production of a fraction enriched in cholesterol and cholesterol esters from the residues of the processing of fats, in which the residues in emulsified form in an aqueous culture medium are broken down with a microorganism of the species *Nocardia paraffinica, Nocardia salmonicolor, Nocardia opaca, Candida lipolytica,* or *Corynebacterium petrophilum* at a temperature of 25° to 55°C and a pH value of 4.0 to 8.0, and the enriched sterols obtained are then separated from the culture solution by solvent extraction.

Residues from the processing of fats of animal origin, especially residues from the distillation of fatty acids after cleavage of fats are mostly used for the process. These fatty residues consist essentially of fatty acids, fatty acid mono- and diglycerides, polymeric products of fats and fatty acids, oxidation products of fats and fatty acids, and combustion products of fats and fatty acids, and tarry constituents. These fats, fatty acids and fatty residues are individual substances or mixtures thereof and generally are higher fatty substances.

In order to stimulate the microorganism growth, it is advisable to provide an additional metabolizable source of carbon, for example, paraffins, glycerin, lower carboxylic acids, starch, dextrin, sucrose, glucose, fructose, maltose and sugar-containing waste materials. Suitable sources of nitrogen or growth substances are ammonium salts, nitrates, urea, peptones, corn steep liquor, soybean meal or cake, distiller's mash and fish meal.

Further, inorganic salts, for example, alkali metal salts of hydrogen phosphates such as sodium, potassium or ammonium hydrogen phosphate, or the alkaline earth metal salts such as calcium salt or magnesium salt, the manganese salt and the iron salts, as well as fermentation accelerators, such as yeast extracts, meat extracts and vitamins are suitably added to the nutrient medium.

The residues to be broken down are added to the nutrient medium in emulsified form in a concentration of about 1 to 20% by weight, preferably 2 to 10% by weight. If the residue is used as the only source of carbon, this is added at the beginning of the culture process and, if necessary, is further continuously added during the period of microorganism growth. It is also possible to cultivate the cultures first of all on the usual substrates, for example, sugar-like substances or other carbon sources, so that the highest possible cell densities are obtained, and then to insert the residue to be decomposed, whereby higher concentrations may be obtained.

The emulsification of the fatty residues in the culture solution is carried out by means of known emulsifiers, in particular, nonionic emulsifiers, such as, for example, fatty acid sorbitol esters and their adducts with 20 to 200 mols of ethylene oxide, preferably adducts with 20 to 80 mols of ethylene oxide.

Before starting the cultivation, the culture medium used is suitably sterilized by heating. The incubation temperature is preferably 27° to 39°C. The pH value of the nutrient solutions is preferably 5.0 to 7.0. The culturing procedure is chiefly carried out under aerobic conditions by shaking or stirring and aerating, and generally requires a period of 1 to 5 days.

After the microbial decomposition of the fatty residues is finished, the whole culture solution is extracted with a lipophilic solvent, without the cells having to be previously separated. The extraction is preferably carried out in the culture vessel. Hydrocarbons such as saturated hydrocarbons having 4 to 10 carbon atoms, for example, hexane, cyclohexane, or benzine mixtures, or alcohols such as lower alkanols, for example, butanol or hexanol, or ketones such as lower alkanones, for example, methyl isobutyl ketone, or ethers such as di- lower alkyl ethers, for example, diethyl ether, or chlorinated hydrocarbons such as chlorinated lower alkanols, for example, chloroform or other usual solvents are examples of suitable extraction agents.

After removal of the solvent, a residue enriched with cholesterol or cholesterol esters remains. This residue may either be further processed as such, or it may be worked up for the production of the pure sterols, for example by recrystallization or chromatography.

Example: *Nocardia paraffinica* ATCC 21198 was first cultivated at 30°C for 24 hours with shaking in a culture medium which contains 1.56% of special peptone, 0.28% of yeast extract, 0.56% of NaCl and 0.1% of D(+)-glucose and was adjusted to a pH value of 6.8. 10 ml of this starter culture was then injected into 100 ml of the main culture and incubated at 30°C in a 500 ml Erlenmeyer flask on a shaking machine. The composition of the nutrient medium of the main culture was as follows: 0.05% $NaH_2PO_4 \cdot 2H_2O$, 0.18% K_2HPO_4, 0.06% NH_4NO_3, 0.06% $MgSO_4 \cdot 7H_2O$, 0.02% $MnCl_2 \cdot 4H_2O$, 0.01% $FeSO_4 \cdot 7H_2O$, 0.01% $CaCl_2 \cdot 2H_2O$, 0.50% corn steep liquor, 0.50% soy meal, 0.05% Tween 80, and 2% fatty acid distillation residues.

The pH value of the nutrient solution was adjusted to 6.5. The cultures were harvested after 72 hours incubation. The whole culture was extracted with chloroform. After distilling off the solvent, a faintly yellowish residue remained which consisted of 80 to 95% by weight of cholesterol ester.

Similar results were obtained when the decomposition was carried out with the species *Nocardia salmonicolor* ATCC 19149 or *Nocardia opaca* DSM 363 (German collection of microorganisms, Inst. für Mikrobiologie, Göttingen) or *Candida lipolytica* DSM.

CARBOHYDRATES

DEXTROSE

Mixed Immobilized Enzymes

K.N. Thompson, R.A. Johnson and N.E. Lloyd; U.S. Patent 4,011,137; March 8, 1977; assigned to Standard Brands Incorporated describe a process for converting starch to dextrose by the use of an enzyme system comprising immobilized glucoamylase and α-amylase selected from the group consisting of soluble α-amylase, immobilized α-amylase and mixtures thereof.

In this process, the partially hydrolyzed starch may be prepared either by an enzyme or acid treatment. In the case of enzyme treatment, the partially hydrolyzed starch should have a DE (dextrose equivalent) in the range of from 10 to 60. When a partial acid hydrolysate is used in this process, the DE thereof should be in the range of from 10 to 30. At higher DE, substantial amounts of reversion products are present which are not acted upon by this enzyme system. The pH of the partial hydrolysate being treated may be from 3.5 to 6.5 and preferably will be from 4 to 6. The most preferred temperature range is 50° to 60°C.

This process may be performed by a number of techniques. For instance, soluble or immobilized α-amylase and immobilized glucoamylase may be used concurrently or sequentially. It is preferred that they be used concurrently as, for example, when partially hydrolyzed starch is contacted with a mixture of immobilized glucoamylase and immobilized α-amylase. Of course, it will be realized that the α-amylase and glucoamylase may be immobilized on or within the same carrier and results will be obtained which are substantially equivalent to those given by mixtures of α-amylase and glucoamylase immobilized on separate carriers.

In the case where the enzymes are used sequentially, the conversion process will comprise at least three steps in the following sequence: (1) contacting the partial hydrolysate with immobilized glucoamylase; (2) contacting the resulting hydrolysate with a soluble or immobilized α-amylase; and (3) contacting the resulting hydrolysate with immobilized glucoamylase. The last two steps of the sequence may be repeated a number of times depending on the conditions under which the reactions are conducted. The concurrent use of the enzymes results in greater amounts of the partially hydrolyzed starch being converted to dextrose than does sequential use except when the steps employed in sequential use are repeated a large number of times.

The preferred method of preparing the immobilized α-amylase for use in this process is by covalently bonding the α-amylase to carriers such as cellulose, porous ceramic, macroporous synthetic resins, crosslinked dextran and similar materials. The glucoamylase may be immobilized by any of the techniques known in the art, although it is preferred to use glucoamylase which has been immobilized on a cellulose derivative, such as DEAE-cellulose or immobilized covalently to an inert carrier.

A number of different types of α-amylase may be used, although it is preferred that saccharifying or pancreatic type α-amylase be used. Microorganisms such as *Bacillus subtilis* var. *amylosacchariticus* Fukumoto elaborate saccharifying type α-amylase.

The ratio of the activities of the enzymes used in this process should typically be above a certain minimum value to provide optimum catalytic action. In this regard, the amounts of immobilized glucoamylase and of α-amylase which may be used should be sufficient to provide a ratio of dextrinizing activity (hereinafter defined) to glucoamylase activity (hereinafter defined) of at least 0.2 liquefon per glucoamylase unit. Preferably, the amounts of enzymes present will be sufficient to provide at least 1 liquefon per glucoamylase unit, and most preferably, the amounts will be sufficient to provide at least 3 liquefons per glucoamylase unit.

Dextrose Equivalent: Dextrose equivalent (DE) was determined by Method E-26 described in *Standard Analytical Methods of the Member Companies of the Corn Industries Research Foundation,* Corn Refiners Association, Inc., Washington, D.C.

Glucoamylase Activity: A glucoamylase activity unit (GU) is defined as the amount of enzyme which catalyzes the production of 1 g of dextrose per hour at 60°C at pH 4.5 in the procedures described below.

Drum-dried partially hydrolyzed starch was used for the preparation of substrate solutions for glucoamylase activity determinations. A partially hydrolyzed starch solution having a DE of 12, was treated with activated carbon (Nuchar CEE) for 45 minutes at 60°C. The carbon was removed by filtration and the filtrate was treated again with carbon and filtered in the same manner. The filtrate was concentrated to about 50% dry solids and was then dried on a steam-heated drum drier and ground.

The drum-dried partially hydrolyzed starch contained 1.7% moisture and 0.5% ash. Substrate solutions for glucoamylase activity determination were prepared to contain 10 g of the dried hydrolyzed starch and 2 ml of pH 4.5, 1 M sodium acetate buffer per 100 ml of solution.

Dextrinizing Activity of Soluble α-Amylase: The dextrinizing activity of soluble α-amylase preparations was determined by a modification of Standard Test Method, AATCC 103, 1965, "Bacterial Alpha-Amylase Enzymes Used in Desizing, Assay of" published in the 1967 Edition of *Technical Manual of the American Association of Textile Chemists and Colorists*, Volume 43, pp. B-174 and B-175. The method was modified by substituting 10 ml of 1 M sodium acetate buffer, pH 5.0, for the 10 ml of pH 6.6 phosphate buffer solution used in the make-up of the buffered starch substrate. Also, 0.73 g of $CaCl_2 \cdot 2H_2O$ was added per 500 ml of buffered starch substrate. Results were calculated in terms of liquefons where 1 liquefon equals 0.35 bacterial amylase unit.

Dextrinizing Activity of Immobilized α-Amylase: The dextrinizing activity of immobilized α-amylase preparations was determined in the same manner as for soluble α-amylase preparations except that immobilized α-amylase was diluted for assay by suspension in 0.005 M calcium acetate solution at 30°C. A 5 ml aliquot of the suspension was added to the 10 ml of buffered starch substrate and the hydrolyzing mixture so formed was stirred continuously during the 30°C hydrolysis step. At appropriate time intervals, 2 ml aliquots of the hydrolyzing mixture were taken and rapidly filtered and 1 ml of the filtrate added to the 5 ml of dilute iodine solution. Time was counted starting at the instant the 5 ml aliquot of suspension was added to the 10 ml of buffered starch substrate and finishing at the time that the 2 ml aliquot of hydrolyzing mixture was filtered.

Production from Protein-Containing Starch

H. Müller; U.S. Patent 4,069,103; January 17, 1978 describes a process for obtaining dextrose and dextrins from protein-containing starches. The starting product constituting the starch is preferably a vegetable product such as potatoes, arrowroot or manioc (cassava) as well as grain feeds such as maize, sorghum, wheat, rice, rye or barley.

The process comprises subjecting the initial starch product to an acid or enzymatic hydrolysis or a combination of both types of hydrolysis, then circulating the hydrolyzate containing a low viscosity sugar solution and water-soluble high molecular proteins through an ultrafilter to separate the proteins from the sugar solution and recovering the dextrose or, in the case of an incomplete hydrolysis, the dextrins from the filtrate obtained in the ultrafilter.

The advantage of this process is that it is not necessary to start with pure starch. Rather, the hydrolysis can be carried out in conventional form immediately after the wet grinding and dilution of the ground initial material. An additional advantage of the ultrafiltration is the recovery of pure protein from the concentrate of the ultrafiltration. This pure protein constitutes a first class food product for human consumption.

With reference to Figure 5.1, it will be seen that the swollen grains of maize received from the bunker **1** are subjected to a coarse and fine grinding mill **2** (this symbol standing for both types of mills) and preferably to a wet grinding, and are then passed into the stirrer tanks **3** and **3'**. The mass is there subjected to an acid or enzymatic hydrolysis. Preferably there is carried out a hydrolysis successively with two different enzymes obtained from the containers **4** and **4'** such as α-amylase and amyloglucosidase.

Figure 5.1: Process for Obtaining Dextrose

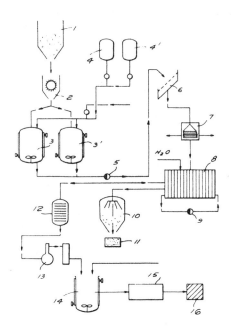

Source: U.S. Patent 4,069,103

The hydrolysis may be carried out with the first enzyme at a temperature of 70° to 80°C and with the second enzyme at a temperature of 50° to 60°C. The pH during the hydrolysis may be adjusted by addition of acid to a range between 3.5 and 4. The time of the hydrolysis may be between 10 and 72 hours depending on the type of raw material and the presence of calcium chloride.

After completion of the hydrolysis the entire solution is passed by means of a pump 5 onto a strainer 6 where the cereal husks and similar hard materials are removed. The mass then passes to a separator 7 where the fatty components including the corn oil are eliminated. Likewise, the precipitate, that is water-insoluble proteins such as zein, are removed from the remaining sugar-containing low viscosity phase. This separator may be in the form of a centrifuge.

The separation of the water-soluble high molecular proteins from the dextrose and the lower molecular dextrins is then effected in the ultrafilter 8. The low viscosity phase may be circulated through the ultrafilter for a time from 15 to 20 hours at a temperature of 20° to 50°C and a pressure of 2 to 6 bars. The circulation in the ultrafilter is effected by a rotary pump 9. The ultrafiltration may be carried out at a pH in the range between 3 and 9. Preferably, the material is washed during the ultrafiltration with relatively small amounts of water for a time sufficient to cause the concentrate which is obtained from the retentate to have a concentration of at least 70 to 80% protein as dry solids. This is accomplished by passing the retentate to a spray dryer **10** whereupon the protein can be discharged into a packaging material **11**.

The sugar solution on the other hand which constitutes the permeate resulting from the ultrafiltration, is passed through a bleaching column **12** which may be an ion exchange column and it may also additionally be treated with activated charcoal. If desired, it may also be subjected to a desalting step. The decolorized, i.e., bleached solution, is then subjected to concentration in an evaporator **13**. If a glucose syrup is to be obtained at this place the concentration is effected to an 85% dry contents.

If it is desired to obtain isomerose, the concentration is carried out only to a dry contents of 40%. The concentrate is in the form of a syrup which may then be treated with magnesium sulfate and cobalt sulfate and glucose isomerase at a pH between 6.6 and 7.5 and a temperature of 60°C.

The syrup may then be subjected to a purification at **15** which can be effected by a cation-anion exchange compound and a subsequent treatment with activated carbon. The product is then discharged into the container **16** as an isomerose syrup or sugar.

Production from Granular Starch

H.W. Leach, R.E. Hebeda and D.J. Holik; U.S. Patent 3,922,197; November 25, 1975; and E.A. Kuske and D.K. Przybylski; U.S. Patent 3,922,198; November 25, 1975; both assigned to CPC International Inc. describe a process for solubilizing starch comprising mixing a granular starch with water, a bacterial α-amylase and glucoamylase, at a temperature between the normal initial gelatinization temperature and the actual gelatinization temperature of starch, and at a pH of from 5.0 to 7.0.

Such a process is useful largely because of the combined action of the bacterial α-amylase and glucoamylase which results in increased solubilization of the starch as well as increased yields of dextrose. Alternatively, these objects may be accomplished by the action of bacterial α-amylase alone, i.e., in the absence of glucoamylase. That is, granular starch, water and a bacterial α-amylase are mixed under the above conditions, to produce a soluble starch hydrolysate.

The starch may be any of those commonly available, including cornstarch, waxy maize starch, tapioca starch, potato starch, white sweet potato starch, wheat starch, sago starch, sorghum starch, high amylose starch, etc. Waxy and the nonwaxy starches are suitable. As indicated, the starch is granular. Corn grits and other raw materials high in starch content may be used satisfactorily.

An important advantage of the process is that it may be carried out in an aqueous slurry at relatively high concentrations. The solids content of the starch slurry is generally within the range of from 5 to 40%, although ordinarily the solids content will be 10 to 30%.

The bacterial α-amylase preferably is one which is active within the pH range of from 4.0 to 7.0 and which possesses appreciable activity at relatively low temperatures, i.e., below the temperature at which a particular starch gelatinizes.

Preferred sources of such α-amylases include certain species of the *Bacillus* microorganism, viz., *B. subtilis, B. licheniformis, B. coagulans* and *B. amyloliquefaciens.* Suitable α-amylases are described in Austrian Patent 4836/70 and in U.S. Patent 3,697,378. Particularly preferred is that α-amylase derived from *B. licheniformis* strain NCIB 8061. A preferred range of concentration of α-amylase is from 1.0 to 10 units per gram of starch (dry basis).

The glucoamylase may be any of the well-known fungal amylase preparations, particularly those derived from members of the *Aspergillus* genus, the *Endomyces* genus or the *Rhizopus* genus. A particularly preferred glucoamylase is that available from the process described in U.S. Patent 3,042,584 whereby a fungal amylase preparation is freed of undesired transglucosidase activity by treatment in an aqueous medium with a clay material. The amount of glucoamylase to be used ranges from 0.05 to 5.0 units per gram of starch (dry basis). Preferably, on an enzyme cost/performance basis, about 0.1 to 0.3 unit of glucoamylase per gram of starch (dry basis) is used.

A particularly preferred embodiment, resulting in the formation of substantial proportions of dextrose, comprises the steps of (1) agitating a mixture of a granular starch, water and a bacterial α-amylase, at a temperature between the normal initial gelatinization temperature of the starch and the actual gelatinization temperature of the starch and at a pH of from 5.0 to 7.0, to convert at least 10% of the starch to a soluble hydrolysate; and (2) adjusting the temperature to 50° to 65°C and the pH to 4.0 to 4.8, and adding glucoamylase to saccharify the soluble hydrolysate. Step (1) generally requires about 12 hours; step (2) on the other hand, ordinarily will require a much longer time, i.e., from 24 to 120 hours. In some instances, it is desirable to add glucoamylase to the mixture in step (1).

FRUCTOSE OR LEVULOSE

Levulose from Granular Starch

In a similar process, *R.E. Hebeda and H.W. Leach; U.S. Patent 3,922,201; November 25, 1975; assigned to CPC International Inc.* describe the procedure of converting starch to levulose comprising mixing a granular starch with water, bacterial α-amylase, glucoamylase, and glucose isomerase at a temperature of from 40° to 70°C below the initial gelatinization temperature of the starch, and at a pH of from 5.0 to 7.0. Such process accomplishes the above objectives largely because of the combined synergistic action of the bacterial α-amylase, glucoamylase and glucose isomerase which results in efficient production of levulose at a single temperature and pH.

The glucose isomerase may be any such enzyme capable of converting dextrose to levulose. Many are known, including principally those elaborated by microorganisms of the *Streptomyces* genus, including *S. bobiliae, S. fradiae, S. roseochromogenes, S. olivaceus, S. californicus, S. vinaceus, S. virginiae, S. olivochromogenes,* and *S. phaeochromogenes.* Glucose isomerases elaborated by microorganisms of the *Arthrobacter* genus likewise are contemplated, e.g., *A.* nov. sp. NRRL B-3724, *A.* nov. sp. NRRL B-3725, *A.* nov. sp. NRRL B-3726, *A.* nov. sp. NRRL B-3727 and *A.* nov. sp. NRRL B-3728. So also, glucose isomerases elaborated by microorganisms of the *Lactobacillus* genus, e.g., *L. brevis, L.*

mannitopens and *L. buchneri,* as well as *Aerobacter cloacae* and *A. aerogenes.* The amount of glucose isomerase to be used ranges from 0.1 to 20 units per gram of starch (dry basis). In the usual, preferred instance, an amount within the range from 0.2 to 2.0 will be used.

Example: A 25% by weight aqueous slurry of granular cornstarch is prepared containing the following ingredients: 125 g of cornstarch; 250 mg of 1.0 N aqueous potassium phosphate buffer, pH 7.5; 5 ml of 1.0 N magnesium sulfate hexahydrate; 5 ml of 0.1 N cobalt chloride heptahydrate; sufficient aqueous calcium chloride to provide 100 ppm of calcium ion; α-amylase *(Bacillus licheniformis),* 5 activity units per gram of starch (dry basis); glucoamylase, 1.0 activity unit per gram of starch (dry basis); and glucose isomerase *(Streptomyces olivochromogenes),* 10 activity units per gram of starch (dry basis).

The above aqueous slurry is maintained at 60°C for 24 hours. The pH is adjusted to 6.0 as necessary by additions of aqueous potassium hydroxide. The conversion mixture is filtered and the filtrate adjusted to a pH of 4.5 by the addition of hydrochloric acid, then boiled to inactivate the enzymes. The solids material comprises 44.9% of the original granular starch, which means that 55.1% of the granular starch is solubilized. This solubilized starch product is found to have a dextrose content of 51.2% and a ketose (levulose) content of 20.6%, both based on solids content.

R.G.P. Walon; U.S. Patent 4,009,074; February 22, 1977; assigned to CPC International Inc. describes the same process, but using *Streptomyces albus* (ATCC 21132) as the preferred microorganism.

Isomerization of Glucose with Bound Glucose Isomerase

K.N. Thompson, R.A. Johnson and N.E. Lloyd; U.S. Patent 3,909,354; Sept. 30, 1975; assigned to Standard Brands Incorporated describe a continuous process for enzymatically isomerizing glucose to fructose, which comprises continuously introducing a glucose-containing solution having a viscosity of from 0.5 to 100 cp, a pH of from 6 to 9, a temperature of from 20° to 80°C and containing from 5 to 80% glucose by weight into a zone containing particles of bound glucose isomerase selected from the group consisting of glucose isomerase bound to anion exchange cellulose and glucose isomerase bound to a synthetic anion exchange resin whereby the particles of the bound glucose isomerase are maintained in suspension and up to 54% of the glucose is converted to fructose.

The color of the converted solution is increased by less than 2 color units and there is no substantial production of psicose and the converted solution is withdrawn from the zone at a rate substantially equivalent to the rate the glucose-containing solution is introduced into the zone, the particles of the bound glucose isomerase being characterized as having a glucose isomerase activity of at least 3 IGIU/cm^3 when packed in a bed and a stability value of at least 50 hours. Various terms and expressions used in the foregoing are defined as follows:

Stability Value: The stability value is determined by placing a sufficient amount of the bound glucose isomerase in a column to obtain from 1,000 to 4,000 IGIU therein. A solution that is 3 M in glucose, at a pH of 6.5, 0.001 M in $CoCl_2$ and 0.005 M in $MgSO_3$ is passed through the column at a rate of from 10 to 200 ml/hr. The column is maintained at a temperature of 60°C. The fraction of glucose converted to fructose in the effluent is determined after 4 hours to in-sure that the bed of bound glucose isomerase is under equilibrium conditions. The activity index of the bound glucose isomerase is calculated using the for-mula: activity index = (R/E) log (0.504)/(0.504-I), where I is the fraction of glucose converted to fructose, R is the flow rate (ml/hr) and E is the number of IGIU initially in the column. The activity index is determined periodically and the time it takes for the activity index to reach one-half the initial value (value after 4 hours) is the stability value in hours.

Color Units: Color was determined spectrophotometrically by measuring the absorbance at 450 μ and 600 μ of an appropriately diluted liquor in a 1 cm cell versus water as a reference. The spectrophotometer was a Beckman DK-2A. The color was calculated by using the following formula:

$$\text{Color units} = \frac{(109)\ (A_{450} - A_{600})}{C}$$

A_{450} = absorbance at 450 μ
A_{600} = absorbance at 600 μ
C = concentration (g/100 ml)

Fructose Content of Isomerized Liquor: Fructose content of the isomerized liquor may be determined by measuring the change in specific rotation which occurred during isomerization. Specific rotations were measured using a Bendix NPL Model 969 Automatic Polarimeter. The rotations were determined at a concentration of 2.5 g/100 ml in a glass cell thermostated at 25°C. The path of the cell was 50 mm. The specific rotations were determined at the beginning of the isomerization reactions after all ingredients in the glucose-containing so-lutions had been combined. To determine change in fructose content, the spe-cific rotation of the isomerized liquor was determined. Change in fructose con-tent was calculated by using the following formula:

$$\text{Percent Fructose} = \frac{100\ (A_1 - A_0)}{-138.9}$$

A_1 = specific rotation of isomerized liquor
A_0 = specific rotation of glucose-containing
       solution before isomerization

In the formula, the factor -138.9 is the change in specific rotation which occurs when glucose is converted completely to fructose.

IGIU: This is the abbreviation for International Glucose Isomerase Unit and is that amount of enzymes which will convert 1 μmol of glucose to fructose per minute in a solution initially containing 2 mols of glucose per liter, 0.02 mol of $MgSO_4$ per liter and 0.001 mol of $CoCl_2$ per liter at a pH of 6.84 to 6.85 (0.2 M sodium maleate) and a temperature of 60°C.

Maintaining the particles of bound glucose isomerase in suspension may be accomplished in any convenient manner. For instance, particles of bound glu-cose isomerase may be placed in a suitable vessel having a top outlet and bot-tom inlet. Positioned in the outlet and inlet may be a screen or other suitable porous means. The glucose-containing solution is continuously introduced into the vessel through the bottom inlet at a rate and under conditions whereby the particles of the bound glucose isomerase are fluidized and maintained in suspension in the glucose solution. A portion of the glucose in the solution is thereby converted to fructose and the fructose-containing solution is withdrawn through the top outlet.

Forming the bound glucose isomerase may be accomplished in any convenient manner so long as the bound glucose isomerase has the characteristics set forth above. It is contemplated that glucose isomerase may be bound to a host of inert carriers. For instance, glucose isomerase may be bound to DEAE-cellulose (diethylaminoethyl cellulose) or like material and excellent results will be obtained. Of course, to effectuate binding the glucose isomerase must be removed from the cells and there must be no interfering substances present during binding.

The binding may be accomplished in an aqueous medium or in a sugar solution, e.g., corn syrup. Also, the glucose isomerase may be bound on inert carriers either along with cellular material or in a relatively pure state. Various polymeric materials may be suitable for this purpose but, of course, the porosity of such materials must be such that allows the glucose to contact the glucose isomerase.

Isomerization of Glucose Under Oxygen-Limiting Conditions

R.P. Cory; U.S. Patent 3,910,821; October 7, 1975; assigned to CPC International Inc. has found that the enzymatic isomerization of aldose sugar to ketose sugar, and particularly the enzymatic isomerization of glucose to levulose, can be substantially improved by conducting the isomerization under nonoxidizing conditions. One preferred mode involves bubbling nitrogen, argon, or some other inert gas through the isomerizing mixture of sugar plus enzyme preparation simultaneously with mild agitation of the mixture, to exclude oxygen (air), while establishing, at the same time, an inert atmosphere above the isomerization mixture.

The process can be employed, so far as it is known, with all types of glucose isomerase enzyme preparations. These enzyme preparations can be derived from a large number of different microbial sources. Each enzyme preparation seems to have its own particular characteristics. The characteristics of each enzyme preparation, such as, for example, its optimum pH, its optimum temperature, the required metal ions, its Michaelis constant, and the mechanism of levulose formation, seem to be somewhat different from one enzyme preparation to another. However, the process seems to be applicable to all of these known glucose isomerase enzyme preparations. The isomerization may be conducted by adding the enzyme preparations, in soluble form, to a sugar solution such as, for example, a glucose-containing syrup.

Example: *Inoculum Development* — Spores from a slant of *Streptomyces olivochromogenes* ATCC 21114 were inoculated into several 500 ml Erlenmeyer flasks each containing 100 ml quantities of a sterile medium composed of the following ingredients:

Xylose	5.0 g
Cornstarch	5.0 g
Corn steep liquor	40.0 g
$MgSO_4 \cdot 7H_2O$	5.0 g

(continued)

$CoCl_2 \cdot 6H_2O$	0.24 g
Distilled water	1,000 ml

The pH of the culture medium was adjusted to 7.1 with sodium hydroxide prior to sterilization. The flasks were inoculated and incubated for 60 hours at a temperature in the range from 28° to 30°C on a reciprocating shaker.

Production of the Enzyme Preparation — Aliquot portions of 10 ml each were taken from the inoculum flasks. These aliquots were used to inoculate 1,000 ml Hinton modified Erlenmeyer flasks each containing a quantity of 200 ml of sterile medium of the following ingredients:

Xylose	10.0 g
Cornstarch	10.0 g
Corn steep liquor	40.0 g
Brewers' yeast extract	2.5 g
$MgSO_4 \cdot 7H_2O$	0.5 g
$CoCl_2 \cdot 6H_2O$	0.24 g
Distilled water	1,000 ml

The pH was adjusted to 7.1 with sodium hydroxide prior to sterilization. The inoculated flasks were incubated for 48 hours at a temperature in the range of from 28° to 30°C on a rotary shaker.

In preparing either of the media described above, xylan hydrolysate can be used in place of xylose with substantially equivalent results. The hydrolysate is an inexpensive source of xylose.

Harvesting of the Enzyme Preparation — After fermentation, the contents of the flasks were pooled. The liquor was centrifuged at 10,000 times gravity for 15 minutes. The cell pack was then separated and frozen for storage. For use, the cell pack or a proportionate part of it was brought back to its original volume with distilled water, and the cells were resuspended. When reconstituted, the cell suspension was found to contain 1.67 units per ml.

Conversion of the Cell Suspension to Solubilized Form — A portion of the cell suspension, that had been reconstituted from a frozen portion of the cell pack, was treated with crystalline lysozyme (General Biochemical, 3X crystalline, 4 μg/ml) for 24 hours at 25°C. The released enzyme was then adsorbed on DEAE (diethylaminoethyl) cellulose (500 units per gram of cellulose).

The DEAE cellulose was then washed with 0.2 M sodium chloride which removed very little of the adsorbed enzyme. The cellulose was then eluted with 0.35 M sodium chloride, which eluted most of the isomerase activity. The resulting enzyme preparation, a solution of glucose isomerase, contained 20 units per ml.

Isomerization Procedure — A standard substrate was prepared and was used in each of the following isomerization demonstrations. This substrate was made up as described as follows: 95 DE starch hydrolysate—60 g/100 ml dry basis; cobaltous chloride—enough to make the substrate 0.001 M; and magnesium sulfate—enough to make the substrate 0.01 M.

This standardized substrate was then subjected to a uniform isomerization procedure in the following examples, so that the results obtained are comparable.

The portion of the substrate to be subjected to enzymatic isomerization was maintained at 70°C in a jacketed vessel. The pH was maintained at 6.25 by means of additions, as necessary, of small amounts of a saturated solution of sodium bicarbonate. During the isomerization, the substrate was stirred within the vessel by an internal magnetic stirring bar. The dosage of enzyme activity employed was 0.6 unit/gram of dry substance in the standardized substrate. In terms of the enzyme preparation in solubilized form, 7.2 ml of the enzyme preparation was used for each 400 ml of the standardized substrate.

Several different gases were utilized in different isomerization reactions, for comparative purposes. For each isomerization, the gas was injected into the isomerizing mixture through a simple capillary tube. The rate of injection was sufficient to cause some agitation of the mixture, and to maintain a blanket of the gas above the mixture.

Following the procedure just described, isomerization reactions were conducted using sweep gas injections in which the gas was air, nitrogen, and argon. For comparative purposes, the isomerization was conducted using the same equipment and procedure, but with no gas injection. The observations made during the several demonstrations are reported below.

Percent Isomerization of 95 DE Hydrolysate to Ketose

Injection Gas Used	. . . . . . Isomerization Period (Hours)			
	21	44	67	67*
Air	14.7	20.2	21.2	19.7
N$_2$	17.5	28.2	35.2	33.4
Argon	15.5	29.0	36.4	34.3
None	17.3	27.0	30.0	28.4

*% isomerization to levulose.

As the data in the table demonstrate, the exclusion of air (oxygen) has the result of permitting a greater amount of ketose formation from the dosage of the enzyme preparation employed. In addition, the data also demonstrate that it is the existence of nonoxidizing conditions, or, in other words, the exclusion of air (oxygen), that is the cause of the effect observed.

Treatment of Cellular Material Containing Glucose Isomerase

J.H. Littlejohn and R.G. Dworschack; U.S. Patent 3,909,355; September 30, 1975; assigned to Standard Brands Incorporated describe the treatment of cellular material containing glucose isomerase which has been heated to fix or stabilize the glucose isomerase in or on the cellular material under suitable conditions with a sufficient amount of a proteolytic enzyme whereby a liquid will more readily pass through a bed or column of the material as compared to the passage of the liquid through a bed or column of cellular material not so treated.

The conditions under which the cellular material containing glucose isomerase is treated with proteolytic enzymes may vary widely, but, of course, must not be such that would deleteriously affect the glucose isomerase. Temperatures up to about 70°C provide satisfactory results, although it is preferred that the treatment be performed at a temperature of from 25° to 45°C. The pH of the aqueous suspension of cellular material during the treatment may also vary widely, for instance, from 4 to 10, although a pH of from 5.5 to 7.5 is preferred. The period during which the treatment is performed is dependent upon the other conditions of treatment, e.g., pH, temperature, etc., and the amount of proteo-lytic enzymes used. Generally, under the preferred conditions, the treatment may be effected in from 10 minutes to 3 hours.

Exemplary of proteolytic enzymes which may be used in this process are ficin, bromelain, papain, alkaline protease and mixtures thereof. The preferred proteo-lytic enzyme is papain. The preferred cellular material containing glucose iso-merase is derived from microorganisms of the *Streptomyces* genus. Particularly preferred is cellular material derived from *Streptomyces* sp. ATCC 21175 and *Streptomyces* sp. ATCC 21176.

Carrier for Glucose Isomerase

R.A. Messing; U.S. Patent 3,868,304; February 25, 1975; assigned to Corning Glass Works describes a method of making fructose which comprises contacting a glucose-containing solution with an immobilized enzyme composite comprising glucose isomerase adsorbed within the pores of porous alumina body having an average pore diameter between 100 and 1000 A. The porous alumina carrier preferably has a particle size between 25 and 80 mesh, U.S. Standard Sieve, and enzyme composites using the porous alumina particles are preferably placed in a flow-through column through which a glucose-containing solution is continu-ously passed under isomerizing conditions.

The incubation temperature is preferably between 50° and 70°C. The reaction proceeds well with a variety of buffer systems and a pH range of 7.2 to 8.2 is preferred with a pH of 7.4 to 7.8 being especially preferred. Since the carrier consists of porous alumina, the resulting enzyme composite has excellent alka-line durability at the pH range within which the bonded glucose isomerase dem-onstrates optimum activity.

Example: A crude glucose isomerase preparation consisting of about 444 IGIU/g derived from a *Streptomyces* organism was used for the adsorption step. About 5 g of the glucose isomerase preparation was added to 28.7 ml of 0.1 M mag-nesium acetate solution in a 50 ml beaker. The slurry was stirred for 25 min-utes at room temperature and then filtered through filter paper. The residue on the paper was washed with 14.3 ml of 0.1 M magnesium acetate followed by washes with 14 ml of 0.5 M $NaHCO_3$ and 4.3 ml of 0.5 M $NaHCO_3$. The washes were collected directly into the original enzyme filter. The total volume of the enzyme-wash solution (filtrate) was 50 ml.

500 mg of porous alumina bodies were placed in a 50 ml round bottom flask. 10 ml of the above glucose isomerase solution was then added to the flask. The flask was attached to a rotary evaporator. Vacuum was applied to the apparatus and the flask was rotated in a bath maintained between 30° and 45°C over a 25 minute interval. An additional 10 ml of glucose isomerase solution was then added to the flask, and evaporation was continued over the next 35 minute in-terval under the same conditions. This addition and evaporation was carried out over a 4 hour period under the same conditions.

The procedure was repeated two more times with separate 10 ml glucose isomerase solutions. A final 7 ml aliquot of glucose isomerase solution was then added to the flask and evaporation was continued for an additional 1 hour and 10 minutes at 45°C. The flask and contents were removed from the apparatus and placed in a cold room over the weekend. A total of 47 ml of glucose isomerase solution had been added to the alumina.

50 ml of buffer (0.01 M sodium maleate, pH 6.8 to 6.9, containing 0.001 M cobalt chloride and 0.005 M magnesium sulfate) was added to the composite and the sample was extracted over the next hour at room temperature. The extract, volume 50 ml, was saved for assay. The composite was then washed with 200 ml of water, followed by 10 ml of 0.5 M sodium chloride. The final wash was performed over a fritted glass funnel with 50 ml of water. The composite was then transferred to a 50 ml Erlenmeyer flask and stored in buffer at room temperature with periodic assays of the total sample over a 106 day interval.

Results — Enzyme extract assay (50 ml): 39.8 IGIU/ml (enzyme activity recovery in extract, 90%). Composite average assay value: 130 IGIU/500 mg sample (enzyme activity recovery on composite, 6%). Average loading value: 260 IGIU/g. The storage stability determinations over the 106 day interval involved 21 periodic assays in which 16 assays on the same sample indicated an activity between 100 and 150 IGIU/500 mg or 200 to 300 IGIU/g.

Continuous Isomerization of Dextrose

M. Tamura, S. Ushiro and S. Hasagawa; U.S. Patent 3,960,663; June 1, 1976; assigned to CPC International Inc. describe a method of isomerizing dextrose continuously with immobilized dextrose isomerase prepared by having the isomerase adsorbed on MR (macroreticular) type or porous type strongly basic anion exchange resin.

The Enzyme: Preferred microorganisms for producing suitable isomerase for use in this process are the members of the *Streptomyces* genus. Particularly preferred species among this genus are *S. venezuelae* and *S. olivochromogenes*. Cultures of preferred strains of these organisms have been deposited in the American Type Culture Collection, Washington, D.C., and added to its permanent collection of microorganisms. They have been assigned the following identification: *S. venezuelae* ATCC 21113 and *S. olivochromogenes* ATCC 21114.

The most preferred microorganisms are mutant strains of *Streptomyces olivochromogenes*, especially, *S. olivochromogenes* ATCC Nos. 21713, 21714, 21715 and their equivalents. These microorganisms form appreciable quantities of isomerase when cultivated in nutrient media free of xylose and xylose-supplying material and free of added cobalt.

Cells from a known volume of culture broth are resuspended in 0.05 M phosphate buffer (pH 7.5). The suspension is then sonified using a Branson Sonifier Model 185-1 (20 kc) until the microbial cells of the same are sufficiently disrupted so that the isomerase enzyme is substantially all liberated. Holding the sample tube in an ice bath during sonification prevents overheating and enzyme inactivation. The resulting enzyme preparation is a solution of solubilized isomerase.

The Resin Carriers: By porous-type ion exchange resins are meant those ion exchange resins having many pores. Such ion exchange resins are generally adapted for adsorption. Macroreticular-type ion exchange resins are commonly designated as the MR type, and have relatively large pores. By strongly basic anion exchange resins is meant resins having $-N-(CH_3)_3X$ as the ion exchange group and these are distinguished from those having $-N-(C_2H_4OH)(CH_3)_2X$ as the ion exchange group.

Amberlite IRA-904 and IRA-938 are commonly available as MR-type strongly basic anion exchange resins, and Dianion PA-308 and PA-304 as porous-type strongly basic ion exchange resins.

The resin or polymer particles are most desirably used in the form of granules or beads and are further, usually, in the range of 16 to 100 mesh particle size (U.S. Standard Screen Series) and more preferably in the form of 20 to 50 mesh beads. For convenience of use the beads are placed or arranged in a column.

Preparation of Immobilized Isomerase: The isomerase is used in the form of solutions: 0.01 to 0.1 M tris-HCl or phosphate buffer solutions; salt solution such as $(NH_4)_2SO_4$, $MgCl_2$ or KNO_3 solution, all adjusted to pH 6 to 9 (preferably pH 7 to 8); or simply in water or in dextrose solutions; all having concentrations ranging from 3 to 50 units per ml.

The ion exchange resin is packed in a column of the proper size and then equilibrated by passing the same solution as used for dissolving the isomerase through the column at a flow rate between 1 and 3 SV for 5 to 10 hours (the term SV is an abbreviation for substrate velocity, and refers to the flow rate in bed volumes per hour; one bed volume is the volume of substrate per hour that is equivalent to the column volume that is taken up by the resin in the column). Then, an amount of the dextrose isomerase solution corresponding to 10 to 100 units (preferably 50 units) the enzyme per gram of wet resin is passed through the resin column at a flow rate between 1 and 3 SV. After all of the enzyme solution has passed through the column, water is passed through the column to wash away unadsorbed isomerase.

Isomerization: Examples of dextrose to be used as materials are: crystalline dextrose (dextrose content: above 99%); powdered dextrose (dextrose content: around 90%); glucose syrup (40 to 90% dextrose); and hydrol (50 to 60% dextrose). These kinds of dextrose are each dissolved at a concentration between 30 and 70% (preferably around 60%) and mixed with 0.001 to 0.01 M $MgCl_2$. Then, the resulting sugar solution is adjusted to a pH value of about 8.0 with NaOH or KOH. Here $MgCl_2$ plays the role of an activator of the isomerase. Meanwhile, immobilized isomerase, prepared on an MR-type or porous-type strongly basic anion exchange resin, as described above, is packed in a column. With the column kept at temperatures between 60° and 70°C, the dextrose-containing solution is passed through it.

Column Reactivation for Continuous Isomerization: First, a dextrose isomerase is dissolved in the same glucose solution as that which is used for isomerization. Then, an amount of the resulting enzyme solution, containing 10 to 50 units (preferably 25 to 50 units) of isomerase per gram of wet resin, is passed through the columns at the same flow rate as that at which the glucose solution to be isomerized is passed. As it passes through the column, more enzyme is adsorbed, at the same time that the glucose is being isomerized.

In this manner, the resin columns regain their original rate of isomerization (52%). Moreover, the immobilized isomerase, which has been reactivated, shows the same half-life as the original one (17 to 22 days). If this procedure is repeated, it is possible to carry out the isomerization without repacking columns as long as the ion exchange resins last. Further, the levulose content in the effluent from resin columns can be held constant at a given isomerization rate up to about 52% (the normal equilibrium value) in the following way. When the isomerization rate is found to drop, as can be detected by measurement with a polarimeter, the resin column is immediately reactivated in the above manner, or, alternatively, the flow rate is gradually reduced to keep the isomerization rate constant, and then the column is reactivated at a proper time.

Example: *Continuous Isomerization with the Enzyme Immobilized on Amberlite IRA-904* — 50 g of moist Amberlite IRA-904, which is an MR-type strongly basic anion exchange resin, was packed in a column (2.5 x 20 cm). With the resin equilibrated with a 0.01 M $MgCl_2$ solution adjusted to pH 8.0, 3,000 ml of a solution of crude dextrose isomerase in the same solution (containing 5,000 units of isomerase) was passed through the column at a flow rate of SV 3. After all the isomerase solution had been passed through the column, a 60% dextrose solution containing 0.01 M $MgCl_2$ and adjusted to pH 8.0 was passed through the column at a flow rate of SV 1 at 70°C.

The rate of isomerization was measured with a polarimeter and expressed as the percentage of levulose of the solid substance in the effluent from the column. The rate remained 52% (equilibrium value) for the first 12 days. After that, however, it lowered gradually and dropped to 26% on the 17th day. At this time, 1,500 ml of a solution of isomerase in a 60% dextrose solution containing 0.01 M $MgCl_2$ and adjusted to pH 8.0 (corresponding to 2,500 units of isomerase) was passed through the column at a flow rate of SV 1. Then, the isomerizing reaction was continued on by passing the same dextrose solution (without isomerase) through the column under the same conditions as those used for the isomerizing reaction.

As the result of this procedure, the resin column regained its original rate of isomerization of 52%, and retained this value for further 12 days. In the succeeding 5 days, however, the rate of isomerization dropped to 26%. At this time, the resin column was again reactivated in the same manner. It was again returned completely to its initial activity.

Removal of Heavy Metal Ions

S. Enokizono, N. Kamata and S. Kanno; U.S. Patent 4,100,025; July 11, 1978; assigned to CPC International Inc. have found that the isomerization ratio is greatly influenced by certain impurities in the starting material. When dextrose is isomerized with glucose isomerase, if certain heavy metal ions such as zinc are present, its isomerizing power is considerably inhibited. The isomerization ratio is also influenced by the purity of the glucose isomerase used.

It has been found that the interfering heavy metal ions may be substantially completely removed by passing the dextrose-containing solution through a bed containing a chelating resin or other type resin capable of removing heavy metal ions. These include the chelating resins, complex adsorbing exchanger resins, and selectively adsorbing cation exchange resins.

In general, the useful resins of this process are different from those resins useful for conventional treatments in terms of production method, functional groups and affinity for adsorbable ions. For example, Dowex A-1, is produced by the addition of iminodiacetate to styrene divinylbenzene polymer (R), and its partial structure is:

$$
\begin{array}{c}
\text{CH}_2-\text{C} \overset{\displaystyle O}{\underset{\displaystyle O-\text{H}^+}{\big\langle}} \\
\text{R}-\text{N} \\
\text{CH}_2-\text{C} \overset{\displaystyle O-\text{H}^+}{\underset{\displaystyle O}{\big\langle}}
\end{array}
$$

However, styrene, phenol, and methacrylic acid are used as the monomers for the production of general ion exchange resins. The functional groups are such acid radicals as $-SO_3H$, $-COOH$, $-OH$ and $-PO_2H_2$.

The cation exchanger on which heavy metals are efficiently adsorbed can be used in any form, for instance in hydrogen or salt forms. It is preferable to utilize the hydrogen form. The resin can be made into the hydrogen form by passing a mineral acid such as H_2SO_4 or HCl through a tower packed with the resin. It is preferred to pass a proper amount of HCl of a proper concentration. For example, twice the volume of the ion exchanger of 5% HCl may be passed to produce the hydrogen form. Further, as the conditions at this time, SV (space velocity) 2 to 10 or favorably 3 to 5, and room temperature to 70°C or favorably 20° to 50°C, are preferred.

Next, when the solution containing dextrose is passed through a tower packed with the abovementioned cation exchange resin, the heavy metals in the solution are adsorbed onto the ion-exchange resin. In passing the dextrose-containing solution through a tower packed with the abovementioned cation exchanger, it is advantageous to pass a solution of pH 1 to 8, or preferably 3 to 6, of a concentration of 10 to 60% (weight %) or preferably 30 to 50% and at room temperature to 70°C at SV 1 to 8.

Example: Cornstarch was liquefied with an α-amylase liquefying enzyme, Kleistase L-1 (Daiwa Kasei Co.), and saccharified with a glucoamylase saccharifying enzyme, Sumizyme 800 (Shinnihon Kagaku Co.), by a conventional method. The saccharizates obtained were filtered on a filter paper using diatomaceous earth as filter aid under reduced pressure.

The filtered saccharizate was purified in the usual way; the liquor was successively passed through (1) a single bed resin column packed with 1,000 ml of a strongly acidic cation exchange resin, IR-120B (Tokyo Yukikagaku Kogyo Co.), which had been regenerated with hydrochloric acid, (2) a single bed resin column packed with 1,200 ml of a weakly basic anion exchange resin, IRA-93, which had been regenerated with sodium hydroxide, and (3) a mixed bed resin column packed with 150 ml of a strongly acidic resin, Amberlite-200, which had been regenerated with hydrochloric acid, and 300 ml of a strongly basic anion exchange resin, IRA-411, which had been regenerated with sodium hydroxide. The purified saccharizate was then concentrated. The quality of the purified saccharizate, so obtained, was as follows:

> Sugar Concentration—51.2° (Brix)
> Dextrose Equivalent—95.7
> Dextrose—93.2 (%)
> Color Value (OD at 427 μ)—0.018 (1 cm light path)
> Total Salts—60.3 ppm (as $CaCO_3$)
> pH—5.4

The heavy metal ions in this liquor were determined by the chelate titration method and found to be 1.6 ppm as zinc ion.

Then, 10 liters of the purified saccharizate were further treated with a chelating resin, Lewatit TP-207. 50 ml of the resin was packed in a glass column (height: 25 cm, diameter: 2.1 cm) and the resin was regenerated by passing 200 ml of 10% hydrochloric acid through a column at 30°C at a flow rate of 5 bed volumes per hour, followed by washing with 1,000 ml of deionized water. Then, the purified saccharizate was passed through this column by the descending method at 30°C and a flow rate of 5 bed volumes per hour. The quality of this chelate resin treated liquor was as follows:

> Sugar Concentration—51.2° (Brix)
> Dextrose Equivalent—95.7
> Dextrose—93.2 (%)
> Color Value (OD at 420 μ)—0.016 (1 cm light path)
> Total Salts—59.9 ppm (as $CaCO_3$)
> pH—3.1

The heavy metals contained in this chelate resin treated liquor were determined by the abovementioned method and no heavy metal ions were detected. Next, this chelate resin treated liquor was isomerized by the batch method. That is, to 5 liters of this chelate-resin-treated dextrose liquor, 5 g of $MgCl_2 \cdot 6H_2O$ were added and stirred, the pH of the mixture being adjusted to 6.5 with sodium hydroxide. To this mixture, a glucose isomerase solution was added, at a dosage of 2 units per gram of dextrose contained in the mixture.

Glucose isomerase was extracted from *Streptomyces olivochromogenes*, a microorganism producing glucose isomerase in the following way: *S. olivochromogenes* was cultured in a liquid medium for about 50 hours and the cultured cells were separated from the culture medium by centrifugation. The cells obtained were digested with lysozyme. Centrifugation of the cell digest gave a supernatant liquid, containing glucose isomerase. Isopropanol was added to the supernatant liquid to precipitate the enzyme. The precipitates obtained on centrifugation of the solution were redissolved in water containing $MgCl_2$ and used as the isomerizing enzyme preparation.

The isomerization reaction was continued for 48 hours in a 15 liter reactor equipped with a heater and stirrer while the saccharified solution was stirred slowly. During the reaction, the pH was adjusted with a 5% $NaHCO_3$ solution so that it remained between 6.3 and 6.7. After 48 hours of isomerization the isomerization rate was found to be 43.2%.

For comparison, the concentrated purified saccharizate not treated with the abovementioned chelating resin was isomerized under the same conditions as above. The isomerization rate was found to be 19.2% after 48 hours.

Fatty Acid Esters of Fructose

T. Suzuki and S. Ito; U.S. Patent 3,909,356; September 30, 1975; assigned to Kyowa Hakko Kogyo Co., Ltd., Japan describe a process whereby fatty acid esters of fructose, particularly α-branched-β-hydroxy fatty acid esters of fructo-furanose, are obtained by culturing a fructose-utilizing microorganism belonging to the genus *Arthrobacter, Corynebacterium, Nocardia* or *Mycobacterium* and capable of producing fatty acid esters of fructose in a medium containing fructose as a carbon source, forming the fatty acid esters of fructose intracellularly and recovering the esters from the microbial cells.

The fatty acid ester of fructose obtained according to this process consists of 1 mol of fructofuranose having the following structure:

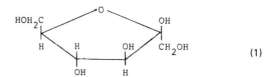

(1)

and 1, 2, 4 or 5 mols of an α-branched-β-hydroxy fatty acid represented by the general formula:

$$\underset{\substack{\| \quad | \\ O \quad R'}}{HO-C-CH-CH-R''} \quad \overset{OH}{\underset{|}{}}$$

(2)

where R' is an alkyl group having 8 to 22 carbon atoms and R'' is an alkyl group having 15 to 61 carbon atoms.

The fatty acid esters of fructose of this process are glycolipids which are used for biochemical experiments. The product has a surface activity and may be used as a surfactant.

The microorganisms applicable in this process are found widely among the microorganisms of the genera *Arthrobacter, Corynebacterium, Nocardia* and *Mycobacterium*. Examples thereof are as follows:

>*Arthrobacter paraffineus* ATCC 15591
>*Arthrobacter hydrocarboglutamicus* ATCC 15583
>*Corynebacterium hydrocarboclastus* ATCC 21628
>*Corynebacterium pseudodiphtheriticum* ATCC 10701
>*Nocardia paraffinica* ATCC 21198
>*Nocardia globerula* ATCC 13130
>*Nocardia convoluta* ATCC 4275
>*Mycobacterium rubrum* ATCC 14346
>*Mycobacterium paraffinicum* ATCC 12670

(continued)

Mycobacterium smegmatis ATCC 21293
Mycobacterium smegmatis ATCC 607

Either a synthetic culture medium or a natural nutrient medium may be used for the culturing of the microorganisms so long as the medium properly contains a carbon source, a nitrogen source, inorganic materials and other nutrients which may be required by the specific strains employed. It is essential that the medium contains fructose, or a crude substance containing fructose, such as molasses, as the carbon source.

As the nitrogen source, ammonium salts such as ammonium chloride, ammonium sulfate, ammonium nitrate, ammonium acetate and ammonium phosphate, ammonia and urea may be used. Further, natural substances containing nitrogen, such as corn steep liquor, yeast extract, meat extract, peptone, casamino acid, etc., may be used. These substances may be used either singularly or in combination of two or more. As inorganic materials, potassium phosphate, magnesium sulfate, iron, manganese salts, calcium chloride, sodium chloride and zinc sulfate may be used. When the microorganism used has a requirement for some nutrients such as amino acids and vitamins, the nutrients must, of course, be supplemented to the medium.

Fermentation is carried out under aerobic conditions at 25 to 40°C. During the fermentation, the pH of the fermentation liquor is adjusted to 4 to 9, preferably 6 to 8, with a urea solution, an aqueous ammonia or an aqueous ammonium carbonate solution. Usually, fermentation is complete in 1 to 7 days. When the yield of the desired product reaches a maximum, fermentation is discontinued. The product is formed predominantly in the microbial cells. After the completion of the fermentation, the microbial cells are separated from the fermentation liquor, for example, by centrifugation. The cells are subjected to extraction with a mixture of chloroform and methanol in a volume ratio ranging from 50:100 to 100:50. The solvent used in the resulting extract is distilled off under reduced pressure, for example, at 20 to 30 mm Hg.

The resulting residue is subjected to extraction with an appropriate solvent such as chloroform, hexane and ethyl acetate. The solvent of the extract is removed. The residue is again dissolved in a small amount of a solvent such as chloroform, hexane and ethyl acetate and subjected to filtration. The filtrate is poured into a silica gel column. First, nonpolar substances such as pigments and free fatty acids are eluted with chloroform. Then, elution is further carried out with a chloroform-methanol mixture, while stepwise varying the volume ratio of chloroform to methanol in the mixture from 99:1 to 95:5. The eluate containing the desired product is recovered and the solvent is removed. The resulting residue is dissolved in warm acetone. The solution is allowed to stand still in a cold place, e.g., at –15°C for 5 to 24 hours. The resulting precipitates are recovered and dried, whereby a white powder of the product is obtained.

OTHER MONOSACCHARIDES

Galactose

P. Galzy and G.J. Moulin; U.S. Patent 3,981,773; September 21, 1976; assigned to Agence Nationale de Valorisation de la Recherche (ANVAR), France describe a process for the treatment of solutions containing lactose, in particular, lacto-

serum or milk, with microorganisms to produce, on the one hand, a mass of microorganisms which can be used in the feeding of animals and, on the other hand, a solution of galactose which can be directly used after clarification for the preparation of beverages.

According to the process, galactose or a beverage containing galactose is prepared from a solution containing lactose, in particular lactoserum or milk, by treating the lactose-containing solution with a mutant, nonpathogenic, prototrophic, microorganism which contains a β-galactosidase and which has a gal⁻ character, i.e., is incapable of fermenting galactose, subsequently separating the microorganisms with the solution containing galactose, if desired after fermentation, and, if necessary, extracting the galactose from this solution.

In one of the preferred embodiments, the microorganism used is a bacteria obtained by mutation and selection from a wild strain. This bacteria is preferably selected from the *Enterobacteriaceae* or the *Lactobacteriaceae*, in particular *Escherichia coli*. The *Escherichia coli* strain is preferably the CBS 6051 B strain lodged in Delft.

In another preferred embodiment, the microorganism used is a yeast. This yeast is obtained by mutation and selection, preferably from a haploid strain, the wild strain preferably being selected from the *ascomycetes*:

> *Debaryomyces cantarellii*
> *Debaryomyces castellii*
> *Debaryomyces hansenii*
> *Debaryomyces morana*
> *Debaryomyces tamarii*
> *Hansenula capsulata*
> *Kluyveromyces aestuarii*
> *Kluyveromyces bulgaricus*
> *Kluyveromyces cicerisporus*
> *Kluyveromyces fragilis*
> *Kluyveromyces lactis*
> *Kluyveromyces wikerhamii*
> *Lipomyces lipofer*
> *Lipomyces starkeyi*
> *Pichia farinosa*
> *Pichia polymorpha*
> *Pichia pseudopolymorpha*
> *Pichia scolyti*
> *Pichia stipitis*
> *Schwanniomyces castellii*
> *Wingea robertsii*

Preferred strains are in particular the *Kluyveromyces fragilis* or *Kluyveromyces lactis*. The strain may also be selected from the *basidiomycetes*:

> *Leucosporium frigidum*
> *Leucosporidium scottii*
> *Bullera alba*
> *Bullera strigea*
> *Sporobolomyces singularis*

Two different types of mutation and selection can be used for obtaining bacterial strains suitable for use in the process. The first method, which starts with a strain of a collection having the gal⁻ character and a β-galactosidase, this strain being nonpathogenic, but auxotropic to certain substances, comprises treating this strain in such a way as to reverse it for its various auxotrophs until a prototrophic strain which has retained its other characteristics is obtained. On an industrial level, a prototrophic strain has the advantage of not necessitating the addition of growth factors which are necessary to the development of an auxotrophic strain.

The following technique is one example of a process for reversing an auxotrophic strain. Starting with a culture of the auxotrophic strain, prepared naturally in a medium containing the growth factors, 10^8 to 10^9 cells are spread over the surface of a minimum medium (one of the characteristics of a prototrophic strain, of course, is that all its constituents can be synthesized in an exclusively mineral synthetic medium known as the minimum medium), from which a growth factor X necessary to the growth of the starting strain is missing. The reversed mutant which has become prototrophic for the factor X then grows.

The cultures are then spread over the surface of a medium accommodated in a Petri dish. The colonies appearing are isolated and sampled. They have become prototrophic for the factor X and can be used in another stage of the process in which they will be reversed for another of the auxotrophs of the initial strain.

It is also possible to obtain from a wild strain a bacterial strain which can be used in the process by another mutation-selection process. The description of this process, given below with reference to *Escherichia coli*, can, of course, be used for other wild strains, in particular for strains of *Enterobacteriaceae* or *Lactobacteriaceae*. The wild strain of *Escherichia coli* is treated with a physical or chemical mutagen in order to increase mutation frequency for the genes involved in the metabolism of galactose. Examples of physical mutagens include ultraviolet rays, x-rays and gamma rays, while chemical mutagens include ethyl methanesulfonate, nitrous acid, alkylating agents, nitroguanidine, acriflavin or other compounds well-known as mutagens.

This treatment is followed by enrichment in mutant bacteria having the gal⁻ character. To this end, the population which has undergone mutation is transferred to another medium allowing growth in which the only carbon source is galactose and which additionally contains an antibiotic, such as penicillin. The antibiotic kills the cells in the process of vegetative multiplication, but has no effect upon the mutants incapable of multiplying through their inability to metabolize galactose. After a certain time, only those bacteria which have a gal⁻ character are left. The isolated strains have to be subjected to another selection because some of the strains obtained, although incapable of using galactose as carbon source, cannot be used in the process. These strains are, for example, strains which do not possess the permease.

In order to eliminate these undesirable strains, the strains obtained in the preceding stage are cultured in a Petri dish containing a complete medium based on glucose as carbon source. It is advisable to spread approximately 100 living cells per dish. The colonies which appear are replicated by replicaplating (for culture by replication) in a medium based on galactose. The required gal⁻ mutants give colonies in the first medium based on glucose, but not in the medium based on galactose. Among the mutant gal⁻ lactose⁺ clones, it is advisable to select, through suitable tests, those which are able to hydrolyze the lactose while rejecting the galactose in the medium. In every case, it is possible to make acellular preparations and to dose the enzymatic activities in order to determine what the missing enzyme is for each clone.

In cases where it is desired to use a yeast as microorganism, the process of mutation and selection of the yeast is the following, the following description being given with reference to the strain *Kluyveromyces fragilis*, but is also applicable to other yeast strains.

The strain *Kluyveromyces fragilis* occurs in nature in diploid form. However, the search for mutants for a given gene can only be made with reasonable chances of success from an individual haploid. The process for obtaining a strain suitable for use in the process is, therefore, in two main parts. The first is the search for individual haploids, while the second is the mutation and selection of the haploid strains obtained.

In order to obtain haploid clones, the strain is grown in a rich medium (yeast extract, glucose, peptone). The culture in its exponential growth phase is transferred to a sporulation medium which consists of a mixture of equal parts of a 0.4 M potassium acetate solution and Sorensen M/15 phosphate buffers (pH 7).

After 48 to 72 hours, 80% of the cells have sporulated. The asci burst very rapidly and release the haploid spores into the medium. The standard technique using Fontbrune's micromanipulator cannot be used in this case because the spores are 99% lethal. Heat treatment has to be applied in order to obtain haploid clones because the vegetative cells are less resistant to heat than the spores, and treatment for 2 minutes at 60°C destroys 99% of the vegetative diploid forms.

Following this treatment, colonies are spread over the surface of a rich medium in a Petri dish at the rate of 100 to 200 colonies per dish. The dishes are observed after 48 to 72 hours incubation in an oven at 26°C and the cellular clones which appear are removed. Each clone is cultured in a rich medium and then tested in a sporulation medium. Each clone which sporulates can be eliminated as being diploid. Those clones which do not sporulate are assumed to be haploid. In order to confirm the haploid character of each clone, it is necessary to attempt to mark them by auxotrophic mutations.

One quick and simple method is to look for the clones which lead to auxotrophic mutants for an amino acid or a base during a mutagenesis. The technique involved comprises allowing a physical or chemical mutagen, such as those described for the mutation of bacteria for example, on a population in the stationary phase. In this case, it is preferred to use ethyl methanesulfonate. The culture is then transferred to a medium in which only the prototrophs grow, and on which a fungicide is allowed to act after one to two generations. The mycostatin kills the cells in the growth phase which correspond to the prototrophic clones that have not undergone mutation.

After dilution, 100 cells per dish are spread over a complete medium in which the auxotrophs are able to develop, after which the dishes are incubated for 48 to 72 hours. By the replicaplating technique, it is then possible to identify those clones which have developed in a complete medium and have not produced colonies in the minimum medium in which only the prototrophs develop. After incubation for 48 hours, it is possible to mark the auxotrophs. This technique enables haploid clones to be obtained. The strains thus obtained and marked are used in the search, in the following stage, for mutants for one of the genes preventing the utilization of galactose.

The technique of mutation and selection intended for obtaining gal⁻ mutants is the same as that described above for obtaining auxotrophic mutants. Only the culture media are different. The complete medium allowing growth will be a medium containing glucose, while the minimum medium will contain galactose. When, after replicaplating, a certain number of clones do not develop in the minimum medium, they are removed and tested in liquid medium in the following media: yeast extract + galactose, and lactoserum.

Of all the clones which do not grow on galactose, some cannot be used in the process. For example, a strain which has mutated for permease will not develop on lactose because galactose can only penetrate by osmosis. Nevertheless, this strain will consume the lactose following the hydrolysis of lactose, because hydrolysis takes place inside the cell. It is possible to determine the lost enzyme by an enzymatic study. In the same way as for bacteria, when a gal⁻ mutant is marked for an auxotroph, it is necessary to attempt reversion by a technique such as that indicated earlier. Although the yeast strains thus obtained can be used in the process, it is generally preferred to convert these strains into diploids, because diploid strains have a better growth than the haploids.

A certain number of techniques have already been developed with a view to converting haploid strains into diploids. Cells can be crossed individually using a micromanipulator, or alternatively to auxotrophic clones, if necessary complementary cells, can be crossed while, at the same time, sifting out the prototrophs. It is, of course, necessary to cross clones with opposite sexual signs. The methodology described can be used for all the cells capable of metabolizing lactose and galactose provided that they are diplobiontic and heterothallic. In the case of haplobiontic yeasts, mutagenesis can be applied directly to collective cells or to cells isolated in nature.

Example 1: The process is carried out with a bacteria of the *Escherichia coli* type lodged at the Centraal Bureau voor Schimmelcultures in Delft under the No. CBS 6051 B obtained by mutation and selection from a strain from the Pasteur Institute in Paris named B 112.21 and catalogued under the No. 115.15. This strain has the following genotype: Thi⁻ gal⁻ (transferase), Hfr P 0.00.

After having obtained reversion of the auxotroph for thymine by the process described above, growth on lactoserum was studied. The precultures were made in a mineral medium complemented with thiamine with the following composition:

K_2HPO_4	7 g/l
KH_2PO_4	3 g/l
$(NH_4)_2SO_4$	1 g/l
$MgSO_4 \cdot 7H_2O$	0.10 g/l
Sodium citrate·$2H_2O$	0.5 g/l
Iron citrate	0.3 mg/l
$MnSO_4 \cdot H_2O$	0.17 mg/l
Glucose or lactose	5 g/l
Thiamine	1 mg/l

Culturing proper is carried out on lactoserum complemented with thiamine in a quantity of 1 µg/ml at a temperature of 34°±0.5°C, being accompanied by agitation in order to aerate the medium and to keep the cells in suspension (80 oscillations per minute, oscillation amplitude 7 cm). The culture is inoculated with the preculture in a mineral medium. A population giving around 1.5 to 3.5 mg of dry material per ml of lactoserum is obtained under these conditions. Quantitative analysis of the remaining sugars is carried out by chromatography on Whatman No. 1 paper by Partridge's method. All the chromatographs show the presence of galactose in the medium at the end of growth. It was not possible to detect any trace of glucose. After growth for 48 hours on lactoserum containing 35 g/l of lactose, almost all the lactose had disappeared.

This strain thus effects the required operation and quantitatively hydrolyzes the lactose in the lactoserum into galactose which accumulates in the medium, and into glucose which is consumed.

Example 2: In this example the process is carried out with a yeast *Kluyveromyces fragilis* SG 11 registered at Delft under the No. CBS 6498 and obtained from a *Kluyveromyces fragilis* strain with the name SG 1 lodged under the No. CBS 6497, to which the mutation and selection treatment described above has been applied.

The cultures are made in 5 liter Erlenmeyer flasks filled to $^1/_{10}$ of their volume with crude lactoserum without any growth factor. The temperature is kept at 26°±0.5°C. The medium is agitated at 80 oscillations per minute, oscillation amplitude 7 cm. In this medium, the strain gives populations of the order of 300×10^6 cells per ml after 120 hours' growth. This strain consumes all the lactose in a lactoserum containing 35 g/l of lactose. Disappearance of the lactose and the presence of galactose were determined in the medium by the chromatographic method described above for bacteria. The growth of this strain is complete after 120 hours, but shows a particularly long latency phase of 48 hours.

D-Ribose

K. Sasajima and M. Yoneda; U.S. Patent 3,919,046; November 11, 1975; and K. Sasajima, M. Doi, T. Fukuhara, A. Yokota, Y. Nakao and M. Yoneda; U.S. Patent 3,970,522; July 20, 1976; both assigned to Takeda Chemical Industries, Ltd., Japan describe a method for producing D-ribose, which comprises cultivating a D-ribose-producing microorganism belonging to the genus *Bacillus*, which lacks sporulation ability or has high 2-deoxy-D-glucose-oxidizing activity or has both of these properties and also lacks at least one of transketolase and D-ribulose phosphate 3-epimerase, in a culture medium containing assimilable carbon and nitrogen sources as well as nutrients necessary for the growth of the strain, thereby causing the strain to elaborate and accumulate D-ribose and, then, recovering the D-ribose thus accumulated from the resultant culture broth.

Among examples of the strains employable are *Bacillus pumilus* No. 911 (IFO 13566), No. 1027 (IFO 13585) and No. 1083 (IFO 13620) and *Bacillus subtilis* No. 957 (IFO 13565), No. 941 (IFO 13573), No. 1054 (IFO 13586), No. 1067 (IFO 13588) and No. 1097 (IFO 13621) and so on. The numbers following IFO mean the accession numbers at the Institute for Fermentation, Japan.

Referring to the nutrients that are used as the constituents of a medium for the cultivation of microorganisms according to this process, the carbon sources include, among others, D-glucose, D-fructose, D-mannose, sorbitol, D-mannitol, sucrose, molasses, starch hydrolyzates, starch, acetic acid and ethanol. The nitrogen sources include organic nitrogen sources such as corn steep liquor, cottonseed refuse, yeast extract, dried yeast, fish meal, meat extract, peptone, casamino acid, etc., inorganic nitrogen compounds such as aqueous ammonia, ammonia gas, ammonium sulfate, ammonium nitrate, ammonium chloride, ammonium carbonate, ammonium phosphate, sodium nitrate, etc., and organic nitrogen compounds such as urea, amino acids, etc. Also incorporated in the medium, in addition to the carbon and nitrogen sources, are various metals, vitamins, amino acids and other substances which may be essential to the growth of the particular microorganism, the proportions of which are optional.

The cultivation is conducted aerobically, for example by shaking culture or submerged culture under sparging and stirring. The incubation temperature is usually selected from within the range of 20° to 45°C, depending upon the temperature suited for the particular organism to grow and accumulate D-ribose. The pH of the medium is preferably somewhere between 5 and 9.

To maintain the pH within the optimum range throughout the cultivation period, one may incorporate from time to time such a neutralizer as hydrochloric acid, sulfuric acid, aqueous ammonia, ammonia gas, an aqueous solution of sodium hydroxide, calcium carbonate, slaked lime, etc. Ordinarily, a substantial amount of D-ribose accumulates in the medium in 2 to 5 days.

The D-ribose thus accumulated can be easily recovered, for example, by the following procedure. The culture broth is first filtered or centrifuged, whereby the cells can be removed with great ease. Then, the filtrate is desalted and decolorized by treatment with activated carbon and ion exchange resin and then concentrated. To the concentrate is added an organic solvent such as ethanol, whereupon D-ribose crystals separate.

Example: A transketolase-lacking, asporogenous mutant No. 911 of *Bacillus pumilus* (IFO 13566), which had been derived from *Bacillus pumilus* (IFO 12113) by irradiation with ultraviolet rays, (transketolase activity: not more than 0.01 μmol/min/mg protein; sporulation frequency: 2×10^{-8}) was used to inoculate 10 liters of a medium comprising 2.0% of sorbitol, 2.0% of corn steep liquor, 0.3% of dipotassium hydrogen phosphate, 0.1% of potassium dihydrogen phosphate, 100 μg/ml of tyrosine and 100 μg/ml of phenylalanine and the inoculated medium was incubated at 36°C for 24 hours.

The entire amount of the resultant culture was transferred to 100 liters of a culture medium comprising 15.0% of D-glucose, 1.0% of dried yeast, 0.5% of ammonium sulfate, 2.0% of calcium carbonate, 50 μg/ml of tryptophan, 50 μg/ml of phenylalanine, in which it was cultivated at 36°C for 60 hours, whereupon D-ribose accumulated at the rate of 64 mg/ml. From this D-ribose fermentation broth, the cells were removed by filtration and the filtrate was concentrated to half the original volume. Then, about one-quarter of its volume of ethanol was added and the precipitate was discarded. The residue was desalted with cation and anion exchange resins and then decolorized on a column of activated carbon. The decolorized solution was concentrated and about 4 times its volume of ethanol was added, whereby 5.0 kg of crystalline D-ribose was obtained.

L-Sorbosone

In a related process, *S. Makover and D.L. Pruess; U.S. Patent 3,912,592; Oct. 14, 1975; assigned to Hoffmann-La Roche Inc.* have found that L-sorbosone can be produced from L-sorbose by the microbiological oxidation of L-sorbose from a microorganism selected from the following genera:

> *Gluconobacter*
> *Pseudomonas*
> *Acinetobacter*
> *Bacillus*
> *Sarcina*
> *Streptomyces*
> *Serratia*
> *Aerobacter*
> *Mycobacterium*
> *Paecilomyces*

The compound L-sorbosone has the formula:

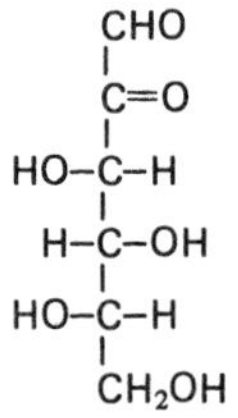

$$
\begin{array}{c}
\text{CHO} \\
|\\
\text{C=O} \\
|\\
\text{HO--C--H} \\
|\\
\text{H--C--OH} \\
|\\
\text{HO--C--H} \\
|\\
\text{CH}_2\text{OH}
\end{array}
$$

Any microorganism of the aforementioned genera can be utilized. Among the preferred strains are included: *Acinetobacter calcoaceticus* (ATCC 10153) and *Bacillus* sp. strain TA (ATCC 27860); *Streptomyces cellulocae* (ATCC 3313), *Serratia* sp. (ATCC 93), *Serratia marcescens* (ATCC 27857), *Aerobacter aerogenes* (ATCC 27858), *Sarcina lutea* (ATCC 9341), *Pseudomonas putida* (ATCC 21812), *Gluconobacter melanogenus* (IFO 3293), *Mycobacterium phlei* (ATCC 355); and *Paecilomyces varioti* (ATCC 26820).

The process can be carried out by culturing the microorganism in a medium containing appropriate nutrients and L-sorbose. On the other hand, the process can be carried out by culturing the microorganisms and then, after culturing, bringing the whole cells or the cell-free extract prepared from the culture into contact with L-sorbose.

In the case where the microorganism is cultured in a medium containing L-sorbose and appropriate nutrients, the microorganism may be cultured in an aqueous medium in an aerated fermentor. The cultivation should be conducted at pH values of from 5 to 9, with pH of from 6.5 to 7.5 being preferred. Especially preferred is utilizing a pH of 7.0 to 7.2. A preferred temperature range for carrying out this cultivation is from 20° to 45°C, with temperatures of from 25° to 30°C being especially preferred. While the time for cultivation varies with the kind of microorganisms and nutrient medium to be used, 1 to 10 days' cultivation usually brings about most preferable results. Concentration of L-sorbose in the media varies with the kind of microorganism, but it is generally desirable to be from 1 to 150 g/l, most preferably from 2 to 70 g/l.

It is usually required that the culture medium contains such nutrients for the microorganism as assimilable carbon sources, digestible nitrogen sources and preferably inorganic substances, vitamins, trace elements, other growth promoting factors, etc.

In the case where after cultivation, the cells collected from the culture are brought into contact with the L-sorbose, cultivation of the microorganisms is carried out under similar conditions described above. Substances mentioned above can, if desired, also be used for nutrients for this cultivation.

On the other hand, no nutrients need be present in cultivating the microorganisms. The whole grown culture is then utilized to convert L-sorbose to L-sorbosone. This conversion can be simply carried out in an aqueous medium under submerged conditions utilizing a pH of from 5 to 9. In this conversion, no additional nutrients need be present. Generally, from 1 to 3 days culture is preferable for obtaining the most effective cells for the conversion of L-sorbose to L-sorbosone. The reaction can be stopped by freezing the reaction medium, i.e., cooling the reaction medium to a temperature of 0°C or below.

MALTOSE

Production from Starch Using a *Streptomyces* Amylase

C. Yomoto, T. Adachi, Y. Nakajima, H. Hidaka, T. Yoshida and F. Sugawara; U.S. Patent 3,998,696; December 21, 1976; assigned to Meiji Seika Kaisha Ltd., Japan describe a method of producing maltose, wherein a liquefied starch is saccharified by an amylase produced by a strain of *Streptomyces* and a β-amylase. The particular strains of *Streptomyces* which are used for the production of the amylase are shown below:

Strain	FRI Deposit No.	ATCC Deposit No.
Streptomyces tosaensis SF-1085	601	21723
Streptomyces hygroscopicus SF-1084	602	21722
Streptomyces viridochromogenes SF-1087	603	21724
Streptomyces albus SF-1089	604	21725
Streptomyces flavus	605	–
Streptomyces aureofaciens	606	–
Streptomyces hygroscopicus var. *angustomycetes*	607	–

FRI is an abbreviation of Fermentation Research Institute, Agency of Industrial Science and Technology of the Ministry of International Trade and Industry of Japan, Inage, Chiba City, Japan, and ATCC is that of American Type Culture Collection, Washington, D.C.

The production medium used for the cultivation of a strain of the *Streptomyces* to accumulate the amylase may contain one or more of starch, soluble starch, glucose and corn meal, etc., as the carbon sources and one or more of defatted soybean meal, defatted cottonseed meal, wheat embryo, peanut meal, ferma media, fish meal, dried yeast, skimmed milk, casein, malt extract, yeast extract, sodium nitrate and potassium nitrate, etc., as the nitrogen sources. In addition, it is possible to use and incorporate in the culture medium one or more of inorganic salts such as potassium dihydrogen phosphate, magnesium sulfate, manganese sulfate, ferric sulfate and calcium carbonate, etc., as well as trace elements in order to promote the growth of the microorganism and to enhance the production of the enzyme, if required.

The process may be carried out in a known manner and under the conventional culturing conditions which are usually employed for the cultivation of *Streptomyces*. However, liquid cultivation and particularly liquid cultivation under submerged aerobic conditions is most preferred. When one of the abovementioned strains of the *Streptomyces* is incubated at a temperature of 25° to 37°C and at a pH in the range of a weak acidity to a weak alkalinity under submerged aerobic conditions with aeration and agitation, the production of the amylase reaches a maximum in 3 to 5 days of incubation.

For the recovery of the amylase, the culture medium or culture broth in which the incubation of the *Streptomyces* has been carried out may be treated in a known manner to separate the amylase therefrom. Thus, the culture broth may be filtered to remove the mycelium cake, and the resulting filtrate may then be treated either through a salting-out technique by adding a water-soluble inorganic salt such as ammonium sulfate, etc., or through a precipitation method by adding a water-miscible organic solvent such as ethanol, methanol, isopropanol, acetone, etc., or through an adsorption-elution method with an ion-exchange resin, etc. In this way, the amylase may be separated from the incubated culture medium. The amylase thus separated may be treated further by spray-drying, hot-air drying, vacuum-drying, freeze-drying or lyophilizing to give a crude powder preparation of the amylase.

An electrophoretically homogeneous and pure preparation of the amylase may be obtained by purifying the abovementioned crude powder preparation in a conventional manner which is known for the purification of enzymes, as described in the French Patent 71-38545.

Following is described a process in which maltose is produced according to the action of the abovementioned *Streptomyces* amylase following the treatment with β-amylase. Starch must be liquefied prior to the action of β-amylase. As the starch materials, those used commonly for the production of starch hydrolyzate, such as cornstarch, potato starch, sweet potato starch, tapioca starch, rice starch, wheat starch, and their α-inverted ones are available for use in this process. As the procedure for liquefaction, a mechanical process accompanied with heating as well as conventional enzymatic processes using α-amylase or the *Streptomyces* amylase are available.

At the liquefaction step, the extent of the degradation of the chain of the starch molecule is desirably small. Generally, the value of DE is favorably less than 4, though a value below 10 does not deviate from the purpose of this process.

After liquefaction, the abovementioned liquified solution is cooled to the temperature range in which β-amylase can react actively, e.g., 45° to 65°C, then the pH is adjusted to the range of 4.5 to 6.0, and β-amylase is added. The reaction period is dependent on the amount of the enzyme used, the reaction temperature, and the reaction pH. The action of β-amylase may either be allowed to proceed to its maximum extent for degradation or it may be stopped a little before it. The content of maltose at this step is generally around 60%.

Next, the degradation of starch in the solution is further progressed by addition of the *Streptomyces* amylase. At this step, β-amylase in the solution may either be inactivated prior to the addition of the *Streptomyces* amylase or be left as it is. The appropriate amount of the *Streptomyces* amylase added is generally 200 to 1,500 units per gram of starch, and the reaction temperature and pH are kept at 45° to 65°C and 5.0 to 7.0, respectively.

Units of activity of the abovementioned enzyme are determined as follows. A mixture composed of 1 ml of enzyme solution, 2 ml of 2% soluble starch solution and 2 ml of McIlvaine buffer at pH 5.5, was incubated at 40°C for 3 minutes. The reaction was stopped by adding 1 ml of the reaction mixture into Somogyi's reagent. The amount of the produced reducing sugars is determined by the Somogyi's titration method, and the measured amount of the reducing sugars produced is calculated as maltose in the whole 5 ml of the reaction mixture. One unit is designated as the amount of the enzyme capable of producing 1 mg of maltose in 60 minutes.

A practical period for saccharification is within 72 hours in total and the content of maltose in the solution at this step is 80 to 90%. After the completion of saccharification, the solution is decolorized and purified using active carbon and ion-exchange resin and concentrated to the prescribed concentration according to the conventional method to obtain a final product.

Example: 100 kg of potato starch containing 18% of water was made up to a 30% aqueous suspension, adjusted to pH 6.0, and liquefied by liquifaction process at 85°C with 200 units per gram of starch of the *Streptomyces* amylase. The resulting liquefied solution was steamed for inactivating the amylase, cooled to 65°C and saccharified for about 2 hours at pH 6.0 with 15 units per gram of starch of β-amylase to obtain a saccharified solution consisting of 62.0% maltose, 0.3% maltotriose, and 37.7% limit dextrin. The solution was cooled to 55°C, 1,500 units per gram of starch of the *Streptomyces* amylase were added and was further saccharified for 48 hours while maintaining the pH at 6.0.

The resulting solution was decolorized and purified according to the conventional method, and concentrated to 75% to obtain the final product. According to a gas liquid chromatographic analysis, the sugar composition of the product was 8.0% glucose, 84.5% maltose, and 5.0% maltotriose.

Use of β-Amylase

S. Sakai and N. Tsuyama; U.S. Patent 4,032,403; June 28, 1977; assigned to Kabushiki-Kaisha Hayashibara Selbutsukagaku Kenkyujo, Japan describe a process for the production of saccharified starch products wherein maltose is the predominant constituent, characterized in that the maltotriose content of the products is reduced while the maltose purity is improved by subjecting a starch hydrolysate simultaneously or successively to the action(s) of an enzyme(s), possessing a maltotriose-decomposing activity versus maltose-decomposing activity ratio of 2.5 or higher and which is derived from higher plant and/or from a bacterial culture, and the action(s) of maltogenic enzyme(s) during the production of saccharified starch products wherein maltose is the predominant constituent.

Any starch is employable regardless of origin, i.e., cereal, tuber or root starches, and amylose/amylopectin proportion. First, a starch suspension is gelatinized or liquefied. The liquefaction is carried out with acid and/or enzyme to a DE (dextrose equivalent), preferably, to not more than 25. Thereafter, the liquefied starch is saccharified by subjecting the starch to the action of known maltogenic enzyme(s), e.g., β-amylase or β-amylase and starch-debranching enzyme. In general, enzymes derived and prepared from wheat bran (cf Specification of Japanese Patent Publication 70-18937), soybean and sweet potato are used as β-amylase.

For starch-debranching enzyme, microbial enzymes, such as those derived from any culture broth of genera *Escherichia intermedia* (ATCC 21073), *Aerobacter aerogenes* (ATCC 8724), *Pseudomonas amyloderamosa* (ATCC 21262), *Corynebacter sepedonicum* (IFO 3306), *Aeromonas hydrophyla* (IFO 3820), *Flavobacterium esteroacromaticum* (IFO 3751), *Vibro metschnikovii* (IFO 1039), *Actinoplanes philippinesis* (ATCC 12427), and *Streptosporangium roseum* (ATCC 12428) as described in Japanese Patent Publications 68-28939, 69-8070, 70-9229, 70-16788, 71-28151, and 73-18826 are employable.

Depending on the method used for the process of production, the resultant saccharified starch hydrolysate will have a maltotriose content, in general, of 5 to 25% and its maltose purity limit would be about 50 to 93%. By allowing an enzyme(s) derived from higher plants and/or culture of bacteria and which have the specified activity ratio during saccharification along with conventionally-known maltogenic enzyme(s), or allowing the former enzyme to saccharify on the saccharified starch hydrolysates upon completion of saccharification with such known maltogenic enzyme(s), the maltotriose content, present in the initial saccharified starch hydrolysate and which prevents further improvement in maltose purity, substantially decomposed and the maltose purity of the final saccharified starch product can be improved by an additional approximate 3 to 15%.

A typical example of saccharification conditions will be described. The maltose purity of a saccharified starch hydrolysate, prepared by either saccharifying a 20 to 40% aqueous acid- or enzyme-liquefied starch hydrolysate with a DE of 5 to 20, or saccharifying a starch hydrolysate with a DE not more than 5 at a temperature of about 60°C using malt amylase, will be at the highest 50 to 60%, thus the production of starch hydrolysates with higher maltose purity has so far been rendered extremely difficult.

By adding an enzyme(s) with the specified activity ratio of 2.5 or higher which is derived from higher plants and/or obtained by culture of bacteria in an amount of 0.05 unit or more per gram saccharified starch hydrolysates, dsb, to the above saccharified starch hydrolysates during or after its saccharification process and incubating the mixture at a temperature from 30° to 55°C and pH 4.0 to 8.0, an attainment of a maltose purity of 60 to 75% in the resultant saccharified starch products was easily possible.

The preparation of bacterial enzyme with the specified activity ratio of 2.5 or higher will be described. The cultivation of various bacteria are usually carried out at 20° to 35°C under a stationary or aeration-agitation conditions for 1 to 5 days on a liquid culture medium containing carbon, nitrogen and inorganic sources and traces of growth factors, sterilized at a temperature of 120°C for 10 to 30 minutes and inoculated with a bacterium. The culture broth may be used intact as enzyme solution, or if necessary, its supernatant obtained by removal of cells with filtration or centrifugation.

Use of α-1,6-Glucosidase

K. Sugimoto, M. Hirao, M. Kurimoto and E. Miyake; U.S. Patent 4,016,038; April 5, 1977; assigned to Hayashibara Company, Japan describe a process for economically preparing highly pure maltoses at high concentrations from various starch slurries at 10% or higher concentrations, which comprises gelatinizing such a slurry evenly at an elevated temperature of not lower than 100°C and then rapidly cooling the gelatinized solution and, at the same time, subjecting it to the action of any of various α-1,6-glucosidases and β-amylase.

Example: Purified starch was adjusted to a concentration of 15 to 35% thereby to obtain a starch emulsion, then the pH of the emulsion was adjusted. The emulsion was forced into a continuous multiblade agitating tank in which the temperature inside the tank was adjusted to 150° to 165°C by the feed of steam to liquefy the starch emulsion. The decomposition rate of liquefied solution was below 1 DE. The liquefied solution was sprayed into a tank with a normal pressure and cooled to 100°C. Then the resulting mixture was sprayed into a vacuum tank and cooled to 55° to 65°C, while β-amylase at a rate of 70 units per gram of the starch was introduced therein, and the mixture was quickly mixed together.

The resulting solution with a reduced viscosity was withdrawn from the lower part of the tank and cooled to 55°C while the pH of the cooled solution was adjusted to 7.0. Thereafter it was divided into 5 equal parts, and to each of them one of five kinds of enzymes prepared from their corresponding genus of *Lactobacillus* was added at a rate of 30 units per gram of the starch. Each of the resulting mixtures was reacted at 50° to 55°C for 40 hours, each enzyme added was deactivated by heat. Colorless maltose solution was obtained in each case by passing the aforementioned reaction product solution through a decoloring activated carbon and an ion-exchange apparatus.

The resulting colorless maltose solution was concentrated to 90% of concentration and it was allowed to stand until crystallization of the maltose occurred. Colorless and sublimated product maltose of 90% yield was obtained. Each product maltose was chromatographically fractionated and determined and the results are shown below.

	. . . . . .Enzyme Sources of α-1,6-Glucosidase				
	Lacto-bacillus brevis IFO 3345	*Lacto-bacillus brevis* IFO 3960	*Lacto bacillus bulgaricus* IFO 3533	*Lacto-bacillus gayonii* ATCC 8289	*Lacto-bacillus plantarum* ATCC 8008
Concentration of starch emulsion	20	25	25	23	23
DE of saccharified solution	62.1	60.1	60.0	61.0	60.2
Sugar components of the product					
glucose	0.5	0.6	0.6	0.5	0.6
maltose	91.5	90.3	90.0	91.0	90.0
maltotriose	5.0	4.2	4.3	4.1	5.0
oligosaccharides	3.0	4.9	5.1	4.4	4.4

Two-Enzyme Process

Y. Takasaki and Y. Takahara; U.S. Patent 3,992,261; November 16, 1976; assigned to Agency of Industrial Science & Technology, Japan describe a method for the manufacture of maltose which comprises causing bacteria of genus *Bacillus* capable of producing both β-amylase and α-1,6-glucosidase at the same time to be cultured under conditions permitting production of the enzymes, collecting from the resultant culture the produced β-amylase and α-1,6-glucosidase, adding the collected β-amylase and α-1,6-glucosidase to starch or a starch derivative (solubilized starch, dextrin, etc., and hereinafter referred to as starch), and maintaining the system under conditions permitting the starch to be hydrolyzed.

When starch or a starch derivative is treated with the combined β-amylase and α-1,6-glucosidase, the β-amylase functions to sever maltose units of amylopectin present in starch from the nonreducing ends of the chain and, as the severance has advanced to approach the branching linkage, the α-1,6-glucosidase functions to sever the α-1,6-glucosidase linkage of the branch and the β-amylase once again functions. Since the two enzymes, i.e., β-amylase and α-1,6-glucosidase, alternately hydrolyze with maltose units of amylopectin from the nonreducing ends of the chain, maltose is produced directly by a one-step process from starch in yields higher than is attainable by the conventional method.

The microorganisms of the process have been deposited with the American Type Culture Collection and assigned the following numbers: *Bacillus cereus* var. mycoides (ATCC 31102), *Bacillus* sp. YT 1002 (ATCC 31101), and *Bacillus* sp. YT 1003 (ATCC 31103). These species of bacteria invariably produce both β-amylase and α-1,6-glucosidase when they are cultured in a medium incorporating carbon sources and nitrogen sources of the type generally adopted for microorganism culture.

Further, it was found that when the culture is carried out in a medium incorporating rapeseed, rapeseed cake or an extract (a substance extracted from rapeseed by heating in water or a liquid having alkalinity of pH 9 to 10), the amount of α-1,6-glucosidase produced is increased to a notable extent. When the medium contains 4% of rapeseed cake, for example, the amount of α-1,6-glucosidase produced is 2 to 3 times as large as when the addition is omitted.

The culture of bacteria is carried out under aeration at 20° to 40°C for 24 to 72 hours. During the culture, the pH value of the medium varies in the range of from 5.5 to 9.5. It is desirable that the pH value of the medium be maintained in the range of from 6 to 8 throughout the entire course of culture.

The culture broth is filtered or centrifuged to remove the microorganism cells, then added to starch solution. For higher yields of maltose, it is desired to use starch or soluble starch in low DE. Where a highly concentrated starch is offered to be hydrolyzed, the starch is liquefied by the action of α-amylase and the resultant liquefied starch is put to use. Dextrin and other similar substances are also usable as the substrate. The concentration of the substrate is from 5 to 60%, more ordinarily from 10 to 40%. Where necessary, calcium ions are added.

The mixture is maintained at pH 5 to 8, preferably 6 to 7, at a temperature from 20° to 60°C, preferably from 40° to 55°C, and then the culture broth containing the two enzymes, α-amylase and α-1,6-glucosidase, is added to the mixture. The precipitate containing the two enzymes or activated carbon having the two enzymes adsorbed thereon can be added to the reaction solution in place of the culture broth containing the two enzymes. It is also permissible for the two enzymes to be separated from the culture broth independently by virtue of adsorption or extraction and to be added to the reaction solution at the same time. The amount of β-amylase to be added generally ranges from 200 to 400 units and that of α-1,6-glucosidase ranges from 10 to 50 units per gram of starch on a dry basis. Under these conditions, the starch is hydrolyzed into maltose in a yield of 80 to 90% after 48 to 72 hours of reaction.

SUCROSE

Production Using α-Galactosidase-Producing Molds

α-Galactosidase is known as an enzyme capable of hydrolyzing raffinose into sucrose and galactose. It has successfully been produced commercially from the mold, *Mortierella vinacea* var. raffinoseutilizer (ATCC 20034) (U.S. Patent 3,647,625). It has subsequently been learned that an enzyme having high α-galactosidase activity is produced from the mold belonging to genus *Absidia* (U.S. Patent 3,832,284). The enzymes have since been used in the beet sugar industry.

In the enzymes obtained by culturing these molds, however, not only α-galactosidase but also invertase is present. When beet juice or beet molasses (hereinafter referred to briefly as molasses) is treated by use of such enzymes, the raffinose contained in the molasses is decomposed by the α-galactosidase so as to lessen the hindrance to sucrose crystallization by raffinose in the course of beet sugar production and also enhance the yield of sucrose. However, the invertase which is present therein in conjunction with the α-galactosidase has been found to function adversely in that it causes undesired decomposition of sucrose as well.

Because it is economically difficult to accomplish separation of this invertase from the enzymes and further because the invertase activity is weak in contrast to the α-galactosidase activity, the decomposition of sucrose by the invertase has heretofore been tolerated as an incurable innate defect. When a large volume of molasses is treated, however, the volume of sucrose decomposed by the invertase increases proportionally to a level too great to be overlooked.

S. Narita, H. Naganishi, A. Yokouchi and I. Kagaya; U.S. Patent 3,957,578; May 18, 1976; assigned to Hokkaido Sugar Co., Ltd., Japan describe a method for the manufacture of an enzyme having notably high α-galactosidase activity and absolutely no invertase activity by use of molds belonging to genus *Absidia*, i.e., *Absidia griseola* (ATCC 20430) and *Absidia griseola* var. iguchii (ATCC 20431).

Carbon sources usable for the medium are starch, glucose, glycerin, maltose, dextrin, sucrose and invert molasses, for example. Suitable nitrogen sources are, for example, soybean flour, peanut powder, ground cottonseed, corn steep liquor, meat extract, peptone, yeast extract, nitrates and ammonium salts. Inorganic salts usable include, for example, common salt, potassium chloride, magnesium sulfate, manganese sulfate, iron sulfate, phosphates and calcium carbonate. If occasion demands, vitamins may be incorporated in the medium. As inducers for effective production of α-galactosidase in the mycelia, there are used such well-known substances as lactose, raffinose, melibiose and galactose.

The culture of the molds is carried out aerobically. Specifically, the molds are inoculated to the medium described above and subjected to shaken culture or aerobic culture at a temperature of about 30°C for a period of from 40 to 70 hours, with the pH value controlled in the range of 5 to 8.

When the molds are cultured as described above, the α-galactosidase activity is produced to a notably high level and absolutely no invertase activity is produced in the cultured mycelia. The mycelia thus cultured are obtained and preserved in the form of pellets through a known process comprising the steps of separation from the culture broth by filtration, washing with water, centrifugal dehydration, drying, etc.

The cells thus obtained in the form of pellets are packed in or added to a vertical reaction vessel or horizontal reaction vessel and the molasses is allowed to flow through the reaction vessel at pH 4 to 6 at a temperature of 30° to 60°C. While the molasses is in contact with the mycelia, the raffinose present in the molasses is hydrolyzed into sucrose and galactose. Because no invertase is contained in the enzyme obtained, the enzymatic hydrolysis is effected only on raffinose and is not at all effected on sucrose. Thus the process permits the yield of sucrose to be enhanced efficiently.

Example: A medium having a composition of 0.5% of lactose, 0.7% of glucose, 0.1% of $(NH_4)_2SO_4$, 0.4% of KH_2PO_4, 0.2% of $MgSO_4 \cdot 7H_2O$, 0.06% of $(NH_2)_2CO$, 0.2% of NaCl and 1.2% of corn steep liquor was adjusted with NH_4OH to pH 5.5. In a jar fermentor having an inner volume of 20 liters, 15 liters of the medium thus prepared was placed and, with 800 ppm of cocolin added thereto as an antifoam agent, sterilized for 30 minutes under pressure of 1.2 kg/cm². The medium was then cooled. To the cooled medium, a spore suspension obtained by culturing *Absidia griseola* var. iguchii (ATCC 20431) in potato-glucose agar medium at 27.5°C for 10 days was inoculated in an amount to give 1 x 10⁵ spores/cc of medium and cultured at 30°C for 18 hours, with aeration given at the rate of ¼ v/v/m and agitation at the rate of 250 rpm. The mycelia thus propagated were used as the seed.

A medium having a composition of 1.5% of lactose, 0.5% of glucose, 0.6% of $(NH_4)_2SO_4$, 1.0% of corn steep liquor, 0.4% of KH_2PO_4, 0.2% of $MgSO_4 \cdot 7H_2O$ and 0.2% of NaCl was adjusted to pH 5.5 and 15 liters of the medium were placed in a jar fermentor having an inner volume of 20 liters and, with 1,280 ppm of cocolin added thereto as an antifoam agent, sterilized for 30 minutes under 1.2 kg/cm². To the medium thus prepared, 500 ml of the seed was inoculated and cultured at 30°C with aeration given at the rate of ¼ v/v/m and agitation at the rate of 350 rpm.

The culture was stopped at the time at which the sugar content in the medium reached 0.1% (as indicated in terms of glucose by Somogyi method) (44 hours of culture). The relation between the culture time and the amount of α-galactosidase formed was as shown below. The assay of the culture broth revealed total absence of invertase.

Culture Time (hr)	pH	Residual Sugar as Glucose (g/100 ml)	Dry Mycelia (g/100 ml)	. . . α-Galactosidase Activity . . .	
				Total Activity (unit/ml)	Specific Activity (unit/g*)
0	6.2	1.37	–	0	–
16	5.4	–	–	0	–
20	6.4	–	–	37,200	–
24	6.5	–	–	84,300	–
28	6.4	–	–	109,000	–
32	5.9	–	–	155,400	–
36	5.7	–	–	226,700	–
40	5.4	–	–	250,900	–
44	5.8	0.08	1.29	257,000	$1,992 \times 10^4$

*Dry basis

In a similar process, H. Suzuki, H. Yoshida, Y. Ozawa, A. Kamibayashi, M. Sato, A. Mori and M. Endo; U.S. Patent 3,992,260; November 16, 1976; assigned to Agency of Industrial Science & Technology, Japan describe a method for readily increasing the yield of sucrose by overcoming the various drawbacks which are involved in the production of sucrose from beets by the conventional processes.

This process aims to recover sucrose additionally from the sugar solution which has undergone the treatment of purification and concentration or centrifugation. It accomplishes this recovery by allowing α-galactosidase to act upon the sugar solution thereby hydrolyzing the raffinose contained therein into sucrose and galactose, returning the resultant hydrolyzate to the stage following the process in which the sugar solution has been withdrawn and subjecting it to the sugar-boiling treatment.

The molasses upon which the α-galactosidase is to act is first diluted with water to a suitable concentration (20° to 60° Brix). Then, the α-galactosidase is added to the diluted molasses. The beet juice (the sugar solution which is in a stage ranging from purification process to the process prior to sugar-boiling treatment) generally has a concentration in the range of from 10° to 60° Brix. Therefore, the α-galactosidase may be added directly to the beet juice without any necessity for diluting or concentrating the juice. If the beet juice is concentrated, it is suitably diluted, as required, prior to the addition of the α-galactosidase.

The α-galactosidase to be added to the aforementioned molasses or beet juice may be produced by any method available. It is preferred to manifest little activity of invertase. If the α-galactosidase is of a type having high invertase activity, it should have its invertase activity inactivated suitably prior to use.

Examples of the α-galactosidase suitable for the process are α-galactosidases extracted from such plant seeds as coffee bean, Vicia sativa, and Vicia faba, and α-galactosidases produced from such microorganisms as brewer's yeast, Aspergillus oryzae, Aspergillus niger, Penicillium paxilli, Calvatia cyathiformis, Mortierella vinacea var. raffinoseutilizer, Streptomyces olivaceus var. raffinoseutilizer, Streptomyces fradiae, Streptomyces roseopinus, E. coli, Aerobacter aerogenes, Streptococcus bovis, Bacillus Delbrückii, Bacillus circulans, and Pseudomonas eisenbergii. Of these enzymes, most suitable is one obtained by culturing Mortierella vinacea var. raffinoseutilizer (ATCC 20034).

The hydrolysis ratio of raffinose by α-galactosidase is variable with the concentration of sugar solution, the nature of sugar solution, the activity of the enzyme to be used, and such factors as duration, pH and temperature selected for the action of the enzyme. Generally, this hydrolysis ratio of raffinose increases in inverse proportion to the concentration of sugar solution and in direct proportion to the activity of the enzyme and the duration and temperature selected for the enzymatic action.

At temperatures exceeding the level of 65°C, the enzyme is frequently inactivated and cannot withstand a continued use for a long time. When the concentration of sugar solution exceeds 65° Brix, the hydrolysis ratio is lowered to the extent of making the operation unprofitable. In the case of molasses, for example, conditions which are suitable for practicing the hydrolysis vary with the concentration of molasses, the activity of enzyme to be used and so on. To be more specific, 50 to 90% of the raffinose contained in the molasses can be hydrolyzed when α-galactosidase is allowed to act for a period of 2 to 3 hours on the molasses in relative amounts such as to give a ratio of 2,500,000 to 5,500,000 units of the enzyme to 1 g of the raffinose, with the temperature fixed in the range of from 45° to 55°C, the concentration in the range of from 15° to 60° Brix, and the pH value in the range of from 4.8 to 5.2.

The unit of the activity of α-galactosidase is defined to be such that will produce 1 μg of free glucose after the enzyme has been left to act for 2 hours on melibiose having a final concentration of 0.015 mol at 40°C and pH 5.2.

Example: As the source of a α-galactosidase, there were used pellet-shaped cells containing α-galactosidase (hereinafter referred to as enzyme-containing cells) and showing very little invertase activity, obtained by inoculating a culture medium containing 1% of lactose, 1% of glucose, 1% of corn steep liquor, 1.0% of ammonium sulfate, 0.1% of urea, 0.2% of sodium chloride, 0.3% of potassium primary phosphate, 0.2% of magnesium sulfate and 1% of calcium carbonate with *Mortierella vinacea* var. raffinose-utilizer and aerobically culturing the microbe therein at 30°C for 3 days.

The molasses selected for the hydrolysis was the fourth molasses obtained in a beet sugar production plant involving five sugar-boiling treatments and utilizing the deionization process using ion-exchange resins. The composition of this molasses was as follows: 76.0° Brix concentration; 53.5% sucrose concentration; 7.05% (9.28% on Brix) raffinose concentration; and 70.4% sucrose purity.

The molasses of the abovementioned composition was diluted with water to 30° Brix and adjusted to pH 5.0. To this molasses, the enzyme-containing cells were added in an amount such as to give 4,500,000 units of α-galactosidase potency per gram of raffinose. The mixture was agitated at 50°C for 2.5 hours to allow the enzyme to act upon the raffinose in the molasses. At the end of the agitation, the hydrolysis ratio of raffinose was found to have reached 69.5%. When the mixture was filtered to remove the enzyme-containing cells therefrom, the resultant molasses was found to have the following composition: 30.0° Brix concentration; 22.2% sucrose concentration; 0.85% (2.83% on Brix) raffinose concentration; and 74.0% sucrose purity.

Sugar solutions of the compositions described above, each in an amount to 20 kg of raw molasses, were separately subjected to sugar-boiling and centrifugal treatments to obtain sugar and waste molasses of the compositions shown below.

	Molasses Undergone .....Raffinose-Hydrolysis...			Molasses not Undergone .....Raffinose-Hydrolysis....		
	Brix	Sucrose Purity (%)	Yield (kg)	Brix	Sucrose Purity (%)	Yield (kg)
Sugar	95.8°	94.2	8.03	95.9°	95.3	6.15
Waste molasses	85.2°	51.2	9.19	79.2°	54.6	11.73

The molasses which had not undergone the raffinose-hydrolyzing treatment gave inferior results in the centrifugal treatment. Thus, the sucrose purity of the sugar had to be heightened by using about 1 liter of hot water.

The data of the preceding table indicate that 5.62 kg of sucrose was recovered in the run involving no raffinose hydrolysis and 7.24 kg of sucrose was recovered in the run involving raffinose hydrolysis each from 20 kg of molasses. This means that there was 1.62 kg of increase in the yield of sucrose. This increase of sucrose yield is noted to far exceed 0.67 kg of sucrose which was formed in consequence of the hydrolysis of raffinose by α-galactosidase.

AMYLOSE

Porous Amylose Powder

M. Kurimoto and K. Sugimoto; U.S. Patent 3,881,991; May 6, 1975; assigned to Hayashibara Company, Japan describe production of powdery amyloses having excellent solubility, absorption and other properties, by hydrolyzing a gelatinized starch solution with isoamylase to a straight-chain amylose solution, slowly cooling the solution thereby forming a crystalline precipitate, and then drying the precipitate by spraying at a low temperature.

Manufacture of straight-chain amylose from starch is accomplished, as disclosed in detail in U.S. Patent 3,730,840, by liquefying a starch slurry either at a high temperature of 160°C or with the use of a liquefying enzyme to a low dextrose equivalent (in the range of 0.5 to 2), and then allowing an enzyme, e.g., the isoamylase produced by the bacteria of *Pseudomonas amyloderamosa* (ATCC 21262) or the pullulanase produced by *Aerobacter aerogenes* (ATCC 8724), that hydrolyzes the α-1,6-glucoside bonds of amylopectin to act in such a manner as to hydrolyze the branches of the branched structure of amylopectin into straight molecules of amylose type. In this way, a low-molecular amylose composed mainly of a polymer having mean degree of polymerization of about 30 to 20 is obtained.

It is essential to separate amylose from water and scour and purify the amylose in the form of crystals, thereby decreasing the water content of the product. This was realized in two ways. In one approach, an amylose-butanol complex was formed and precipitated by adding butanol, an organic acid or the like to amylose. To be more exact, a 5 to 15% solution of amylose was heated to upwards of 80°C for uniform dissolution, and then butanol was added to a saturation point. The mixture was slowly cooled over a period of 20 hours and the butanol complex was precipitated. Thus a crystalline precipitate having a small water content and which was readily separable from water was obtained. After the removal of the supernatant fluid by a centrifuge, a pure amylose complex was left behind.

The other method consisted of the following steps. An amylolyzed starch solution at a concentration of 10 to 20% was heated to more than 100°C and a thoroughly dispersed, homogeneous solution was prepared. It was cooled as slowly as possible, and the resulting crystalline precipitate was centrifuged, and the supernatant fluid was heated again for dissolution, and then the solution was slowly cooled and centrifuged before complete gelling took place. The resultant had a water content of less than 80% and had such low viscosity that could facilitate subsequent treatment. The formation of crystalline precipitate by this method was particularly easy when the isoamylase produced by bacteria of the genus *Pseudomonas* was used.

The amylose precipitates formed by both of the procedures abovedescribed were sprayed by means of a high-pressure pump from the top of a drying column through a nozzle or via a rotary disk. The drying air temperature was at less than 100°C and the amylose temperature was maintained under the temperature of 45°C to avoid gelatinization or dissolution.

The powders thus obtained were white starch powders with very small apparent specific gravity, and their water contents were not more than 10%. X-ray diffraction showed that the products were porous, crystalline powders having considerable crystallinity. Especially the dry powder of the butanol complex appeared to retain much of the helical crystalline structure of the original amylosic substance.

As regards moisture absorption in the air, tests indicated that the powders absorb water at relative humidities of over 80%, and they begin to absorb odors as well. Because of these properties the powders appear to be effective as porous absorbents. As for solubility in water, the powders are faster to disperse and dissolve in water than those obtained by usual processes and are easier to form homogeneous solutions for greater convenience in dissolution and gelatinization. Ordinary products dried as by hot rolls are very hard and difficult to dissolve. They have great specific gravity and have very poor flavor-retaining quality and solubility. Moreover, high heat used for the drying purpose causes partial decomposition and degradation of amylose and makes it impossible to form pure products of uniform quality.

The amylose powders prepared by spray-drying in accordance with this process are porous solids free from the foregoing disadvantages. With relatively large surface areas, the powders according to the process, take the form best suited for the synthesis of amylose derivatives through vapor phase reaction.

Example: Waxy cornstarch at a concentration between 5 and 10% was rapidly heated to 155°C with stirring to form a completely homogeneous gelatinized solution. It was sprayed into vacuum at 50°C for quick cooling. Immediately upon cooling, isoamylase (pullulanase as described in the specification of U.S. Patent 3,622,460) was added for a reaction at pH 6.0 for 30 hours. Then, the mixture was heated to 120°C with stirring to a homogeneous solution, and the solution was cooled slowly over a period of 12 to 24 hours. The low-molecular-weight amylose contained deposited as crystalline powder and could be centrifugally separated. A slurry of the precipitate having a water content of 75% was sprayed by means of a high-pressure pump from the upper part of a drying column through a nozzle. The drying column had an effective height of 15 meters, and dry air at an inlet temperature of not higher than 100°C was forced in a downward stream into the column. The output temperature was kept below 60°C and the material temperature below 40°C. The powder obtained was fine, porous and lightweight, practically globular in shape. The product was readily soluble in warm water and absorbed little water but had a flavor-retaining quality.

Nitrogen-Containing Amylose Derivatives

M. Yoshida and S. Yuen; U.S. Patent 3,870,704; March 11, 1975; assigned to Hayashibara Company, Japan describe a process relating to nitrogenous short chain water-soluble amylose derivatives produced by subjection of starch to the actions of α-1,6-glucosidases, and reaction of urea or urea derivatives with the resultant short chain amyloses with polymerization degrees (DP) of 15 to 50.

The process is characterized in the production of nitrogen-containing short chain amylose derivatives by reacting the short chain amylose and urea or urea derivatives according to either of the following methods. (a) Reacting the mixture at above or below the melting point of urea in the presence or absence of catalysts, such as potassium acetate. (b) Reacting the mixture in solvents, such as dimethylformamide, formamide, etc., under anhydrous conditions with or without the employment of a catalyst at a temperature above or lower than the melting point of urea.

Such amylose carbamates may be formed into water-soluble methylol compounds by methylolation with aldehydes, e.g., formaldehyde, under slight alkaline conditions, and such methylol compounds can be derivatized in turn into water-insoluble and nitrogen-containing short chain amylose derivatives by heating the substance. Moreover, such carbamates have various useful applications, including in materials for the production of coating agents, binders and sizing agents for paper and fabrics.

The amylose employable in this process may be prepared by debranching the α-1,6-glucosidic bonds of amylopectin present in starches with the employment of α-1,6-glucosidases or debranching enzymes derived from cultivation of strains of *Pseudomonas amyloderamosa* (ATCC 21216) or *Escherichia intermedia* (ATCC 21073), etc.

Accordingly, starch comprising 70 to 80% by weight of short chain amylose, DP about 15 to 50, which corresponds with the branched chain length of amylopectin, and 20 to 30% by weight of long chain amylose, is obtainable by treating native common starches (cornstarch, sweet potato starch, potato starch, tapioca starch, and other varieties of starches). Subsequently, the resulting product is cooled gradually whereupon long chain amylose (20 to 30% by weight of the resultant) precipitates. After the long chain amylose is isolated from the reaction mixture, the remainder, 70 to 80% by weight of the starch, will be short chain amylose, DP of 15 to 50, which is employable for the processes described above.

Example: A suspension prepared by admixing 20 g of short chain amylose, which has a relatively high water solubility, 6 g of urea and 20 ml of water, was converted into fluid state by heating in an oil bath at about 110°C and then water was removed by evaporation. When the temperature of the reaction solution reached about 110°C, the temperature of the oil bath was elevated gradually and reaction was carried out for 1 hour maintaining the temperature of the reaction solution at 135° to 140°C under agitating conditions.

By heating, the suspension became a solution. Until nearly complete evaporation of water, the solution exhibited a temperature of about 102°C, the temperature elevated when the lack of moisture content reached an appreciable degree. (Ammonia evolved during the reaction.) The reaction product was a slightly brown colored resinous material which solidified upon cooling. The resultant product was cooled to room temperature after completion of the reaction and dissolved in 50 ml of cold water.

To the aqueous solution of the product was added about 50 ml of methanol and sticky precipitates formed. These procedures were repeated three times and finally to the precipitates were added absolute methanol and pulverized, then a nearly white powder was obtained. The yield was 16.1 g. The nitrogen content of the product was 1.6% and degree of substitution (DS) of urethane, 18 mol %.

To 50 ml of 10% aqueous solution of the powder was added 4.5 ml of 10% formalin. The mixture was then adjusted to pH 10, with sodium hydroxide, and reacted at 50°C for 3 hours with stirring; thus methylolated short chain amylose-urethane was obtained. Such methylolated short chain amylose-urethane was a complete fluid in which no precipitate formed. To the aqueous methylolated short chain amylose solution was added 3% of $Zn(NO_3)_2$. Cotton fabric, and mixed yarn fabrics comprising cotton and synthetic fibers were immersed in the mixture, and heated at 120°C for 30 minutes resulting in the formation of water-insoluble and firm films within the fabrics. The fabrics thus treated maintained their initial starch coating stage even after several repetitions of vigorous washing in hot water.

GLUCAN

Water-Insoluble Glucan

K. Yokobayashi, T. Ikeda and A. Misaki; U.S. Patent 4,072,567; February 7, 1978; assigned to KK Hayashibara Seibutsu Kagaku Kenkyujo, Japan have found that highly-branched, water-insoluble glucan comprising α-(1→3), (1→4) and (1→6) glucosidic linkages (hereinafter referred to as the water-insoluble glucan) is producible in high yield from *Streptococcus salivarius* TTL-LP₁ FERM-P No. 3310, derived from the saliva of healthy persons.

The bacterium of genus *Streptococcus*, wherein the aforementioned strain is present, is cultivated on a medium containing sucrose as the major carbon source, nitrogen source, minerals and other nutrients necessary for the cell growth. Although the culture medium may be in solid or liquid form, a liquid medium is generally used. The water-insoluble glucan may be produced by static culture, but shaking culture or aeration-agitation culture results in higher glucan yield, up to 30 to 45% against material sucrose, w/w, dry solid basis (hereinafter referred to as dsb).

Sucrose is most suitable as carbon source and the desirable concentration range is 1 to 30%, w/v. Synthetic compounds, such as nitrates, ammonium salts, urea and natural organic substances such as polypeptone, corn steep liquor, yeast extract and amino acids may be used freely as nitrogen source. If necessary, inorganic salts such as phosphates, sulfates, potassium salts, calcium salts, magnesium salts, manganese salts and ferrate are employable. Vitamins, nucleic acids and their analogs may be also added to the culture medium as growth factors. The initial pH, when the microorganism starts growing and producing the water-insoluble glucan, is generally in the range of 6.0 to 8.0. The cultivation is carried out until maximum glucan production is attained, which usually requires 24 to 96 hours.

The crude glucan thus obtained can be purified by dissolving the crude product in, for example, 0.5 to 1.0 N aqueous alkali metal hydroxide solution, centrifuging the resulting solution, and then neutralizing the supernatant thereof. The purification will provide a water-insoluble glucan product. Repetition of the procedures and thorough water-washing yield the purified water-insoluble glucan of this process.

Example: A culture medium comprising polypeptone 0.6%, w/v, sodium acetate 1.0%, w/v, K_2HPO_4 0.05%, w/v, KH_2PO_4 0.05%, w/v, NH_4Cl 0.3%, w/v, yeast extract 0.1%, w/v, $MgSO_4 \cdot 7H_2O$ 0.5%, w/v, $MnSO_4 \cdot 4H_2O$ 0.04%, w/v, sucrose 7%, w/v, and tap water was sterilized at 120°C for 15 minutes, inoculated with strains of *Streptococcus salivarius* TTL-LP$_1$ FERM-P No. 3310 and then cultivated at 35°C for 72 hours under aeration and agitation.

Subsequent of the cultivation, the resultant was subjected to water-washing and sieving to obtain crude glucan at a yield of about 43% dsb, against sucrose. The nitrogen and ash contents of the thus obtained crude glucan were respectively 0.25% and 0.17%. The crude glucan was then dissolved in a 1 N-NaOH solution and centrifuged at 10,000 g for 15 minutes. The recovered filtrate was neutralized with 0.5 N-HCl and then the insolubilized glucan was collected by centrifugation. After repetition of the procedures four times, the collected glucan was washed with water thoroughly until no chloride was detected in the wastewater and then dried to form white purified glucan powder at a yield of about 34% w/w dsb against sucrose.

β-1,3-Glucan Derivatives

T. Takahashi, Y. Yamazaki and K. Kato; U.S. Patent 4,075,405; February 21, 1978; assigned to Takeda Chemical Industries, Ltd., Japan describe a process relating to β-1,3-glucan derivatives which comprises reacting a water-insoluble β-1,3-glucan with cyanogen halide, and which provides a water-insoluble carrier of high reactivity useful for the production of water-insoluble enzymes and of columns for affinity chromatography.

The water-insoluble β-1,3-glucan to be employed is exemplified by the polysaccharides elaborated by microorganisms belonging to the genus *Alcaligenes* or the genus *Agrobacterium*. More particularly, there may be mentioned the polysaccharide produced by *Alcaligenes faecalis* var. *myxogenes* 10C3K [*Agr. Biol.*

Chem., vol. 30, pages 196 et seq. (1966) by Harada, et al.], the polysaccharide produced by the mutant strain NTK-u (IFO 13140, ATCC 21680) of *Alcaligenes faecalis* var. *myxogenes* 10C3K (U.S. Patents 3,754,925 and 3,822,250) (hereinafter referred to as PS-1), the polysaccharide produced by *Agrobacterium radiobacter* (IFO 13127, ATCC 6466) or its mutant strain U-19 (IFO 13126, ATCC 21679) (U.S. Patents 3,754,925 and 3,822,250) (hereinafter referred to as PS-2) and pachyman which occurs in the crude drug known as *Poria cocos [Agr. Biol. Chem.,* vol. 32, No. 10, p 1261 (1968)].

As the cyanogen halide which is an activating agent, there may normally be employed the bromine, chlorine or iodine compound or a mixture of such compounds. The alkali to be employed is exemplified by caustic alkali, such as sodium hydroxide or potassium hydroxide, etc., and, normally, is preferably used in 1 to 5 N aqueous solution. Desirably the alkali is added until a pH value of 9 to 13 is established, a pH of 11 being particularly preferred. The alkali is added gradually so as to prevent dissolution of the water-insoluble β-1,3-glucan, generally the rate of addition being preferably in the range of from 0.2 to 0.5 pH unit/min. The following is a specific example of the activation reaction of a water-insoluble β-1,3-glucan with a cyanogen halide.

One part of a powdery water-insoluble β-1,3-glucan was suspended in 20 volume parts of water, followed by the addition of 20 volume parts of water containing 0.1 to 3 parts of a cyanogen halide. Under stirring at an optional temperature of 0° to 50°C, the pH of the reaction mixture was increased to pH 11 by the dropwise addition of a 2 N solution of sodium hydroxide at a rate that would not cause dissolution of the glucan (about 0.5 pH unit per minute). The reaction mixture was then maintained at pH 11 for 15 minutes, whereby the activation reaction was carried to completion. Following the reaction, the solid fraction was recovered by filtration and rinsed with 100 volume parts of water.

The above procedure provided a powdery activated β-1,3-glucan. This activated β-1,3-glucan is insoluble in water and alkali solutions; is thermally ungelable and hydrophilic; and is comprised of particles of such size and strength as will give an adequate flow rate when packed into columns. Therefore, this β-1,3-glucan can be processed by procedures such as the following to obtain water-insoluble enzymes with excellent properties or carrier-ligand products suitable for affinity chromatography. The production of such a water-insoluble enzyme or a carrier-ligand product for affinity chromatography is carried out, e.g., by reacting the above activated β-1,3-glucan with a substance containing primary or secondary amino groups, e.g., an enzyme, protein, peptide, amino acid, an enzyme substrate or inhibitor, antigen, antibody, hormone, etc., preferably in weakly alkaline aqueous solution and at optional temperature in the range of 0° to 50°C.

An enzyme which is bound to an activated β-1,3-glucan can be packed into a bed and be utilized as a reactor for chemical reactions. In this connection, a solution of a substrate can be passed through the enzyme bed to convert this substrate into a valuable product. The reaction is automatically stopped when the solution leaves the bed. The solid enzyme can be used for a long time in a continuous operation. Another valuable example is antibodies bound to water-insoluble β-1,3-glucans. Such products can be combined specifically with the corresponding antigen, e.g., for the analytical determination of the latter. A water-insoluble enzyme of α-amino acid ester hydrolase can be prepared by binding covalently an α-amino acid ester hydrolase as elaborated by a microorganism to a cyanogen halide-activated water-insoluble β-1,3-glucan.

MALTOPENTAOSE

Production from Amylose

T.J. Pankratz; U.S. Patent 4,039,383; August 2, 1977; assigned to E.I. DuPont de Nemours and Company describes a multistep process for obtaining a high yield of maltopentaose which comprises the steps of:

(a) dissolving amylose in a first organic solvent capable of dissociating and dissolving the amylose to form a first solution;

(b) mixing the first solution with an aqueous acid solution capable of partially hydrolyzing the amylose, thereby forming a second solution;

(c) heating the second solution to partially hydrolyze the dissociated amylose into lower molecular weight components;

(d) mixing the second solution containing the partially hydrolyzed amylose with a second organic solvent capable of stripping the first organic solvent from the hydrolyzed amylose to form a third solution, and allowing a precipitate containing partially hydrolyzed amylose to form;

(e) collecting, washing and drying the precipitate to form soluble amylose;

(f) mixing amylase from *Bacillus licheniformis* and soluble amylose in buffered aqueous solution to form a fourth solution;

(g) incubating the fourth solution for a time sufficient to produce a fifth solution containing a high percentage of maltopentaose;

(h) deactivating the amylase in the fifth solution; and

(i) fractionating the fifth solution to obtain a sixth solution predominant in maltopentaose.

The sixth solution can be dried to obtain a solid maltopentaose-containing product. In the preferred embodiment, the first organic solvent is dimethyl sulfoxide, (DMSO), the aqueous acidic solution is an aqueous sulfuric acid solution, and incubation of the fourth solution is accomplished at an elevated temperature.

Example: *Preparation of Soluble Amylose* — A 10-liter graduated cylinder with a contained stirring bar is placed on a Cole-Parmer magnetic stirrer. 4.73 liters of DMSO are added to the graduated cylinder, and the stirring bar is rotated at moderate speed. 1 kg of amylose is added to the graduated container so that the amylose mixes evenly in the DMSO and no clumps are formed. The stirring bar is adjusted to maximum speed and this first solution is allowed to stir overnight or until the amylose goes completely into solution. Any concentration of amylose in the DMSO will work, but, in the preferred embodiment, this first solution contains between about 15 and about 20 w/v of amylose in the DMSO.

60 ml of 10% sulfuric acid (v/v) are then added to the graduated cylinder and mixed thoroughly with the contents of the graduated cylinder to form a second solution. The graduated cylinder with its contents is then placed in an autoclave and heated for 55 min at 121°C to partially hydrolyze the amylose in the second solution. Again, any concentration of acid in the solution will produce some

hydrolyzation, but, in the preferred embodiment, the second solution contains between about 0.05 and 0.20% of H_2SO_4.

15 liters of methanol and 7.5 liters of acetone are brought to a temperature of 4°C and then mixed in a 38 liter container. After the graduated cylinder with its contents has cooled to 100°C, the contents of the graduated cylinder are poured into the cold methanol-acetone solution and stirred. The resulting solution, referred to as the third solution, is then left for a time sufficient to allow a precipitate to form.

By allowing the mixture to stand overnight, most of the precipitated soluble amylose settles from the third solution. The liquid residue is decanted off and discarded, and the precipitate is filtered using a Buchner funnel and coarse filter paper. The precipitate is washed in 3.79 liters of methanol and then removed from the methanol by filtration, again using a Buchner funnel and filter flask. In both collection stages, care must be taken not to allow moisture in the air to dissolve the amylose.

The precipitate is placed in a lyophilization tray and dried overnight in a Hull lyophilizer to provide dry soluble amylose.

Preparation of Maltopentaose — An aqueous buffer solution is prepared using 5 liters of Trisma buffer to which is added 12.1 g of Trisma base and sufficient acetic acid to bring the pH of the buffer solution to 8.0±0.2. This buffer solution is placed in a 20 liter stainless steel container and autoclaved for 15 min at 121°C.

When the buffer is removed from the autoclave, and while the temperature is greater than 90°C, 0.5 ml of amylase obtained from *Bacillus licheniformis* is added to the buffer solution and stirred with magnetic stirring bar. One such amylase is Thermamyl 60 (Novo Chemical Company).

After the amylase is dispersed in the buffer solution, 1,000 g of soluble amylose prepared according to the procedure set forth above is added to the buffer so that no large aggregates of soluble amylose are formed. This fourth solution is autoclaved for 15 min at 121°C. The solution is then removed from the autoclave while it is still hotter than 90°C, and 0.5 ml of the same amylase is again added to the buffer solution. When the temperature of the buffer solution drops to 85°C, an additional 1.0 ml of amylase is added, and 1.0 additional ml of amylase is added at every 5° decrease in temperature, down to and including 55°C, the total amount of amylase added being 8.0 ml.

This fourth solution is incubated overnight at 50° to 55°C using a heated stirring plate. A fifth solution containing a high maltopentaose concentration is formed.

After about 24 hr, a 1 ml sample of this fifth solution is removed and the carbohydrate composition of the sample is determined by standard chromatographic techniques. If, on the basis of this analysis, it is determined that the maltopentaose fraction of the total carbohydrate is greater than about 25%, then, the entire solution is placed in an autoclave for 15 min at 121°C, to deactivate the amylase enzyme. If the maltopentaose fraction of the carbohydrate in the fifth solution is below about 25%, then additional amylase is added to the solution and the incubation continues until the proper maltopentaose content is achieved.

Once the amylase activity in the solution has been destroyed by autoclaving, the solution is cooled to about 50°C, and all particulate matter is removed from the solution by filtration or centrifugation. The supernate is decanted, and the solid impurities are discarded.

The supernate is then separated using a separation column filled with –400 mesh P2 gel (Bio-Rad). Using conventional techniques, fractions eluting from the column containing primarily maltopentaose are collected. These fractions, referred to as sixth (maltopentaose-rich) solution when pooled, contain about 95 to 98% of maltopentaose, with small concentrations of maltotetraose and maltohexaose. The pooled fractions are then lyophilized by conventional techniques to produce a solid maltopentaose-rich product.

Production Using *Streptomyces myxogenes*

K. Iwamatsu, S. Omoto, T. Shomura, S. Inoue, T. Niida, T. Hisamatsu and S. Uchida; U.S. Patent 4,151,041; April 24, 1979; assigned to Meiji Seika Kaisha, Ltd., Japan describe a process to produce maltopentaose and maltohexaose comprising the successive steps of:

   (1)   cultivating *Streptomyces myxogenes* SF-1130 (ATCC 31305) under aerobic conditions in a culture medium containing assimilable carbon and nitrogen sources at a temperature of 25° to 38°C

   (2)   filtering the culture broth under acidic conditions

   (3)   passing the filtrate through a column of strongly acidic ion-exchange resin

   (4)   passing the effluent through a column of active carbon

   (5)   washing the carbon column with water, followed by elution successively with aqueous solutions containing ethanol at different concentrations

   (6)   collecting the given fractions of the eluates

   (7)   concentrating the fractions to dryness to afford a crude powder comprising a mixture of maltopentaose and maltohexaose

   (8)   passing the crude power taken up in water through a column of active carbon which is then washed with water and eluted with aqueous solutions of ethanol

   (9)   collecting the eluates in fractions

  (10)   subjecting the fractions to paper chromatography developing with a mixed solvent of n-butanol-pyridine-acetic acid-water

  (11)   collecting the fractions which give a single spot characteristic of maltopentaose, followed by concentration to dryness to yield maltopentaose in the form of colorless pure powder and collecting the further fractions which give a single spot characteristic of maltohexaose, followed by concentration to dryness to yield maltohexaose also in the form of colorless pure powder.

It is to be noted that no antibacterial activity of maltohexaose and maltopentaose in themselves has been observed. Both enzymes stimulate the host defense system.

Example: The strain *Streptomyces myxogenes* SF-1130 (identified as FERM-P 676 or ATCC 31305) was inoculated to 20 liters of a liquid culture medium (pH 7.0) comprising 5.0% maltose syrup, 2.5% soybean meal, 1.0% wheat embryo and 0.25% sodium chloride. The inoculated medium was incubated under aeration and agitation at 28°C for 66 hr in a jar-fermenter. At the end of the incubation, the resultant culture broth was filtered under acidic conditions (pH 3) to give 15 liters of a broth filtrate.

The filtrate was passed through a column of 2 liters of a strongly acidic ion-exchange resin (Dowex 50WX2) and the effluent from the column was then passed through a column (8 x 50 cm) of 2.5 liters of active carbon. The carbon column was well washed with water and eluted successively with aqueous solutions of 10, 15, 20, 25 and 30% ethanol.

The fractions (10 liters) obtained from the elution with the 20 and 25% ethanolic aqueous solutions were concentrated to dryness to give 20 g of a mixture of crude maltopentaose and maltohexaose in the form of yellowish brown powder.

15 g of the powder was dissolved in 100 ml of water and the solution was passed through a column (4.0 x 33 cm) of 350 ml of active carbon to adsorb the desired substances on the active carbon. The column was well washed with water, followed by elution successively with aqueous solutions of 10, 15, 20 and 25% ethanol. The eluate was collected in 15 ml-fractions and each of the fractions was subjected to paper chromatography developed with a mixed solvent of n-butanol-pyridine-acetic acid-water (6:4:1:3 by volume). The fractions 261 to 340 which gave a single spot at R raffinose = 0.41 (calculated with assumption that the Rf value of raffinose is 1.00) colored by a reagent of silver nitrate were combined together and concentrated to dryness to yield 5.2 g of colorless powder of maltopentaose, then recrystallized from ethanol to yield 4.8 g of pure maltopentoase.

On the other hand, the fractions 396 to 500 which gave a single spot at R raffinose = 0.25 were treated by the procedure as described above to give 3.1 g of pure maltohexaose as colorless powder.

PULLULAN

Production from Saccharides

K. Kato and M. Shiosaka; U.S. Patent 3,912,591; October 14, 1975; assigned to Hayashibara Biochemical Laboratories, Incorporated, Japan describe a process for the production of pullulan by microorganisms in a culture medium containing a saccharide as a principal carbon source. Pullulan is a polysaccharide consisting of α-1,4-linked maltotriose units which are connected by α-1,6-linkages between the terminal glucosidic residues of the trisaccharide. The pullulan-producing microorganisms employable in the process include *Pullularia fermentans var fermentans* IFO 6401, *Pullularia fermentans var fusca* IFO 6402, *Pullularia pullulans* AHU 9553, *Pullularia pullulans* IFO 6353, and *Dematium pullulans* IFO 4464.

Suitable known carbon sources include invert sugar, isomerized sugar, fructose, and glucose, but partly hydrolyzed starch gives very satisfactory results.

Generally, ammonium salts, nitrates, and peptone may be used as nitrogen sources.

Appropriate amounts of phosphate and metal ions, such as magnesium and ferrous ions, may be added to the culture medium. After heat sterilizing of the culture medium and pH adjustment, the microorganism is cultured in the medium with aeration or agitation, at 25° to 30°C, preferably at 27°C for about 7 days. About 3 days after the start of the fermentation, accumulation of a considerable amount of pullulan is observed, and the viscosity of the culture mixture increases.

The residual sugar in the medium is determined at certain intervals, and cultivation is discontinued when the amount of residual sugar approaches a minimum. The microbial cells are removed from the liquid medium by centrifuging, and the cell-free liquid medium is decolorized with active carbon, if desired.

Preferably, the liquor is mixed with a hydrophilic organic solvent, such as methanol or ethanol, to precipitate the accumulated pullulan which is recovered by centrifuging. If desired, the recovered pullulan is dissolved in warm water and again precipitated by solvent. The pullulan obtained after drying is a whitish powder which very readily dissolves in water to form a viscous solution. The molecular weight of the pullulan obtained varies depending on the culture conditions between 50,000 and 4,500,000. The yield of pullulan based on the sugar employed as a carbon source also varies from 20 to 75% depending on the culture conditions.

At a pH of 5.5, the viscosity of the broth was over 1,000 cp, whereas at pH 7.0 to 8.0 the viscosity was merely 24 to 31 cp. However, a pH of 7.0 to 8.0 results in a yield above 50%, while the yield of pullulan based on the initial sugar was 30% at pH 5.5.

The mean molecular weight of the pullulan was about 3,000,000 at pH 6 of the culture medium, whereas it was about 80,000 at pH 7.0.

The higher the pH value of the culture medium, the quicker the consumption of sugar and the formation of pullulan, and the higher the yield of pullulan.

It has further been found that the concentration of phosphate ion in the medium affects the culture period and the yield of pullulan. For example, 0.1 to 0.5 g/dl phosphate at pH 5.5 results in the production of pullulan having a molecular weight from 1,500,000 to 3,000,000, and the molecular weight of the pullulan decreases with increasing phosphate concentration.

Method of Purification

K. Kato and T. Nomura; U.S. Patent 4,004,977; January 25, 1977; assigned to K.K. Hayashibara Seibutsukagaku Kenkyujo, Japan describe a method for purifying pullulan which comprises cultivating a strain of *Aureobasidium pullulans* in a liquid medium, removing the cells from the culture broth, precipitating pullulan present in the broth with at least one solvent selected from the group consisting of alcohols, esters and ethers respectively with three or more carbon atoms and ketones with four or more carbon atoms to recover, and then dehydrating and/or drying the recovered pullulan.

Suitable solvents employable in the precipitation are those with relatively low hydrophilicity, more particularly, alcohols with three or more carbon atoms such as propyl alcohol, isopropyl alcohol, n-butyl alcohol and sec-butyl alcohol, esters with three or more carbon atoms such as methyl acetate, ethers with three or more carbon atoms such as tetrahydrofuran, dioxane, ethylene glycol monoethyl ether, ethylene glycol monobutyl ether and diethylene glycol monomethyl ether, and ketones with four or more carbon atoms such as methyl ethyl ketone, diacetone alcohol and acetylacetone.

The solvents enumerated above effect precipitation and purification of pullulan with a small amount and with ease. Especially, methyl ethyl ketone, methyl acetate and sec-butyl alcohol have lower water solubility, about 20 to 25%, and a portion thereof dissolves and is dissolved mutually in and with water to form two layers. A one-third to one-fourth, v/v, addition of such solvent to culture broth brings saturation of the resultant mixture and effects complete precipitation of pullulan. The efficacy is sufficiently realized with an addition of an amount of less than one-tenth against that of acetone or low molecular alcohol.

In this case, most of the residual sugars such as mono- and oligosaccharides, pigments and other impurities are eluted satisfactorily in the water layer. In addition, slight amounts of the impurities which remain with water in the precipitated pullulan layer can be eluted and removed by adding again a small amount of the solvent to the layer. Thus, products with higher purities than those obtained by effecting precipitation with methanol or acetone, can be obtained.

STARCHES

Production of Starch Hydrolysates

N.C. Holt, C. Bos and K.V. Rachlitz; U.S. Patent 3,910,820; October 7, 1975; assigned to DDS-Kroyer A/S, Denmark describe a method of making starch hydrolysates which comprises the step of subjecting a slurry of a milled crude starch-containing product to a pretreatment before the addition of enzymes, the pretreatment comprising the step of heating the slurry to a temperature of between 50 and 65°C for a period of time of from 1 to 5 hr.

In an enzymatic liquefaction of starch, the starch slurry containing added liquefaction enzyme is normally quickly heated through the gelatinization range of the starch to a temperature of 80 to 90°C in order to effect a rapid enzymatic decomposition of the starch and thus to reduce the viscosity of the gelatinized starch slurry.

In the pretreatment in this method, no enzymes are added and the starch slurry is heated to a temperature below the gelatinization temperature of the starch and maintained at that temperature for a considerable period of time. During the pretreatment the starch absorbs water and swells but no significant gelatinization takes place.

As a result of the water absorption obtained during the pretreatment, the attack of the starch molecules by the liquefaction enzymes during the subsequent liquefaction step at temperatures of between 80° and 105°C is facilitated. Consequently, the necessary liquefaction time can be reduced and the well-known retrogradation which is effected when starch and especially the amylose fraction thereof is heated over long periods of time can be avoided. Such a retrogradation results in the formation of products which subsequently cannot be broken down and therefore cause difficulties in the subsequent filtration and reduce the yield of the final product.

The elimination of these retrogradation products and the obtaining of a complete liquefaction makes it possible to filter the liquefied product. Such a filtration of the liquefied product has previously been considered to be undesirable partly due to the presence of the retrogradation products which make such a filtration difficult to perform and partly because the partially converted product or a product containing free starch has a higher viscosity than the conversion product formed after subjecting the liquefied product to further treatment, for example a saccharification.

As set forth above, the liquefaction enzyme preparation is added when the pretreatment has been completed. This is due to the fact that the activity of alpha-amylase may be reduced if alpha-amylase is present in the starch slurry during the pretreatment. Thus, it appears that Ca ions, which normally are added to the starch slurry in order to protect the alpha-amylase against inactivation, form complexes with the proteins during the pretreatment. Furthermore, it appears the crude starch-containing materials contain some inhibitors for the alpha-amylase which inhibitors are active at temperatures of between 50° and 65°C.

The filtered liquefied product may be subjected to an enzymatic saccharification so as to obtain a hydrolysate of a high DE value. If necessary, the product is finally refined, for example by treatment with active carbon and by ion exchange.

Example 1: Dry milled maize grits are slurried with hot water (60°C). The slurry is kept for 1½ hr at 60°C and is continuously agitated. The concentration is 30% w/w. The pH is kept between 6.4 and 6.6 with dilute sodium carbonate solution. Also $CaCl_2$ solution is added at a concentration of 0.5 g calcium ions per liter slurry.

After 1½ hr retention time alpha-amylase is added into the slurry at a concentration of 1.2 g per kg dry substance just before the slurry is pumped through a converter tube where the temperature is raised to 89°C. After the converter tube, an additional amount of alpha-amylase is added, 1.4 g per kg dry substance, and the slurry is kept at 85°C for an additional 3 hr.

The pH is then adjusted to 4.8 and the slurry filtered. The hydrolysate obtained is crystal clear and has a DE value of between 28 and 32.

Example 2: A hydrolysate prepared as described in Example 1 and having a concentration of 28% dry substance w/w is passed through ion-exchange columns and/or active carbon and then concentrated to a concentration of 80% dry substance w/w syrup. This product is then spray-dried, to produce a white free-flowing nonhygroscopic powder.

Example 3: A filtered hydrolysate prepared as described in Example 1 is put in saccharification tanks, the temperature is adjusted to 60°C and saccharification enzymes (amyloglucosidase) are added at a concentration of 2.4 g/kg dry substance, and the solution continuously agitated for 48 to 72 hr. The product which then has a DE of 97 to 99 is then refined by means of ion-exchange and active carbon, and concentrated to 74% dry solids w/w and put in crystallizers, where dextrose monohydrate crystallizes out. The crystals are washed and dried to form pure dextrose monohydrate. Alternatively, after refining the dextrose liquor, it is concentrated to 85 to 92% dry substance w/w and spray crystallized to produce a total sugar having a DE value of between 97 to 99 and a moisture content of less than 1%.

Lipophilic Starch Derivatives

C.N. Richards and C.D. Bauer; U.S. Patent 4,035,235; July 12, 1977; assigned to Anheuser-Busch, Incorporated describe a spray-dried enzyme converted starch reaction product which may be used in the form of an aqueous dispersion with an oil to form a stable emulsion. The oil-in-water emulsion may be dried and later reconstituted to provide a stable emulsion.

When resuspended in water, the spray-dried product causes a cloud effect. The cloud effect is an opaqueness in the fluid which is used in certain types of drinks made from dried flavorings.

The starch used is waxy maize, tapioca, dent, potato, wheat, rice, or other starches. The preferable starch is waxy maize. The starch is treated with a substituted cyclic dicarboxylic acid anhydride of the following structural formula:

$$
\begin{array}{c}
O \\
\parallel \\
C \\
O \diagdown \diagup \diagdown R\text{--}R' \\
C \diagup \\
\parallel \\
O
\end{array}
$$

wherein R represents a dimethylene or trimethylene radical and wherein R' is the substituent group, which is a hydrophobic group (ordinarily a long chain hydrocarbon radical). Substituted cyclic dicarboxylic acid anhydrides falling within the above structural formula are the substituted succinic and glutaric acid anhydrides.

The hydrophobic substituent group R' may be alkyl, alkenyl, aralkyl, or aralkenyl and should contain 5 to 18 carbon atoms. R' may be joined to the anhydride moiety R through a carbon-to-carbon bond (as in alkenyl succinic anhydride), or through two carbon-to-carbon bonds (as in the adduct of maleic anhydride with methylpentadiene, or as in the cyclo-paraffinic cyclo-dicarboxylic acid anhydrides, such as for example, cyclohexane-1,2-dicarboxylic acid anhydride), or may be linked through an ether or ester linkage (as, for example, in octyloxy succinic anhydride or in capryloxy succinic anhydride).

The products formed by the reaction of starch with any of the above-listed reagents are the acid esters of the substituted dicarboxylic acids and, more specifically, they are the acid esters of either substituted succinic or glutaric acid. These acid esters may be represented by the following structural formula:

$$
\begin{array}{c}
COOH \\
\vert \\
Starch\text{--}OOC\text{--}R\text{--}R'
\end{array}
$$

wherein R is a dimethylene or trimethylene radical and R' is the substituent hydrophobic group (this being an alkyl, alkenyl, aralkyl, or aralkenyl group containing from 5 to 18 carbon atoms). The hydrophilic group in all cases is the remaining free carboxyl group (COOH) resulting from the esterification of only one of the carboxyl groups.

In making the starch derivative of this process at least 2%, and preferably at least 3% (based on the weight of dry starch), of n-octenyl succinic anhydride must be used in order to give a product which has the desired emulsifying, coating, and clouding properties. Varying amounts of other anhydrides, depending on the molecular weight thereof, are used. Generally, from 0.1% to about 10% by weight anhydride is added to the starch, based on the dry weight of starch.

The pH of the starch-substituted cyclic dicarboxylic acid anhydride reaction should be adjusted to the range of 5 to about 11. Preferably the pH is adjusted to 7 to 8 by the addition of an alkaline solution of sodium hydroxide, sodium carbonate, or any alkali metal base.

After the pH of the starch slurry is adjusted as mentioned, from 0.1 to about 100% (by weight based on the weight of dry starch) of substituted cyclic dicarboxylic acid anhydride is added. Preferably, n-octenyl succinic anhydride is used in the amount of 2 to about 5%.

After the pH is stabilized, the slurry is kept in a range of 60° to about 130°F, preferably 75°F, for about 1 to 10 hr to complete the reaction. The reacted starch is then washed and filtered. The water washing removes impurities and provides for a better flavored product. The preferred starch reaction product has about 2 to 3% n-octenyl succinyl substitution. The percent substitution depends upon the amount of reactant and the efficiency of the reaction.

The modified starch preferably is resuspended in water at about a 35% solids concentration. The concentration can be from about 10 to 50% solids. A starch conversion enzyme, preferably alpha-amylase (liquifying enzyme) is added to the starch suspension. The amount of enzyme used depends on activity, time of reaction, pH, temperature, degree of conversion desired, use of activators, etc.

The enzyme treatment decreases the viscosity of the starch suspension to the desired level for the final product. The enzyme treatment is at a temperature of about 60° to about 100°C, or higher for very heat stable enzymes. This temperature is above the temperature at which the starch starts to swell. The swelling allows the enzyme to attack the granular starch to depolymerize the starch ester. The starch is partly cold water soluble because of the breaking up of the starch polymers by the enzyme. A molecular dispersion of the starch occurs after depolymerization of the enzyme.

Deactivation of the alpha-amylase is needed to control the reaction to prevent the desired viscosity of the starch from changing. The enzyme is deactivated by the application of heat and/or addition of sodium hypochlorite. A small amount of oxidation of the starch molecule may occur from the sodium hypochlorite. The sodium hypochlorite may also purify the starch by killing bacteria.

The modified starch is then spray-dried using an inlet temperature of about 175° to 700°F (preferably about 500°F), and an outlet temperature of about 150° to 350°F (preferably 225°F). Inlet and outlet conditions are selected so that desired drying is effected and no undesirable damage (browning, charring, etc.) occurs to the product. Any drying technique, such as passing over heated drums may be used, but spray drying is a preferred method for waxy starch-n-octenyl succinate.

Example: One hundred parts of waxy maize starch is slurried to a Baumé of 21.0°. Three parts of n-octenyl succinic anhydride is added to the starch slurry and the pH is kept between 7.0 and 8.0 with a basic solution (70 g NaOH-150 g Na_2CO_3 per liter). The temperature is kept at 75°F. When the pH is stabilized, the starch is washed and filtered. The modified starch (10,000 parts) then is resuspended in water to a 35% solids suspension.

One part of an alpha-amylase, Rhozyme 86L, is added to the suspension. The temperature is raised to 80°C until a Dudley viscosity of about 120 seconds at 80°C is achieved. Then 5.0 parts of sodium hypochlorite is added and the temperature is raised to 94°C for 30 min. The starch is then spray dried using an inlet temperature of 500°F and an outlet temperature of 225°F.

The product, when mixed with coconut oil in a 1:1:2 starch:coconut oil:H_2O ratio and emulsified, forms stable emulsions.

VARIOUS POLYSACCHARIDES

Production from Glucides Using Ammonium Phosphate

J.C. Campagne; U.S. Patent 4,154,654; May 1979; assigned to Rhone-Poulenc Industries, France describes a process for the production of polysaccharides by fermentation of glucides in the presence of selected microorganisms and effective nitrogen-containing compounds.

The most preferred microorganism is *Xanthomonas campestris,* while the preferred inorganic nitrogen-containing compound is ammonium phosphate, most preferably diammonium phosphate. The glucide may be selected from any of a number of effective compounds, the sugars being most preferred.

The fermentation mixture is maintained within a pH range of from about 6 to 7.5 preferably from about 6.5 to 7.2, by a buffer agent such as dipotassium phosphate. Alternatively, if the medium is not buffered, a pH regulator may be employed to introduce requisite amounts of an alkaline reagent, such as sodium hydroxide, potassium hydroxide, or lime, which may or may not be in solution, into the medium.

Upon conversion of the glucide to the polysaccharide, the latter may be appropriately extracted as by, for example, precipitation, washed, dried and ground to a usable form for storage. Subsequently, by adding the ground polysaccharide to an aqueous solution, gels of the appropriate consistency may be produced, which gels exhibit both a high viscosity and good filterability.

Example: A 500 cc Erlenmeyer flask is charged with 75 cc YM broth which is subsequently inoculated, by means of a platinum loop, with a culture of *Xanthomonas campestris* maintained on agar in a tube. The broth culture medium, obtained from DIFCO Chemical Company in dehydrated form, has the following composition:

Component	Amount
Yeast extract for bacteriology	3 g
Malt extract	3 g
Soya peptone for bacteriology	5 g
Pure glucose	10 g

This mixture is allowed to incubate at from about 28° to 30°C for about 48 hr and the resultant contents are employed to inoculate a sterile medium (6 liters), contained in a 10 liter laboratory fermenter and having a composition of:

Component	Amount
Glucose	20 g/l
Diammonium phosphate	1.5 g/l
Dipotassium phosphate	3 g/l
Magnesium sulfate heptahydrate	0.25 g/l
Antifoaming agent	2 cc/l
Water (qsp)	6,000 cc

The dipotassium phosphate serves to buffer the pH of the composition within the range of from about 6.9 to about 7.5.

The composition is allowed to ferment, under stirring and aeration, for 64 hr at a temperature within the range of from about 28° to about 30°C. After that period, no glucose is found in the medium. The viscosity is measured on a Brookfield LVT viscometer at 30 rpm with a No 4 needle and found to be 4,800 cp. The polysaccharide content is found to be 14.4 g/kg, corresponding to a yield of 72%.

Production Under Carbon-Limiting Conditions

R.C. Righelato and T.R. Jarman; U.S. Patent 4,110,162; August 29, 1978; assigned to Tate & Lyle Limited, England describe a process for the production of a polysaccharide consisting of a partially acetylated variable block copolymer of 1,4 linked D-mannuronic and L-guluronic acid residues. The process comprises subjecting to continuous cultivation a bacterium of the species *Azotobacter vinelandii* under aerobic conditions in an aqueous culture medium containing as essential ingredients at least one monosaccharide or disaccharide as carbon source and sources of phosphate, molybdenum, iron, magnesium, potassium, sodium, calcium and sulfate.

The medium contains a fixed source of nitrogen and/or the aerating gas contains nitrogen. The concentration of the saccharide carbon source in the medium is limiting on the growth of the bacterium, and during the cultivation the pH in the medium is maintained within the range of 6.0 to 8.2.

Any strain of *Azotobacter vinelandii* can be used in the process. However, particularly valuable strains which give especially good yields of the polysaccharide are those bearing the culture collection numbers NCIB 9068 and NCIB 8789, and that bearing the culture collection number NCIB 8660 (National Collection of Industrial Bacteria) and are described in the catalogue of the collection.

Of the various mono- and disaccharides suitable as carbon sources in the production of polysaccharide by *Azotobacter vinelandii,* it was found that sucrose is preferable.

Apart from the use of limiting conditions for the monosaccharide or disaccharide carbon source, the medium can otherwise contain any of the usual components used in the production of polysaccharide from *Azotobacter vinelandii*.

It was however, found that the rate of oxygen supply is related to the efficiency with which the monosaccharide or disaccharide carbon source is converted into polysaccharide. In continuous culture conditions, it is the relationship of the oxygen level to the cell concentration which is critical and not the oxygen concentration in the medium as such. High oxygen levels promote the respiration of the microorganism, thus converting the sucrose (or other carbon source) into carbon dioxide.

On the other hand, oxygen-limiting conditions restrict the formation of polysaccharide. Careful control of the oxygen supply, can however, give a sucrose conversion figure of about 30% or even over 40%, thus providing a much more efficient utilization of the carbon source. Typically the oxygen uptake is restricted to from 5 to 20 mmol O_2/hr/g cell.

Example: *Alginate Production Under Carbon-Limiting Conditions in Continuous Culture* — *Azotobacter vinelandii* NCIB 9068 was grown continuously in a continuous culture of the chemostat type (Herbert, Elsworth and Telling 1956, *Journal of General Microbiology*, 14, 601) using *Azotobacter vinelandii* NCIB 9068 in a continuous culture apparatus with a culture volume of 1.0 liter, with the phosphate decreased to the concentrations stated. The medium was pumped into the culture at 150 ml/hr and the culture broth overflowed via a standpipe weir into a receiver vessel. The temperature was controlled at 30°C. The pH was controlled at 7.4 by automatic addition of 1 M NaOH. Air was sparged into the fermenter at 1 liter/min, the impeller speed being adjusted so that oxygen uptake rates in the range 10 to 30 mmol/g cell/hr were obtained.

Samples were taken from the fermenter at daily intervals and assayed for cell mass and polysaccharide concentration. To each 40 ml sample, 0.8 ml of 0.5 M EDTA plus 0.8 ml of 5 M NaCl was added. The samples were then centrifuged at 25,000 g for 40 min. The cell pellet obtained was resuspended in distilled water, centrifuged at 25,000 *g* for 40 min and the supernatant decanted. The preweighed tube containing the sediment was dried at 105°C for 12 hr and weighed. The polysaccharide was precipitated from the supernatant of the first centrifugation by adding three volumes of propan-2-ol. The precipitate was collected by filtration, dried in vacuo at 45°C for 24 hr and weighed.

The culture medium, the final composition of which is given in Table 1, was prepared and added to the culture in two batches. Batch (1) was autoclaved at 1 kg/cm^2 for 1 hr in 2 parts which were combined aseptically after cooling. One part contained sucrose, KH_2PO_4 and K_2HPO_4 in 14 liters, the other part contained $MgSO_4$, NaCl and trace elements other than Ca and Fe in 4 liters. Batch (2) contained $CaCl_2$ and $FeCl_2$ in 2 liters; the $CaCl_2$ was autoclaved in 1.9 liter and the $FeCl_2$ was filter sterilized and then added to the bulk of Batch (2). Batches (1) and (2) were added to the culture through separate lines, (1) being added at a rate of 135 ml/hr and (2) at 15 ml/hr.

Carbon limitation was demonstrated by increasing the sucrose concentration in the medium to 40 g/l. Within 5 hr, the cell concentration increased by more than 10%. The level of sucrose was determined by GLC.

Table 1

Constituent	Amount (g/l culture medium)
Sucrose	8 (24 mM)
KH_2PO_4	0.064
K_2HPO_4	0.25
$MgSO_4 \cdot 7H_2O$	1.6
NaCl	1.6
Na_2MoO_4	0.008
$CaCl_2 \cdot 2H_2O$	0.34
$FeCl_2 \cdot 2H_2O$	0.017
H_3BO_4	23×10^{-3}
$CoSO_4 \cdot 7H_2O$	9×10^{-3}
$MnCl_2 \cdot 4H_2O$	0.7×10^{-3}
$ZnSO_4 \cdot 7H_2O$	9×10^{-3}
$CuSO_4 \cdot 7H_2O$	0.8×10^{-3}

The amount of polysaccharide produced per g cell under carbon limitation and the amount of residual sucrose are shown in Table 2. The residual sucrose level under conventional conditions may be as high as 15 mM.

Table 2

Concentration of limiting substrate in culture medium (mM)	24
Cell concentration (g/l)	1.3
Respiration rate (mmol O_2/hr/g cell)	16
Polysaccharide concentration (g/l)	2.2
Conversion efficiency* (%)	28
Residual sucrose in culture medium (mM)	0.5
Polysaccharide yield per g cell (g)	1.7
Dilution rate (h^{-1})	0.15
Productivity** (g/l/hr)	0.33

*Utilized substrate into polysaccharide produced.
**Polysaccharide concentration x dilution rate.

Production Under Phosphate-Limiting Conditions

In a similar process, *R.C. Righelato and L. Deavin; U.S. Patent 4,130,461; Dec. 19, 1978; assigned to Tate & Lyle Limited, England* describe a method of producing a polysaccharide from *Azotobacter vinelandii,* the concentration of phosphate in the medium being limiting on the concentration of the bacteria and being at least 1.0 millimolar.

It was found that under conventional continuous culture conditions, for example, those using a chemostat, particularly advantageous phosphate levels are from 2.0 to 3.0 millimolar, giving a cell concentration of about 4 to 4.7 g/l at a dilution rate of 0.15 h^{-1} or about 14 g/l at 0.05 h^{-1}. Use of these levels has been found to give a polysaccharide concentration of up to about 27 g/l. No upper limit to the phosphate concentration can be set, except that dictated by the practical considerations governing the continuous fermentation process.

Heteropolysaccharide from Fermentation of Methanol

A.L. Tannahill and R.K. Finn; U.S. Patent 3,878,045; April 15, 1975; and R.F. Finn, A.L. Tannahill and J.E. Laptewitz, Jr.; U.S. Patents 3,923,782; December 2, 1975; and 3,932,218; January 13, 1976; all assigned to Cornell Research Foundation, Inc. describe the preparation of a heteropolysaccharide by fermentation which comprises culturing a heteropolysaccharide-producing strain of microorganism of the genus *Methylomonas* on a culture medium containing methanol as the sole source of assimilable carbon. The improvement comprises initiating the fermentation reaction with a methanol concentration of between about 0.5 to 2% by volume on a suitable salts medium and subsequently providing additional methanol in the range of from about 1 to 4% by volume after the aqueous fermentation broth displays an optical density of between about 1.0 and 5.0.

Fermentation is carried out at a pH between about 6.0 and 7.8 and preferably within the range of 6.2 to 7.5, and at a temperature between about 25° and 33°C. A further feature of the process comprises providing adequate levels of available iron in the medium, either as ferrous or ferric ions, to secure enhanced yields of biopolymer. The heteropolysaccharide produced by this process exhibits improved thickening and drag reducing properties when employed in dilute aqueous solutions.

The present organism, since it does not fall within any previously described species, has been designated as *Methylomonas mucosa* (NRRL B-5696).

The broth medium utilized for growth of the microorganism consists of the following:

Material	Grams per Liter
KH_2PO_4	3.75
Na_2HPO_4	2.50
$NaNO_3$	2.50
$MgSO_4 \cdot 7H_2O$	0.40
$Ca(NO_3)_2 \cdot 4H_2O$	0.005
$FeSO_4 \cdot 7H_2O$	0.005
$ZnSO_4 \cdot H_2O$	0.005

Generally, the cultivation procedure involves adding the salts to water in the above listed order, followed by pasteurization or autoclaving, if desired, since this is not essential. Methanol is then added to the salt solution at approximately room temperature. When solid medium is desired to carry the organism, 1.75% Bactoagar may be added to the salts before pasteurization or autoclaving. Growth is carried out preferably at about 30°C in a shaker-incubator utilizing, for example, 1 liter shaker flask filled with from about 200 to 400 ml of broth. Growth times are normally from about 48 to 96 hr, after which a highly viscous slime has been produced in the culture broth.

Similarly, recovery of the fermentation product may be accomplished in a conventional manner utilizing acetone, methanol, propanol, quaternary ammonium salts, etc. as precipitants, with acetone being preferred. For example, the broth culture may be first centrifuged for a period of from 10 to 20 min to remove some of the bacterial cells, and the clarified, slightly yellow supernatant decanted from the centrifuge to leave a cell pellet behind. To this supernatant is added from 1.2 to 2 parts by volume of acetone and the mixture is well mixed. The biopolymer, which is obtained as a light colored cottony precipitate, is then drained, placed into from 2 to 3 parts of fresh acetone to remove as much water as possible, and finally, is dried at room or elevated temperature to give a dry, powdery product.

Example: 350 ml of a culture broth having the following composition are sterilized for 10 min at 121°C:

Material	Grams per Liter
KH_2PO_4	3.75
Na_2HPO_4	2.50
$NaNO_3$	2.50
$MgSO_4 \cdot 7H_2O$	0.40
$Ca(NO_3)_2 \cdot 3H_2O$	0.025
$FeSO_4 \cdot 7H_2O$	0.005
$ZnSO_4 \cdot H_2O$	0.005

The above broth was dispensed into a 1 liter indented shake flask and was mixed with 1½% by volume of methanol, inoculated with a strain *Methylomonas mucosa* and was cultivated with shaking at 30°C. The pH is metered constantly and appropriate amounts of KOH or H_2SO_4 are added to maintain the pH at 7.0. At the end of 24 hr, the optical density of the fermentation broth is 2.5, and at this point an additional 1½% by volume of methanol is added to the broth. Fermentation is conducted for an additional period of 24 hr.

At the end of the fermentation time, the broth culture is centrifuged at 20,000 *g* for 15 min to remove some of the bacterial cells. Then acetone is added in an amount of 1.5 volumes per volume of broth supernatant obtained by decanting the clear, slightly yellow liquid from the centrifuge tube, leaving the cell pellet behind. The acetone-broth combination is mixed well, and cell-free polymer is recovered as a light-colored cottony precipitate. The polymer is drained, then placed into 2 parts of fresh acetone to remove as much water as possible, and is finally dried at room temperature for 24 hr. 9.8 g of the crude solid polymer are obtained.

Alkali Metal Glycerophosphate in Methanol-Containing Medium

In an improvement on the previous process, *J.G. Savins; U.S. Patent 4,006,058; February 1, 1977; assigned to Mobil Oil Corporation* describes the production of a heteropolysaccharide by fermentation of a methanol-containing culture medium with a microorganism of the strain *Methylomonas mucosa*. This improvement comprises incorporating into the culture medium an alkali metal salt of glycerophosphoric acid to provide a readily assimilable source of phosphate. The alkali metal glycerophosphate is the predominant source of the assimilable phosphate in the culture medium and preferably is present in an amount within the range of 0.3 to 3.0 weight percent.

There is also incorporated into the culture medium a chelating agent for iron selected from the group consisting of the alkali metal salts and mixed ferric/alkali metal salts of ethylenediaminetetraacetic acid. In a still further improvement, tricine, N-tris(hydroxymethyl)methyl glycine, is employed in the culture medium as a hydrogen ion buffer.

While any alkali metal glycerophosphate may be used, sodium glycerophosphate is preferred because of its commercial availability. This material is easily incorporated into the culture medium and readily assimilated by the microorganism to provide the phosphate needed for synthesis of biomass and biopolymer. An especially suitable phosphate source is the disodium pentahydrate (α and β mixture) of glycerophosphate.

Emulsifying Agent

J.E. Zajic and E. Knettig; U.S. Patent 3,997,398; December 14, 1976; assigned to Canadian Patents and Development Limited, Canada describe a process for the production of an emulsifying agent of microbiological origin comprising the step of cultivating by an aerobic fermentation in aqueous solution and with paraffinic hydrocarbon substrate as principal source of assimilable carbon a microorganism of species *Corynebacterium hydrocarboclastus* of the type UWO 419 or NRRL-B-5631 until the fermentation medium contains at least 0.1% by weight of an active emulsification agent consisting of an extra-cellular polymer formed as a result of the fermentation.

The extra-cellular polymer comprises a polysaccharide component including galactose, glucose and mannose.

The nutrient medium consists of a solution in tap water of the following by weight:

	Percent by Weight
$NaNO_3$	0.5
K_2HPO_4	0.5
Yeast extract	0.3
KH_2PO_4	0.2
$MgSO_4$	0.2
NaCl	0.1

Under laboratory conditions about 15 liters of this solution were added to a 24 liter fermentation vessel, together with about 5% by volume of culture inoculum and about 1.5% by weight of kerosene. The pH of the contents was adjusted to 6.5 to 6.8. The temperature was controlled at $28°\pm0.5°C$, aeration was supplied by bubbling air therethrough at 5 l/min, while the contents were agitated vigorously by an impeller operating at 500 rpm. A standard plot of cell growth against time gives a sigmoid curve with the exponential phase completed in 60 hr.

The fermentation broth itself may be used as an emulsifying agent, and may be freeze-dried directly for storage purposes, or the active principle comprising an extracellular polymer that accompanies the cell formation may be separated from the broth. In a typical separation any unused kerosene is removed from the broth by extraction with ether or chloroform, and thereafter the cellular component is removed by centrifuging (e.g., at 10-12,000 rpm in a Sorvall centrifuge for 30 to 60 min). The cellular mass is dried overnight and weighed to determine yield.

The impure polymer also is effective as an emulsification agent and may be recovered from the supernatant liquid of the centrifugate by precipitation at pH 3.0 to 6.5 with 2 or more volumes of alcohol, acetone or similar solvents. The polymer is soluble in water and weak or strong acids or bases. The precipitation starts at pH 5.0 and no precipitation occurs at pH 7.0.

If desired the polymer can be purified by redissolving in water, if necessary, dialysis with water and subsequent removal of low molecular weight contaminants, particularly those with molecular weight less than 20,000 by first filtration through Millipore filters of 1.2 μ and 0.45 μ pore size and then by use of an ultra filter with Diaflo membrane PM-30 or XM-50. The purified agent is reprecipitated and the solvent removed under vacuum at slightly elevated temperature, or by use of dry heat at temperatures less than 130°C.

The fermentation broth, the precipitated polymer and the isolated polymer are all usable and highly effective as emulsifying agents for long chain paraffinic hydrocarbons (e.g., of carbon content greater than 10), particularly fuels such as kerosene and Bunker C fuel oil.

Nonfibrous Polysaccharides

A. Bouniot; U.S. Patent 3,988,313; October 26, 1976; assigned to Rhone-Poulenc SA, France; has found that it is possible to produce fermentation polysaccharides in the form of a powder which consists of small granules and which consequently has good flow characteristics. To achieve this result, it is necessary to use an aqueous organic washing liquid which contains neither too much nor too little water for washing the polysaccharide produced by precipitation from the fermentation medium.

It has been found that if the washing liquid contains a proportion of the organic liquid below a certain level (referred to herein as a percentage of organic liquid y% above the minimum percentage of the organic liquid at which the polysaccharide is insoluble), the polysaccharide fibers, after washing, are very soft, difficult to dry, and tend to agglomerate with one another.

On the other hand, if the washing liquid has more than a certain proportion of the organic liquid (referred to herein as a percentage of organic liquid z% above the minimum percentage of the organic liquid at which the polysaccharide is insoluble) the product, after it has been washed, dried and ground, has a fibrous and not a granular structure.

Between these two limits (i.e., if the washing liquid contains a proportion by weight of organic liquid from y% to z% above the minimum percentage of the organic liquid at which the polysaccharide is insoluble), the polysaccharide fibers, after washing, are easy to dry, and the product, after drying and grinding, is granular and not fibrous.

It is necessary to determine beforehand, for a given type of operation, the correct range of proportions of organic liquid in the washing liquid, as defined above, as well as, where appropriate, the exact optimum proportion, because the values of these working factors are not the same for all washing liquids nor for all working conditions.

For a given organic washing liquid, in order to find the optimum range of proportions by weight of this agent (between y% and z% above the proportion which corresponds to the solubility limit of the polysaccharide in the washing liquid), it is advisable to carry out prior tests. It is convenient and effective to carry out these tests in the manner described below.

If an organic washing liquid, which can also be used to precipitate the polysaccharide from the fermentation liquid, is to be used, the solubility limit of the polysaccharide is first determined by making a gel, for example, containing approximately 1% by weight of the polysaccharide in distilled water, and then introducing, over the course of less than 5 min, the organic washing liquid under investigation into a volume of approximately 2 liters of this gel, which is kept at the desired temperature and is stirred, until the polysaccharide precipitates.

It is then easy to calculate the concentration of water relative to the total of water plus organic liquid when precipitation occurs. A second similar test is preferably carried out thereafter, which makes it possible to determine the concentration of organic liquid with more precision.

In a second stage, parallel tests of washing the freshly drained, precipitated polysaccharide are carried out, using, as the washing liquids for the various parallel tests, mixtures of water and the organic washing liquid. These mixtures contain gradually increasing proportions of the organic liquid, for example, proportions of 5, 10, 15, 20, and 25% and the like respectively (or, alternatively, gradually decreasing proportions of water relative to the total of water plus organic agent, which makes it possible to know the corresponding proportions of the latter), above the proportion by weight of organic washing liquid corresponding to the solubility limit as determined by the test above.

These washing processes are effected using the working conditions (temperature, concentration of dry material, period of contact and the like) which it is proposed to use in practice. The product obtained after each washing test is isolated, drained, dried, ground and optionally screened, all of these processes being carried out under the conditions and with the type of equipment which it is proposed to employ in practice. The particles obtained in each test are examined under a microscope and their flow characteristics are measured. The optimum working conditions to be employed in the washing process, as a function of the other conditions of the entire process for the production of the polysaccharides, are thus determined.

Addition of Sodium Hydroxide to *Xanthomonas* Culture

J. T. Patton; U.S. Patent 3,964,972; June 22, 1976; has found that heteropolysaccharides prepared by the fermentation of carbohydrates with bacteria of the genus *Xanthomonas* and subsequent reaction of the fermentation product with a base such as sodium hydroxide produce marked increases in the viscosities with little or no decrease in the clarity of brine and similar solutions to which they are added in low concentrations. Such heteropolysaccharides are stable for long periods at elevated temperatures and are not substantially degraded by salts normally found in oil field brines. They are not absorbed to a significant extent upon subsurface formations.

These and other properties of the substituted heteropolysaccharides of the process render them eminently suitable for thickening brines to be used in oil field secondary recovery processes and in a variety of other applications that require a highly stable thickener which is effective at low concentrations.

Xanthomonas begoniae, Xanthomonas campestris, Xanthomonas incanae and *Xanthomonas pisi* are preferred for purposes of the process.

Organisms of the *Xanthomonas* genus act upon a wide variety of carbohydrates to produce the heteropolysaccharides. Suitable carbohydrates include glucose, soluble starch, corn starch and the like. Fermentation studies have shown that the carbohydrates employed need not be in a refined state and may instead be utilized in the form of crude materials derived from natural sources.

The heteropolysaccharides are normally produced from carbohydrates by employing an aqueous fermentation medium containing 1 to 5 weight percent of the carbohydrate. From 0.1 to about 0.5 weight percent of dipotassium acid phosphate and from 0.1 to about 10 weight percent of a nutrient containing suitable trace elements and organic nitrogen sources is usually added to the carbohydrate solution to complete the fermentation medium. The nutrient employed will normally be a by-product material such as distillers' solubles or the like. A mixture containing 2 weight percent raw sugar and 0.4 weight percent distillers' solubles has been found to yield excellent results.

Fermentation of the medium thus prepared to produce the heteropolysaccharides is carried out by first sterilizing the medium and then inoculating it with bacteria of the genus *Xanthomonas.* The fermentation reaction is conducted under aerobic conditions and hence, sterilized air is bubbled through the medium as it ferments. The medium is maintained at a temperature between 70° and about 100°F, preferably between 75° and about 85°F, for a period from 1 to about 3 days.

As the fermentation reaction progresses, the viscosity of the medium increases rapidly due to formation of the heteropolysaccharide.

The rate of fermentation is controlled to some extent by the pH of the fermenting medium. In general, fermentation takes place most rapidly at pH values between 6.0 and about 7.5. Control of the pH at a level between 6.5 and about 7.2 is preferred. Sodium hydroxide or a similar alkaline material may be added to the medium continuously or at intervals in amounts sufficient to maintain the pH levels within the desired time range.

After the viscosity of the medium has reached a value of about 70 cp or higher as determined by testing the fermentate in 1:6 dilution with distilled water with a Brookfield viscometer at 80°F, the reaction may be halted. In a well controlled fermentation process, this point is normally reached after about 48 hr. The crude heteropolysaccharide produced by fermentation can then be separated from the bacterial cells by centrifugation or filtration, if desired. Precipitation with methanol, ethanol, acetone or a similar reagent permits isolation of the relatively pure heteropolysaccharide.

Monokaryotic Mycelium of *Coriolus versicolor* as Polysaccharides Source

The usefulness of the polysaccharides obtained from extraction of *Coriolus versi-*

color (Fr.) Quél. or culture thereof as a base component for preparation of medical drugs or foods and drinks has become acknowledged recently, and various techniques for producing such Basidiomycetes by artificial culture in a high yield have been proposed. Nevertheless, there is not yet available an advantageous method capable of propagating the Basidiomycete in a high yield.

C. Yoshikumi, Y. Omura, T. Wada, H. Makita, T. Ando, N. Toyoda and K. Matsunaga; U.S. Patent 4,159,225; June 26, 1979; assigned to Kureha Kagaku Kogyo KK, Japan have found that when this Basidiomycete is subjected to submerged culture while performing a mechanical treatment such as grinding or shearing in a liquid medium, the Basidiomycete loses clamp connection, which is its intrinsic morphological characteristic, and is changed into a monokaryotic mycelium, and that the thus formed monokaryotic mycelium is stable and also has a special characteristic which is a very high propagation rate as compared with the known dikaryotic mycelia.

The monokaryotic mycelium such as stated above can be produced by subjecting the dikaryotic mycelium of *Coriolus versicolor* (Fr.) Quél. to a mechanical (physical) treatment such as grinding or shearing in a liquid medium, or by subjecting the dikaryotic mycelium to submerged culture while mechanically treating the mycelium. More specifically, this method may be accomplished in the following ways.

(1) In case of subjecting the dikaryotic mycelium of *Coriolus versicolor* (Fr.) Quél. to shaking culture, the mycelium is ground by adding inert solid granular materials such as glass beads.

(2) In case of subjecting the dikaryotic mycelium to continuous submerged culture, such culture is performed while shearing the mycelium with an agitating element.

(3) In case of subjecting the dikaryotic mycelium to submerged culture, the mycelium is sheared or ground by a homogenizer to such an extent that no pellet-shaped mycelium is noted by external observation.

In the formation of the monokaryotic mycelium under the abovementioned conditions, it is preferable to apply additionally the following techniques.

(a) An enriched nutritious condition is provided by using a medium with 1.5 to 3 times as high concentration as the ordinary medium, for example, a glucose-yeast extract medium containing 5% glucose and 0.75% yeast extract.

(b) The atmosphere of submerged culture is maintained under reduced oxygen partial pressure which is provided by keeping the fermenter airtight, or by flowing an inert gas such as nitrogen gas or carbon dioxide gas into the fermenter.

(c) Submerged culture is carried out continuously while additionally supplying the liquid medium.

The dikaryotic mycelium of *Coriolus versicolor* (Fr.) Quél. can be easily converted into the monokaryotic mycelium by employing the abovementioned methods (1) through (3) either singly or in suitable combination with the techniques (a) through (c). In case a sufficient amount of monokaryotic mycelia can not be obtained by one run of culture, the abovementioned operation is continued after homogenizing the culture until the desired amount of monokaryotic mycelia is obtained. Culture is usually practiced at a temperature of 25±5°C for a period of 3 to 15 days.

The monokaryotic mycelium of this process was named *Coriolus versicolor* (Fr.) Quèl. GX-101-3 and deposited under FERM-P No. 3686.

This monokaryotic mycelium can be applied for the same purposes of use as the dikaryotic mycelium of *Coriolus versicolor* (Fr.) Quèl. For instance, it is possible to obtain a nitrogen-containing polysaccharide from its extraction with an aqueous medium (such as water, dilute alkaline solution or dilute acid solution), and such nitrogen-containing polysaccharide can be used for preparation of a pharmaceutical preparation having utility as an antitumor agent, immunity activating agent, antiviral drug, antifungal agent, antileprous drug, appetite promoting drug, etc.

Gel-Forming Polysaccharide from *Arthrobacter carbazolum*

K. Hisatsuka, S. Ishiyama, A. Inoue, O. Tsumura and M. Sato; U.S. Patent 4,153,508; May 8, 1979; assigned to Idemitsu Kosan Company Limited, Japan describe a process whereby a polysaccharide characterized by gel formation can be produced by cultivating a microorganism in a medium containing assimilable carbon sources and nitrogen sources until the polysaccharide is substantially accumulated in the culture, and recovering the accumulated polysaccharide therefrom.

The microorganisms which can be employed in this process belong to the genus Arthrobacter, for example, *Arthrobacter carbazolum,* which was isolated from soil. This strain was deposited under FERM-2574, and also deposited under ATCC-31258.

All strains belonging to the genus Arthrobacter which possess an ability to produce a polysaccharide having gel-forming property can also be used as well as artificial or natural mutants of the strain described above.

In the case of cultivating this strain, such generally utilized substances as saccharides (for example, glucose, starch, etc.) can be used as carbon sources. In addition, acetic acid and such derivatives from petrochemistry as ethanol, glycerol, propanediol, butanediol, ethylene glycol, etc., can also be used as carbon sources. As a nitrogen source, potassium nitrate, ammonium nitrate, ammonium sulfate, ammonium chloride, ammonium phosphate, polypeptone, carbazole, etc., can be used. As inorganic salts, dipotassium hydrogen phosphate or disodium hydrogen phosphate and potassium dihydrogen phosphate, etc., can be used, and as a source of trace element metal, magnesium sulfate, ferrous sulfate etc., can be used.

A medium prepared by adding the abovedescribed components into tap water was inoculated with the strain, and cultivated in an aerobic condition by reciprocal shaking.

After completion of the cultivation, the polysaccharide accumulated in a culture broth can be harvested by various methods according to the purpose of use. For example, a water-soluble solvent such as methanol, ethanol, isopropanol, acetone, tetrahydrofuran, etc., is added to the highly viscous broth to precipitate the polysaccharide with the cells, and after drying this precipitate can be used as a crude product.

When a solid material in the broth such as microbial cells is disadvantageous to the purpose of use, an appropriate amount of water is added before refining the broth, and the broth is heated. Subsequently, the polysaccharide is recovered from the solution obtained by centrifugation of the broth.

In refining, such routine methods for separation of the polysaccharide from impurities as condensation, precipitation with the abovedescribed water-soluble solvent, precipitation with ammonium sulfate, washing, centrifugation, column chromatography, extraction with solvent, dialysis, etc., can be used singly or in combination.

For example, after cultivation, an appropriate amount, preferably 3 to 10 times the amount based on the amount of the broth, of water is added into the broth, and the resultant mixture is heated to 45° to 65°C. The supernatant fluid is obtained by centrifugation of the broth. Subsequently, an aliphatic quaternary ammonium salt such as cetyltrimethylammonium bromide, etc., is added into this supernatant to precipitate the polysaccharide. The polysaccharide separated is washed with 85 to 95% methanol or ethanol saturated with sodium chloride or potassium chloride, etc. Thereafter, the polysaccharide is swelled by addition of water. After washing with acetone, etc., the polysaccharide is subjected to drying.

On the other hand, the polysaccharide can also be obtained by the addition of from 5 to 10 times the amount of water-soluble solvent such as acetone to precipitate the polysaccharide, or by spray-drying after desalting of the above supernatant by ultrafiltration or with an ion exchange resin such as Amberlite IR-120 B, Amberlite IRA-410, etc., and concentration. When a polysaccharide with higher purity is required, such procedures as deproteinization using a mixture of chloroform and isoamyl alcohol (4:1) or dialysis can be applied in addition to the abovedescribed procedures.

Example: *Arthrobacter carbazolum* (FERM-2574, ATCC-31258) was inoculated with platinum loop in a 500 ml Sakaguchi flask containing ethanol (1.56%), carbazole (1%), K_2HPO_4 (0.2%), $MgSO_4 \cdot 7H_2O$ (0.025%) and $FeSO_4 \cdot 7H_2O$ (0.001%) in tap water (pH 8.0), and the cultivation was carried out for 4 days at 30°C by reciprocal shaking.

After the cultivation, an amount of hot water (60°C) corresponding to nine times the amount of the cultured broth was added to the cultured broth, and cells were removed by centrifugation (10,000 rpm, 30 min). One third the amount (by volume) of cetyltrimethylammonium bromide (2% solution) was added to the supernatant with stirring to precipitate a polysaccharide. The polysaccharide precipitated was filtered, and washed twice with 90% ethanol saturated with sodium chloride. Subsequently, one fourth the amount (by volume) of water based on the amount of the cultured broth was added to swell, and then washed with five times the amount of acetone. After drying, crude polysaccharide (4 grams per liter) was obtained. After dissolving the crude polysaccharide in water (0.1% concentration), cations were removed by using Amberlite IR-120 B, and then one third amount (by volume) of a mixture of chloroform and isoamyl alcohol (4:1) was added. After mixing thoroughly, a supernatant was obtained by centrifugation. The supernatant obtained was dialyzed for 3 days at low temperature in water. After dialysis, vacuum drying was applied and thus purified polysaccharide (3.6 g/l) was obtained.

Heteropolysaccharide from *Erwinia tahitica*

K.S. Kang, G.T. Veeder, III and D.D. Richey; U.S. Patent 3,933,788; January 20, 1976; assigned to Kelco Company describe a process for producing a heteropolysaccharide by bacterial fermentation of a selected carbon source under controlled conditions. The heteropolysaccharide of this process is a high molecular weight polysaccharide containing primarily carbohydrate residues and a minor amount of protein. In the following description it will sometimes be referred to as Heteropolysaccharide 10.

This compound may be prepared by fermentation of a suitable nutrient medium with an organism which has been named *Erwinia tahitica*. A deposit of this organism employed in making the heteropolysaccharide was made with the American Type Culture Collection on August 11, 1971 under Accession No. ATCC 21711.

A suitable nutrient fermentation medium is inoculated with heteropolysaccharide-producing strain of *Erwinia tahitica* and permitted to incubate at a temperature of 33° to about 37°C, preferably about 35°C, for a period of 45 to about 60 hr. The carbon source required by the bacteria in order to produce the polysaccharide is an oligosaccharide containing from about 3 to about 10 monomer units at a concentration of 1 to about 5% by weight, and preferably 2 to 4% by weight.

A further ingredient which is present in the fermentation medium is a source of magnesium ions. The magnesium salt content of the fermentation medium may range from about 0.005 to about 0.02% by weight. Suitable sources of magnesium ions include water-soluble magnesium salts, such as magnesium sulfate heptahydrate, magnesium acetate, magnesium chloride, magnesium nitrate and magnesium acid phosphate which may be deliberately added or present as an impurity in the carbon source or the water used.

The pH of the fermentation medium is important to suitable growth of the bacteria. It was found that the optimum pH for production of Heteropolysaccharide 10 is in the range of about 6.0 to 7.5, and preferably about 6.0 to 6.5.

At least a trace quantity of phosphorus, generally in the form of a soluble potassium salt, is also present in the fermentation medium.

In order to obtain a rapid fermentation, it was found that it is essential to have a sufficient quantity of oxygen available for the growing *Erwinia tahitica* culture. The liquid medium should contain 5 to 10% of the amount of oxygen that can be dissolved in the medium, when the oxygen is added as air.

A source of nitrogen is also present in the fermentation medium. When utilizing an organic nitrogen source in the fermentation medium, an amount ranging between 0.01 and 0.07% by weight of the fermentation medium is satisfactory.

Also, if desired, an inorganic nitrogen source, such as ammonium nitrate, ammonium sulfate, ammonium citrate or ammonium acetate may be present in the fermentation medium. The amount of such a salt which may be employed can range from 0.02 to about 0.15% by weight and preferably from about 0.045 to about 0.1% by weight of the fermentation medium.

On completion of the fermentation, the desired Heteropolysaccharide-10 may be recovered by treatment of the fermentation beer with a miscible solvent which is a poor solvent for the heteropolysaccharide and does not react with it. In this way the heteropolysaccharide is precipitated from solution. The quantity of solvent employed generally ranges from 2 to about 3 v/v of fermentation beer. Among the various solvents which may be employed are acetone and lower alkanols such as methanol, ethanol, isopropanol, n-butanol, sec-butanol, tertiary butanol, isobutanol, and n-amyl alcohol. Isopropanol is preferred.

Precipitation of the desired heteropolysaccharide is facilitated when the fermentation beer is first heated to a temperature of 70° to about 90°C for a short time, e.g., about 5 to 10 min, and then cooled to about 30°C or lower before addition of the solvent. Thus, this is a preferred method of precipitating the heteropolysaccharide from the fermentation beer. The solid is recovered by separating it from the liquid, as by filtering or straining, and then drying at elevated temperature.

Heteropolysaccharide–10 imparts viscosity to an aqueous medium when dissolved in water in low concentration. Because of this, its sensitivity to shear, its pseudoplasticity, its stability with salts, and because of its overall rheology, Heteropolysaccharide–10 is useful as a thickening, suspending and stabilizing agent in aqueous systems. More specifically, it is useful as an additive to textile printing pastes or in formulating low drift aqueous herbicidal compositions. It is also of value as a thickening or suspending agent in salad dressings, in forming thickened puddings, and as a thickener in adhesive compositions.

Production from *Azobacter indicus*

K.S. Kang and W.H. McNeely; U.S. Patents 3,915,800; October 28, 1975; 3,960,832; June 1, 1976; the former assigned to Kelco Company describe a heteropolysaccharide which is designated as Heteropolysaccharide-7, and which contains glucose, rhamnose and galacturonic acid in an approximate ratio of 6.6:1.5:1, and is obtained by fermentation of an appropriate nutrient medium with bacteria designated by *Azotobacter indicus var. myxogenes.*

A deposit of a strain of the bacteria of the process was made in the American Type Culture Collection on August 19, 1969 and the accession number of the deposit is 21423.

Heteropolysaccharide-7 is produced by growing the *Azotobacter indicus var. myxogenes* organism in an aqueous nutrient medium at a temperature of from about 25° to 35°C, and preferably at about 30°C until substantial Heteropolysaccharide-7 is elaborated. The fermentation time is normally from about 35 to 60 hr, and preferably from 37 to 48 hr.

The aqueous nutrient medium, i.e., the fermentation medium, contains an appropriate source of carbon and nitrogen as well as a source of low levels of magnesium and phosphorus. The carbon source is a carbohydrate at a concentration of about 1 to 5% by weight, and preferably about 2 to 3% by weight. Suitable carbohydrates include, for example, dextrose, sucrose, maltose, fructose, mannose, starch hydrolysate or corn syrup. Preferably, the carbohydrate source employed is dextrose (glucose). Crude sugars may be used, such as deionized molasses, or a product such as Hydrol-E-081 (Corn Products Refining Company).

A further ingredient which is present in the fermentation medium is a source of magnesium ions. The magnesium salt content of the fermentation medium is in the range of 0.005 to about 0.02% by weight. The source of magnesium ions is not critical, and suitable sources include water-soluble magnesium salts, such as magnesium sulfate heptahydrate, magnesium acetate, magnesium chloride, magnesium nitrate, and magnesium acid phosphate.

At least a trace quantity of phosphorus, generally in the form of a soluble potassium salt, is also present in the fermentation medium. Larger quantities of phosphorus, such as about 0.65% by weight of the fermentation medium, calculated as dipotassium acid phosphate can, however, also be used without adverse effects.

A further ingredient which is present in the final fermentation medium is a source of nitrogen. The nitrogen source may be organic in nature as, for example, soy protein.

When utilizing an organic nitrogen source in the fermentation medium it may be present in an amount ranging between 0.01 and 0.07% by weight of the fermentation medium.

Also, it has been found desirable to have present in the fermentation medium an inorganic nitrogen source, such as ammonium nitrate, ammonium chloride, ammonium sulfate or ammonium acetate. The amount of ammonium salt which may be employed can range from about 0.02 to 0.15% by weight and preferably from about 0.045 to 0.1% by weight of the medium.

The pH of the fermentation medium is important for suitable growth of the bacteria and elaboration of Heteropolysaccharide-7. It was found that the optimum starting pH for production of colloid is within the range of about 7±0.5. Sodium hydroxide should be added so as to maintain a pH of at least about 6.5.

It is important that a dissolved oxygen level of 5 to 10% be maintained at least during the first 20 to 40 hr of the fermentation. Thus, the liquid medium should contain 5 to 10% of the amount of oxygen that can be dissolved in the medium, when the oxygen is added as air.

The course of the fermentation may be followed by determining the residual sugar content of the fermentation medium. For best results, the fermentation is continued until the residual sugar content of the medium is in the order of about 0.3% by weight, and preferably in the order of about 0.1% by weight or less.

When the fermentation is completed, Heteropolysaccharide-7 may be recovered from the fermentation liquor by known techniques, and preferably by solvent precipitation. Thus, the fermentation beer is treated with a water-miscible solvent which does not react with the heteropolysaccharide and in which the product is only slightly soluble. The product is thus precipitated and may be recovered by accepted and known techniques and dried.

Typical organic solvents which may be used for this purpose are straight or branched chain lower alkanols, i.e., methanol, ethanol, isopropanol, butanol, tert-butanol, isobutanol, n-amyl alcohol of which isopropanol is the preferred alcohol; lower alkyl ketones, such as acetone, may be employed. In some cases the precipitation is improved if the fermentation medium is first heated to a temperature of about 70° to 90°C for a short period of time and then cooled to about room temperature before addition of the solvent.

Heteropolysaccharide-7 obtained as described above is a high molecular weight polysaccharide that functions as a hydrophilic colloid to thicken, suspend and stabilize water based systems.

Cyclodextrin

K. Horikoshi and N. Nakamura; U.S. Patent 4,135,977; January 23, 1979; and K. Horikoshi; U.S. Patent 3,923,598; December 2, 1975; both assigned to Rikagaku Kenkyusho, Japan describe a process for producing cyclodextrin which comprises reacting a cyclodextrin glycosyl transferase with starch at a pH of 6.0 to 10.5, adding a glycoamylase to the resulting reaction mixture liquid to decompose unreacted starch, concentrating the liquid to form a concentrate containing cyclodextrin at a content of at least 40%, and adding a small amount of cyclodextrin as a seed to the concentrate to precipitate cyclodextrin.

The cyclodextrin prepared according to the process has the following structures:

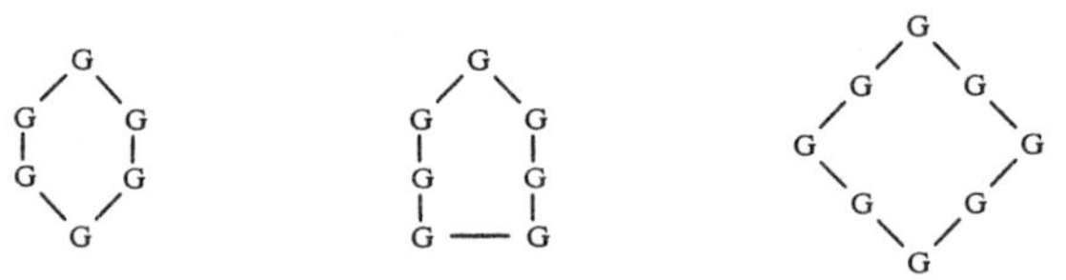

```
      G                  G                      G
    /   \              /   \                  /   \
  G       G          G       G              G       G
  |       |          |       |            /           \
  G       G          G       G          G               G
    \   /            |       |            \           /
      G              G ----- G              G       G
                                              \   /
                                                G
```

Such dextrin is valuable as a sweetening substance or a substitute for gum arabic and so on.

The enzyme to be used in the process is a cyclodextrin glycosyl transferase having an optimum pH on the alkaline side, which is a fermentation product of a microorganism growing only in an alkaline medium.

The strains identified as *Bacillus* sp. No. 38-2, *Bacillus* sp. No. 135, *Bacillus* sp. No. 169, *Bacillus* sp. No. 13 and *Bacillus* sp. No. 17-1 were deposited with the American Type Culture Collection (ATCC) as ATCC access numbers 21783, 21595, 21594, 31006 and 31007, in unrestricted condition permitting the public to have full access to the cultures.

Either solid or liquid media can be used for culturing these strains, but it is indispensable that the culture medium should be an alkaline medium containing a carbonate. More specifically, a medium formed by adding a carbonate to a medium comprising components necessary for growth of microorganisms, such as a carbon source, a nitrogen source and inorganic salts, for example, a culture medium formed by adding a carbonate to a medium comprising soluble starch, peptone, yeast extract, K_2HPO_4, $MgSO_4 \cdot 7H_2O$ etc., is used in this process.

Culturing of the abovementioned microorganisms is performed aerobically under agitation with air current. For example, the microorganism is cultured under shaking at 30 to 37°C for 24 to 96 hr, and after the culturing cells are removed and the resulting enzyme is precipitated or salted out by an organic solvent or a salt such as ammonium sulfate after or without neutralization of the added carbonate with acetic acid or a similar acid. The recovered enzyme is dehydrated and dried to obtain a powder of a crude enzyme for production of cyclodextrin.

Example: The pH of 15 liters of water containing 4% (w/v) of potato starch (amount: 600 g) was adjusted to 10 by sodium hydroxide, and the liquid was heated at 125°C for 30 min to gelatinize the starch. The gelatinized starch was cooled to 50°C and 600 mg of an alkaline amylase produced from the abovementioned *Bacillus* sp. No. 38-2 (ATCC 21783) was added to the starch and reaction was conducted at 50°C for 30 hr.

After the reaction, the reaction mixture was heated at 100°C for 5 min to deactivate the enzyme, and the mixture was cooled to 55°C and the pH was adjusted to 5.0 by hydrochloric acid. Then, 900 mg of a glucoamylase GAS-1 (Amano Seiyaku K.K.) having an activity of 2,000 units per gram, was added to the reaction mixture and reaction was carried out for 20 hr. Then, the reaction mixture was decolorized by active carbon and filtered according to customary procedures. The reaction mixture was then concentrated so that the solid content was elevated to 65%, and a small amount of cyclodextrin was added to the concentrate and it was allowed to stand overnight in a cold chamber.

The resulting precipitated cyclodextrin was recovered by filtration and dried under reduced pressure to obtain 280 g of cyclodextrin.

MISCELLANEOUS CARBOHYDRATES

Soluble Flavonoid Glycosides

J. Fukumoto and S. Okada; U.S. Patent 3,878,191; April 15, 1975; assigned to Kyowa Hakko Kogyo KK, Japan describe flavonoid glycosides containing at least two glucoses which are sweetening agents of high solubility. These compositions are prepared by the action of α-amylase on a mixed solution of a flavonoid monoglycoside (e.g., dihydrochalcon monoglucoside) and starch which acts as a sugar donor whereby sugar is transferred from the sugar donor to the flavonoid monoglycoside. A decomposed product of the starch may also be used as the sugar donor.

The flavonoid monoglycosides which may be employed are exemplified by hesperetin-7-glucoside, citronin-7-glucoside, naringenin-7-glucoside (prunin), sakuranin-7-glucoside, isosakuranin-7-glucoside, and the like. Generally those glucosides which belong to flavonoid monoglycosides may be utilized. It is also possible to use both chalcon and dihydrochalcon compounds of the abovementioned flavonoid monoglucosides.

As the amylase enzymes any and all α-amylases may be used. It is preferred to use, for example, bacterial saccharifying α-amylases, α-amylases of fungi and α-amylase obtained from *Bacillus macerans* etc. because of their high conversion efficiency. However, bacterial liquifying α-amylases of malt, etc. are less efficient, and are generally not practical for use in the process.

The process is advantageously carried out in the following manner. Flavonoid monoglycoside is added to a 1 to 30% solution of a sugar donor, i.e., starch or starch decomposate such as cyclodextrin, liquefied dextrin (when not completely dissolved, dissolved previously in ethanol). The amount of the monoglycoside is adjusted to give a ratio of starch (or decomposate thereof) to glycoside of at least 1:1; however, preferably the amount of starch (or decomposate thereof) exceeds the amount of glycoside.

α-Amylase, preferably diluted with water, is added to the mixture and the reaction is carried out for one to two days at a pH of 4.0 to 8.0 and at a temperature of 30° to 60°C. The reaction mixture is heated to from about 80° to 90°C for 10 min to inactivate the enzyme and the product is then dried to yield a sweetening agent, i.e., flavonoid glycosides containing two or more glucoses and having a high solubility.

β-Amylase may be additionally used to act on the reaction mixture described above. With such additional reaction the glycoside having a high polymerization degree is decomposed and a flavonoid glycoside having two or three glucoses is solely obtained. The reaction and inactivation of β-amylase are carried out in a similar manner to that described in relation to the α-amylase.

The flavonoid glycosides obtained by the process have almost the same degree of sweetness as that of the original monoglycosides, but they tend to exhibit somewhat milder effects.

Example: 2 g of hesperetin dihydrochalcon-7-glucoside was dissolved in 20 ml of methanol with heating. Separately, 5 g of soluble starch was dissolved in 50 ml of water with heating and was thereafter decomposed by adding a bacterial liquefying α-amylase obtained from *Bacillus subtilis*. When the decomposition rate became about 18%, the starch-α-amylase mixture was heated to give a digested starch solution having a decomposition rate of about 20%. Subsequently, both solutions were combined and 30 ml of amylase obtained from *Bacillus macerans* was added thereto. The mixture was reacted for 16 hr at 40°C and at pH 5.6.

The resultant liquor was heated for 5 min at 90°C to inactivate the enzyme and was then spray-dried to yield 6 g of powdered flavonoid glycosides. The product was in the form of a mixture of glucose, maltose and maltotriose as well as of mono, di-, tri-, tetra- and pentaglucosides. The ratio of respective mono-, di-, tri-, tetra- and pentaglucosides was 1:2:2.5:1.5:0.7.

VITAMINS

2-KETO-L-GULONIC ACID

Production from L-Sorbosone

The compound 2-keto-L-gulonic acid, which has the formula:

$$
\begin{array}{l}
\text{COOH} \\
| \\
\text{C=O} \\
| \\
\text{HOCH} \\
| \\
\text{HCOH} \\
| \\
\text{HOCH} \\
| \\
\text{CH}_2\text{OH}
\end{array}
$$

is an important and valuable intermediate in the production of L-ascorbic acid (vitamin C). In the past, 2-keto-L-gulonic acid has been produced microbiologically by the fermentation of L-sorbose. However, this process has suffered from the disadvantage that yields have been on the order of 5%. These low yields preclude the commercialization of such a procedure. Therefore, it has long been desired to provide a direct method for producing 2-keto-L-gulonic acid in high yields.

S. Makover and D.L. Pruess; U.S. Patent 3,907,639; September 23, 1975; assigned to Hoffmann-La Roche Inc. have found that 2-keto-L-gulonic acid can be produced in one step by the microbiological oxidation of L-sorbosone.

$$
\begin{array}{l}
\text{CHO} \\
| \\
\text{C=O} \\
| \\
\text{HOCH} \\
| \\
\text{HCOH} \\
| \\
\text{HOCH} \\
| \\
\text{CH}_2\text{OH}
\end{array}
$$

By this process, 2-keto-L-gulonic acid has been produced in yields as high as 80%.

Among the microorganisms suitable for the process are included members of the following genera: *Acetobacter, Pseudomonas, Escherichia, Serratia, Bacillus, Staphylococcus, Aerobacter, Alcaligenes, Penicillium, Candida* and *Gluconobacter.*

Any microorganism of the aforementioned genera capable of converting L-sorbosone to 2-keto-L-gulonic acid can be utilized. Among the preferred strains are included *Gluconobacter melanogenum IFO* 3293; *Bacillus subtilis* (NRRL 558); *Candida albicans* (NRRL 477); *Penicillium digitatum* (ATCC 10030); *Pseudomonas putida* (ATCC 21812); *Aerobacter aerogenes; Staphylococcus aureus* (ATCC 6538P); *Pseudomonas aeruginosa; Escherichia coli; Alcaligenes* species (ATCC 10153) and *Serratia marcescens.*

The abovementioned microorganisms can be induced from mutants having superior ability relative to the parent wild strains of producing the enzyme system capable of transforming L-sorbosone to 2-keto-L-gulonic acid. Such mutation can be caused by treating a wild strain with a mutagen such as ultraviolet irradiation, x-ray irradiation or contact with nitrous acid, or by isolating a clone occurring by spontaneous mutation. These means for inducing the desired mutation on a wild type strain may be effected in any of the ways well known for this purpose by one skilled in the art.

The production of 2-keto-L-gulonic acid is effected by the cultivation of one of the 2-keto-L-gulonic-acid-producing organisms in an aerated deep tank, i.e., under submerged fermentation. In order to obtain high yields, certain conditions should be maintained. The fermentation should be conducted at pH values of from about 5 to about 9 with pH values of from about 6.5 to about 7.5 being preferred. It is particularly preferred to carry out the process at a pH of about 7.2. Although the temperature is not critical, best results are usually obtained utilizing temperatures of from 20° to 45°C, with temperatures of from about 25° to 35°C being particularly preferred. In general, about 1 to 10 days are required to obtain the best results and from about 4 to 7 days is found most suitable.

The process can be carried out by culturing the microorganism in a medium containing L-sorbosone and other appropriate nutrients. On the other hand, the process can be carried out by culturing the microorganism and then after culturing, the whole cells or the cell-free extract collected from the culture are brought into contact with L-sorbosone.

It is usually required that the culture medium contains such nutrients for the microorganism as assimilable carbon sources, digestible nitrogen sources and preferably, inorganic substances, vitamins, trace elements, other growth promoting factors, etc.

2-keto-L-gulonic acid can be converted to L-ascorbic acid by application of any of the known methods, if desired. L-ascorbic acid is generally synthesized by esterifying 2-keto-L-gulonic acid in the presence of a mineral acid such as sulfuric acid, hydrochloric acid or strongly acidic cation exchange resin, as a catalyst, followed by enolizing the ester and subsequently lactonizing the enol compound. The resulting 2-keto-L-gulonic acid, in the reaction mixture, need not be separated, but the reaction mixture can directly be esterified, enolized and lactonized by procedures well known in the art.

The following example is illustrative of the process. All temperatures are in degrees centigrade. The *Pseudomonas putida* utilized in the following example was *Pseudomonas putida* ATCC 21812.

Example: A *Pseudomonas putida* culture was grown overnight on a rotary shaker at 28°C in a 125 ml Erlenmeyer flask containing 20 ml of the following medium:

	Grams per Liter
Glycerol	2.5
Sodium citrate	5.0
K_2HPO_4	10.0
KH_2PO_4	5.0
Na_2SO_4	2.5
$(NH_4)_2SO_4$	0.28
$MgCl_2 \cdot 6H_2O$	0.20
$CaCl_2 \cdot 2H_2O$	0.016
$FeCl_3 \cdot 6H_2O$	0.001
$ZnCl_2$	0.0005
$CuCl_2 \cdot 2H_2O$	0.0005
$MnCl_2 \cdot 4H_2O$	0.0005

One-half of the resulting suspension culture was inoculated into a 50 ml Erlenmeyer flask containing 9.5 ml of fresh medium supplemented with 5 g/l L-sorbosone (L-xylo-hexogulose). The flask was incubated on a rotary shaker at 28°C. The production of 2-keto-L-gulonic acid was monitored by paper electrophoresis at pH 2.6 and by paper chromatography in a solvent system consisting of pyridine:ethyl acetate:acetic acid:water (5:5:1:3 parts by volume). The paper strips were treated with aniline phthalate to develop the 2-keto-L-gulonic acid produced.

Production from 2,5-Diketo-D-Gluconic Acid

T. Sonoyama, B. Kageyama and T. Honjo; U.S. Patent 3,922,194; November 25, 1975; assigned to Shionogi & Co., Ltd., Japan describe a method for producing 2-keto-L-gulonic acid which comprises contacting a 2-keto-L-gulonic-acid-producing strain selected from the genera of: *Brevibacterium, Arthrobacter, Micrococcus, Staphylococcus, Pseudomonas* or *Bacillus* or any product obtained by treating cells of the strain with 2,5-diketo-D-gluconic acid or any salts thereof.

The 2-keto-L-gulonic-acid-producing strains of microorganisms employed in the method include: *Brevibacterium ketosoreductum* nov. sp. ASM-1005 (deposited with the Fermentation Research Institute as FERM-P 1905, and with the American Type Culture Collection as ATCC-21914), *Arthrobacter simplex* ASM-10 (FERM-P 1902, ATCC 21917), *Bacillus megaterium* ASM-20 (FERM-P 1903, ATCC 21916) and *Staphylococcus aureus* ASM-30 (FERM-P 1904, ATCC 21915).

In addition to the above, available microorganisms include strains which are preserved in any public depository (culture collection) for delivery to anyone upon request, such as the Institute of Fermentation, Osaka (IFO) and include *Micrococcus dinitrificans* IFO 12442, *Micrococcus rubens* IFO 3768, *Micrococcus roseus* IFO 3764 and *Pseudomonas chlororaphis* IFO 3904.

Any means known for incubating microorganisms may be adopted although the use of aerated and agitated deep-tank fermenters are particularly preferred. A preferred result may be obtainable from an incubation which utilizes a liquid broth medium.

As regards the nutrient medium available for the incubation of the microorganism, although no special restriction is imposed on its class, a medium suitably includes carbon sources, nitrogen sources, other inorganic salts, a small amount of other nutrients and the like.

Although the concentration of 2,5-diketo-D-gluconic acid in the medium may also be varied with the generic character and the like of the employed strain, a concentration of about 1 to 200 g/l is generally applicable and, inter alia, a concentration of about 1 to 50 g/l is preferred.

An incubation temperature of about 20° to 35°C and a pH value of the medium of about 4 to 9 may preferably be maintained. Normally, an incubation period ranging from 10 hours to 100 hours may be sufficient and the formation of the intended product in the medium reaches its maximum value within such period.

Any methods may be employed for the separation of the intended product from the medium unless they deteriorate the product. For instance, the separation may be performed in any suitable combination or repetition of the following unit processes: (a) removal of the cells of the microorganisms from the fermented broth by filtration, centrifugation or treatment with active charcoal, (b) precipitation of the intermediate crystals by concentrating the filtered broth, (c) recovery of the precipitated crystals by filtrating or centrifugating the concentrated broth, (d) recrystallization of the intermediate crystals, (e) extraction with solvent, and (f) fractionation by chromatography.

Example: A sterilized medium (600 ml) containing 1.5% of calcium 2,5-diketo-D-gluconate, 0.3% of glycerol, 0.1% of polypeptone, 0.1% of yeast extract, 0.1% of monopotassium phosphate and 0.02% of magnesium sulfate ($7H_2O$) and having a pH value of about 6.3 to 7 is introduced into a small fermenter of 1.5 liters, and a suspension (20 ml, in sterilized water) of *Brevibacterium ketosoreductum* which is previously cultured on a bouillon agar at 30°C for 2 days, is inoculated.

An aerated (1 v/v/min) culture while being stirred (300 rpm) is performed at 30°C. At each given time during the incubation, samples are withdrawn from the broth to confirm the formation of 2-keto-L-gulonic acid as a pink spot on a paper partition chromatogram which utilizes a solution of phenol:water:formic acid (75:25:4) as a developing solvent.

A quantitative determination by means of gas-liquid chromatography (column, silicone gum; SE-52; sample, silylated, gives the following results:

Incubation time (hr)	36	48	72
2-Keto-L-gulonic acid (γ/ml)	200	620	1,890

In a similar process, *T. Sonoyama, H. Tani, B. Kageyama, K. Kobayashi, T. Honjo and S. Yagi; U.S. Patent 3,959,076; May 25, 1976; assigned to Shionogi & Co., Ltd., Japan* describe a method for producing 2-keto-L-gulonic acid which comprises contacting a 2-keto-L-gulonic acid-producing microorganism strain selected from the genus of *Corynebacterium* or any product obtained by treating cells of the strain with 2,5-diketo-D-gulonic acid or any salts thereof.

The 2-keto-L-gulonic acid-producing strains of microorganism employed in this process include: *Corynebacterium* sp. ASM 3311-6 (ATCC 31081), *Corynebacterium* sp. ASM-20A-77 (ATCC 31090), *Corynebacterium* sp. ASM-T-13 (ATCC 31089), and *Corynebacterium* sp. ASM-K-106 (ATCC 31088).

In another similar process, *T. Sonoyama, H. Tani, B. Kageyama, K. Kobayashi, T. Honjo and S. Yagi; U.S. Patent 3,963,574; June 15, 1976* describe a method whereby the 2-keto-L-gulonic-acid-producing microorganism includes strains which belong to the species of the *Brevibacterium* nov. sp. ASM-856-4, ATCC 31083.

Production from D-Glucose

T. Sonoyama, H. Tani, B. Kageyama, K. Kobayashi, T. Honjo and S. Yagi; U.S. Patent 3,998,697; December 21, 1976; assigned to Shionogi & Co., Ltd., Japan describe a method for producing 2-keto-L-gulonic acid which comprises contacting a microorganism strain capable of producing 2,5-diketo-D-gluconic acid from D-glucose which belongs to the genera of *Acetobacter, Acetomonas* and *Gluconobacter* or any product obtained by treating cells of the microorganism having an enzymatic activity for the same effect, and a microorganism capable of converting 2,5-diketo-D-gluconic acid into 2-keto-L-gulonic acid which belongs to the genera of *Brevibacterium* and *Corynebacterium* or any product obtained by treating cells of the microorganism having an enzymatic activity for the same effect with a medium containing D-glucose.

This is done in a manner wherein both microorganisms or products thereof co-exist together in the medium during at least part of the entire process in a condition sufficient to produce and accumulate 2-keto-L-gulonic acid or any salts thereof in the medium. These are recovered from the resultant mixture.

The microorganism strain capable of producing 2,5-diketo-D-gluconic acid from D-glucose (designated Strain A), specific examples of which are shown below, belongs to the genera of *Acetobacter, Acetomonas* and *Gluconobacter,* although these three genera are not specifically discriminated with respect to each other in the description of *Bergey's Manual of Determinative Bacteriology* (7th Ed.).

The microorganism strain capable of converting 2,5-diketo-D-gluconic acid into 2-keto-L-gulonic acid (Strain B) employed herein belongs to the genera of *Brevibacterium* and *Corynebacterium.* Specific examples are shown below.

Strain A

Acetomonas albosesamae FERM-P No. 2439, ATCC No. 21998.
Acetobacter melanogenum IFO No. 3293.
Gluconobacter rubiginosus IFO No. 3244.

Strain B

Brevibacterium ketosoreductum ASM-1005, FERM-P No. 1905, ATCC No. 21914.

(continued)

Brevibacterium nov. sp. ASM-856-4, FERM-P No. 2686, ATCC No. 31083.
Brevibacterium sp. ASM-3356-31, FERM-P No. 2685, ATCC No. 31082.
Brevibacterium testaceum IFO No. 12675.
Corynebacterium sp. ASM-3311-6, FERM-P No. 2687, ATCC No. 31081.
Corynebacterium sp. ASM-20A-77, FERM-P No. 2770, ATCC No. 31090.
Corynebacterium sp. ASM-T-13, FERM-P No. 2771, ATCC No. 31089.
Corynebacterium sp. ASM-K-106, FERM-P No. 2769, ATCC 31088.

The "mixed culture" can be conducted by means of a variety of methods, for instance, a method wherein both kinds of microorganism strains are simultaneously inoculated in the medium at the initiation of the culture, a method wherein Strain A is inoculated first and Strain B is subsequently inoculated after a period of incubation, and a method wherein each strain is inoculated separately in its respective medium and then either is added to the other broth all at once, portionwise or continuously after some period of incubation, followed by another period of incubation.

Any means per se known as a method in connection with the incubation technique for microorganisms may be adopted though the use of aerated and agitated submerged fermenters is particularly preferred. A preferred result may be obtainable from an incubation which utilizes a liquid broth medium.

As regards the nutrient medium available for the incubation of the microorganism, although no special restriction is imposed on its class, an aqueous nutrient medium suitably including carbon sources, nitrogen sources, other inorganic salts, small amounts of other nutrients and the like, which can be utilized by the microorganism is desirable for the advantageous incubation of the microorganism. Various nutrient materials which are generally used for the better growth of microorganisms may suitably be included in the medium.

Although the concentration of the starting material, D-glucose, in the medium may also be varied with the generic character and the like of the employed strain, a concentration of about 20 to 200 g/l is generally applicable and, inter alia, a concentration of about 20 to 100 g/l is preferred.

The conditions of the incubation may also vary with the species and generic character of the strain employed, the composition of the medium and other attendant factors, and may, of course, be selected or determined in accordance with the particulars of the individual cases in order to yield the intended product most efficiently, although an incubation temperature of about 20° to 35°C and a pH value of the medium of about 4 to 9 may preferably be maintained. Normally, an incubation period ranging from 10 hours to 100 hours may be sufficient and the formation of the intended product in the medium reaches its maximum value within such period.

Production from L-Gulonic Acid

D.A. Kita; U.S. Patent 4,155,812; May 22, 1979; assigned to Pfizer Inc. describes the conversion of L-gulonic acid as the calcium or sodium salt to 2-keto-L-gulonate by means of a growing culture of selected strains of species of the genus *Xanthomonas*. Yields of 90% or greater are obtained with concentrations of L-gulonate up to 15%.

Conversion of L-gulonate to 2-keto-L-gulonate was obtained by the following *Xanthomonas* cultures obtained from The American Type Culture Collection and The New Zealand Reference Culture Collection.

		Culture No.
Xanthomonas		
amaranthiocola	ATCC	11645
begoniae		8718
begoniae		11725
begoniae		11726
campestris		6402
campestris		13951
holcicola		13461
incanae		13462
juglandis		11329
malvacearum		9924
malvacearum		12131
malvacearum		12132
malvacearum		14981
malvacearum		14982
malvacearum		14983
malvacearum		14984
malvacearum		14985
malvacearum		14986
melhusi		11644
papavericola		14179
phaseoli		11766
phaseoli		17915
pruni		15924
translucens		9000
translucens		9002
translucens		10731
translucens		10768
translucens		10769
translucens		10770
translucens		10771
translucens		10772
begoniae	NZRRC	191-65
begoniae		193-62
begoniae		194-66
begoniae		75-65
incanae		574-63
incanae		573-63
campestris		1686-65
campestris		2385-68
campestris		3984-74
campestris		4013-74

The preferred microorganism is *Xanthomonas translucens* ATCC 10768 which quantitatively converts L-gulonate to 2-keto-L-gulonate.

In the process, an aqueous nutrient medium containing an assimilable source of carbon and nitrogen is inoculated with a suitable L-gulonate-converting strain of *Xanthomonas*. After aerobic propagation for about 24 to 48 hours at 24° to 34°C, preferably 28° to 30°C, an aliquot is transferred to a fermenter containing an aqueous nutrient medium comprising a carbohydrate, an assimilable source of nitrogen and trace elements. The aqueous nutrient medium also contains about 8% (the solubility limit) of calcium gulonate. The sodium, potassium and ammonium salts of L-gulonic acid are inhibitory at this concentration to cell growth of the inoculum.

One of these salts, preferably the sodium salt, is added at a level of 40 g/l of fermentation broth at the fermentation time of about 48 hours and an additional 30 g/l at about 72 hours. L-gulonic acid and L-gulono-1,4-lactone are not suitable conversion substrates for the production of 2-keto-L-gulonic acid. The calcium and sodium salts of L-gulonic acid are prepared from L-gulonic acid obtained by the hydrolysis of L-gulono-1,4-lactone which may be synthesized by the method described in *Chem. Pharm. Bull.,* 13, 173 (1965) and *Rec. Trav. Chim. des Pays-Bas,* 74, 1365 (1955).

The conversion fermentation is conducted at a temperature of about 28° to 30°C with mechanical stirring at about 1,700 rpm and an aeration rate of about 0.75 volume of air per volume of broth per minute. The optimum pH conversion range is fairly narrow (6.5 to 7.5). The fermentation is allowed to continue until a yield of at least 85% (based on L-gulonate) 2-keto-L-gulonate is obtained.

2-keto-L-gulonate may be isolated and recovered as the free acid or in salt form by methods well known to those skilled in the art or by hydrolyzing the 2-keto-L-gulonate in the fermentation broth to yield ascorbic acid.

VITAMIN B_{12}

Production from Arthrobacter Strain

I. Kojima, H. Sato and Y. Fujiwara; U.S. Patent 4,119,492; October 10, 1978; assigned to Nippon Oil Company, Ltd., Japan describe a process for producing vitamin B_{12} by a fermentation technique that comprises cultivating a vitamin B_{12}-producing microorganism of the genus *Arthrobacter* under aerobic conditions in a culture medium containing at least one compound selected from the group consisting of alcohols with 2 or 3 carbon atoms and ketones, and recovering vitamin B_{12} from the culture broth.

Examples of the known vitamin B_{12}-producing microorganisms of the genus *Arthrobacter* include *Arthrobacter simplex* (ATCC 6946), *Arthrobacter tumescens* (deposited at Institute for Fermentation, Osaka, Japan under deposit number IFO 12960), and *Arthrobacter globiformis* (ATCC 8010). The strain *Arthrobacter hyalinus* was deposited in Fermentation Research Institute, Agency of Industrial Science and Technology, Japan under deposit number FERM-P 3125. This strain was also deposited in American Type Culture Collection under deposit number ATCC 31263.

Cultivation is carried out under aerobic conditions, for example, with aeration and stirring. The cultivation temperature is generally about 20° to about 40°C, and the pH is about 4 to about 9.5. The cultivation time is usually from about 2 to 8 days, and can be suitably changed according to changes in the other cultivation conditions.

Separation of vitamin B_{12} from the fermentation product can be performed in the same way as in the prior art. Since vitamin B_{12} builds up mainly within the cells of the microorganism, it is desirable first to centrifuge the culture broth so as to obtain the cells. When it is desired to separate it as a cyano-type vitamin B_{12}, a cyanogen ion is added to the cells, and after adjusting the pH to 5 with an acid such as sulfuric acid, the mixture is maintained at 80° to 100°C in an aqueous medium. Where it is desired to separate it from the culture broth as a coenzyme-type vitamin B_{12} (5,6-dimethyl benzimidazole cobamide coenzyme) and hydroxyl-type vitamin B_{12}, the cells may be extracted in a customary manner with a solvent such as methanol, ethanol, acetone or pyridine in an aqueous medium in a dark place. The extracting temperature may be room temperature, but the extraction system may be heated to about 100°C.

Vitamin B_{12} extracted from the culture broth can be purified by a suitable combination of known means such as extraction with phenol, adsorption with activated carbon, or column chromatography using an ion exchange resin or cellulose.

Example: *Arthrobacter hyalinus* was inoculated in three 500 ml conical flasks each containing 100 ml of a sterilized culture medium containing 10 ml of isopropanol, 3 g of peptone, 1 g of yeast extract, 3 g of NH_4NO_3, 0.4 g of KH_2PO_4, 1.5 g of $Na_2HPO_4 \cdot 12H_2O$, 0.5 g of $MgSO_4 \cdot 7H_2O$, 10 mg of $FeSO_4 \cdot 7H_2O$, 10 mg of $ZnSO_4 \cdot 7H_2O$, 5 mg of $Co(NO_3)_2$, 50 μg of $CuSO_4 \cdot 5H_2O$, 10 μg of MoO_3, and 5 g of $CaCO_3$, and culture broth obtained was reserved for use as a seed.

Five liters of a culture medium composed of deionized pure water, and per liter of the water, 10 ml of isopropanol, 3 g of peptone, 1 g of yeast extract, 9 ml of corn steep liquor, 12 g of NH_4NO_3, 1.5 g of $Na_2HPO_4 \cdot 12H_2O$, 1.9 g of KH_2PO_4, 0.5 g of $MgSO_4 \cdot 7H_2O$, 10 mg of $FeSO_4 \cdot 7H_2O$, 10 mg of $ZnSO_4 \cdot 7H_2O$, 5 mg of $Co(NO_3)_2$, 50 μg of $CuSO_4 \cdot 5H_2O$, 10 μg of MoO_3, 6 g of $CaCO_3$ and 0.5 ml of an antifoamer was prepared in a 10 liter fermentation tank, and sterilized. Then, 300 ml of the seed prepared as set forth above was inoculated in the culture medium, and cultivated at 32°C for 60 hours with stirring at 600 rpm while passing aseptic air at a rate of 2.5 liters/minute.

During the course of fermentation, the cells were grown for 30 hours in the logarithmically growing phase while adding isopropanol so that its concentration in the culture broth did not exceed 3.0 ml/liter. Subsequently, in the vitamin B_{12}-producing phase, the amount of isopropanol added was controlled so as to maintain the pH of the culture broth at 7 to 8. The concentration of vitamin B_{12} produced reached 1,100 μg/l. The amount of isopropanol consumed was 33 ml per liter of the culture broth.

5.3 liters of the resulting culture broth was centrifuged at 10,000 g to separate the cells. Extraction of vitamin B_{12} from the cells and its purification were performed in a customary manner. Specifically, isopropanol was added to the cells in an amount four times the amount of the latter. The mixture was allowed to stand overnight at room temperature in a dark place. The resulting suspension of the cells in isopropanol was centrifuged at 10,000 g to separate the isopropanol extract.

The isopropanol was evaporated off from the extract, and the residue was extracted with an 80% aqueous solution of phenol to transfer vitamin B_{12} to the lower phenol layer. The phenol layer was washed once with water, and a mixture of equal volumes of ether and water was added to transfer vitamin B_{12} to the aqueous layer. The resulting aqueous solution of vitamin B_{12} was purified by column chromatography using TEAE cellulose. Fractions containing vitamin B_{12} were concentrated, and acetone was added to afford 2.8 mg of vitamin B_{12} crystals. This vitamin B_{12} was a coenzyme-type vitamin B_{12}.

Synchronization of Bacteria Population

B. Johan, L. Szemler, T. Szontagh, L. Kuti, E. Simonovits, J. Bekes, D. Szekely, J. Kiss, E. Kovacs and A. Hargitai; U.S. Patent 3,979,259; September 7, 1976; assigned to Richter Gedeon Vegyeszeti Gyar Rt., Hungary describe a process for the anaerobic septic fermentation of a broth to produce vitamin B_{12} in the presence of baterial nutrients, methanol and a methane-producing bacteria capable of fermenting methanol to produce vitamin B_{12} wherein portions of the fermenting broth are periodically removed and replaced by a nutrient broth containing nutrient components in a normal concentration to sustain fermentation.

The improvement comprises the combination of the following steps:

(a) Adding methanol in daily amounts of 0.4 to 0.7 volumes per volume percent for 2 to 4 days to the fermenting broth;

(b) Thereafter monitoring the pH of the fermenting broth and upon the pH falling to 5.6 to 5.8 removing 5 to 15 volume percent of the fermenting broth and replacing it with an equal volume of concentrated bacterial nutrient broth which contains the nutrient components in 5 to 12 times higher concentration than the normal concentrations, and repeating this step once during each interval of 5 to 12 days;

(c) Between repetitions of Step (b) and each day for periods of 5 to 10 days removing 5 to 15 volume percent of the fermenting broth daily and replacing it with an equal volume of the bacterial nutrient broth with the nutrient compositions in the normal concentrations together with 0.4 to 1.5 volume percent of methanol; and

(d) Interrupting the removal of fermenting broth in Step (c) for a period up to 2 days and adding only 0.4 to 1.5 volume percent of methanol to the fermenting broth during the interruption.

Advantageously, the methanol is added in Step (a) for substantially 3 days, the fermenting broth is removed in Step (b) and replaced with an equal volume of concentrated bacterial nutrient broth about every 10th day, the fermenting broth removal is interrupted in step (d) for a period of substantially 1 day, and the concentrated nutrient broth is added in Step (b) only on the day that fermenting broth is removed while on all other days of the period of Step (b) only methanol is added.

The main advantages of the process can be summarized as follows:

(1) The process provides, under certain conditions on a laboratory scale, high vitamin B_{12} values that could not be attained so far with septic fermentation processes and can hardly be attained even with artificially activated monocultural fermentation processes. The high vitamin B_{12} level, about 3.5 to 4

times higher than that of the average industrial scale processes (about 10,000 $\mu g/l$) can be accomplished in laboratory scale with the synchronized periodical procedure of
the process with a 2- to 2.5-fold nutrient requirement and
3-fold methanol requirement compared to the average industrial values.

(2) In the laboratory process, reproduced several times, vitamin
B_{12} levels of 30,000 to 35,000 $\mu g/l$ could be maintained for
about a one month semicontinuous fermentation process
even in an iron fermenter of 50 m^3 capacity.

Of the additives utilized in the fermentation process, the amount of methanol
introduced has an outstanding significance. There is a unanimous and linear
correlation between the amounts of introduced methanol and the biogas formed
therefrom. When the volume of the biogas formed under unit time is plotted
against the time, the curve consists of three stages: a steep ascending stage, a
horizontal stage (plateau) and a steep descending stage. The total amount of the
gas formed from the introduced methanol can be determined from the area under
the curve by graphical integration [E. Siminovits, T. Szontagh: XI Biochemical
Conference (Sopron, 1970) lectures Nos. 59 and 60; T. Szontagh, E. Simonovits,
I. Bekes: XII Biochemical Conference (Pecs, 1971), p. 219 to 225; T. Szontagh,
E. Simonovits, I. Bekes, E. Ezer, A. Lauko: XIII Biochemical Conference
(Szombathely, 1972), lecture No. 35].

The height of the plateau (expressed as produced gas, ml/5 l of fermentation
broth times hours) is a good measure of the actual biochemical activity of the
given fermentation broth, therefore this figure was used to determine the amount
of methanol which can be consumed by the population within 24 hours from a
certain time. These values should always be taken into account in the control
of the fermentation, because a higher increase in the amount of nonconsumed
methanol during the process exerts a toxic effect on the fermentation broth.

The amounts of methanol which may be supplied at or above the indicated
minimum plateau heights are given in the table below:

Plateau Height (ml/5 l of fer-mentation broth x hours)	Supplied Methanol (v/v % fer-mentation broth)
560	0.5
700	0.6
820	0.7
930	0.8
1,050	0.9
1,160	1.0
1,280	1.1
1,400	1.2
1,520	1.3
1,630	1.4
1,740	1.5

Example: 5 liters of an industrial fermentation broth produced by semicontinuous fermentation with methane-producing bacteria (vitamin B_{12} content: 9,936
$\mu g/l$) are introduced into two laboratory-scale fermenters with 5 liter working
capacities, equipped with gas removal outlets. The fermenters are thereafter

thermostatically heated to 32° to 34°C. On the first three days of the experiment no fermentation broth is removed, and only 25 ml of methanol are added to each of the fermenters. On the fourth day 500 ml of fermentation broth is removed, and 500 ml of a concentrated broth with the following composition are supplied:

Corn steep liquor	24.0 g
Hydrolyzed brewer's yeast (20%)	5.0 ml
Succinic anhydride	0.03 g
Cobalt chloride	0.05 g
5,6-Dimethyl-benzimidazole	0.015 g
Magnesium chloride	0.5 g
Molasses	10.0 g
Sulfite waste liquor	8.0 g
Liver extract (10%)	5.0 ml
Glycine	0.03 g
Ammonium hydroxide (24%)	5.5 ml
Diammonium hydrogen phosphate	10.0 g
Ammonium bicarbonate	15.0 g
Ammonium sulfate	7.5 g
Sodium bisulfate	0.1 g

In the following period of fermentation the above amount of the concentrated broth is fed into the fermenters on days 10, 17, 25, 33, 40, 46, 52, 60, 66 and 72 of the fermentation; prior to the administration of the broth 10% of the fermentation broth is removed. No fermentation broth is removed on days 5, 6, 7, 8, 9, 11, 12, 13, 14, 15, 16, 18, 19, 20, 21, 22, 23, 24, 26, 34, 41, 47, 53, 61, 67, 73 and 74 of the fermentation; on these days only methanol is introduced. On the remaining days of the fermentation 10% of fermentation broth is removed, and an equal volume of usual broth containing the same components as listed above for the concentrated broth but in 10 times less concentrations, is introduced. The maximum amount of methanol is 75 ml (1.5 v/v %); the actual amount is chosen according to the table above.

The average vitamin B_{12} level of the fermentation broths removed from the two parallel experiments during the semicontinuous process, i.e., from the 25th to the 75th day of fermentation, was as follows: A fermenter, 43,268 µg of vitamin B_{12}/liter of fermentation broth; and B fermenter, 43,543 µg of vitamin B_{12}/liter of fermentation broth.

Control of Nitrogen/Carbon Ratio

B. Johan, L. Szemler, J. Fülöp, T. Szontagh, E. Simonovits, J. Bekes, L. Kuti, R. Toròs, D. Szekely, L. Szabo, K. Seitz, T. Vajda, E. Kovacs, A. Hargital and K. Polinszky; U.S. Patent 3,964,971; June 22, 1976; assigned to Richter Gedeon Vegyeszeti Gyar Rt., Hungary describe a method for increasing the vitamin B_{12} production of fermentation processes performed in a known way with a mixed population of mesophilic methane-producing bacteria under anaerobic conditions, in which in an enrichment period of preferably 4 to 7 days a nutrient concentrate containing mainly inorganic ammonium compounds as nitrogen source and mainly methanol as carbon source and having a N:C weight ratio of 1:10 to 1:20, preferably 1:11 to 1:15 is added in daily portions to a fermentation broth obtained from a usual vitamin B_{12} fermentation process and containing living bacteria populations so as to increase the total concentration of assimilable nitrogen by a factor of maximum 4 until the end of the enrichment period. Thereafter the fermentation is terminated and the obtained fermentation broth with an increased vitamin B_{12} content is processed, or the fermentation is continued by periodically removing a portion of the fermentation broth and supplementing it with the same volume of fresh nutrient medium containing mainly inorganic ammonium compounds as the nitrogen source and mainly methanol as the carbon source and having a N:C weight ratio of 1:10-1:20, preferably 1:11-1:15.

Example: 9,400 ml of a fermentation broth containing 10,500 μg/l of vitamin B_{12}, produced by a known semicontinuous industrial fermentation utilizing a methane-producing mixed bacterium population, are introduced into a laboratory-scale fermenter with a working capacity of 10 liters. The fermentation broth is heated to 28° to 30°C, and a nutrient mixture of the following composition is added:

Ammonium bicarbonate	4.5 g
Ammonium hydroxide (24%)	1.5 ml
Diammonium hydrogen phosphate	0.45 g
Ammonium sulfate	1.5 g
Brewer's yeast extract (20%)*	1.0 ml
Cobalt chloride	0.010 g
Succinic acid	0.006 g
o-Xylidine	0.020 g
Magnesium chloride	0.1 g
Methanol	60.0 ml
Water qs 100.0 ml	

*Dry material content: 20–30%; total nitrogen content: 43–47%

In this nutrient mixture the weight ratio of the inorganic ammonium nitrogen related to the methanol carbon is about 1:11. Thereafter a mixture of 7.0 g of corn steep liquor (dry material content: 55 to 60%, total nitrogen content: 3 to 4%) and 0.006 g of 5,6-dimethylbenzimidazole is added to the broth.

After the introduction of the nutrients the fermentation broth is thoroughly mixed, and then the fermenter is covered with a rubber plate and placed into a thermovessel heated to 32° to 34°C. The broth is incubated at this temperature for 6 days, and during this period 100 ml portions of a nutrient mixture with the above composition are added daily to the broth. Thus, on the seventh day of incubation the volume of the fermentation broth reaches 10 liters.

From this time on the fermentation is continued in semicontinuous manner, i.e., 10% of the fermentation broth is removed daily in a single portion for processing and replaced by the same volume of a fresh nutrient broth. The composition of the fresh nutrient broth should be adjusted so that the assimilable nitrogen and carbon removed with the separated portion of fermentation broth and in the form of methane is replaced completely.

The total nitrogen concentration of the removed fermentation broth (determined, e.g., by the Kjeldahl method) is 2,025 mg/l, consequently 2.025 g of nitrogen are removed from the fermenter with the separated 1 liter of fermentation broth. Thus the daily supplement of 1 liter of nutrient broth should contain about

2.025 g of nitrogen in the form of inorganic ammonium compounds, and about 12 times greater amount of carbon in the form of methanol. Accordingly, the tenth volume of fermentation broth, removed daily from the fermenter, is replaced by 1 liter of a fresh nutrient broth with the following composition:

Ammonium bicarbonate	6.0 g
Ammonium hydroxide (24%)	2.0 ml
Diammonium hydrogen phosphate	0.6 g
Ammonium sulfate	2.0 g
Hydrolyzed brewer's yeast (20%)	1.0 ml
Cobalt chloride	0.025 g
Succinic acid	0.006 g
o-Xylidine	0.020 g
Magnesium chloride	0.1 g
Methanol	80.0 ml
Water qs 1,000.0 ml	

In this nutrient broth the ratio of ammonium nitrogen and methanol carbon (g N:g C) is 1:12. Furthermore, a mixture of 9 g of corn steep liquor and 0.015 g of 5,6-dimethylbenzimidazole (precursor) is added daily to the fermentation broth.

The removal of 1 liter of fermentation broth and the addition of 1 liter of nutrient broth with the above composition are repeated daily for any desired time. The portion of fermentation broth removed daily contains, with a negligible fluctuation, about 20,460 μg/l of vitamin B_{12} and about 200 mg/l of factor III, biologically equivalent with vitamin B_{12}. Accordingly, the vitamin B_{12} production increases to 195% as compared to the initial value of 10,500 μg/l.

OTHER VITAMINS

Vitamin B_1

L. Šilhánková; U.S. Patent 4,186,252; January 29, 1980; assigned to Vysoka skola chemicko-technologicka, Czechoslovakia describes a process for preparing vitamin B_1 by fermentation in an economic manner as a by-product of industrial alcohol fermentation, such as during the fermentative preparation of alcoholic beverages such as beer, wine, fruit wines, cider, sorghum, sake, quass, kifir, kumis, etc. Strains *Saccharomyces cerevisiae* Hansen DBM 159 and *Saccharomyces uvarum* Beijerinck (syn. Sacch. Carlsbergensis Hansen) DBM 189 are particularly suited for this purpose.

Cultivation of the strains employed herein may be effected either in a synthetic culture medium or a natural medium, the prime requirement being that the medium contain the essential nutrients for the growth of the strain employed. Such nutrients are well known in the art and may typically be selected from among a carbon source, a nitrogen source, inorganic compounds and small amounts of organic growth factors.

Carbon sources found suitable for this purpose include sugars such as glucose, fructose, sucrose, maltose, maltotriose and the like. The source of these sugars may be molasses, malt extract, fruit mash, starch syrup, cellulose hydrolysates, sulfite liquors and the like.

The nitrogen sources employed may be chosen from among various ammonium salts such as ammonium sulfate, ammonium phosphate, liquid ammonia, etc., or natural substances containing nitrogen such as malt extract, yeast extract, casein hydrolysates, corn steep liquor, etc.

The media employed also include inorganic compounds which must be present to assure growth of the strains employed. Typical of such inorganic compounds are potassium phosphate, magnesium sulfate, iron sulfate, the chlorides of iron or sodium, etc. Traces of other inorganic ions are usually present in sufficient amounts in the raw materials employed to aid in the growth of the strains.

The organic growth factors employed may be biotin, pantothenate, etc. These compositions may be furnished by adding natural substances such as malt extract, yeast extract, cornsteep liquor and the like.

In the operation of the process, the conditions normally employed in culturing and fermentation, for example, in alcoholic fermentation, baker's yeast production and the production of beer, wines, ciders and the like may be used. Typically, fermentation temperatures range from 6° to 10°C for bottom beer type fermentation and from 30° to 32°C for ethanol production.

Following fermentation, the vitamin may be recovered from the fermentation medium by conventional means such as ion exchange techniques, adsorption chromatography, gel filtration or the like.

Example: *Saccharomyces cerevisiae* Hansen, strain DBM 159 was propagated from a laboratory culture in a molasses fermentation medium acidified to a pH within the range of 4.2 to 4.5 and supplemented with ammonium sulfate and ammonium phosphate. Fermentation was effected at temperatures of 30° to 32°C. After the removal of the yeast, vitamin B_1 was obtained from the fermented liquor by passing it through a column of an ion exchanger. The liquor, in this way deprived of the vitamin, was then processed in accordance with conventional distilling techniques. The vitamin, accumulated in an Ionex column, was then eluted and isolated from the eluate by precipitation with silver nitrate and decomposition of the precipitate with diluted hydrochloric acid. Vitamin B_1 yield was approximately 10 mg/l of fermented liquor, the ethanol yield being essentially unaffected.

Riboflavin Purification

A. Epstein, G. Graham and W.A. Sklarz; U.S. Patent 4,165,250; August 21, 1979; assigned to Merck & Co., Inc. describe a process wherein riboflavin fermentation broth or riboflavin fermentation solids are upgraded to an intermediate product either usable as such as an animal feed supplement or suitable for further purification for pharmaceutical use.

The starting material for the process is a riboflavin fermentation broth. Briefly, a nutrient medium is sterilized and inoculated with an organism capable of producing riboflavin. When the fermentation yield approaches or is at about the maximum the broth is heated to a temperature of from about 50°C to about 65°C for from about 15 to about 45 minutes, preferably for from about 25 to about 35 minutes and the riboflavin recovery begins. This heating serves to lyse the cells and to decrease broth viscosity thus enhancing the effectiveness of subsequent recovery and purification steps. Heating beyond about 45 minutes is undesirable as it increases rather than decreases broth viscosity.

The broth is then cooled and diluted with a predetermined quantity of water. The quantity of water chosen is insufficient to dissolve suspended solids in the broth but sufficient to optimize centrifugal separation by both diluting the previously dissolved solids in the broth and enhancing separation of solid suspended particles having a density less than that of crystalline riboflavin. Typically, this added quantity of water is from about 25 to about 100 volume % of the volume of the fermentation broth, and preferably is from about ⅓ to about ½ the volume of the fermentation broth.

A proteolytic enzyme may be present during the heating or may be added following the heating. The enzyme is allowed to digest proteinaceous matter for several hours, generally for from about 1 to about 5 hours, preferably for from about 3 to about 4 hours. During enzyme treatment the pH is adjusted to a level at which the enzyme functions most effectively, generally at a pH of from about 6.0 to about 9.0. The broth is then cooled, and the pH, if alkaline, is adjusted to about pH 7.0.

The diluted broth either with or without enzyme treatment next is converted to a sludge by centrifugation. The sludge is then resuspended with a predetermined quantity of water. The quantity of water chosen is insufficient to dissolve suspended solids in the resuspended sludge but sufficient to optimize separation of solid particles having a density less than that of crystalline riboflavin. Typically this quantity of water is equal to from about 1 volume to about 3 volumes per volume of sludge, and preferably is about twice the volume of the sludge. On a solids basis, the resuspended sludge contains from about 15 to about 30 weight percent solids. The resuspended sludge is then centrifuged to yield a centrifugate usable as such as an animal feed supplement.

Examples of enzymes suitable for use are alkaline proteases such as *B. subtilis* protease, *B. ligninoformis* protease, and *B. amylofaciens* protease, or neutral proteases such as neutral *B. subtilis* proteases. Such enzymes are commercially available. For example, a suitable alkaline protease enzyme is Rhozyme P-62 and a suitable neutral protease enzyme is Enzeco Bacterial Protease.

Example: A 553 ml portion of pasteurized riboflavin fermentation broth is inoculated with 10 ml of *B. subtilis* (National Collection Type Culture No. 3610) nutrient broth seed culture. The fermentation is allowed to continue 48 hours at 37°C. Suspended solids decrease 72% during fermentation. The *B. subtilis* fermented broth is heated, diluted to 750 ml with water and centrifuged. The centrifugate is suspended with 200 ml water and centrifuged again. The centrifugate is 53% more pure than a non-*B. subtilis* fermented control, and 4.4 times more pure than dried fermentation broth.

Biotin

S. Ogino, S. Fujimoto, H. Wada and Y. Tanigawa; U.S. Patent 3,859,167; Jan. 7, 1975; assigned to Sumitomo Chemical Company Ltd., Japan describe a process for producing biotin of the formula shown on the following page.

This is done by oxidizing cis-tetrahydro-2-oxo-4-n-pentyl-thieno[3,4-d]imidazoline of the formula,

(which will be referred to as A hereinafter) by cultivating a microorganism belonging to the genera *Pseudomonas, Corynebacterium, Arthrobacter, Brevibacterium, Mycobacterium, Nocardia, Candida, Cunninghamela, Cladosporium, Gibberella, Penicillium* and *Mucor* in a culture medium containing a proper nutrient and preferably under an aerobic condition, adding the compound A at a proper concentration from the beginning of or after the growth of the microorganism and continuing the cultivation to convert the compound into biotin; or producing biotin from the compound A by using resting cells previously grown in an adequate medium in the presence or absence of a proper amount of the compound A, and then separating biotin from fermented broth or incubated fluids with resting cells.

The advantages of the process are that complicated steps of producing biotin required as in the conventional method for the synthesis of biotin can be simplified and that the amount of production of biotin is high (e.g. several grams per liter) as compared with the conventional fermentation method wherein the yield of a biotin active substance is at most 20 mg/l.

More particular examples of these microorganisms are as follows:

> *Pseudomonoas mutabilis* B-3252, FERM-P No. 1429, ATCC 31014
> *Corynebacterium primorioxydans* B-321, FERM-P No. 1427, ATCC 31015
> *Corynebacterium primorioxydans var. forte* B-323, FERM-P No. 1428, ATCC 31016
> *Arthrobacter paraffineus*, ATCC 21003
> *Brevibacterium ketoglutamicum*, ATCC 21004
> *Mycobacterium smegmatis*, IFO 3083

Nocardia corallina, IFO 1954
Candida arborea, IAM 4147
Cunninghamela blakesleeana, IFO 4443
Cladosporium herbarum, IFO 4458
Gibberella fujikuroi, ATCC 14842
Penicillium chrysogenum, ATCC 7326
Mucor microsporus, ATCC 8541

The compound A may be obtained by Grignard-reacting N-substituted-cis-tetra-hydrothieno[3,4-d]imidazoline-2,4-dione of the formula,

with n-pentyl bromide, then dehydrating and hydrogenating the reaction product and removing the N-substituents.

The nutrients to be used for the cultivation of the microorganism may be any usual ones which are commonly used for the cultivation of microorganisms. Thus, for the carbon source, e.g., n-paraffin, acetic acid, kerosene or combination thereof is preferable although glucose, starch, glycerine or sorbose may be used and, for the nitrogen source, e.g., peptone, corn steep liquor, soybean powder, ammonium chloride, ammonium nitrate, ammonium sulfate, urea or mixtures thereof may be used.

Further, for the inorganic salt, sodium chloride or a phosphate may be used and, for the trace elements, a sulfate or hydrochloride of magnesium, iron, manganese, cobalt, zinc or copper may be used. The pH of the culture medium may be about 6 to 8, preferably about 7 when a bacteria is used and may be about 4 to 7, preferably about 5 to 6 in the case of using mold or yeast. The cultivation temperature may be about 25 to 37°C, preferably about 25° to 30°C for bacteria and may be about 22° to 30°C, preferably about 25°C in the case of yeast or mold. In the culture, aeration and agitation give favorable results.

In carrying out the process, any of the abovedescribed microorganisms is grown in a culture medium containing the abovementioned nutrient source. The compound A as dissolved in a proper organic solvent such as ethanol, methanol or dimethyl formamide is added from the beginning of or after the growth of the microorganism and the cultivation is continued for 2 to 7 days. The concentration of the compound A in media may be 0.05 to 0.3%, preferably 0.1 to 0.2%.

After the cultivation biotin produced in the culture may be separated and purified. For this purpose a process generally used for extracting a certain product from the culture may be applied by utilizing various properties of biotin. Thus, for example, the cells are removed from the culture fluids, the desired substance in the filtrate is adsorbed on active carbon, is then eluted out and is purified with an ion exchange resin. Alternatively the culture filtrate is purified by being treated directly with an ion exchange resin and after the elution the desired product is recrystallized from water or alcohol. Further, in some cases, the culture filtrate may be concentrated without using an ion exchange resin and the deposited crystals are collected by adjusting the liquid pH near the isoelectric point.

Optical Resolution of α-Tocopherol

α-Tocopherol and the ester thereof are broadly used as vitamin E for human and veterinary medicines or as various additives for foods. Most of these α-tocopherols and esters thereof are produced and used as 2-DL-racemic compound. However, the direct use, as pharmaceutical agent, of a racemic mixture containing nonnatural type optical isomers, typically amino acids, which are biologically nonactive or low-active, has lately been questioned.

T. Kawamura, S. Kato, Y. Ikeda and N. Kuwana; U.S. Patent 4,022,664; May 10, 1977; assigned to Eisai Co., Ltd., Japan describe a process for the separation of α-tocopherol or the ester thereof which has been prepared by the condensation reaction of trimethylhydroquinone with natural phytol or racemic isophytol, into 2-D-isomer and 2-L-isomer.

Such optical resolution ability exists in microorganisms such as *Candida sp., Mucor sp., Rhizopus sp., Aspergillus sp., Pseudomonas sp.* and the like.

When the microorganisms used in the process are cultured, there may be used a culture medium which contains suitable nutrition sources used conventionally for culturing these microorganisms, i.e., carbon sources, nitrogen sources, inorganic salts, etc.

Cultivation is carried out wholly aerobically. Therefore, shaking culture method or bubbling an agitating culture method may be used. Culture temperature is suitably within the range of from 25° to 32°C, and preferably within the range from 27° to 30°C depending upon the species of the microorganisms. Suitable culture period is usually in the range of from 48 to 72 hours, though there may be cases where sufficient activities are provided in the period of 12 to 24 hours or less.

The resulting culture broths of the microorganisms are used for the enzymatic reaction, in a form of broths per se, in a form of either whole cell or culture filtrate which are obtained after filtering the medium, or in a form of the treated products thereof.

As the esters of 2-DL-racemic compound of α-tocopherol used as the substrate, there can be used any of the esters of straight or branched chain, unsubstituted or substituted with a halogen or carboxyl group, saturated or unsaturated aliphatic acid having 2 to 20 carbon atoms, such as acetic acid, monochloracetic acid, propionic acid, butyric acid, isobutyric acid, valeric acid, isovaleric acid, caproic acid, caprylic acid, lauric acid, myristic acid, palmitic acid, stearic acid, linolic acid, succinic acid and the like. The esters of monochloracetic acid and of stearic acid are most suitable. These substrates are added directly to the culture filtrate, or to the aqueous solution containing microorganism bodies or their various crude enzymes, such as M/15 phosphate buffer solution. Concentration of the substrate is preferably on the order of 1 to 100 mg/ml, particularly in the range of from 3 to 15 mg/ml. The reaction is carried out by stirring sufficiently or in a form of an emulsion which has been prepared by adding about 0.5 mg/ml to 2.0 mg/ml of surfactant, such as polyoxyethylene nonylphenyl ether (Emulgen 905), because these substrates are generally insoluble in water.

The pH value in the reaction is generally in the range of pH 3 to pH 9, preferably in the range of pH 5 to pH 7, though it is varied depending upon the properties of the microorganisms or the enzymes produced by the microorganisms. The reaction is carried out by purging the space above the liquid in the reactor with nitrogen gas, allowing to stand, shaking or stirring the reaction mixture at the temperature ranging from 10° to 60°C, preferably from 25° to 35°C, for 0.5 to 60 hours.

The separation and purification of the free α-tocopherol and the unreacted α-tocopherol ester from the enzymatic reaction solution may be achieved by any conventional method for separation and purification, as follows: After the completion of the reaction, the reaction mixture is extracted with petroleum ether or benzene, the extracts are combined, and the solvent therein is removed by distillation under a reduced pressure. The residue is dissolved in a small amount of benzene or n-hexane, and subjected to silica gel chromatography according to a conventional method.

Example: A 30 liter jar fermenter was charged with 12 liters of culture medium containing 10 ml of kerosine, 1 ml of polyoxyethylene sorbitan monostearate (Tween 60), 5 g of ammonium sulfate, 0.4 g of potassium dihydrogen phosphate, 0.2 g of magnesium sulfate, 0.03 g of ferrous sulfate and 0.01 g of manganese sulfate per liter of the culture medium (hereinafter referred to as KT medium). The mixture was heated at 120°C for 10 minutes to sterilize it.

On the other hand, the KT medium was inoculated with 3 liters of culture seeds of *Candida cylindracea* ATOC 14830 which had been cultured for 24 hours in a culture medium containing 10 g of glucose, 5 g of peptone, 3 g of the yeast extract, and 3 g of the malt extract per liter of the medium (hereinafter referred to as MY medium). The cultivation was achieved under the conditions as follows: culture time, 48 hours; culture temperature, 27°C; volume of aeration, 30 liters/minute; and revolution of stirring, 300 rpm.

After the cultivation was over, the microorganism was filtered and removed to provide 15 liters of culture filtrate which was then concentrated to 750 ml at the bath temperature of 40°C. The concentrated solution was used as the crude enzyme solution in enzymatic reaction for optical resolution of 2-DL-racemic compound.

Thus, a 1 liter glass reactor was charged with 800 ml of the substrate suspension containing 5.6 g of 2-DL,4'-DL,8'-DL-α-tocopherol stearate, and 200 ml of the crude enzyme solution were added to the suspension. After purging with nitrogen gas, the reactor was sealed and stirred at 30°C for 48 hours, and then 800 ml of ethanol were added to terminate the reaction. The resulting mixed solution was extracted once with 900 ml of petroleum ether and three times with 350 ml quantities of the same solvent. A total of 2,170 ml of the petroleum ether layers was combined, and the amount of α-tocopherol therein was determined.

The content of about 23 g of the free α-tocopherol was confirmed, which corresponds to a hydrolysis ratio of 66.4%.

The extract was concentrated and evaporated to dryness, and the residue was dissolved in 40 ml of benzene. The benzene solution was charged on a column that had been filled with 270 g of silica gel, and subjected to column chromatography by using benzene as developing fluid. The extract was fractionated into fractions of 19 ml each. It was found that unreacted substrate, α-tocopherol stearate, was extracted in the fraction Nos. 27 to 38.

These fractions, Nos. 27 to 38, were combined together, and then concentrated and evaporated to dryness, and 1.9 g of waxy residue were obtained. This product exhibited a single spot on thin layer chromatograph, and this spot was identified with that of α-tocopherol stearate. After hydrolyzing the product by an alkali, it was subjected to the potassium ferrocyanide treatment and the angle of optical rotation was determined. The specific rotation was +21.98. It was found that product was 2-D,4'-DL,8'-DL-α-tocopherol stearate having about 93% of optical purity.

On the other hand, the fraction Nos. 52 to 84 were combined together, and then concentrated and evaporated to dryness, to obtain about 2.6 g of an oily residue which is identical with the free α-tocopherol. When the oily product was subjected to the potassium ferrocyanide treatment, the specific rotation of the resulting solution was −11.04. As the result of this determination, it was found that the oily product was a mixture of 2-D and 2-L isomers in the ratio of 28.5:71.5.

VARIOUS ORGANIC COMPOUNDS

METHANE

Anaerobic Digestion at Superatmospheric Pressures

A worldwide energy shortage has prompted a search for nonfossil sources of energy. One obvious alternate source of energy is the production of methane gas through anaerobic digestion of organic solid wastes. Methane gas has been produced from the digestion of sewage sludge for at least fifty years in both Europe and the United States. Methane has also been produced from animal wastes and used for running automobile engines and heating homes as well as for running engines and compressors at sewage plants.

Thus-produced methane has two important impediments to full utilization: (1) it contains 30 to 40% carbon dioxide, which reduces its heating value to about 600 Btu/ft^3 (compared to 1,000 Btu natural gas); and (2) it also normally contains a small amount of hydrogen sulfide which makes it highly corrosive. The corrosive element makes the gas usable only in specifically-designed more expensive engines and makes it unsuitable for blending with pipeline-quality natural gas.

To make the gas fully compatible with pipeline gas or usable in standard engines, it may be scrubbed. There are a number of commercial processes available for removing acid gases (CO_2 and H_2S), but they are all relatively expensive. Also, they do not lend themselves well to scrubbing the relatively small quantities of gas which might be produced from organic solid wastes from cities or even the wastes from a large beef cattle feedlot operation.

J.E. Ort; U.S. Patent 3,981,800; September 21, 1976; assigned to ERA, Incorporated describes a system which employs (a) digestion of solid wastes at a temperature of approximately 95°F (35°C), (b) mixing the contents of digesters, and (c) recirculating sludge to provide seed and favorable population dynamics. All of these practices improve digestion efficiency and decrease to about 10 days the time required to achieve complete digestion.

Whereas conventional anaerobic digestion is effected under a pressure of at most only a few inches of water, this digestion process is conducted under a pressure of from 2 to 5 atmospheres (approximately 30 to 75 psig). For aboveground operations digestion tanks operated at such pressures are economically constructed of steel. Alternatively-employed underground digestion tanks can be made of other material. Pressurization by use of a hydrostatic head is readily employed for underground digestion tanks (digesters).

Henry's Law indicates that the solubility of a gas is directly proportional to pressure. For instance, at 30°C and standard atmospheric pressure, the solubility of carbon dioxide in water is 655 ml/l, and that of methane is 27.6 ml/l. If the pressure is increased by one atmosphere, solubility of both gases is doubled. Each additional atmosphere of pressure adds 665 ml of solubility to a carbon dioxide-water system at 30°C. In typical digestion practice a pressure of 5 atmospheres gauge would increase solubility of carbon dioxide and methane to yield a gas with 66 to 68% methane. A pressure of nearly 20 atmospheres gauge would be required to yield nearly pure methane.

In addition to pressure, the volume of water is an important variable in such a system. A large water volume entails large tanks and high tanks; dilution is detrimental to population dynamics and raises energy costs (for heating the water to 35°C). To provide a large effective volume of water without these problems, recirculation is used. By combining a pressure of from 2 to 5 atmospheres gauge with a recirculation rate of from 4 to 10 times the daily feed volume, high quality methane is produced efficiently at little additional construction or operating cost, compared to conventional practice. Alternatively, digestion tanks are periodically depressurized; this, too, results in an increase in effective volume for carbon dioxide absorption.

In modern digesters which use external sludge heating units, approximately 2 to 4 times the feed volume is circulated through the sludge heater daily. This makes adaptation of a carbon dioxide stripping unit very simple. On this same recirculation loop at 5 atmospheres gauge, the entire sludge flow, after heating, may be degassed. This heating (normally about 6°C) decreases carbon dioxide solubility approximately 13% and improves degassification efficiency. When the digesters are operated at lower pressures, a greater recirculation volume is required and only a portion of the degassing loop is heated.

After degassing, either by mixing, sheet flow, and/or bubbling warm gas (optionally from a compressor exhaust through the sludge), a pump repressurizes the sludge back into the stage 1 digester (withdrawal is from the stage 2 digester as shown in Figure 7.1). In effect, the operation consists of withdrawing nearly saturated (with respect to carbon dioxide at operating temperature and pressure) sludge from stage 2, heating at least part of the flow, degassing and then reintroducing the unsaturated sludge into the stage 1 digester. This maintains the sludge within the digesters in an unsaturated state and essentially all carbon dioxide, as well as hydrogen sulfide, is kept in solution in the sludge. Even under pressure and recirculation, no more than 4% of all methane generated remains in solution and is subject to loss during degassing.

If in-tank degassing is used, a reduction in pressure to atmospheric is required for 15 to 20 minutes periodically. Frequency of degassing depends upon operating pressure. Higher pressures require less frequent depressurization.

Figure 7.1: System Design for Anaerobic Digestion of Sludge to Produce 0.4 Million Cubic Feet Methane Gas per Day

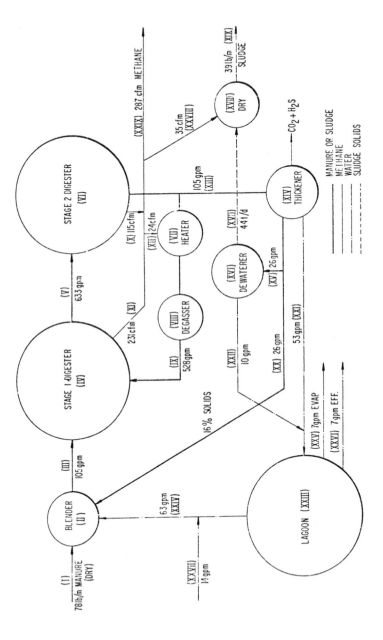

Source: U.S. Patent 3,981,800

By subjecting relatively conventional digestion to a pressure of from 2 to 5 atmospheres gauge and incorporating pressurized recirculation in a practical application of Henry's Law, relatively pure methane gas is produced without incurring scrubbing costs. In effect scrubbing is accomplished internally; it requires digesters which are designed to operate at the noted superatmospheric pressures, but such does not greatly add to either construction or operation cost as would scrubbing.

Two-Phase Digestion Process

S. Ghosh and D.L. Klass; U.S. Patent 4,022,665; May 10, 1977; assigned to Institute of Gas Technology describe a method and system for producing methane gas by anaerobic digestion of organic wastes wherein a separate acid-forming phase leads to improved liquefaction and denitrification of liquid wastes; and a separate methane-forming phase results in improved methanation of the organic wastes, as well as methanation of carbon dioxide which is formed as a by-product.

The microorganisms which ferment the organic wastes under anaerobic conditions require few conditions for adequate activity. Such requirements include the usual nutritive salts, carbon dioxide, a reducing agent, a single oxidizable compound suitable for the organism, and a source of nitrogen. Several species of methane-producing bacteria have been reported, including *Methanobacterium omelianskii, Mb. formicicum, Methanosarcina barkerii, Mb. sohngenii, Ms. methanica,* and *Mc. mazei.*

A wide variety of substrates are utilized by the methane-producing bacteria, but each species is believed to be characteristically limited to the use of a few compounds. It is therefore believed that several species of methane-producing bacteria are required for complete fermentation of the compounds present in sewage. In fact, mixed cultures are required for complete fermentation. For example, the complete fermentation of valeric acid requires as many as three species of methane-producing bacteria. Valeric acid is oxidized by *Mb. suboxydans* to acetic and propionic acids, which are not attacked further by this organism. A second species, such as *Mb. propionicum*, can convert propionic acid to acetic acid, carbon dioxide, and methane. A third species, such as *Methanosarcina methanica*, is required to ferment acetic acid.

An operative mixed culture is capable of maintaining itself indefinitely as long as a fresh supply of organic materials is added because the major products of the fermentation are gases, which escape from the medium leaving little, if any, toxic growth-inhibiting products.

Reference is made to Figure 7.2 which is a highly schematic block diagram illustrating representative embodiments of the process. The dotted lines represent alternative embodiments. A source of organic waste feed **2** is delivered along line **3** to an acid phase digester **4**. The organic feed may be waste such as manure, municipal refuse, raw sewage, primary sludge, activated sludge, or any combination. The organic feed may also be a biomass of land or water base plants, such as trees, grass, kelp, algae and the like. The term organic feed shall refer to both organic waste and organic biomass. The mode of operation provides continuous feed along line **3** into the acid phase, and continuous or intermittent agitation of the organic waste in the acid phase digester.

Figure 7.2: Two-Phase Digestion Process

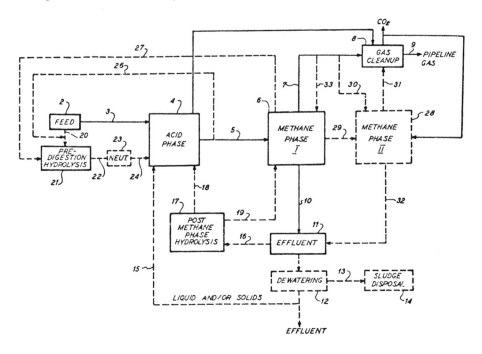

U.S. Patent 4,022,665

The organic waste is kept in the acid phase for a detention time of from ½ to 1½ days at a pH of from 5 to 6. The organic waste is continuously loaded into the acid phase at a rate of 1 to 10 pounds of total organics per cubic foot per day.

The liquid effluent from the first phase is then delivered along line **5** to a separate methane phase digester **6**, and such delivery may be intermittent or continuous at a loading rate of from 0.1 to 0.5 pound of total organics per cubic foot per day. The acid waste is gently agitated in the methane phase and the detention time of such waste in the methane phase is from 2 to 7 days over a pH range of from 6.8 to 7.4. The formed gases consisting principally of carbon dioxide and methane are moved along line **7** to a gas clean-up zone **8**, and thereafter conveyed to pipeline **9** for delivery to a fuel utilization means.

The effluent is moved through line **10** to a collection zone **11**. Such effluent may be processed by dewatering at **12** and conveying the sludge cake along line **13** to a sludge disposal station **14**. In the alternative, the effluent may be conveyed along line **15**, with or without dewatering, for recycling to the acid phase **4**. Such effluent can comprise liquid alone, solids alone, or a mixture of liquids and solids.

In another alternative embodiment, the effluent from station **11** can be conveyed along line **16** to a hydrolysis zone **17** where mild hydrolysis can occur by acid or alkaline treatment. The posthydrolyzed effluent can then be recycled along line **18** to the acid phase **4**. In the alternative, the hydrolyzed effluent from the methane digester can be delivered along line **19** back to the methane phase digester **6**. A neutralization station (not shown) may be interposed in lines **18** and **19** prior to return to their respective phase digesters.

In the preferred practice, the feed from source **2** is moved in a stream along line **20** to a predigestion hydrolysis zone **21** where mild acid or alkaline hydrolysis occurs. The hydrolyzed feed is then moved along line **22** to a neutralization zone **23** where either the acid or alkaline phase is neutralized; and the stream is then moved along line **24** to the acid phase digester **4**. The stream of acid waste may be taken from line **5** and moved along line **26** for recycling to the prehydrolysis zone **21** along line **20**. Similarly, effluent from the methane phase digester **6** can be returned along line **27** to the predigestion zone **21**. Recycling the acid phase and methane phase effluents results in successive neutralization, enhanced hydrolysis, and dilution, as well as improved acidification in the acid phase and gasification in the methane phase as a result of a second pass.

A second methane phase digester is indicated at **28**, and the methane phase effluent is moved to the second methane phase along line **29**. As previously stated, improved gasification occurs in the second methane phase from preferential action by microorganisms, including methanation of carbon dioxide gas which can be returned to the second methane phase along line **30**. The gas production from the second methane phase can be delivered to the gas clean-up zone along line **31** and the effluent can be conveyed along line **32** to the effluent zone **11**. A sulfide neutralization salt such as ferric chloride can be introduced from sources (not shown) to one or both of the methane phase digesters to neutralize sulfide formation and reduce hydrogen sulfide which occurs in the digester gases.

The following table presents a range of optimum conditions which lead to an improved two-phase digestion process. The organic waste which is treated can be primary sludge, activated sludge, a mixture of both, manure, solid waste, or industrial waste. The following data represent a delivery of organic waste to an acid phase digester, maintenance and detention of the waste in the digester as set out in the following table, and then delivery of the lower molecular weight acid and intermediate products to a methane phase digester, as well as collection of methane gas in the gasification process in the methane phase digester.

Nature of Feed	Acid Phase	Methane Phase (Acid Phase Effluent)
Feed consistency, % total solids	2-10	1-7
Temperature, °C	20-40	20-40*
Culturing mode	continuous	intermittent or continuous
Recycling, % of influent	0-50**	0-40**
Mixing	continuous moderate agitation	intermittent or continuous gentle agitation
Residence time, days	0.5-1.5	2-7
Loading, lb total organics/ft^3/day	1-10	0.1-0.5
pH	5-6	6.8-7.4
ORP, mV	200-300	425-550

*Can be operated in thermophilic range, 50° to 65°C. In this case, the loading can be increased up to 2.0 lb of total organics/ft^3/day.
**Concentrated or raw effluent.

The foregoing two-phase digestion process, under the conditions recited, results in an increased methane content of the product gas, and the waste processing capacity is increased for a given detention time relative to conventional digestion.

Control of Volume-to-Interface Ratio

J.E. Ort; U.S. Patent 4,040,953; August 9, 1977; assigned to RecTech, Inc. describes a process for producing methane gas which involves an anaerobic digestion process where formation of gaseous carbon dioxide is suppressed and greatly minimized without effect upon methane production and indeed to the benefit of methane production. Through this process there is a recognition that gaseous methane has a coefficient of diffusion which is 4.53 greater than that of gaseous carbon dioxide. Thus, by carrying anaerobic process at predetermined volume-to-interface (V/I) ratios, the transfer of carbon dioxide across the interface within the digestion tank is minimized while the transfer of methane gas is maximized. The transfer across the interface takes place for a desired liquid retention time (LRT), temperature and pH.

It will be seen from Figure 7.3 that organic material is slurried or pulped in a wet pulping tank where it is mixed with recycled water and fresh water into a slurry with a minimum of 4% solids by weight. The slurrying or pulping tank is of steel construction and the organic solids and water within, are impacted at high speed with a rotor having a hardened steel surface. The rotor is mounted on a vertical shaft which gives the rotor blades a speed of about 5,000 fpm. A perforated bottom plate in the tank, in conjunction with rotating member, produces a rasp and sieve effect. Drive requirement for the rotor is approximately 40 horsepower for each ton per hour of organic feed material (drive basis). The unit operates much as a scaled-up home garbage disposal unit. Wet pulpers are commercially available.

The slurry is then passed from the wet pulping tank through a liquid cyclone for removal of grit and inorganic heavy solids. The liquid cyclone is of steel construction and operates on as little as 4 feet of hydraulic head. An optimal operating range is approximately 10 to 15% underflow. Commercial units of the liquid cyclone tank are readily available. The underflow from the cyclone tank is dewatered and wasted. Overflow from the cyclone tank next passes through a strainer which dewaters the slurry to a minimum solids content of about 10%. The strainer unit is capable of retaining most material which is larger than 0.10" in particle size. Commercial strainer units are also readily available.

Excess water from the strainer is recycled to the wet pulping tank. The dewatered slurry is then transmitted to a carbon dioxide stripping and feeding tank which is a progressing cavity pump well that will hereinafter be termed as the feeding tank. The slurry is well-blended in the feeding tank with recirculation sludge from the stage 2 digester, which sludge has been acidified, nutrient-balanced and carbon dioxide stripped. In the feeding tank there is a progress-cavity type pump that is capable of pumping thick slurries under a variety of conditions and is commercially available.

The blended material in the feeding tank is then pumped into the stage 1 digester which is constructed of reinforced concrete or steel.

Figure 7.3: Anaerobic Digestion Process

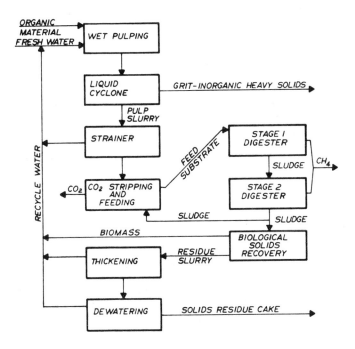

Source: U.S. Patent 4,040,953

The interior of the stage 1 digester is coated with a corrosion-protective material, especially near the upper edges which will be adjacent the intended liquid-gas interface in the tank. Mechanical mixing capability is provided by auger-type mixers so that the contents of the tank will be well mixed. It is intended that there be an auger, propeller or axial flow impeller which will pump a total of approximately 12 times the tank volume in a 24 hour period assuming no hydrostatic head to overcome and a rather nominal dynamic head due to friction.

Such an impeller or pump requires approximately a 1 horsepower drive for each 30,000 gallons of tank capacity in the range of tanks having volumes of 100,000 to 1,000,000 gallons where the tank is either spherical or square cylindrical design (where D = h). The tank should also be baffled in ways well known to those skilled in the art, to prevent short-circuiting. Feed into a spherical tank should be tangential at the equator of the tank and discharge from the bottom.

It is preferred that the feed into the stage 1 digester have a carbon-to-nitrogen ratio of 5:1 and a carbon-to-phosphorus ratio of 20:1. The nutrients may be conveniently supplied as required, by adding ammonia and phosphoric acid to the sulfuric acid which is fed to adjust pH for carbon dioxide stripping. Unadjusted effluent will have a pH of 6.8 to 7.0 which is adjusted downwardly to 6.4 to 6.5. A pH as low as 6.0 may be tolerated. A chemical solution feeder suitable for strong acid handling is used. However, where organic solids such as manure are being fed, then usually the nitrogen addition can be dispensed with or considerably reduced. However, with municipal waste, the nitrogen addition is most desirable.

The sludge from the stage 1 tank is transferred into the stage 2 tank. The stage 2 tank is identical to the stage 1 tank except mixing is continuous in the stage 1 tank, but mixing is intermittent in the stage 2 tank.

It has been recognized that the rate of digestion peaks at from $35°$ to $40°C$ (mesophilic) and then drops at slightly higher temperature. However, the rate then climbs to a still higher point at from $55°$ to $60°C$ (thermophilic) and a second, but higher peak is achieved. Thus, operation at about $60°C$ achieves a higher digestion rate, such that the solids retention time (SRT) is 4 to 5 days in the stage 1 tank and 4 to 5 days in the stage 2 tank. This compares favorably with an SRT of 12 to 15 days total for the two tanks when operating at $40°C$.

The liquid retention time (LRT) for the two tanks combined should be approximately 48 hours, although as little as 24 hours may suffice and as many as 72 hours may be necessary depending upon the gas quality desired and the volume-to-interface ratio (V/I) used. LRT is calculated by determining the total volume of liquid in the stage 1 and stage 2 tanks and dividing by the volume of liquid fed back into the stage 1 tank. Thus, if the total volume of liquid in stage 1 and stage 2 tanks is 200,000 gallons and if 100,000 gallons of liquid is fed back into the stage 1 tank in a 24 hour period, there is an LRT of 48 hours or 2 days.

With a 48 hour LRT, a gaseous product of approximately 90% methane can be produced at a V/I of approximately 400 (gal/ft² of interface). As the LRT changes, the quotient of LRT divided by the V/I should remain approximately 0.12. With appropriate changes in LRT, a V/I range of 200 to 500 is most desirable. If the LRT becomes longer, then the appropriate V/I ratio must be greater.

The volume-to-interface ratio can be varied most readily by changes in operating levels in existing tanks or by variations in tank geometry in new facilities. The following table shows how the V/I ratio changes on a spherical tank for various operating levels.

Volume-to-Interface Ratios for Spherical Tanks at Various Depths

Depth (%)	Volume (%)	V/I (gal/ft^2)
100.00	100.00	Infinite
96.67	99.67	1,112.5
93.33	98.65	592.3
90.00	97.13	405.0
86.67	95.14	308.8
83.33	92.59	249.3
80.00	89.60	210.1
76.67	86.21	180.5
73.33	82.46	157.4
70.00	78.40	139.0
66.67	74.08	124.1
63.33	69.53	111.7

From Figure 7.4 it can be seen that at a 24 hour LRT and a V/I ratio of about 220 virtually all methane will be transferred while only 21% of the carbon dioxide will be transferred. At a 48 hour LRT and a V/I ratio of 410, all of the methane and 26% of the carbon dioxide will be transferred. At an LRT of 72 hours and a V/I ratio of 640, all of the methane and about 30% of the carbon dioxide will be transferred. From an economic standpoint the lower LRT values are preferred.

In view of the above, it can be seen that the stage 1 and stage 2 tanks should be operated at a 24 hour LRT under conditions that the tank is filled for 80% of its depth which calculates out to a filled volume of 89.6% of the total volume of the tank in accordance with figures taken from the table above. This gives a V/I ratio of 210. From Figure 7.4 it can be seen that at a 24 hour LRT and a V/I ratio of 210, virtually all of the methane will be transferred and only about 21% of the carbon dioxide will be transferred. The most effective operating pH range is from 6½ to 7 or slightly on the acid side. The sludge discharged from the stage 2 tank is then treated for carbon dioxide stripping.

Figure 7.4: Effect of Volume-to-Interface Ratio on the Transfer of Carbon Dioxide and Methane

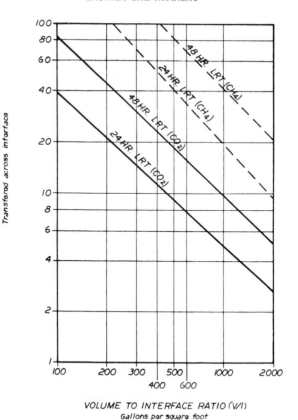

Source: U.S. Patent 4,040,953

Production Using Solar Energy

O.W. Boblitz; U.S. Patent 4,057,401; November 8, 1977; assigned to Bio-Gas Corporation describes a method for the manufacture of methane gas which basically comprises bacterially decomposing sewage sludge or other organic waste in an aqueous slurry at a pH of from 6 to 8, heating the slurry for periods up to 10 days to a temperature between 100° and 140°F in major part by in-direct heat exchange with hot air produced in a solar energy heater, collecting methane gas produced by bacterial action on the heated slurry, withdrawing portions of the slurry as the process proceeds, dewatering such withdrawn por-tions and recovering the dewatered solid residue for use as fertilizer.

The objects are accomplished using a methane gas-manufacturing apparatus which basically comprises a series of digester tanks; a substantially airtight en-closure surrounding the tanks; a solar heater to heat a stream of air; conduit means to recycle air through the solar heater from the tank enclosure and re-turn it to the enclosure in heated condition; and plumbing to pass organic waste slurry in, through and out the digester tank system.

The solar heater comprises a rectangular container with a bottom and four sides, a transparent panel covering the top of the container, a rectangular section of black wire screen positioned in the container parallel to the transparent panel, a hinge attached to one edge of the container to permit it to be moved between a position of about 30° to 80° of horizontal and an accordion pleated unit at-tached to the bottom of the container to provide a closed chamber beneath the container which can vary in size as the container is moved between the 30° to 80° positions.

The methods and equipment provide a 10 day continuous process that can ex-tract about 80% of the methane from sewage sludge and other organic waste. About 11 ft³ of methane gas (STP) can be extracted for each pound of organic solids, equal to about 7,000 to 9,000 Btu. Most of the digester tank heat, in view of the construction of the equipment, will be obtained by the solar heat collectors which can be adjusted at right angle to the sun. This solar heating system can reduce the cost of producing methane gas by about 40%. Moreover, the solid waste that is recovered may be sold and used as fertilizer providing additional production cost savings of about 15%.

Use of Electrolysis

H. Switzgable; U.S. Patent 4,053,395; October 11, 1977; assigned to Alpha Systems Corporation describes a method of producing methane gas which util-izes a plurality of tanks. Each of the tanks is covered with a floating hood which seals the contents of the tanks with respect to the ambient. Inside each tank is a slurry solution which consists essentially of a composite of waste ma-terials being partly solid and partly liquid. The volume of the slurry in each tank is controlled such that a region exists under the hood for collecting gas-eous materials.

One of the tanks is provided with a cathode which is mounted inside the tank such that it protrudes into the slurry solution. Another tank is provided with an anode which also extends into the slurry solution in that tank. The anode tank and the cathode tank are joined by a conduit which connects the tank at a point in the tank wall below the level of the slurry in each tank such that although the slurry solution in the two tanks is interconnected the gaseous portion of each tank is sealed with respect to each other and with respect to the ambient.

The floating hood of each tank is provided with a means for withdrawing the gas collected therein whenever the hood is vertically moved to a predetermined position. Preferably, at this predetermined height a compressor will be activated which will pump the collected methane or other gas out of the hood and the compressor will cease operation when the hood returns below a predetermined level. This compressor system can be utilized in each tank. Each tank is provided with an auxiliary inlet and an auxiliary outlet for the introduction of additional new waste material and for the elimination of unprocessible and processed waste material.

The advantage of the apparatus of this process is that the cathode tank will be sealed with respect to the ambient and with respect to the anode and thereby prevent any oxygen from entering this tank and inhibiting the growth of the anaerobic bacteria. Therefore, the anaerobic bacteria will be able to grow at will and increase the rate of production of methane. Also, in the cathode tank in accordance with standard electrolysis reaction, hydrogen gas will be produced. This hydrogen gas will aid in the operation of this tank in two ways. First, any free carbon in the cathode tank will tend to associate with the hydrogen thereby producing additional methane. Second, the hydrogen produced in the cathode tank has been shown to be an effective element to aid in the growth and reproduction of certain anaerobic bacteria.

In the anode tank in accordance with the electrolysis reaction oxygen will be produced. This oxygen introduced into the gaseous volume of the second tank is useful in aiding in the growth of aerobic bacteria. It is also desirable that this oxygen is separated from the cathode tank since it would inherently inhibit the growth of anaerobic bacteria therein. Alternatively, in the anode tank the oxygen could be continually pumped off and thereby allow for the growth of anaerobic bacteria in the anode tank, or, of course, the oxygen could be contained therein to aid in the growth of aerobic bacteria.

The two tanks are interconnected below the slurry level by a conduit which allows the free flow of ions due to the electrolysis reaction. The location of this conduit, however, eliminates any possibility of fluid flow communication between the gaseous sections of the cathode tank and the anode tank which would obviously inhibit the bacterial action in each tank.

In addition, a method of removing the unprocessible floating slurry from the tanks is shown. A skimming system can be utilized to eliminate the floating sludge and thereby facilitate contact between the tank atmosphere and the tank slurry. Also, the floating sludge material can be more evenly distributed throughout the slurry solution by the use of agitation and then strained and expelled from the tank. Each tank can be provided with resistance heater elements to increase the temperature above room temperature to facilitate bacterial growth.

In the embodiment illustrated in Figure 7.5, a two tank system is employed, referenced generally as tank **2** and tank **4**. Both tanks are shown containing the normal amount of slurry **6** contained therein.

Figure 7.5: Process for Methane Production Using Electrolysis

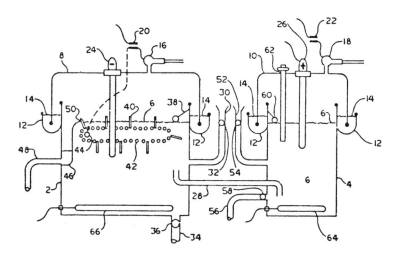

Source: U.S. Patent 4,053,395

The slurry consists of a large variety of waste materials, the exact composition of which is determined by the particular commercial or household application involved.

Tanks 2 and 4 are provided with the vertically movable hoods 8 and 10, respectively. These hoods are shown suspended in channels 12 which contain therein an oil solution 14. The channel 12 extends around the entire periphery of the tank so that by utilizing the sealing properties of the oil channel the internal environment of the tanks will be hermetically sealed with respect to the ambient. In this manner elements can be excluded from the tank environment which may be detrimental to the growth of the particular bacteria being cultivated therein.

Tank 2 is provided with a compressor 16 for removing gas accumulated within hood 8 of tank 2. Similarly hood 10 is equipped with compressor 18 for withdrawing gas accumulated under hood 10. Associated with compressors 16 and 18 are switches 20 and 22, respectively, which serve to activate the compressors whenever the hoods 8 and 10 move vertically in response to an accumulation of gasses lighter than air therein to a predetermined vertical position such that a switch is closed. Upon closing of either of these switches the associated compressor is activated to withdraw the gas accumulated within the particular hood and thereby allow the associated hood to again move downward. The moving downward thereby releases the switch and thereby ceases the operation of the associated compressor.

Tank 2 is provided with a cathode element 24 which is mounted such that it protrudes into the slurry solution when filled at the normal level. Similarly, tank **4** is provided with an anode **26** which is mounted such that it protrudes into the slurry solution located in tank **4**. Tanks **2** and **4** are interconnected by conduit **28**. This conduit provides fluid flow communication between the slurry located in each tank. The conduit **28** is positioned in the walls of tanks **2** and **4** such that it is completely below the normal level of the slurry to prevent any flow communication of the gasses located in the upper portions of tanks **2** and **4**. The conduit **28** is of sufficient size to permit the flow of ions between cathode **24** and anode **26**, which allows the electrolysis action to take place whenever direct current or alternating current is applied to electrodes **24** and **26**.

Tank **2** is provided with an inlet **30** and an inlet valve **32** to allow and control an input of additional waste material into the slurry solution located within tank **2**. Additionally, tank **2** is provided with an outlet **34** and outlet valve **36** for the exhaust of processed and unprocessible slurry matter. Valves **32** and **36** can be associated with sensing element **38**. Sensing element **38** is responsive to the level of the slurry solution located in tank **2** such that when the level of the solution becomes sufficiently low valve **32** admits additional fresh slurry. Sensing element **38** is also provided such that if the level of slurry solution in tank **2** becomes sufficiently high valve **36** will open and thereby exhaust part of the slurry solution.

The maintenance of a constant level of slurry solution is important due to the action of wiper elements **40**. Wiper elements **40** are mounted on chain drive element **42**. Chain drive element **42** is driven by motor **44**. Whenever motor **44** is actuated chain drive element **42** operates in a counterclockwise manner such that wiper elements **40** skim along the surface of the slurry solution of tank **2**. Part of the cross section of wiper element **40** is above the top surface of the liquid slurry solution and part of the wiper element moves along below the level of the liquid slurry solution when moving toward trough **46**. In this manner wiper element **40** picks up all the processed and unprocessible floating sludge on the surface of the slurry solution.

The wipers which are partly located above the surface as they move to the left push this floating sludge into trough **46** and it is thereby exhausted through drain **48**. As the wipers each contact trough **46** at protruding edge **50** they wipe the floating surface sludge into the trough by being urged against the leading edge **50**. After an individual wiper passes the leading edge **50** it is then returned back to the right edge of the tank along a path slightly submerged from the surface as shown in Figure 7.5. These wipers **40** are preferably made with disposable rubber sections which form the contact with leading edge **50** of trough **46** during each revolution of the chain **42**.

The formation of sludge is usually in direct proportion to the amount of methane gas given off from the slurry solution. To prevent the consumption of excess energy by motor **44** it is connected to its power source through switch **20**. Therefore, the motor **44** rotates chain **42** in a counterclockwise direction only during the intermittent periods when compressor **16** is also activated. As discussed before this occurs whenever the hood **8** moves to a predetermined vertical height. In addition, whenever the switch **20** is opened and compressor **16** thereby ceases action motor **44** will also cease rotating chain **42**.

Tank 2 is also shown equipped with heater element 66 which is used to elevate the temperature of slurry solution 6 above the normal ambient temperature. It has been found that to promote the production of methane the temperatures in the range of 95° to 100°F have been most effective.

Tank 4 is similar to the configuration of tank 2 in that inlet 52 has a valve 54 which is used for the introduction of additional fresh slurry solution. Also, outlet 56 has an outlet valve 58 which is used for expelling processed and unprocessible slurry materials. A sensing device 60 is used to sense the level of the slurry solution and, as in tank 2, maintains the proper level by the control of valves 54 and 58.

Tank 4 is equipped with bubbling element 62 which can be used to introduce gas, such as air or oxygen into the slurry solution in tank 4 to increase the oxygen content thereof. In this manner element 62 can increase the growth of any aerobic bacteria located in tank 4. Tank 4 is equipped with a heating element 64 which can be used to control the temperature of slurry solution 6 located in tank 4.

At this point it should be noted that a power source will be required for various elements in this system such as heaters 66 and 64, the electricity source for cathode 24 and anode 26, power sources for compressors 16 and 18, and other various drives necessary for this system. Preferably these power sources will utilize methane as a fuel or in the alternative will utilize a free energy source such as wind by using a windmill.

Four-Chamber Digester

B.A. McDonald; U.S. Patent 4,100,023; July 11, 1978 describes a digester for controlling the decomposition of organic matter to produce methane gas and a liquid fertilizer. The digester comprises three slurry chambers connected in tandem by pass-through tubes. A fourth chamber used for gas collection is located above the three slurry chambers. An input line is connected to the bottom of the first chamber and an output line is connected to the bottom of an upright tube provided in the third chamber.

A slurry containing raw manure or other organic waste is loaded from a mix tank through the input line to fill the three chambers to a level determined by the top of the upright tube in the third chamber. The process of aerobic digestion and a limited form of anaerobic digestion is carried out in the first chamber and the process of anaerobic digestion which produces the methane gas is carried out in the second chamber which is heated.

As an input load of slurry is fed daily from the mix tank into the first chamber, the digested slurry in the third chamber in addition to being fed out through the upright tube as a liquid fertilizer has a small portion thereof recycled back into the mix tank to condition the input load. The gas collected in the fourth chamber is compressed and stored.

Such gas, in addition to being available for general usage, provides for heating the water used to heat the second chamber and for agitating the slurry in the three slurry chambers.

ZEAXANTHIN

Cultivation of Genus *Flavobacter*

A.J. Schocher and O. Wiss; U.S. Patent 3,891,504; June 24, 1975; assigned to Hoffmann-La Roche Inc. describe a fermentation process for obtaining zeaxanthin (3,3'-dihydroxy-β-carotene). Zeaxanthin is a naturally occurring pigment useful for imparting a yellow color to a wide variety of foodstuffs, such as poultry products, including egg yolks, as well as to cosmetic preparations.

In accordance with the process, zeaxanthin is obtained by cultivating a bacterial species selected from the group consisting of *Flavobacter* ATCC 21,588 and *Flavobacter* ATCC 21,081, in an aqueous nutrient medium containing assimilable sources of carbon and nitrogen. The resulting biomass of this process contains zeaxanthin in high yields.

Among the particularly preferred media for the cultivation of *Flavobacter* ATCC 21,588 and *Flavobacter* ATCC 21,081 to obtain zeaxanthin are nutrient solutions, containing readily assimilable carbon and nitrogen as well as conventional inorganic salts and trace elements, of the following compositions:

Nutrient Solution 1

Yeast extract	5 g
Glucose	5 g
Tris-(hydroxymethyl)-aminomethane	1 g
Sodium chloride	1 g
Magnesium sulfate·$7H_2O$	100 mg
Potassium chloride	100 mg
Sodium nitrate	100 mg
Calcium chloride·$2H_2O$	100 mg
Sodium glycerophosphate	100 mg
Cobalamin in cyanocobalamin	1 mg

Trace Elements

Iron, e.g., iron sulfate·$7H_2O$	0.5 mg
Zinc, e.g., zinc chloride	0.3 mg
Boron, e.g., boric acid	0.1 mg
Cobalt, e.g., cobalt nitrate·$6H_2O$	0.1 mg
Copper, e.g., copper sulfate·$5H_2O$	0.1 mg
Manganese, e.g., manganese sulfate	0.1 mg
Molybdenum, e.g., sodium molybdate	0.1 mg
Distilled water qs	1,000 ml
pH	7.5

Nutrient Solution 2

Yeast extract	1 g
Tryptone	10 g
Glucose	10 g
Sodium chloride	10 g
Tap water qs	1,000 ml
pH	7.5

Nutrient Solution 3

Yeast extract	10 g
Tryptone	10 g
Glucose	1 g
Sodium chloride	26 g
Magnesium chloride	3.1 g
Magnesium sulfate·7H$_2$O	2.2 g
Calcium sulfate	1.2 g
Potassium chloride	0.7 g
Magnesium bromide	0.07 g
Tap water qs	1,000 ml
pH	7.5

Cultivation of the microorganisms to form zeaxanthin according to the process may be carried out in any conventional manner. In accordance with the preferred manner of carrying out this process, cultivation is carried out submergedly in an aqueous media. The fermentation is suitably carried out at a temperature between 10° and 35°C with a temperature of between 20° and 28°C being preferred.

According to a particularly preferred embodiment of the process, the microorganism is submergedly and aerobically cultivated in an agitated nutrient solution initially at a temperature of 28° to 35°C until a sufficient amount of culture is formed, and then cultivation is carried out between 20° and 28°C. Formation of pigment increases in proportion to the growth of the culture, with the maximum pigment formation being reached after about 1 to 4 days.

Natural light and artificial light with the spectral range of daylight are preferably utilized to promote the formation of zeaxanthin. Cultivation is also preferably carried out by employing certain yield-promoting additives in the media. Among the preferred additives are included emulsifiers, defoaming agents, and promoters of pigment formation. The increase in yield achievable in a particular case can be seen from the following table in which some particularly preferred additives are set out.

Increase of Yield of Zeaxanthin*

Additive	Amount (%)	Increase in Formation of Zeaxanthin (%)
Tween 80**	1-5	140-150
Tween 60***	1	157
Tween 80 + soy oil	1 + 2	150
Tween 80 + kerosene	1 + 1	128
Tween 80 + kerosene + soy oil	1 + 1 + 2	138
Isonicotinic acid hydrazide	0.001	120
Sorbitol	0.5	133
Palmitic acid methyl ester	0.5	286
Palmitic acid methyl ester + sorbitol	0.5 + 0.5	270
Lactic acid	0.5	140

*When using nutrient solution 1 or 3 in the presence of the specified additives.
**Polyoxyethylenesorbitan monooleate.
***Polyoxyethylenesorbitan monostearate.

In carrying out the cultivation, it is further preferred to provide a high carbo-hydrate content of from 1 to 10% by weight of the nutrient media and a high phosphate content of from 0.1% up to a maximum of 1% by weight of the nutrient media. In this cultivation, it is especially preferred that the high con-tent of carbohydrate comprise glucose and/or that the high content of phosphate comprise potassium orthophosphate. It is particularly preferred to carry out the cultivation under partial pressures of oxygen greater than ambient at a temper-ature of 24°C.

After completion of the cultivation of the microorganism, the cell-mass is iso-lated by centrifugation or filtration. The pigment-containing cell-mass can be advantageously utilized to color foodstuffs without the necessity for isolating pure zeaxanthin pigment. On the other hand, the intracellular zeaxanthin can be separated from the cells in a conventional manner.

Example: A *Flavobacter* ATCC 21,081 agar-slant culture is suspended in a 500 ml Erlenmeyer flask which contains 100 ml of Nutrient Solution 1 and incubated with shaking for 24 hours at 22°C. The preculture is used as the inoculum for 8 liters of a nutrient solution of the same composition which, after inoculation, is aerated with shaking at 22°C for 36 hours. The fermenta-tion solution is subsequently centrifuged. In dried state, the cell-mass can be directly used for coloring purposes.

For the analytical determination of content, a portion of the cell-mass is digested in acetone. The acetone solution is dried over sodium sulfate and evaporated. The residual zeaxanthin is isolated in pure form by preparative thin-layer chro-matography.

For the preparative production of pure zeaxanthin, the moist cell-material is dried under careful conditions and exhaustively extracted with ethyl acetate. The extract is then concentrated and adsorbed on silica gel. The pure zeaxan-thin is obtained by elution with diethyl ether.

Addition of Sulfur-Containing Amino Acids and Bivalent Metal Ions

D. Shepherd and J. Dasek; U.S. Patent 3,951,742; April 20, 1976; assigned to Societe d'Assistance Technique pour Produits Nestle SA, Switzerland describe a process for the preparation of zeaxanthin which comprises culturing a micro-organism of the genus *Flavobacter* producing this pigment, in a nutrient medium containing at least one carbohydrate as assimilable carbon source, at least one source of assimilable amino nitrogen containing free amino acids, mineral salts, oligoelements and vitamins. The spectrum of amino acids of the medium is mod-ified by increasing the content in the medium of at least one of the sulfur-con-taining amino acids, methionine, cystine and cysteine, in relation to the content in the medium of other amino acids, while limiting the quantity of methionine and/or cystine and/or cysteine added to the medium to a value not exceeding 1 mg/ml, and continuing culturing until a substantial quantity of intracellular zeaxanthin has been obtained.

In the practical application of the process, the source of assimilable carbon can be glucose or saccharose used in a proportion of from 0.1 to 15% by weight of the culture medium, while the source of assimilable amino nitrogen can be a yeast extract and/or a corn steep liquor and/or a protein hydrolysate used in a proportion of from 0.1 to 8% by weight of the medium. Magnesium sulfate can also be included in this medium in a proportion of from 0.1 to 2% by weight. All of the aforementioned substances are diluted in tap water, for example, which makes up the balance to 100% by weight.

The spectrum of amino acids of the medium can be modified by adding methionine and/or cystine and/or cysteine in a given quantity in L- and/or D- and/or DL form. The methionine and/or cystine and/or cysteine content of the medium is preferably adjusted to from 3 to 10% of the content of other amino acids in the medium.

An inoculum of a microorganism of the genus *Flavobacter* which produces zeaxanthin can then be added to the nutrient solution thus prepared. The fermentation process can take place with agitation and aeration, at suitable pH values and temperatures and over the period of time required to produce an appreciable quantity of intracellular zeaxanthin. Thereafter, the culture may be dried optionally after concentration. Zeaxanthin may be extracted from the cells with a polar organic solvent such as acetone, ethyl alcohol or a chlorine-containing solvent such as chloroform. Alternatively, the biomass after separation from the culture medium, by centrifugation, decantation or filtration, may be used as such, for example, as an additive in poultry feeds.

It has been found that the concentration of zeaxanthin in the biomass thus obtained is distinctly higher (an increase of approximately 50%) than the concentration obtained by growing the same microorganism in a conventional nutrient medium.

It is worth noting here that a phenomenon similar to the effect mentioned above has been observed in the case of an addition to the starting nutrient solution of at least one of the bivalent metal ions Fe^{++}, Co^{++}, Mn^{++} and Mo^{++} without the addition of at least one of the aforementioned sulfur-containing amino acids, this phenomenon being reflected by an increase in the concentration of zeaxanthin in the biomass of as much as 50%, and by an increase in the biomass of again as much as 50%, i.e., by double the output of zeaxanthin in relation to that obtained by growing the same microorganism in a conventional nutrient medium.

Addition of Pyridoxine

In a similar process *D. Shepherd, J. Dasek and M.S.C. Carels; U.S. Patent 3,951,743; April 20, 1976; assigned to Societe d'Assistance Technique pour Produits Nestle SA, Switzerland* describe the preparation of zeaxanthin which comprises culturing a microorganism of the genus *Flavobacter* producing this pigment in a nutrient medium containing at least one carbohydrate as assimilable carbon source, at least one source of assimilable amino nitrogen containing free amino acids, mineral salts, oligoelements and vitamins, the composition of the medium being modified by adding pyridoxine to it, and continuing culturing until a substantial quantity of intracellular zeaxanthin has been obtained.

The composition of the medium can be modified by adding pyridoxine to it in a quantity of 0.1 to 1 μg/ml. The pyridoxine is preferably added in a quantity of 0.1 to 0.3 μg/ml.

MISCELLANEOUS ORGANIC COMPOUNDS

Optically Active Cyclohexane Derivatives

W. Boguth, H.G.W. Leuenberger, H.J. Mayer, E. Widmer and R. Zell; U.S. Patent 3,988,205; October 26, 1976; assigned to Hoffmann-La Roche Inc. describe a process for producing optically active cyclohexane derivatives by fermentatively hydrogenating ketoisophorone of the formula:

(1)

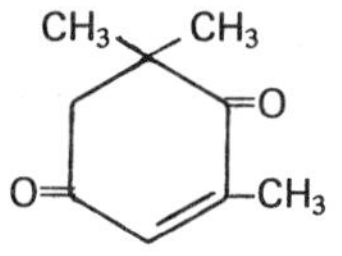

in an aqueous medium to produce [6R] -2,2,6-trimethyl-1,4-cyclohexanedione of the formula:

(2)

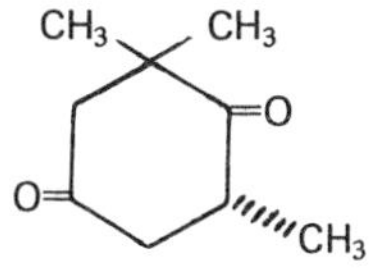

which is reduced to produce [4R,6R] -4-hydroxy-2,2,6-trimethylcyclohexanone of the formula:

(3)

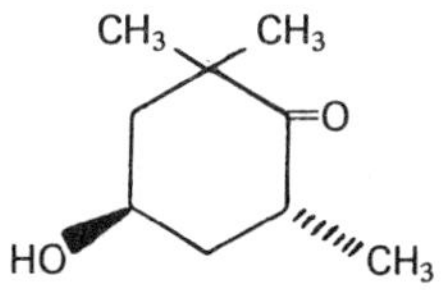

The compound of formula (3) can, if desired, be converted to known optically active carotenoids useful as food coloring agents.

The substituents in the structural formulae given in this specification are characterized by the notation ➤ insofar as they lie in front of the plane of the molecule and by the notation ⦀ insofar as they lie behind the plane of the molecule. The fermentation of the compound of formula (1) to a compound of formula (2) can be carried out utilizing any conventional means.

The pH at which the fermentation is carried out is preferably within the range of from 2 to 10, preferably from 3 to 8, and this is generally achieved without special additives. If desired, the pH can be regulated by using buffers; for example, phosphate, phthalate or tris-buffer [tris-(hydroxymethyl)-aminomethane]. The temperature at which the fermentation is carried out can vary within wide limits (e.g., between 4° and 50°C). A temperature of 15° to 35°C, especially 25° to 35°C, is preferred.

In order to obtain optimal yields it is preferred that the ketoisophorone be present in the fermentation medium in a concentration of from 0.1 to 2.0% by weight, preferably 0.5 to 1.2% by weight. After completion of the fermentative hydrogenation, fresh ketoisophorone can be added in a preferred concentration of from 0.5 to 1% by weight. This procedure can be repeated several times until the microorganism becomes inactivated. In a preferred fermentation procedure using press-yeast as the microorganism, there can be fermented, with periodic educt addition, up to 10%, preferably 6 to 8% by weight of ketoisophorone in the same batch. The fermentation temperature in the case of this periodic educt addition is advantageously 15° to 25°C. The effective fermentation time depends on the microorganism used, but normally varies between 10 and 200 hours.

The nonspecificity of the required microorganism is exemplified in that any microbially infected soil and water samples from nature are capable of being used successfully as microorganism-providers in the fermentative hydrogenation process. The fermentation is carried out aerobically, preferably with stirring, shaking or by means of any aeration process. In order to control foam, the usual antifoaming agents such as silicon oils, polyalkylene glycol derivatives, soy bean oil and the like can be added. Having regard to the nonspecificity of the required microorganism, the fermentation has the advantage that it need not be carried out under sterile conditions.

After termination of the fermentation, the [6R]-2,2,6-trimethyl-1,4-cyclohexanedione is isolated from the fermentation broth in the usual manner. Extraction with a water-insoluble organic solvent is preferably used.

The reduction of the oxo group in the 4-position of the resulting [6R]-2,2,6-trimethyl-1,4-cyclohexanedione of formula (2) to the hydroxy group, i.e., the compound of formula (3), proceeds in good yields with stereospecific selectivity, i.e., not only with retention of the oxo function in the 1-position but also with formation of the R,R-trans configuration for the two substituents in the 4- and 6-position (hydroxy or methyl). In carrying out this procedure, any conventional method of reducing an oxo group to a hydroxy group can be utilized.

In accordance with one embodiment, the reduction can be carried out advantageously using an organoaluminum compound, especially a β-branched aluminum tri(lower alkyl) (e.g., triisobutylaluminum) or a corresponding halo-substituted derivative thereof (e.g., isobutylaluminum dichloride). In order to obtain optimal yields of the desired [4R,6R]-4-hydroxy-2,2,6-trimethylcyclohexanone of formula (3), the aluminum compound and the [6R]-2,2,6-trimethyl-1,4-cyclohexanedione of formula (2) are to be used in approximately equimolar amounts. Other reducing agents which may be used are organic alkali metal aluminum hydrides such as sodium dihydrobis(2-methoxyethoxy)-aluminate and alkali metal borohydrides such as sodium borohydride.

The reduction is preferably carried out in an inert organic solvent; for example, n-hexane, n-heptane, benzene, toluene, diethyl ether, tetrahydrofuran, a chlorinated hydrocarbon such as methylene chloride, or chlorobenzene or mixtures of these solvents. A preferred solvent is methylene chloride and a preferred mixture consists of principally n-hexane in admixture with benzene. The reduction is preferably carried out at a temperature between -70°C and room temperature (30°C). The reduction has the advantage that it is completed, especially when an aluminum alkyl or a halo-substituted derivative thereof is used, in a short time (generally in a few minutes at a temperature of 0°C or above). Then, after neutralization of the reduction mixture with acid, the desired [4R,6R]-4-hydroxy-2,2,6-trimethylcyclohexanone can be obtained by purification in the usual manner; for example, by chromatography on silica gel, aluminum oxide, dextran or the like or by extraction using a countercurrent procedure.

The resulting [4R,6R]-4-hydroxy-2,2,6-trimethylcyclohexanone of formula (3) is a key intermediate in the manufacture of optically active carotenoids; for example, for the manufacture of: [3R]-β-cryptoxanthin; [3R,3'R]-zeaxanthin; [3R]-rubixanthin; [3R]-β-citraurin; and [3R]-reticulataxanthin.

Sapogenins from the *Helleborus* Species

The roots and rhizomes of *Helleborus* species contain a saponin mixture whose main sapogenin has the structure of a spiro-5,25(27)-diene-1β,3β,11α-triol with the formula:

O. Isaac; U.S. Patent 4,004,976; January 25, 1977; assigned to Deutsche Gold- und Silber-Scheideanstalt vormals Roessler, Germany has found that this main genin of the sapogenin mixture which is contained in the roots and rhizomes of *Helleborus* species can be recovered in a simpler and quicker manner if the roots and rhizomes of *Helleborus* species or drugs or extracts obtained therefrom are treated with an enzyme mixture whose essential active components are cellulase, hemicellulase or β-glucosidase.

According to the process there can be employed directly the pulverized roots and rhizomes of *Helleborus* species or there can be used an extract recovered in customary manner or a drug from *Helleborus* species which has been worked up and pulverized in customary manner. The treatment can be carried out in solution, in suspension or in form of a mash. Thus, for example, the enzyme can be added to the autolysate of roots and rhizomes of *Helleborus* species. The autolysis or autofermentation can be carried out by pulverizing fresh plant parts and allowing the thus obtained material to stand for a period of time (for example, 2 to 25 days) at temperatures between 18°C to a maximum of 60°C in a given case after adding water and with occasional stirring.

Also, dried plant parts can be stirred with water so long as their fermentative activity is maintained, and be autofermentated. The enzyme can be added either at the beginning of the autofermentation process or also during this process with equally satisfactory results.

However, it is also possible to first produce an extract in known manner from the roots and rhizomes or a drug produced therefrom and treat this, in a given case af-ter a conventional preliminary purification, with the enzyme. For this purpose suitably the roots or rhizomes or drug first is extracted with alcohol (ethyl alcohol) or an alcohol-water mixture containing a maximum of 50% water. The thus ob-tained extract is subsequently extracted by shaking with organic solvents, e.g., water-immiscible organic solvents, for example, aromatic hydrocarbons, e.g., benzene, toluene, xylene, halohydrocarbons, e.g., chloroform, carbon tetrachloride, 1,2-dichloroethane or halohydrocarbon-alcohol mixtures as, for example, chloroform-ethanol. If halohydrocarbon-alcohol mixtures are used the mixing ratio is preferably 2:1. However, it is favorable if the organic extract obtained is prepurified.

Such a preliminary purification can take place in a customary manner. However, it can also take place, for example, by chromatographization of a silica gel whereby there is used a synthetically produced, highly porous, amorphous silica in the form of hard particles with a particle size of 0.15 to 10 mm, preferably 0.15 to 0.30 mm. The water content of this silica can amount to 10%. The specific surface area can be from 300 to 650 m^2/g. Generally the surface area is about 400 m^2/g. The apparent density can be from 400 to 750 g/l. An apparent density (bulk density) of 450 to 500 g/l is propitious.

The prepurified and completely solvent-free saponin fraction is then incubated in water with the enzyme. The incubation can be carried out at a temperature of 20° to 50°C. For protection against bacterial decomposition toluene can be added (mostly in small amounts) to the incubate.

There can be used in the process commercial cellulase, hemicellulase and β-glucosidase preparatives. Likewise, there can be used preparatives of the just named enzymes which have been produced according to known processes for such purpose from fungi, microorganisms (*Trichoderma viride*, for example), protozoa, bacteria, insects, plants or invertebrate animals such as snails (edible snails) and worms [Sumner, et al., *The Enzymes*, Vol. 1, Part 2, pages 729-731 (1951)].

The enzyme, as for example, cellulase, is obtained if the culture medium or the aqueous mycelial extract of fungi is precipitated with alcohol (ethanol), acetone or salt. The crude enzyme can then be further purified, for example, on aluminum oxide. There should be used the freshest possible preparative which has not been stored for long. Stored preparatives should be kept cool and dry. Especially suitable are enzyme preparatives or enzyme concentrates of fungi or *Aspergillus* species such as *Aspergillus niger, Aspergillus oryzae, Aspergillus flavus, Aspergillus fumigatus,* and *Aspergillus nidulans.* Especially favorable are preparatives or enzyme concentrates from *Aspergillus niger.*

Generally, the enzymatic reaction is ended after two days. With enzymes having higher activity, the reaction can be ended even earlier; with enzymes having lower activity, however, the treatment can be carried out a considerably longer time. The enzyme can be mixed either as such or in aqueous solution with the substrate. The optimum pH value for the enzyme treatment is from 4.5 to 4.7. It is advantageous to stir the incubate or to keep it in motion in other ways.

The amount in which the enzyme is added depends on the activity of the enzyme and also on the substrate used. The enzyme, for example, can be added in very large excess. If, for example, a crude saponin fraction is added which is preliminarily purified in the manner given above, there can be added 5 to 100% of the enzyme preparative necessary according to the degree of purity determined by thin layer chromatography. If, for example, the drug is directly reacted or roots and rhizomes of *Helleborus* species are used, then an amount between 1 and 10% based on the drug or the roots and rhizomes is sufficient. The crude sapogenin resulting from the enzyme treatment can be further worked up according to the customary processes described in the literature.

However, especially desirable is a purification (for example, by chromatography) of the crude sapogenin obtained with the help of silica gel with a particle size of 0.2 to 0.5 mm. As elution agents there are preferably suitable lower halogenated hydrocarbons such as methylene chloride and chloroform to which there can be added 1% and up of a lower aliphatic alcohol such as methanol or ethanol. The thus obtained sapogenin can be recrystallized once from propanol-water.

Little was previously known of the physiological activity of the main sapogenin from the roots and rhizomes of *Helleborus* species. It has now been found that the main genin recovered from *Helleborus* saponin has an ulcer healing activity. Furthermore, a muscle relaxing and central nervous system influencing activity has been established.

Example: 100 kg of roots and rhizomes of *Helleborus viridis* were comminuted, defatted with petroleum ether and subsequently exhaustively extracted with 80% aqueous ethanol. The residue of the ethanol extract was taken up in water and the solution extracted by shaking with chloroform/ethanol (2:1 by volume). The chloroform solution portion was chromatographed on a silica gel column with chloroform/methanol (9:1 by volume) and the fractions examined by thin layer chromatography [absorbent: silica gel for thin layer chromatography; flowing agent: chloroform/methanol/water (35:25:10 by volume); and detection: anis-aldehyde/sulfuric acid/acetic acid (1:1:100 by volume)].

The fractions were collected which essentially contained a brown colored material of average Rf-value (about 0.50) which is localized in the chromatogram between desglucohelleborin and helleborin. The residue of the combined fractions (2,621 g) was dissolved under reflux in 2.4 liters of methanol and 2.4 liters of water. After addition of 24 liters of tap water there was distilled off from the solution 4.8 liters of solvent.

After cooling, the solution was treated with 240 g of a commercial cellulase (Rohm and Haas) as well as 100 ml of toluene and aged at 40°C with occasional shaking. After about 48 hours the reaction was complete. The precipitate was filtered off with suction, washed with hot water and dried. Yield, 1,100 g.

The dried precipitate was dissolved in methanol and dichloromethane (50:50 by volume) and the solution chromatographed with dichloromethane/methanol (increasing methanol concentration) on silica gel for column chromatography with a particle size of 0.2 to 0.5 mm.

The sapogenin fraction (736 g) was dissolved in 5 liters of n-propanol and the solution treated with 30 liters of water. After 24 hours the crystals which separated were washed with propanol/water and dried. Formula: $C_{27}H_{40}O_5$; M.P. 238°C; $[\alpha]_D^{20} = 86.93°$ (c = 1.1; pyridine).

3(S or R)-Hydroxy-1-Iodo-1-trans-Octene

C.J. Sih; U.S. Patent 3,868,306; February 25, 1975; assigned to Wisconsin Alumni Research Foundation describes a process for preparing 3(S or R)-hydroxy-1-iodo-1-trans-octene by subjecting 3-oxo-1-iodo-1-trans-octene to the fermentative enzymatic action of microorganisms of the classes *Ascomycetes, Phycomycetes* and *Fungi Imperfecti*. This compound is a key intermediate in the preparation of prostaglandins of the E_2, F_2, A_2 and B_2 series.

General Screening Procedure to Determine Efficiency of Any Specific Organism:
Inoculate the microorganism onto Sabouraud's agar slants or other agar base
media suitable for growth. Place the inoculated slants in an incubator maintained at 25°C and allow to grow for 1 week. Remove the slant and add 15 ml
of sterile distilled water to it. Loosen the spores and vegetative growth from
the agar with a sterile needle. Transfer the suspension to a flask containing
50 ml of the soy-dextrose medium described below and place the flask in a
rotary shaker in an incubator maintained at 25°C at 210 rpm for 24 hours.

After this initial time period (first stage seed), add 5 ml of the submerged growth
to each of duplicate flasks of three types of media, namely, soy-dextrose, Cere-
lose-Edamine and dextrin-cornsteep, the compositions of which are given below.
Place the flasks in the shaker and allow to grow from 24 to 48 hours at 25°C.
The cells were then harvested by centrifugation.

Incubation – The harvested cells were suspended in 50 ml of 0.033 M borate
buffer, pH 8.5 in a 250 ml Erlenmeyer flask. To this was added 25 ml of 3-
oxo-1-iodo-1-trans-octene in 1 ml of acetone. Incubation was carried out at
25°C on a rotary shaker at 290 rpm (1" stroke) for 19 hours.

Extraction – After 19 hours, the cells were removed by centrifugation and the
supernatant solution was extracted with a 25 ml portion of chloroform three
times. The chloroform layer was dried over sodium sulfate and evaporated to
dryness. The residue was dissolved in a few drops of acetone and analyzed via
thin layer chromatography using silica gel G plates. A suitable system was
chloroform. After development, the plates are sprayed with 3% ceric sulfate
in 3 N H_2SO_4 to reveal the compounds as brown spots. The following Rf values
were observed for the compounds of interest.

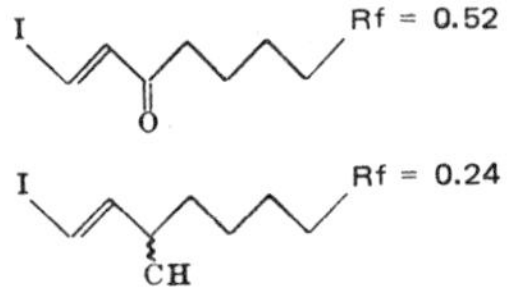

Constitution of illustrative nutrient media suitable for the above screening procedure and for the batch fermentation given below in the examples are as follows:

Soy-Dextrose

Soybean meat	5 g
Dextrose	20 g
NaCl	5 g
K_2HPO_4	5 g
Yeast	5 g
Water	1 liter

pH adjusted to 7.0
    Autoclave at 15 psi for 15 minutes

Cerelose-Edamine*

Cerelose (crude dextrose)	50 g
Edamine*	20 g
Cornsteep liquor	5 ml
Water	1 liter

pH adjusted to 7.0
*Enzymatic hydrolysate of milk protein (Sheffield Farms Co.)

Dextrin-Cornsteep

Dextrin	10 g
Cornsteep liquor	80 g
KH_2PO_4	1 g
NaCl	5 g
Water	1 liter

pH adjusted to 7.0
Autoclave at 15 psi for 30 minutes

Example 1: *Reduction by Penicillium decumbens* – Cells were harvested from 2 liters of culture broth (5 x 400 ml) in 2 liter Erlenmeyer flasks and were suspended in 2 liters of .025 M boric acid-borax buffer solution. The cell suspension was divided into four equal portions, each of which was placed in a 2 liter Erlenmeyer flask. 1 g of 3-oxo-1-iodo-1-trans-octene was dissolved in 40 ml of acetone and the resulting solution was divided equally into the four flasks. The cell suspension was shaken at 290 rpm for 19 hours. The cells were removed by centrifugation (16,000 g, 15 minutes), the supernatant was extracted three times with 2 liters of chloroform.

The removed cells were resuspended in four flasks (those which were used for reduction) by adding 250 ml of distilled water to each flask, after which the flasks were shaken vigorously on a rotary shaker and the emulsion formed by shaking was separated by centrifugation. After removing the chloroform and any ethyl acetate present, both residues were combined and applied to a 2 x 21 cm alumina chromatographic column. The column was eluted with a gradient system consisting of 300 ml of 10% ethyl acetate in benzene followed by 300 ml of 30% ethyl acetate in benzene and 4.0 ml fractions were collected. Fractions 42 through 102 were collected as the reduction product and yielded 104 mg of the desired 3(S) enantiomer; $[\alpha]_D^{27°}$ +7.5 (c, 3.69 in MeOH).

Example 2: *Reduction by Penicillium vinaceum* – The reduction procedure was the same as in Example 1 except that:

(a) the cells were harvested from 2 liters of culture broth (4 x 500 ml in 2 liter flasks) and were suspended in 2 liters of 0.033 M boric acid-borax buffer;

(b) the emulsion formed by shaking was broken by passing through a Celite pad on a Buchner funnel;

(c) the column size was 2 x 20 cm and 5.2 ml fractions were collected. Fractions 27 through 70 were collected as the reduction product and yielded 100 mg of the desired 3(S) enantiomer; $[\alpha]_D^{27°}$ +7.56 (c, 5.86 in MeOH).

Example 3: *Reduction by Aspergillus ustus* — The reduction procedure was the same as in Example 2 except that 6.5 ml fractions were collected. Fractions 16 through 36 were collected as the reduction product and yielded 120 mg of the desired 3(R) enantiomer; $[\alpha]_D^{27°}$ −8.04 (c, 5.97 MeOH).

Sodium Citrate Dihydrate

Having been used in combination with citric acid in food industries, sodium citrate dihydrate has been used as a detergent builder and a desulfurizing agent as well, with its use being constantly enlarged in various applications.

M. Tsuda, Y. Fujiwara and T. Shiraishi; U.S. Patent 3,904,684; September 9, 1975; assigned to Takeda Chemical Industries, Ltd., Japan have found that when the broth resulting from citric acid fermentation, obtained by growing a micro-organism in a medium containing hydrocarbons as principal carbon sources, is subjected to a series of purification procedures at a strongly alkaline environment adjusted to pH 9 to 13 inclusive, the insolubles are discarded, and when this fluid is concentrated at pH 10 to 13 inclusive, the removal of the insolubles is very effective and thorough, and the resultant fluid is completely transparent and the sodium citrate dihydrate can be obtained as colorless crystals in a short time.

The advantages of this method include (1) sodium citrate dihydrate directly crystallizes from a fermentation broth and, therefore, the process is considerably simplified; (2) since the process of purification is conducted on the alkaline side (pH 9 to 13), sodium citrate dihydrate of improved quality is obtained; (3) the mother liquor from which the sodium citrate dihydrate has been recovered contains impurities in a high concentration, which makes it easy to dispose of them at low cost, for example, by removing the aqueous portion by evaporation followed by combustion; and (4) the cells, after drying, can be utilized as a fertilizer.

In this process, use is made of a citric acid fermentation broth which is obtainable by growing a microorganism in a medium containing hydrocarbons as principal carbon source (cf French Patents 1,571,551 and 7,003,025). As the microorganism there is exemplified a yeast of the genera *Candida, Brettanomyces, Debaryomyces, Hansenula, Kloeckera, Torulopsis, Pichia, Trichosporon, Saccharomyces,* etc., and a bacterium of the genera *Corynebacterium, Arthrobacter,* etc.

In the medium there may also be incorporated nitrogen sources which will be digested by the microorganism employed, inorganic salts, amino acids, vitamins, and other nutrients or growth-promoting factors. In conducting the cultivation, the pH of the medium is maintained at from 2 to 10 and preferably from 3 to 8. As the pH regulator, sodium hydroxide, sodium carbonate or sodium hydrogen carbonate is most desirable for the preparation of citric acid fermentation broths to which this purifying process is applied. Addition of calcium to the medium must be by all means avoided. At the end of fermentation, pH of the broth is generally below 9 and, in the broth citric acid dissolves in a free form or as a salt with sodium or ammonium.

When the fermentation broth is adjusted to pH 9 to 13 with a water-soluble inorganic sodium salt and then subjected to a separation process, e.g., centrifugation, there is obtained a clear supernatant from which sodium citrate dihydrate can be caused to crystallize by concentration.

Ergot Alkaloids

G. Wack, L. Nagy, D. Székély, J. Szolnoky, E. Udvardy-Nagy and E. Zsóka;
U.S. Patent 3,884,762; May 20, 1975; assigned to Richger Gedeon Vegyeszeti Gyar RT, Hungary describe a fermentation process for the preparation of ergot alkaloids, mainly ergocryptine and ergocornine, by cultivating a *Claviceps purpurea* strain on a liquid, aerated culture broth containing saccharose, an inorganic nitrogen source and other known additives, in which a *Claviceps purpurea* variant strain deposited with the National Institute of Public Health (Orszagos Kozegeszsegugyi Intezet), Budapest, Hungary, No. 88/1972 is used as microorganism.

This strain, due to its favorable properties, can be used with great advantages for the preparation of ergocryptine and ergocornine. The fermentation process can be carried out on known culture media by methods known in the art, and has several advantages over the hitherto known fermentation process for the preparation of ergotoxine alkaloids. The main advantages are summarized below.

1. The stability of the strain obtained without mutagenic treatment is higher than that of the mutant strains, accordingly the process utilizing this strain is well-reproducible.

2. By excluding the organic nitrogen sources from the culture medium and substituting them by ammonium nitrate, the reproducibility can further be increased and the economy of the process becomes more favorable.

3. The fermentation process requires 5 to 7 days; while using the known processes 8 to 12 days were required for the production of peptide-type alkaloids in appreciable amounts.

4. Due to the decreased pigment production and the low amount of mycelia, the alkaloids can be removed more easily from the culture broth.

5. The ratios of the alkaloids produced in the culture broth are very favorable. The alkaloids belonging to the ergotoxine group are formed in equal amounts and when separating these two alkaloids and supplementing the obtained mixture with ergocristine, a complete ergotoxine composition of the usual component ratios can be obtained.

6. The most important accompanying alkaloid, the water-soluble ergometrine, can easily be removed during the working up procedure, and can be isolated separately.

Example: A typical 30 day old colony of the *Claviceps* strain OKI No. 88/1972 is removed from the surface of SC 101 solid culture medium, and homogenized in 10 ml of sterile water. 100 ml of an S2C culture medium, filled into a 500 ml Erlenmeyer flask, is inoculated with this suspension. The S2C culture medium has the following composition:

Saccharose	200.0 g
Citric acid	15.0 g
Potassium dihydrophosphate	0.5 g
Magnesium sulfate	0.3 g

(continued)

Ammonium hydroxide qs pH   5.2
Water qs   1,000.0 ml

The culture is shaken at 24°C for 6 days, thereafter a 10 ml fraction is removed, and this fraction is used for the inoculation of 100 ml of an SB 101 culture medium, filled into a 500 ml Erlenmeyer flask. The SB 101 culture medium has the following composition:

Saccharose	100.0 g
Succinic acid	10.0 g
Calcium nitrate	1.0 g
Ammonium nitrate	1.0 g
Potassium dihydrophosphate	0.5 g
Magnesium sulfate	0.3 g
Sodium chloride	10.0 g
Ammonium hydroxide qs pH	5.2
Water qs	1,000.0 ml

The culture is shaken at 24°C for 7 days. The dry material content of the culture is 3.42%.

100 ml of the culture obtained as described above is extracted with 50 ml of a 4:1 mixture of chloroform and isopropanol. 10 ml of the extract is evaporated, the residue is taken up in 1 ml of a 1:1 mixture of chloroform and methanol, and the solution is subjected to chromatography on a dry alumina layer. The alkaloid spots are traced in UV light, eluted with 50% aqueous methanol containing 1% of tartaric acid, and the amounts of the respective alkaloids are determined on the basis of UV absorption.

The total alkaloid content of the culture is 1,687 γ/ml, while the ergocornine-ergocryptine content amounts to 1,187 γ/ml. Ergocryptine and ergocornine can be isolated in crystalline form from any of the cultures by known techniques.

2,6-Diaminopyrimidines

G. Greenspan, R.W. Rees and P.B. Russell; U.S. Patent 3,940,393; February 24, 1976; assigned to American Home Products Corporation describe compositions of matter classified in the art of organic chemistry as 2,6-diaminopyrimidines, and processes for their preparation. The 4,5-disubstituted 2,6-diaminopyrimidines are chemically related to the pharmaceutical, pyrimethamine, whose utility as an antimalarial medicament is well-known. The compounds of the process are also useful as antimalarial agents, which activity is evidenced from the results of standard pharmacologic testing. The product of the process is a compound of the formula:

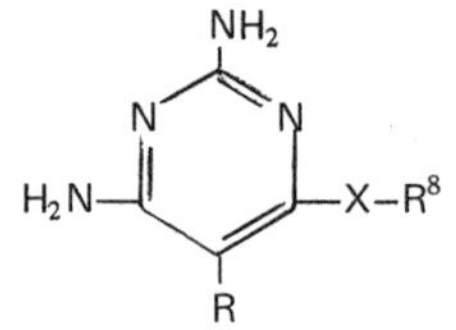

where R is phenyl, 3-chlorophenyl, 4-chlorophenyl, or 3,4-dichlorophenyl, X is $C(OCH_3)_2$, $C(OC_2H_5)_2$, $C=O$, $C=N-OH$, $C=N-NH_2$, $C=N-N(CH_3)_2$, CHOH, or

$C=N^+(CH_3)-O^-$, R^8 is alkyl of from 1 to 6 carbon atoms; and their pharmacologically acceptable acid addition salts.

The principal process aspect comprises the microbiological oxidation of the known antimalarial medicament, pyrimethamine, [2,6-diamino-5-(p-chlorophenyl)-4-ethylpyrimidine] by the fungus *Pellicularia filamentosa* f. sp. *sasakii* IFO 6675. Other fungi of equivalent function are *Pellicularia filamentosa* f. sp. *sasakii* IFO 6258, 6297; *Pellicularia filamentosa* f. sp. *microsclerotia* CBS; *Pellicularia filamentosa* f. sp. *solani* CBS 280.36; *Pellicularia filamentosa* IFO 6476; and the like. Thus, when pyrimethamine is incubated with *Pellicularia filamentosa* f. sp. *sasakii* IFO 6675, the compounds (+)-2,6-diamino-5-(p-chlorophenyl)-α-methyl-4-pyrimidine methanol [(+)I], and [2,6-diamino-5-(p-chlorophenyl)-4-pyrimidinyl] methyl ketone (II) are produced.

The preferred method for carrying out this microbiological oxygenation is to contact pyrimethamine with a buffered suspension of fungal mycelial cells, for example, mycelial cells of *Pellicularia filamentosa* f. sp. *sasakii* IFO 6675, for from 1 to 5 days, preferably from 1.5 to 2.5 days at 28°C. The temperature is not critical, but ideally is maintained between 25° and 30°C. The fermentation is carried out aerobically and with agitation.

It will be obvious to one skilled in the art of organic chemistry that the oxygenated products (II) and [(+)I], may be isolated by standard procedures as, for example, by extraction of the buffered solution with a water-immiscible organic solvent such as methylene chloride or ethyl acetate. Drying and evaporation of the solvent followed by separation and purification of the products, for example, by chromatographic means, yields materials whose elemental analyses, infrared, proton magnetic and mass spectra are in full agreement with the compounds (II) and [(+)I].

In the second process aspect, compound (II) is converted to compound [(+)I] by the action of the fungus *Corynespora cassiicola* IMI 56007. This microbiological process is best carried out by contacting compound (II) with a buffered suspension of mycelial cells of the fungus *Corynespora cassiicola* IMI 56007 for from 1 to 5 days, preferably from 2.5 to 4 days at 28°C. The temperature is not critical, but ideally is maintained between 25° and 30°C. The fermentation is carried out aerobically and with agitation.

The optically active product [(+)I] may be isolated by standard procedures as above, yielding a material whose elemental analysis, infrared, proton magnetic and mass spectra are in full agreement with the structure [(+)I].

7-Amino-Cephem Compounds

T. Matsuda, T. Yamaguchi, T. Fujii, K. Matsumoto, M. Morishita, M. Fukushima and Y. Shibuya; U.S. Patent 3,960,662; June 1, 1976; assigned to Toyo Jozo Kabushiki Kaisha, Japan describe a microbiological process for the production of 7-amino-cephem compounds of the formula

(1)

where X represents hydrogen, hydroxy, acetoxy, or a nucleophilic residual group
from a compound of the formula

(2)

$$R-(CH_2)_3-CONH$$

where R is $-COOH$ or $-COCOOH$ and X has the same meanings as listed above,
by splitting an amide linkage thereof.

It was found that a microorganism belonging to genus *Comamonas* strain SY-77-1
separated from soil samples has the ability to produce 7-amino-cephem compounds. It was also found that *Pseudomonas ovalis* ATCC 950 has the same
ability.

The cultivation of the microorganisms can be advantageously carried out under
aerobic conditions, more preferably by submerged aeration culture. Nutrient
media may include an assimilable source of carbon, an assimilable source of nitrogen and salts. The culturing temperature is preferably 25° to 37°C, and the
culturing period is generally 2 to 10 days and at the time when the N-deacylating
activity reaches a maximum, the cultivation should naturally be terminated.

The thus-obtained microbial culture or preparation thereof can be used for the
N-deacylating reaction on compound (2). The N-deacylating reaction of compound (2) is generally conducted in an aqueous medium, preferably at pH 6 to
8. Alternatively water-insoluble microbial cells or a preparation thereof is used
in the form of a suspension, or in the form of a column wherein compound (2)
is N-deacylated by passing the aqueous solution of compound (2) through the
column.

The reaction period is generally 3 to 30 hours and it should be terminated when
the production of amino compound (1) reaches a maximum. The reaction temperature may be 20° to 45°C, preferably 30° to 37°C. Substrate concentration
is dependent on the N-acylating activity and it may be generally 0.1 to 5%. The
thus-formed amino compound (1) can be isolated and purified by the known
methods such as column chromatography, ion-exchange, precipitation, etc.

Macrocyclic Ketones

*H.W. Bost; U.S. Patent 3,963,571; June 15, 1976; assigned to Phillips Petroleum
Company* describes the oxidation of hydroxy acids contained in fermentation
liquors also containing dicarboxylic acids having from 14 to 22 carbon atoms
per molecule, to produce macrocyclic ketones which are usable as odorants,
particularly in the perfume industry.

The production of dicarboxylic acids having from 6 to 22 carbon atoms produced from C_6 to C_{22} n-paraffins by microbial oxidation in cultures containing
the yeast *Torulopsis bombicola*, n. sp. PRL319-67, obtainable from Prairie Regional Laboratory, National Research Council of Canada, is well-known.

Similarly it is well-known to produce dicarboxylic acids from n-paraffins by use
of yeast mutants of *Torulopsis bombicola*. Such mutants are available from the

U.S. Department of Agriculture, Agricultural Research Center, North Central Region, Northern Regional Research Laboratory, as strain NRRL-Y-7569 and NRRL-Y-7570.

The fermentation of n-paraffins by microbial oxidation employing such yeasts has been found to produce fermentation liquors containing dicarboxylic acids, and, in addition, hydroxy acids, e.g., omega-hydroxycarboxylic acids. It is towards the latter, which are generally considered to be impurities in the fermentation liquor, that this process is directed.

According to this process, hydroxy acids produced by the bacterial oxidation of paraffins are oxidized to the diacid, the diacid is esterified, the diester is subjected to cyclization to produce a cyclic hydroxy ketone and the hydroxy ketone is reduced to produce a cyclic saturated ketone. Alternately, the cyclic hydroxy ketone can be converted directly to a cyclic unsaturated ketone which is then selectively hydrogenated to the cyclic saturated ketone.

In this process, the fermentor liquor or effluent is treated to remove cell residues and other insoluble materials and to produce a fermentation product containing the hydroxy acids and containing from 3 to 30% by weight, and preferably from 10 to 25% by weight, of the crude dicarboxylic acids. Suitable oxidizing agents for treatment of the hydroxy acids contained in these crude dicarboxylic acids are those known for the conversion of primary alcohols to carboxylic acids.

For example, nitric acid (40 to 100% HNO_3) at 40° to 100°C in the presence, or absence, of an oxidation catalyst is a suitable oxidizing agent. The oxidation catalyst, if employed, is usually a compound of vanadium and can be used with or without an added cocatalyst such as copper. Ceric nitrate, with or without copper, is also a suitable oxidation catalyst for use with nitric acid.

Another suitable oxidizing system is dinitrogen tetroxide (N_2O_4) employed at temperatures of –10° to –20°C for up to 50 hours. This reagent can be used in the presence of a paraffinic diluent such as n-heptane. Other suitable oxidizing agents are potassium dichromate and sodium dichromate individually or in combination, in sulfuric acid or in a mixture of sulfuric and acetic acids. Either of the dichromates is usually employed at 10° to 80°C for 2 to 8 hours. Aqueous sodium hydroxide or soda-lime under pressure at 200° to 250°C can also be employed to accomplish the desired oxidation step. The use of air in the presence of potassium hydroxide in hexamethyl phosphoramide is also a suitable oxidizing system. Potassium permanganate in basic aqueous media at 0° to 50°C for up to 20 hours can also be used as the oxidizing agent.

The hydroxy acids are treated with an oxidizing agent to produce diacids and these diacids, combined with those diacids originally in the fermentor liquid product, can be covered by cooling the reaction mixture to precipitate the diacids, if not already insoluble in the reaction medium, and then filtering the diacids from the mixture. The recovered diacids are then washed thoroughly to remove traces of the oxidation reaction system. Usually this washing will result in a color-free (white) diacid product. Occasionally, it may be desirable to use a preliminary washing step employing a dilute solution of a reducing agent in order to destroy any occluded oxidizing agent remaining with the recovered diacids.

The recovered dicarboxylic acids are then esterified with a lower alkanol, e.g., methanol or ethanol to form diesters. The diesters can be treated with a sodium dispersion to perform the cyclization (intramolecular acyloin condensation) to the cyclic α-hydroxyketones (cyclic acyloin) which can be reduced directly to the cyclic ketones by hydrogen iodide or zinc and hydrochloric acid. Alternatively, the cyclic acyloins can be dehydrated to the cyclic unsaturated ketones which can be selectively hydrogenated to the cyclic saturated ketone.

Example: *Diacid* — 14 fermentations were carried out aerobically with n-octadecane as the primary source of carbon according to the following general procedure. A sucrose-containing nutrient was prepared by dissolving the following ingredients in sufficient distilled water to give one liter of solution.

Component	
Sucrose	40 g/l
Commercial yeast extract	5 g/l
Urea	1 g/l
KH_2PO_4	1 g/l
$MgSO_4 \cdot 7H_2O$	3 g/l
Mineral solution	1 ml

The mineral solution used above was prepared by adding the following materials, in the quantities shown, to water to form 1 liter of solution.

Component	Grams per Liter
$CuSO_4 \cdot 5H_2O$	0.06
KI	0.08
$MnSO_4 \cdot H_2O$	0.30
$NaMoO_4 \cdot 2H_2O$	0.20
H_3BO_3	0.02
$ZnSO_4 \cdot 7H_2O$	2.0
$FeCl_3 \cdot 6H_2O$	4.8

A 4 liter sample of the sucrose-containing medium was then inoculated with 500 ml of 5-day old culture of a mutant *Torulopsis bombicola,* specifically with strain NRRL-Y-7570. 50 g of n-octadecane were added to the mixture and the temperature of the composite was adjusted to about 30°C. The mixture was agitated and air was introduced at a rate of about 0.10 vol/min/vol of fermentation liquor. Fermentation was contained for about 137 hours. The fermentor effluent was filtered and the filter cake was air dried and reduced to a powder. It was then extracted with acetone to recover the crude diacid.

The acetone extraction was conducted for about 2 hours using 2 liters of acetone per 200 g of dried cell mixture. The extract was filtered to remove cell residues and other insoluble materials. The filtrate was condensed until a slurry was obtained and the slurry was diluted with n-hexane. The resulting mixture was cooled with ice water to precipitate the extracted crude diacid containing hydroxy acid. This crude acid mixture was dried.

Oxidation of the Crude Diacid Mixture — A run employing 100 g (0.60 mol) of the crude diacid containing the hydroxy acid was carried out in the manner described below. A 1 liter, three-necked, flask equipped with heating and stirring means, reflux condenser, thermometer, and powder addition funnel was charged with 300 ml concentrated nitric acid (70% by weight HNO_3), 0.2 g ammonium vanadate dissolved in 5 ml concentrated HNO_3, and 0.3 g copper shot. The crude diacid mixture was added to the stirred mixture in small portions over a period of 1.25 hours while the temperature was maintained at 55° to 60°C. Several grams of copper shot and two 100 ml portions of concentrated HNO_3 were also added during this period.

After all the crude diacid mixture had been added, the temperature of the reaction mixture was increased slowly to 90°C over a period of 3.5 hours while stirring was continued. The mixture was then cooled and filtered to recover the diacid on the filter. The diacid was stirred vigorously with water at 40°C. The mixture was cooled, filtered, and the washed diacid dried to provide 99.5 g of diacid.

Esterification – A 1 liter flask equipped with heating means, water take-off means (Dean-Stark trap), reflux condenser and thermometer was charged with 97 g (0.31 mol) of the treated diacid, 600 ml toluene, 65 ml ethanol and 1.5 ml concentrated sulfuric acid. The mixture was refluxed about 13 hours during which time 76 ml of a water-ethanol mixture was recovered from the reaction mixture. The residual reaction mixture was washed twice with water and the organic phase recovered and condensed by removal of toluene under vacuum. Fractional distillation under vacuum then provided 88.9 g (0.24 mol) of the diester for a 77% mol yield.

Acyloin Condensation – A 3 liter Morton flask equipped with heating and stirring means, thermometer, reflux condenser, nitrogen gas inlet and addition funnel was charged with 2,500 ml xylene and 20.0 g (0.88 g atom) of sodium. A dispersion of the sodium was formed by heating and stirring. The diester prepared above, (81.5 g, 0.22 mol) dissolved in 100 ml xylene was slowly added over a 2 hour period to the stirred and heated sodium dispersion.

After 1.5 hours of additional stirring with heating, no sodium metal particles could be seen. Treatment of the reaction mixture, which had been cooled from 100° to 70°C, with ethanol (250 ml) and washing the mixture with 400 ml water gave an emulsion with some solid material. This mixture was filtered and the organic phase separated from the aqueous phase. The organic phase was then stripped of xylene and other volatiles under vacuum to give a residue of 52.5 g crude acyloin. Fractional distillation of 49.0 g of the above residue gave 27.2 g of product comprising the cyclic acyloin, α-hydroxycyclooctadecanone, for a yield of 44% based on the diester.

Reduction of Acyloin – A 500 ml flask equipped with heating means and flux condenser was charged with 25 g (0.087 mol) of the acyloin from the acyloin condensation step above, 90 g of HI as a 37% by weight solution in water, and 150 ml glacial acetic acid. The mixture was heated at reflux for 1.5 hours. The reaction mixture was added to a vigorously stirred aqueous solution of NaOH (112 g) and sodium bisulfite (25 g) in water (750 ml).

The organic layer was separated and the aqueous layer was extracted with 200 ml ether. The ether extract was combined with the organic layer and the combination was stripped of volatile materials on a rotary evaporator. The residue was washed with an aqueous solution (50 ml) containing 1 g NaOH and 3 g $NaHSO_3$. The aqueous wash was extracted with 200 ml ether which was combined with the organic layer and stripped of volatiles as before. The residue was washed again with aqueous $NaHCO_3$ until basic to litmus. The aqueous wash was extracted with ether and the ether extract combined with the organic layer and stripped of volatiles to give 26 g of residue. Fractional distillation under vacuum of 23.5 g of the above residue gave 12.8 g of product cyclooctadecanone.

Water-Soluble Polymers

R.I. Leavitt; U.S. Patent 3,989,592; November 2, 1976; assigned to Mobil Oil Corporation has found that a particular strain of microorganism, when grown in the presence of protein, transforms the protein to a water-soluble polymer that exhibits unusual properties in aqueous solutions. A sample of this microorganism, *Pseudomonas fabricans* No. 492 (ATCC No. 21984), has been deposited with the American Type Culture Collection.

The proteins that may be treated include a wide variety of substances derived from vegetable or animal sources. While it is difficult to classify proteins in any simple, completely satisfactory manner, it is believed that the best results are achieved with water-dispersible proteins, and in particular with those types of water-dispersible proteins classified as globulins, which are precipitated from aqueous solution by 50% saturation with ammonium sulfate, and as albumins, which are soluble in pure water, coagulated by heat, and precipitate from solution only at ammonium sulfate concentrations in excess of 50% of saturation. Water-insoluble proteins such as keratins, which are the structural proteins of such things as hair, wool and feathers are not suitable.

Any of the proteinaceous materials described may be suspended in water at a concentration from 0.1 to 15.0%, the mixture adjusted to a pH of from 6.0 to 8.0 with a buffer, and the suspension inoculated with the *Pseudomonas* organism. Other inorganic elements necessary to the fermentation are introduced as salts in a manner and in amounts well-known to those skilled in the art. Subsequently the suspension of protein is incubated aerobically for a period of from 2 to 72 hours, at a temperature from 20° to 40°C. The resultant mixture contains the water-soluble polymer product.

The water-soluble polymers of this process are effective coagulants for a variety of suspensions, including phosphate slimes. Solutions of the water-soluble polymers may be used in the tertiary recovery of petroleum from spent oil wells. Other applications include stabilizing the suspension of solids such as pigments, and stabilizing emulsions of hydrocarbon fluids and fatty oils in water for the preparation of lubricant fluids, for example.

Dihydroxyacetone

W. Charney; U.S. Patent 4,076,589; February 28, 1978; assigned to Schering Corporation describes a process for producing dihydroxyacetone which comprises cultivating *Acetobacter suboxydans* ATCC 621 under aerobic conditions in a nutrient medium containing glycerol, yeast hydrolysate or fish hydrolysate and water at a pH in the range of from 3.3 to 4.3. Advantageously, the process is effected under conditions wherein the glycerol is present at from 5 to 15%, preferably from 9 to 12% (weight to volume), the yeast hydrolysate or fish hydrolysate being present at from 0.2 to 1.0% (weight to volume) with 0.5% being preferred. The fermentation is complete in from 24 to 48 hours, usually in about 30 hours. The product is isolated by methods known in the art, such as, filtration of the whole broth, removal of inorganic cations and/or anions (when present) via ion exchange resin adsorption, concentration of the resin effluent and crystallization of the dihydroxyacetone from the concentrate.

Example: *Inoculum Preparation* — First Stage: The medium contains the following — 0.5% yeast extract, 0.1% potassium dihydrogen phosphate, 6.0% glycerol and 100 ml soft water. Prepare a series (e.g., 10 flasks) of inoculum using 300 ml Erlenmeyer flasks. Sterilize the medium at 121°C for 30 minutes. Inoculate the sterile medium under aseptic conditions with a loopful of *Acetobacter suboxydans* ATCC 621 and incubate under static conditions at 28°C for from 48 to 72 hours.

Second Stage: The medium contains the following — 0.5% Ardamine Z, 0.5% potassium dihydrogen phosphate, 6.0% glycerol and 500 ml soft water. Sterilize a series of flasks containing the second stage medium in 2 liter Erlenmeyer flasks at 121°C for 40 minutes and after cooling to 28°C, add 5.0% (v/v) of the first stage inoculum to the sterile medium. Incubate the flasks at 28°C on a rotary shaker at from 280 to 320 rpm for 48 hours, at which time the turbidometric reading (Klett) is about 250 to 260.

The fermentation medium contains 0.45 kg Ardamine Z, 9.9 kg glycerol, 90 liters soft water, and 50 ml antifoam (GE-60). Prepare the fermentation medium in a 25 gallon (working volume) agitated fermentor. Sterilize the medium for 45 minutes at 121°C. Adjust the pH of the sterile medium to 3.7 using 12 N sulfuric acid. Cool the fermentation medium to 30°C and inoculate with 5.0 liters of the second stage inoculum. Aerate the fermentation mixture at 3.5 ft^3/min while agitating at 350 rpm adding antifoam as required. Maintain the temperature at 30°C and the pH range of 3.3 to 4.3. Monitor the utilization of the glycerol by thin layer chromatography. Stop the fermentation when the glycerol is completely utilized and isolate the dihydroxyacetone by the usual procedure.

Fermentation Products of 8-Chloro-10,11-Dihydrodibenz[b,f][1,4]Oxazepine

J. Jiu and S.S. Mizuba; U.S. Patent 4,086,268; April 25, 1978; assigned to G.D. Searle & Co. describe compounds having utility as pharmaceuticals, or as intermediates in the synthesis of other compounds having useful pharmacological properties which are produced by fermenting 8-chloro-10,11-dihydrodibenz[b,f]-[1,4]oxazepine with *Trichoderma lignorum* NRRL 8138, *Hormodendrum* sp. NRRL 8133, *Cladosporium lignicolum* NRRL 8131, *Hormodendrum cladosporioides* NRRL 8132, *Pullularia pullulans* NRRL 8137, *Penicillium* sp. NRRL 8136, *Mucor* sp. NRRL 8135, *Chaetomium* sp. NRRL 8130, and *Hormodendrum* sp. NRRL 8134 in a suitable growth medium.

The process may be carried out in such a manner so as to produce the various transformation products of 8-chloro-10,11-dihydrodibenz[b,f][1,4]oxazepine as shown below. The process illustrated as (I) is conveniently effected by fermenting *Hormodendrum* sp. NRRL 8133 or enzymes derived therefrom with 8-chloro-10,11-dihydrodibenz[b,f][1,4] oxazepine in a suitable growth medium. One of the resulting products, 8-chloro-10,11-dihydrodibenz[b,f][1,4]oxazepin-11-one, is useful as an intermediate in the synthesis of complex amides of dihydrodibenz[b,f] [1,4] oxazepine-10-carboxylic acids as described in U.S. Patent 3,357,998. These compounds are useful as antihypertensive agents and anti-inflammatory agents. The second product, 2-(2-amino-4-chlorophenoxy)benzyl alcohol, is useful as a smooth muscle antagonist.

The process illustrated as (II) is accomplished by fermenting *Hormodendrum cladosporioides* NRRL 8132 or enzymes therefrom with 8-chloro-10,11-dihydro-dibenz[b,f] [1,4] oxazepine in a suitable growth medium. The isolated product, 8-chlorodibenz[b,f] [1,4] oxazepine is useful as an intermediate in the synthesis of the oxazepine derivatives described in Czechoslovakian Patent 111,215. These compounds possess antihistaminic, antispasmodic, local anesthetic, ataractic, and antidepressive activity. The intermediate product produced in the fermentation medium, 8-chloro-10,11-dihydrodibenz[b,f] [1,4] oxazepin-11-ol, is unstable and not isolated. Its presence, however, is detectable from spectral data.

The process illustrated as (III) may be effected by fermenting *Trichoderma lignorum* NRRL 8138, *Cladosporium lignicolum* NRRL 8131, *Pullularia pullulans* NRRL 8137, *Penicillium* sp. NRRL 8136, *Mucor* sp. NRRL 8135, *Chaetomium* sp. NRRL 8130, or *Hormodendrum* sp. NRRL 8134, or enzymes derived therefrom with 8-chloro-10,11-dihydrodibenz[b,f] [1,4] oxazepine in a suitable growth medium. The resulting product, 2-(2-amino-4-chlorophenoxy)benzyl alcohol, is useful as a smooth muscle antagonist.

The diacetate derivative of 2-(2-amino-4-chlorophenoxy)benzyl alcohol, namely, 2-(2-acetamido-4-chlorophenoxy)benzyl acetate, is additionally useful as a smooth muscle antagonist. This compound is conveniently prepared by contacting 2-(2-amino-4-chlorophenoxy)benzyl alcohol with acetic anhydride and pyridine.

A nutrient medium is required for culture of the organism, that is, a medium containing assimilable nitrogen and carbon. An adequate supply of sterile air should be maintained therein, for example by exposing a large surface of the medium to the air, or preferably by passing it through the medium in quantities sufficient to support submerged growth.

Concentration of the oxazepine substrate in the medium, as also fermentation time and temperature, can vary widely. Such operating conditions are to a certain extent interdependent. A preferred, but a critical, range of concentrations of the substrate is 0.01 to 10.0%, while fermentations of from 2 hours to 10 days duration at temperatures between 24° and 35°C are representative. Obviously, conditions must not be such as to degrade the oxazepine, kill the organism prematurely, or inactivate the involved enzymes.

In a preferred embodiment, a nutrient medium containing about 3% oxazepine of substrate is aerobically incubated at 23° to 25°C with a culture of the desired organism for a period of 1 to 3 days. The desired products are extracted with dichloromethane and isolated by chromatography.

4-Androstene-3,17-Dione Derivatives

A. Weber, M. Kennecke, R. Mueller, U. Eder and R. Wiechert; U.S. Patent 4,100,026; July 11, 1978; assigned to Schering Aktiengesellschaft, Germany describe a process for the preparation of a 4-androstene-3,17-dione compound of Formula (1)

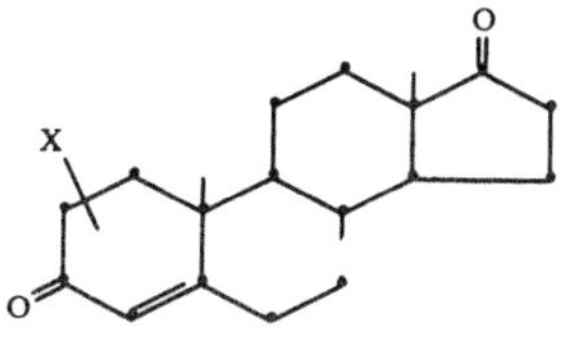

(1)

where X is 1,2-methylene or 1- or 2-methyl, comprising fermenting a sterol of Formula (2)

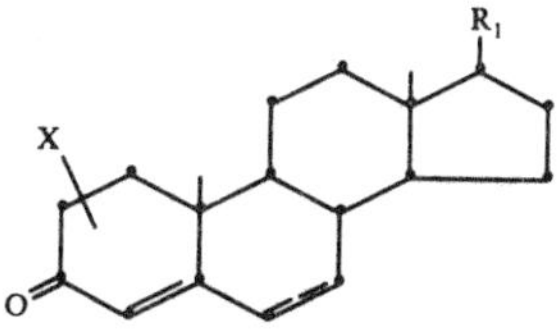

(2)

where X is as above, ----- is a single or double bond, and R_1 is a saturated or unsaturated hydrocarbon side chain of 8 to 10 carbon atoms, with a microorganism culture capable of degrading the side chain.

Exemplary starting compounds for the process are sterols where X is 1α-methyl, 1β-methyl, 1α,2α-methylene, or 1β,2β-methylene. Examples of suitable starting compounds are: 1α-methyl-4-cholesten-3-one; 1β-methyl-4-cholesten-3-one; 1α, 2α-methylene-4-cholesten-3-one; 1α,2α-methylene-4,5-cholestadien-3-one; 1α-methyl-4-stigmasten-3-one; 1α,2α-methylene-4-stigmasten-3-one; 1α,2α-methylene-4,6-stigmastadien-3-one, or the corresponding sitosterol derivatives.

The fermentation is conducted using microorganism cultures customarily employed for the side chain degradation of sterols. Suitable cultures are, for example, those of the genera *Arthrobacter, Brevibacterium, Microbacterium, Protaminobacter,* or *Streptomyces.* Those of the genus *Mycobacterium* are preferred.

Examples of suitable microorganisms are: *Microbacterium lactum* IAM-1640, *Protaminobacter alboflavus* IAM-1040, *Bacillus roseus* IAM-1257, *Bacillus sphaericus* ATCC-7055, *Norcardia gardneri* IAM-105, *Norcardia minima* IAM-374, *Norcardia corallina* IFO-3338, *Streptomyces rubescens* IAM-74 or especially the microorganisms *Mycobacterium avium* IFO-3082, *Mycobacterium phlei* IFO-3158, *Mycobacterium phlei* (Institute of Health, Budapest No. 29), *Mycobacterium phlei* ATCC-354, *Mycobacterium smegmatis* IFO-3084, *Mycobacterium smegmatis* ATCC-20, *Mycobacterium smegmatis* (Institute of Health, Budapest, No. 27), *Mycobacterium smegmatis* ATCC-19979, *Mycobacterium fortuitum* CBS-49566, *Mycobacterium* sp. NRRL-B-3805, and *Mycobacterium* sp. NRRL-B-3683. *Mycobacterium* sp. NRRL-B-3805 is most preferred.

Submerged cultures are grown under conditions customarily employed for these microorganisms, using a suitable nutrient medium with aeration. Then, the substrate, dissolved in a suitable solvent or preferably in emulsified form, is added to the culture and the fermentation is conducted until maximum substrate conversion has been attained.

Suitable solvents for the substrate are, for example, methanol, ethanol, glycol monomethyl ether, dimethylformamide, or dimethyl sulfoxide. The substrate can be emulsified, for example, by adding micronized substrate or substrate dissolved in a water-miscible solvent, e.g., methanol, ethanol, acetone, glycol monomethyl ether, dimethylformamide, or dimethyl sulfoxide, through nozzles under strongly turbulent conditions to preferably decalcified water containing the customary emulsifying agents.

Suitable emulsifying agents include nonionic emulsifiers, for example, ethylene oxide adducts or fatty acid esters of polyglycols. Examples of suitable emulsifiers are surfactants commercially available as Tegin, Tagat, Tween, and Span.

The optimum substrate concentration, time of substrate addition, and duration of fermentation depend on the structure of the substrate employed and on the type of the microorganism utilized. These variables must be determined in each individual case, by preliminary experiments well-known to those skilled in the art.

The 4-androstene-3,17-dione compounds of Formula (1) which can be produced according to the process are valuable intermediates for the synthesis of pharmacologically active steroids, e.g., 17β-hydroxy-1α-methyl-5α-androstan-3-one, 17β-hydroxy-1-methyl-5α-androst-1-en-3-one, 2α-methyl-17β-propionyloxy-5α-androstan-3-one and 1,2α-methylene-17α-hydroxy-4,6-pregnadiene-3,20-dione.

Example: A 2 liter Erlenmeyer flask with 500 ml of a sterile nutrient medium, containing 1% yeast extract, 0.45% disodium hydrogen phosphate, 0.34% potassium dihydrogen phosphate, and 0.2% Tween 80, adjusted to pH 6.7, is inoculated with a suspension of a *Mycobacterium* sp. NRRL-B-3805 dry culture and shaken for 3 days at 30°C at 190 rpm.

20 Erlenmeyer flasks each containing 100 ml of a sterile nutrient medium, containing 2.0% cornsteep liquor, 0.3% diammonium hydrogen phosphate, and 0.25% Tween 80 adjusted to pH 6.5, are inoculated with 5 ml portions of the *Mycobacterium* sp. growth culture and shaken at 30°C for 24 hours with 220 rpm. Then, each culture is combined with 100 mg of $1\alpha,2\alpha$-methylene-4,6-cholestadien-3-one, dissolved in 1 ml of dimethylformamide, and the mixture is fermented for another 96 hours at 30°C.

The combined cultures are extracted with ethylene chloride; the extract is concentrated under vacuum, the residue is purified by chromatography over a silica gel column, and, after recrystallization from diisopropyl ether, 0.9 g of $1\alpha,2\alpha$-methylene-4-androstene-3,17-dione is obtained, melting point 155°C.

Androstane-3,17-Dione Derivatives

In a similar process, *A. Weber, M. Kennecke, R. Mueller, U. Eder and R. Wiechert; U.S. Patent 4,097,334; June 27, 1978; assigned to Schering Aktiengesellschaft, Germany* describe the preparation of an androstane-3,17-dione compound of Formula (1)

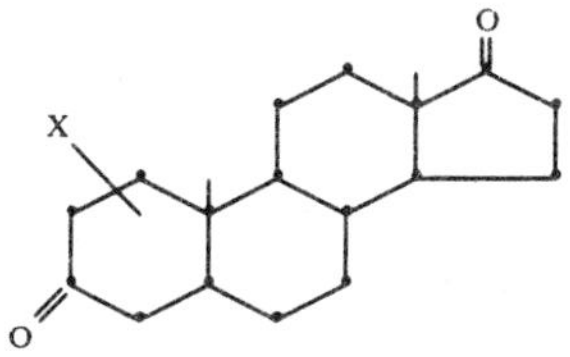

(1)

where X is 1,2-methylene or 1- or 2-methyl, comprising fermenting a sterol of Formula (2)

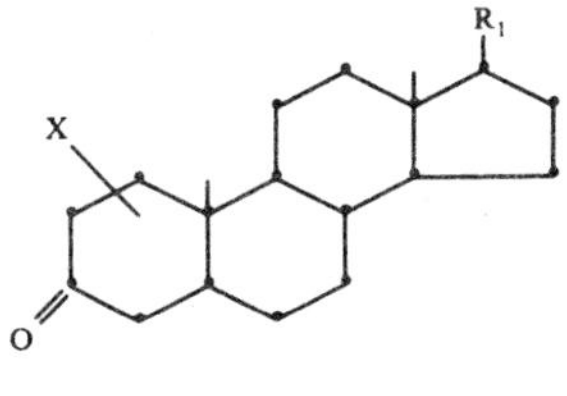

(2)

where X is as above and R_1 is a saturated or unsaturated hydrocarbon sterol side chain of 8 to 10 carbon atoms with a microorganism culture capable of degrading the sterol side chain.

Suitable starting materials for the process are, for example, sterols where X is 1α-methyl, 1β-methyl, $1\alpha,2\alpha$-methylene, or $1\beta,2\beta$-methyl. Examples of suitable starting compounds are: 1α-methyl-5α-cholestan-3-one; 1β-methyl-5α-cholestan-3-one; $1\alpha,2\alpha$-methylene-5α-cholestan-3-one; 1α-methyl-5α-stigmastan-3-one; $1\alpha,2\alpha$-methylene-5α-stigmasten-3-one, or the corresponding sitosterol derivatives. The microorganisms employed are the same as in the previous process.

Spray Drying of Alkali Metal Gluconates

It is well-known that gluconic acid is formed by fermentation of glucose solutions with microorganisms. Sodium or potassium gluconate salts are formed by neutralizing the gluconic acid with a base such as sodium hydroxide or potassium hydroxide. The difficulty in obtaining these salts lies primarily in their recovery.

D. L. Gillenwater; U.S. Patent 3,907,640; September 23, 1975; assigned to Grain Processing Corporation describes a simplified method for recovering sodium or potassium gluconate from fermentation liquors, which method can be carried out relatively rapidly and which involves considerably less equipment and hence expense than is associated with prior recovery methods.

It was found that sodium or potassium gluconate solutions can be successfully spray dried provided that some degree of crystallization or crystal growth has occurred in the solution. In general, a solution of sodium or potassium gluconate can be successfully spray dried when the crystal growth therein ranges from 5 to 50%, preferably 20 to 30%, by volume. Crystal growth in the feed liquor can be achieved by increasing the solids content of the solution to within the range of 45 to 60% by evaporation or adding previously dried product or by cooling the solution or by a combination of these expedients. The sodium or potassium gluconate solution should be kept agitated prior to spray drying to prevent crystal agglomeration which could possibly result in plugging of the nozzles used for introducing the solution into the dryer.

To reduce the possibility of plugging the homogenizing type feed pumps usually employed on commercial spray dryers, it is preferred to employ spray drying equipment having two fluid nozzles.

Drying of sodium or potassium gluconate fermentation liquors can be conveniently accomplished using dryer exhaust air temperatures ranging from 150° to 250°F which correspond to inlet air temperatures ranging from 300° to 500°F. Inlet air temperatures will, of course, vary depending upon the amount of water in the feed unless controlled. The preferred exhaust drying air temperature is from 200° to 250°F.

An unexpected feature of this process is the ability to control particle size of the dried product. Generally speaking, higher drying temperatures and/or lower crystal concentrations tend to result in increasing the particle size of the gluconate product.

Example: A solution (fermentation liquor) of sodium gluconate containing solids in the range of 50 to 60% was transferred to a 100 gallon kettle and cooled by putting water on the jacket. The solution was stirred continuously. As the solution cooled, small crystals of sodium gluconate became apparent in the solution. When the slurry reached a temperature of about 85°F, it was pumped to the dryer and spray dried. Dryer conditions were as follows: nozzle air pressure, 45 psi; nozzle feed pressure, 30 psi; inlet air temperature, 420°F; outlet air temperature, 225°F; and feed pH, 6 to 7.

The product dried easily and was collected via a pneumatic conveying system attached to the bottom of the main drying chambers and the cyclones. Finished product was light colored, free-flowing and contained 0.5% moisture. The loose bulk density was 43.2 lb/ft³ and the packed bulk density was 54 lb/ft³.

Allylic Methylhydroxylated Novobiocins

Novobiocin is an antibiotic useful in the treatment of staphylococcal infections and in urinary tract infections caused by certain strains of *Proteus*. It shows no cross-resistance with penicillin and is active against penicillin-resistant strains of *Staphyloccoccus aureus*. Novobiocin is produced through fermentation by *Streptomycetes*. The methods for production, recovery and purification of novobiocin are described in U.S. Patent 3,049,534. As with any antibiotic it is always highly advantageous to prepare derivatives or analogs since these often lead to new antibiotics with increased potency, fewer and less severe side effects, and/or a different spectrum of antibiotic activity.

O.K. Sebek and L.A. Dolak; U.S. Patent 4,148,992; April 10, 1979; assigned to The Upjohn Company describe a hydroxynovobiocin-type compound of formula:

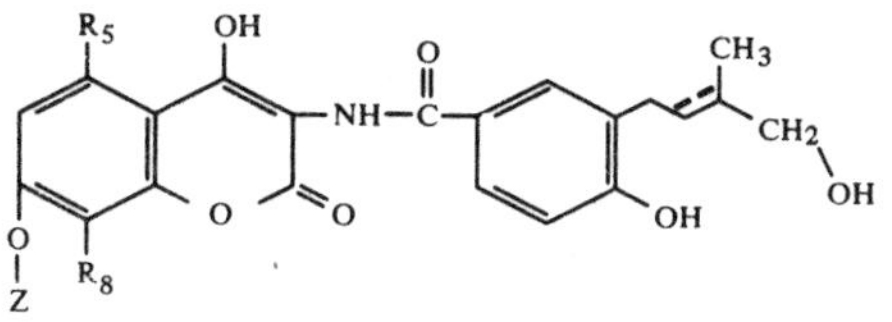

(II)

which comprises (1) cultivating *Sebekia benihana* having the identifying characteristics of NRRL 11,111 and novobiocin-hydroxylating mutants thereof in an aqueous nutrient medium under aerobic conditions; (2) contacting a novobiocin-type compound of the formula:

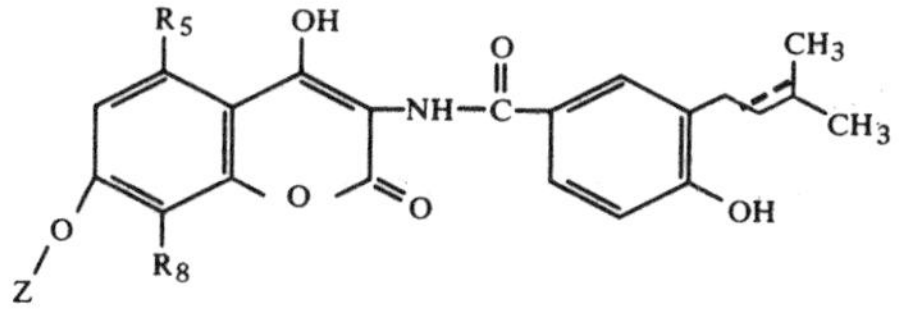

(I)

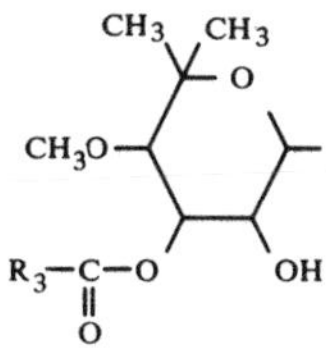

with the *Sebekia benihana* culture; and (3) recovering the hydroxynovobiocin-type compound (II) where R_5, R_8, Z, and - - are defined as follows: R_5 and R_8 may be the same or different and are hydrogen, alkyl of from 1 through 5 carbon atoms, alkenyl of from 1 through 5 carbon atoms, halogen, nitro, cyano, carboxyl, or $-NR_\alpha R_\beta$; R_α and R_β may be the same or different and are hydrogen or alkyl of 1 through 5 carbon atoms; - - is a single or double bond; and Z is hydrogen or where R is amino, 2-pyrryl, 2-(5-methyl)-pyrryl, 2-furyl, and 2-(5-methyl)-furyl.

The microorganism is grown in or on a sterile medium favorable to its development. Sources of nitrogen and carbon are present in the culture medium, the pH is properly adjusted and an adequate sterile air supply is maintained as is well known to those skilled in the art.

The preferred medium for the process is TYG medium. It is utilized for the growth of the microorganism prior to addition of the substrate and during the bioconversion process. The composition of TYG medium is as follows: 0.5% tryptone, 0.3% yeast extract, and 2.0% glucose. TYG medium is adjusted to pH 7.2 in deionized water. The concentrations of the three ingredients in TYG medium may vary somewhat without any problems as is well known to those skilled in the art.

The organism is grown by homogenizing a piece of the mycelium from an agar slant and adding a portion of the suspension of the growth medium (100 ml in a 250 ml Erlenmeyer flask). The organism is grown at a temperature of 20° to 35°C, 25° to 28°C being preferred. The organism is grown with shaking (100 to 500 rpm). Alternatively, the flask may be aerated by bubbling air through it. The growth process takes from 2 to 4 days.

After a suitable period of growth, usually 3 days, (1) the substrate may be added for bioconversion, (2) a portion of the growth may be utilized to inoculate a number of small Erlenmeyer flasks, or (3) the total contents of the shaker flasks along with a number of other small flasks may be added to a larger fermentor. For instance, a fermentor containing 10 liters of fermentation medium may be seeded with the contents of 2 to 10 Erlenmeyer flasks (250 ml) each containing 100 ml of the inoculum.

Usually an antifoam agent such as Ucon (1 to 5 ml) is added to each fermentor of 10 liters capacity. During growth, the fermentation medium is stirred (100 to 400 rpm), aerated (1 to 5 liters air/min/10 liters fermentation medium) and maintained at 20° to 35°C, preferably 25° to 28°C. After a period of 2 to 4 days, usually 3 days, the contents of the 10 liter fermentor may be used to inoculate a larger fermentor or a novobiocin-type compound (I) may be added to the fermentor to undergo bioconversion.

The substrates, the novobiocin-type compound (I), are added to the fermentation medium in their salt form in an aqueous solution. The substrate may be added to give a concentration of as low as 50 μg/ml or as high as 1,500 μg/ml. It is preferred that the concentration of (I) be 100 to 1,000 μg/ml.

The bioconversion takes place at 20° to 35°C, preferably 25° to 28°C, with agitation (100 to 500 rpm) or stirring (100 to 400 rpm) and aeration either by surface contact in a shake flask or 1 to 5 liters/min/10 liters fermentation medium in a fermentor.

As the bioconversion process proceeds, the reaction is monitored by TLC (thin layer chromatography). One TLC system is silica gel with ethyl acetate to methanol ratio of 4:1. As the bioconversion takes place a more polar compound, the product (II), is formed at the expense of the substrate. The time necessary to obtain maximum yields from the bioconversion will range from 3 to greater than

10 days depending on the amount of mycelial growth, the temperature, the aeration, etc., but most importantly on the concentration of the substrate (I).

Following completion of the bioconversion, as measured by TLC, the products are recovered and purified by methods well-known to those skilled in the art. The fermentation beer is adjusted to pH 2 to 5 with an acid such as hydrochloric, sulfuric, phosphoric, etc. The solids are separated by centrifugation or by mixing the fermentation beer with approximately $\frac{1}{10}$ volume of a filter aid such as Dicalite 4,200 or any other diatomaceous earth product.

When using a filter aid, the mixture is then filtered over a bed of the same filter aid. The cake is extracted with an organic aqueous immiscible diluent such as ethyl acetate, chloroform, carbon tetrachloride, benzene, toluene, methylene chloride, SSB (mixture of isomeric hexanes), or mixtures thereof. The filtrate is extracted with the same organic diluent as is used to extract the cake. The combined organic layers are washed with brine, dried with sodium sulfate or magnesium sulfate, filtered, and concentrated under vacuum with or without heat.

An alternative recovery process for the hydroxynovobiocins (II) is utilization of an anion exchange resin as is well-known to those skilled in the art.

Hydroxylated Biphenyl Compounds

R.D. Schwartz; U.S. Patent 4,153,509; May 8, 1979; assigned to Union Carbide Corporation describes a process for producing mono- and dihydroxybiphenyl compounds by enzymatically biotransforming the corresponding biphenyl compound with a microorganism. Preferred biphenyl reactants are those of the formula:

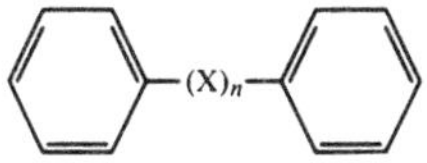

where n is 0 or 1 or 2; X is phenyl, divalent oxygen, sulfur, sulfinyl, sulfonyl, carbonyl, amino or alkylamino or X is a divalent alkylene, alkenylene or alkynylene chain which may optionally include one or more divalent oxygen, sulfur, sulfinyl, sulfonyl, carbonyl, amino or alkylamino moieties in any combination. Illustrative of suitable biphenyl reactants are: biphenyl, 4-hydroxybiphenyl, biphenylmethane, benzophenone, benzil, diphenylacetylene, diphenyl sulfide, diphenyl ether, p-terphenyl, etc.

Illustrative of hydroxybiphenyl derivatives that can be prepared by the process are those of the formula:

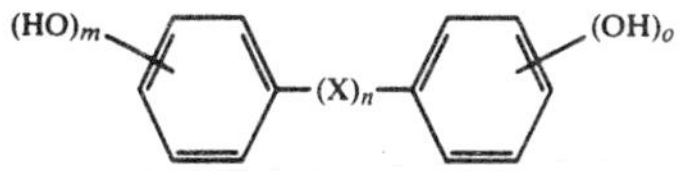

where m and o are individually 1 or 0 with the proviso that both m and o may not be 0, and X and n are as described above.

The process is particularly useful for preparing hydroxylated biphenyl derivatives with hydroxyl groups substituted at one or both p-positions. The microorganisms which are used in the process are *Absidia* and certain species of *Aspergillus* and *Cunninghamella;* preferred for use in the process is *Absidia.* Illustrative of particularly preferred microorganisms are: *Absidia pseudocylindrospora* NRRL 2770; *Absidia ramosa* NRRL 1332; *Absidia glauca* NRRL 1324; *Absidia* sp. NRRL 1341; *Aspergillus niger* NRRL 599; *Absidia spinosa* NRRL 1347; *Cunninghamella echinulata* NRRL 1386; *Cunninghamella elegans* NRRL 1392; and *Cunninghamella elegans* NRRL 1393. *Aspergillus niger* and *Absidia ramosa* are useful for producing the monohydroxy product and the other microorganisms are useful for producing both the dihydroxy and the monohydroxy product.

In a preferred embodiment of the process, the microorganism is cultivated in a suitable medium prior to contacting with the biphenyl reactants. A typical medium will include a carbon source, a nitrogen source, inorganic salts and deionized water.

The pH of the culture medium is not critical and may vary from 4.0 to 8.0, and preferably from 5.6 to 7.2. The cultivation temperature is not critical and may vary from 20° to 40°C.

In carrying out the process any of the abovedescribed microorganisms is grown in a culture medium containing the abovedescribed nutrient sources. An appropriate biphenyl compound is then added at any time between the beginning of microorganism growth and the end of microorganism growth. The biphenyl compound can also be added after the microorganism has been concentrated and redeposited into the culture medium. The biphenyl compound may be added neat or dissolved in a suitable solvent such as ethanol, methanol, dimethylformamide, etc. The concentration of the microorganism required to effect the process is not critical. In preferred embodiments, the concentration of the microorganism is at least 1.0% by weight. The concentration of biphenyl reactant can vary from 0.005 to 10% by weight, based on the total weight of the reaction mixture.

After the addition of the biphenyl reactant, the process is effected for a period of time sufficient to produce the desired hydroxybiphenyl compound. In general, residence times may vary from 1 to 20 days or longer. After the process has gone to completion, the desired product can be collected in pure form by conventional methods. Thus, for example, the mycelium can be removed from the water/product by filtration and the product collected using ion, gas chromatography, extraction, thin layer chromatography, distillation, etc.

The biphenyl compounds prepared in accordance with the process have wide utility and are valuable for a number of useful purposes. For example, monomer 4,4'-dihydroxybiphenyl is extremely useful as a precursor in the preparation of a polymer of high strength and heat resistance. Other of these compounds are useful as heat transfer agents or as precursors in the preparation of photocurable resins.

Epoxides

S. Suzuki, K. Furuhashi and A. Taoka; U.S. Patent 4,106,986; August 15, 1978; assigned to Bio Research Center Company, Ltd., Japan have found that a microorganism that belongs to *Nocardia* genus can produce α-epoxides or α,ω-diepoxides or a mixture of these from α-olefins or α,ω-dienes as a carbon source.

The microorganism employed in this process can be designated as *Nocardia corallina* var. *Taoka* or one affinitive to *Nocardia corallina* by collating its properties with *Bergey's Manual of Determinative Bacteriology,* 8th Edition, (1974). This microorganism was deposited with Agency of Industrial Science and Technology, Fermentation Research Institute of the Ministry of Industrial Trade and Industry, Japan as FERM-P-4094 on June 15, 1977. The microorganism has also been deposited with American Type Culture Collection and bears ATCC No. 31338.

A wide range of assimilable carbon sources can be used as substrate for the microorganism. Preferable sources, however, include the α-olefins having 3 to 20 carbon atoms and/or α,ω-dienes having 4 to 20 carbon atoms. In a medium to which the carbon sources, nitrogen sources and inorganic salts are added, the abovementioned microorganism is inoculated to cultivate by agitation or shaking under aerobic conditions.

The reaction is conducted for 1 to 6 days at 5° to 40°C, or 20° to 38°C, desirably keeping the pH level at 6 to 8, and usually under normal pressure, but it may be carried out under increasing pressure depending on the carbon source used as substrate.

Besides cultivation of the microbe in the abovementioned medium to produce epoxide, this process also includes preculture of the microbe in a medium containing assimilable carbon source, for example, glucose, sucrose, sorbitol, glycol, n-paraffin, α-olefin and propylene, then allowing the resultant grown cells to react aerobically in the medium of the same composition as mentioned above.

Example:

$(NH_4)_2HPO_4$	4 g
$Na_2HPO_4 \cdot 12H_2O$	2.5 g
KH_2PO_4	2.0 g
$MgSO_4 \cdot 7H_2O$	0.5 g
$FeSO_4 \cdot 7H_2O$	30 mg
$CaCl_2 \cdot 2H_2O$	60 mg
$MnCl_2 \cdot 4H_2O$	60 μg
Yeast extract	200 mg

The above components were dissolved into 1 liter of deionized water. The pH of the obtained solution was 7.2. The solution was pipetted, 50 ml into each 500 ml Sakaguchi flask, and was sterilized at 115°C for 15 minutes.

Next, 2 loopsfull of *Nocardia corallina* var. *Taoka*, B-276 (Ferm-P-4094) which had been cultured in a nutrient-glucose agar at 30°C for 24 hours, were inoculated to each solution prepared as described above, and were autoclaved or filtered through a millipore filter.

Carbon sources as substrate were added properly so that the total amount would make 0.5 g; then a cultivation was carried out aerobically. In the case propylene and butene-1 were used as substrate, the Sakaguchi flasks were closed up tight, pressure was reduced, 15 cm Hg and a suitable volume of gas was introduced to put the pressure back to the original one. After 5 days of cultivation at 30°C, the obtained broth was analyzed by gas chromatography in order to identify and quantify the produced epoxides. The results are shown below.

Substrate	Produced 1,2-Epoxides	Yields (g/l)
Propylene	1,2-epoxypropane	0.77
Butene-1	1,2-epoxybutane	0.32
Pentene-1	1,2-epoxypentane	0.02
Hexene-1	1,2-epoxyhexane	0.012
Heptene-1	1,2-epoxyheptane	0.010
Octene-1	1,2-epoxyoctane	0.014
Nonene-1	1,2-epoxynonane	0.016
Decene-1	1,2-epoxydecane	0.016
Undecene-1	1,2-epoxyundecane	0.014
Dodecene-1	1,2-epoxydodecane	0.82
Tridecene-1	1,2-epoxytridecane	2.0
Tetradecene-1	1,2-epoxytetradecane	2.6
Pentadecene-1	1,2-epoxypentadecane	1.4
Hexadecene-1	1,2-epoxyhexadecane	1.0
Heptadecene-1	1,2-epoxyheptadecane	0.81
Octadecene-1	1,2-epoxyoctadecane	0.42
Nonadecene-1	1,2-epoxynonadecane	0.71
Eicosene-1	1,2-epoxyeicosane	0.20
1,13-tetradecadiene	13,14-epoxy-1-tetradecene	0.31

OTHER MICROBIAL PROCESSES

Epoxidation of Linear Hydrocarbons

Many fermentation processes involve chemical reactions carried out by micro-organisms which convert certain organic compounds to other compounds. The process may or may not occur in the presence of air. The microorganisms produce enzymes which serve as catalysts for the chemical reactions. A common characteristic of these fermentations is that the end product of the process is in dilute aqueous solution. Recovery of the product from this solution often contributes significantly to final product cost. Another characteristic shared by many fermentations is that continued biosynthesis of the product is inhibited in the presence of relatively low concentrations of the product itself, or the microorganism responsible for the fermentation may be impaired by the fermentation product. As a result, it is unusual to find fermentations in which the product occurs in high concentration.

C. J. McCoy and R.D. Schwartz; U.S. Patent 4,102,744; July 25, 1978; assigned to Exxon Research & Engineering Co. describe a process for adding cyclohexane to a fermentation broth in order to maintain an inhibiting substance in the fermentation broth at a level which will permit continued fermentation.

In a preferred embodiment, the microorganism responsible for the enzymatic conversion of the substrate is *Pseudomonas oleovorans*, in particular, *Pseudomonas oleovorans* ATCC 29347 and the substrate is selected from the group consisting of 1-alkenes with formula C_nH_{2n}, where $5 \leqslant n \leqslant 12$, dienes with formula $CH_2{=}CH{-}(CH_2)_n{-}CH{=}CH_2$, $1 \leqslant n \leqslant 8$ and normal alkanes with formula C_nH_{2n+2}, $5 \leqslant n \leqslant 12$.

The amount of the solvent added to the fermentation broth may vary over a wide range from 10 to 60% by volume. A preferred range is 15 to 25% by volume. The process should be practiced at a temperature maintained between 15° and 40°C. A preferred temperature range is between 28° and 34°C. The process should be practiced with the pH of the aqueous broth maintained between 6 and 8. A preferred range is 6.8 to 7.4. The period of time required for maximum conversion of the transformed substrate depends on the particular reaction and the initial concentration of the microorganism. In general, one to five days should be sufficient although it is preferable initially to include a high enough inoculum of microorganism so that the period of time for maximum conversion is between one and three days.

Example 1: *Enzymic Epoxidation of 1,7-Octadiene Using the Enzyme System of Pseudomonas Oleovorans* — The mechanism of enzymic epoxidation using an enzyme system in *Pseudomonas oleovorans* ATCC 29347 was studied in this example. In order to study the nature of the products formed from the epoxidation of 1,7-octadiene, it was necessary to synthesize and recover gram quantities of the products 7,8-epoxy-1-octene and 1,2-7,8-diepoxyoctane. Initially, conventional fermentation was used, i.e., an aqueous mineral salts medium containing both octane (1% v/v) and 1,7-octadiene (1% v/v) was inoculated with *P. oleovorans* and incubated for about 30 hr at 30°C. During this time growth occurred at the expense of octane, and the octadiene was epoxidized.

In this system the product yields were at best 1 to 1.2 g of 7,8-epoxy-1-octene/l and 0.3 to 0.4 g of 1,2-7,8-diepoxyoctane/l. One of the limiting factors was the inhibition observed when the concentration of 7,8-epoxy-1-octene reached about 0.8 g/l. The results are included in the following table.

Example 2: *Enzymic Epoxidation According to This Process* — The materials were the same as in Example 1, except for the addition of cyclohexane. Experiments with growing cells were conducted in 300 ml baffled shake flasks containing 100 ml of medium supplemented with 1,7-octadiene and n-octane (1%, v/v each), at 30°C. The medium was modified so as to contain the indicated amount of cyclohexane (v/v). Unless otherwise indicated, the entire contents of the shake flask were centrifuged to separate the phases and the volume of each phase was measured and assayed for epoxides.

Epoxidation of 1,7-Octadiene*

| | . . . . Product Synthesis . . . . | | | |
| | . . Conventional Aqueous . . | | . . . . . Mixed Phase** . . . . | |
	7,8-Epoxy-1-Octene	1,2-7,8-Diepoxyoctane	7,8-Epoxy-1-Octene	1,2-7,8-Diepoxyoctane
Product recovered from				
Aqueous phase, g/l	1.2	0.4	0.19	0.25
Cyclohexane phase, g/l	—	—	6.7	0.27
Cell pellet, g/l	—	—	0.08	0
Molar conversion to				
product, %	14.3	4.2	83.3	5.6

*In conventional aqueous fermentation and combined aqueous phase and solvent phase fermentation. Initial octadiene concentration, 7.32 g/l.
**80% aqueous/20% cyclohexane.

In summary, *P. oleovorans* ATCC 29347, growing at the expense of n-octane and in the presence of 1,7-octadiene, oxidized the octadiene to epoxide products at an efficiency approaching 90 mol % conversion. This was accomplished by incorporating a water-insoluble organic solvent, cyclohexane, into the conventional aqueous fermentation medium. Further, the presence of the cyclohexane resulted in the simultaneous separation and concentration of the products in the organic phase. The modified fermentation results in a five-fold increase in efficiency of conversion of 1,7-octadiene to 7,8-epoxy-1-octene and 1,2-7,8-diepoxyoctane relative to the conversion in conventional aqueous medium.

Recycling of Microorganisms

H Müller; U.S. Patent 4,156,630; May 29, 1979; describes a process of fermentation wherein the microorganisms or enzymes are recycled into the fermentor after separating therefrom the fermentation filtrate.

The filter aides used in this process are particularly those which have a low degree of abrasion. In general the filter aids may be granular or fibrous materials capable of forming filter cakes, and they may be used in amounts between 0.5 and 30 g/l of fermentation broth depending on the type of microorganisms. Preferably, the filter aids are used in an amount equal to the dry mass of microorganisms in the fermentation broth.

Figure 8.1: Apparatus for Recycling of Microorganisms

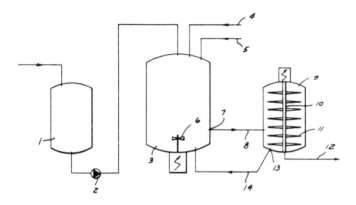

Source: U.S. Patent 4,156,630

The apparatus employed is described with reference to Figure 8.1. In a culture medium tank **1** the nutrient medium is mixed with a source of carbon. The sterilized culture medium is then conveyed from the tank **1** by means of the pump **2** into the fermentor **3**.

At the inception of the fermentation a filter aid is added through the duct **4** and the inoculum is introduced through the duct **5**. The filter aid may be subjected to sterilization prior to the introduction of the inoculum. The fermentation can then be carried out with or without addition of air depending on the type of fermentation. The fermentor is equipped with a stirrer **6** and an outlet **7** is provided for the fermentation product. A passage **8** permits passage of the fermentation product into the filter tank **9**.

The filter, as shown, e.g. in U.S. Patent 3,251,468, is a centrifugal evacuation filter. In its tank there is provided a hollow axle **10** on which the filter elements **11** are disposed. Each filter element comprises a filter web which is stretched above the base portion forming a cavity between web and base. The cavity is in communication with the hollow axle **10**. A discharge duct **12** is provided for the clear filtrate while the concentrate formed in the filter as will be described below is discharged through the outlet **13** to which the recycling passage **14** is connected which leads back to the fermentor **3**. Thus, a feedback is provided between the filter **9** and the fermentor **3**.

The operation of the process is as follows, it being understood that the process may be carried out in a continuous or discontinuous operation. For a continuous operation nutrient medium is continuously passed into the fermentor and the fermentation product is continuously discharged through the outlet **7** and passed through the connecting line **8** in the form of a fermented suspension into the filter **9**. The filtrate, after passing through the web on the filter plates **11**, flows into the hollow axle **10** through inlet apertures provided thereon and is then channeled out of the tank through the duct **12** substantially free of all microorganisms.

The inert filter aid which forms part of the suspension formed in the fermentor, together with the microorganisms, goes into the filter cake which deposits on the web of the plates **11** of the filter. Once the filter cake has reached the desired thickness, the passage of fermentation broth from the fermentor is suspended and the filter cake is loosened by subjecting the filter to a centrifugal action. The detached filter cake containing the microorganisms is then passed back as a slurry through the outlet **13** and passage **14** into the fermentor **3**.

As will be seen the entire operation takes place in a closed circuit and, therefore, the operation can easily be carried out under sterile conditions. In case of the use of free adsorbable enzymes the enzymated product instead of a clear filtrate as obtained with microorganisms will be discharged through the duct **12**.

Example: This example relates to the fermentative oxidation of sorbose to 2-keto-L-gulonic acid, an intermediate product for the vitamin C synthesis. The microorganisms which may be used in this process belong to the genera *Pseudomonas* and *Acetobacter*. A process of this type is described in U.S. Patent 3,234,105.

For the purpose of the fermentation a nutritive base medium was formed in the mixing tank. The medium consisted of 5% sorbitol, 0.5% glucose, 0.5% yeast extract and 1 to 2% calcium carbonate ($CaCO_3$). The mixing and sterilization of the nutrient medium was effected in conventional manner in a mixing tank.

30 liters of the nutrient medium were then pumped into a fermentor of 50 liter capacity which previously had been sterilized by steam at 121°C for half an hour. The fermentor was then inoculated with 500 ml inoculum consisting of *Acetobacter sp.* and simultaneously about 2 to 3 g/l of sterile filter aids were added to the fermentor in the form of a suspension. The fermentation was then carried out at a temperature of 28° to 30°C at 580 to 600 rpm. The aeration was effected with 0.5 to 0.8 vvm.

After about 150 hr the contents of the fermentor was transferred in sterile form and under sterile conditions into the centrifugal evacuation filter. The cells and the filter aid were then separated from the clear solution by the filter web and were caused to deposit as a filter cake on the filter plates while the solution passed into the hollow axle and into the discharge duct. There were obtained

25 to 27 liters of clear filtrate from which 5 g/l of 2-keto-L-gulonic acid were isolated. The filter cake was then ripped off the filter plates by means of a centrifugal action imposed upon the plates and was recycled as a slurry into the fermentor. A reinoculation of the fermentor was not necessary since the microorganisms had retained their full vitality. Other products which may be formed in a similar manner are lactic acid which may be obtained from lactose by means of lactobacilli or the oxidation products of naphthalene which may be obtained by means of *Pseudomonas sp.*

Degradation of Hydroxystyrene Polymers

Generally, synthetic high molecular substances are hardly decomposed by the action of microorganisms and are not susceptible to decay even when buried in soil. This nature would be of great merit for structures for which decay is undesirable. On the other hand, industrial materials not decomposable by the action of microorganisms are out of the natural recurring cycle and thus are treated compulsorily by the combustion method when thrown away as waste materials. When synthetic high molecular substances are treated by the combustion method, they are not effectively utilized and are converted into substantially valueless substances, e.g., CO_2 and H_2O. Further, the treatment of such high molecular substances by combustion is attended by such a disadvantage that the combustion furnace used is susceptible to serious damage so that a specially devised furnace is required.

H. Hatakeyama, T. Haraguchi and E. Hayashi; U.S. Patent 4,154,653; May 15, 1979; assigned to Agency of Industrial Science and Technology, Japan describe a process for the degradation of hydroxystyrene polymers which comprises culturing soil bacteria belonging to *Moraxella* genus or soil fungi belonging to *Penicillium* genus on a medium containing hydroxystyrene polymers having a recurring structural unit of the general formula:

wherein X and Y each stand for a hydrogen atom or an alkoxy group, as carbon source.

The polymers having a recurring structural unit of the above general formula used in this process are known in the literature and prepared by oxidizing lignin obtained from the white liquor of paper mills with oxygen under alkaline conditions to form benzaldehyde derivatives, converting the benzaldehyde derivatives into the corresponding monomeric styrene derivative and polymerizing the monomers.

The B-1 strain used in this process belongs to the *Moraxella* genus and is deposited with Fermentation Research Institute of 5-8-1, Inage-higashi, Chiba-ken, Japan as deposition number FERM-R 3709.

The A-1 strain belongs to the *Penicillium* genus and is deposited with the aforementioned Fermentation Research Institute as deposition number FERM-P 3882.

The media used to degrade the hydroxystyrene polymers may be any of the
known conventional ones so far as they necessarily contain the hydroxystyrene
polymers as the main carbon source, other nutrients required for growth of the
microorganisms such as nitrogen source and inorganic salts and compounds capable of promoting growth of the microorganisms. Utilizable as the nitrogen
source required for growth of the microorganisms are, for example, inorganic
ammonium salts such as ammonium nitrate, ammonium phosphate and ammonium
sulfate as well as other nitrogen-containing compounds such as urea and ammonia per se. Examples of the nutrients include inorganic salts such as potassium
dihydrogen phosphate, dipotassium hydrogen phosphate and magnesium sulfate.
In addition, other metal compounds such as ferrous sulfate are added in a very
small amount.

The culturing treatment is effected by merely using the culture tank in a fixed
position, shaking the tank or agitating the tank by aeration. The hydrogen ion
concentration during the culturing treatment is preferably between pH 7 and 7.5,
inclusive.

Example: A medium which will be referred to hereinafter as PHS medium was
prepared as follows.

Polymer	1.0 g
NH_4NO_3	20.0 g
$CaCl_2$	1.0 g
KH_2PO_4	1.0 g
$MgSO_4 \cdot 7H_2O$	0.3 g
$ZnSO_4$	0.01 g
$FeSO_4$	0.01 g
Water	1 liter
pH	7

100 ml of the medium were placed in a flask and sterilized at 115° to 119°C for
15 min. A very small amount (1 platinum loop) of *Moraxella* B-1 strain which
had been growing on another PHS medium was inoculated on this PHS medium
and cultured at 28°C. The change in the mean molecular weight with the lapse
of time of the polymer contained in the culture solution was analyzed by way
of gel permeation chromatography.

When p-hydroxystyrene polymer was used as carbon source in the medium, the
portion of the polymer corresponding to a molecular weight of 240 (almost a
dimer) was found removed from the original polymers with a mean molecular
weight of about 1,800 (pentadecamer) after culturing for 20 days.

When 3-methoxy-4-hydroxystyrene polymer was used as carbon source in the
medium, the original polymer with a mean molecular weight of about 2,100
(tridecamer) was degraded to a polymer with a mean molecular weight of about
1,600 after culturing for 4 days but showed a tendency of increasing the molecular weight after culturing for 20 days. In this case, however, the proportion of
oligomers not greater than hexamer was increased and the formation of oxalic
acid and maleic acid was detected. Such increase in molecular weight during the
culturing treatment is interpreted as a result of the opening of the benzene rings in
the polymer and subsequent oxidation to carboxylic acid.

The above mentioned polymers showed UV-absorption spectral bands indicating the existence of aromatic rings but the spectral bands were weakened with the lapse of time, thus indicating that the aromatic rings were opened gradually.

The B-1 strain was cultured in a similar manner to that described above, on a film of a p-hydroxystyrene polymer (MW = 2×10^5) for 20 days. The cultured film was then subjected to an IR-absorption spectral analysis whereby a clear absorption band of C=O was found, thus indicating degradation of the polymer.

Biomass Growth Restriction

W.L. Griffith and A.L. Compere; U.S. Patent 4,127,447; November 28, 1978; assigned to U.S. Department of Energy have found that continuous biologically catalyzed anaerobic processes may be conducted in a manner so as to alleviate the problems which exist in the prior art batch processes as well as overcoming problems which inhere in the use of continuous fixed film packed bed upflow reactors. By use of a continuous system the reculturing of appropriate biological catalysts for each batch is not required. The avoidance of reculturing is particularly advantageous when the catalyst used is an anaerobic bacteria because such cultures are well known to be difficult to grow.

Figure 8.2: Anaerobic Upflow Packed Bed Bioreactor

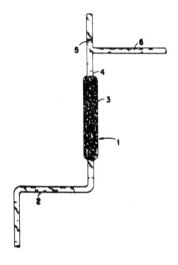

Source: U.S. Patent 4,127,447

The process may be best understood by first referring to the apparatus in which the process is carried out. Such apparatus is shown in Figure 8.2. The apparatus comprises a packed column **1** having a source of influent or reactants **2**. The column is packed with a support **3** to which the biological catalysts are attached. The products of the reaction leave the column via exit **4** such that any gaseous products may be vented or extracted through outlet **5** while liquid products are removed through exit **6**. This type of reactor unit is referred to as an anaerobic upflow packed bed bioreactor.

The term anflow process refers to chemical reactions conducted in such an apparatus. The term itself is an acronym for anaerobic upflow.

An essential part of the apparatus used in carrying out an anflow process is the packing material 3 to which the anaerobes are attached. The packing material may be a conventional material used in distillation tower packings, such as alumina, porcelain, stoneware, or fired clays, or less conventional materials, such as mixed cement compositions. Such packing material may be coated by a technique which involves coating a dense packing material with a polymer which can be crosslinked to form an insoluble hydrophilic loose net film to which microorganisms are attached. The packing can thus be a dense material such as glass, stone, ceramic or plastic which is coated with a hydrogel form material such as gelatin, egg albumin, hide pulp collagen, refined collagen, fibrin, gluten and acrylic acid backbone type polyelectrolytes such as polyacrylamide and polyhydroxy-ethyl methacrylate.

Such polymer can then be crosslinked with an agent selected from the group consisting of gluteraldehyde, ethylchloroformate, formaldehyde, N,N'-methylenebisacrylamide, 1,2-diacrylamide ethyleneglycol, and N,N'-diallyltartardiamide. The anaerobic microorganisms can be attached either before or after cross-linking. Other types of packings which may be used in the anflow apparatus include those which are capable of forming a hydrophilic charged layer on their surface as a function of their composition. Such materials include various charged metals and metal oxides or hydroxides, and substituted plastics, such as ion-exchange resins.

When the process is not employed, the anaerobic organisms which are attached to the packing in the above manner when contacted with the reactant solutions which are to be metabolized tend to expand and grow such that the actual volume of anaerobes (biomass within the column) becomes considerably greater after a short period of operation. This expanded volume of biomass tends to plug the column making further operation impossible. This biomass includes both organisms which are alive and active as well as anaerobes which have died and are no longer contributing to the utility of the process. The essence of this process is the discovery that the volume of biomass can be restricted by limiting the population of anaerobes within the column while simultaneously allowing the anaerobes to catalyze the particular reaction via their metabolic activity.

It has been found that by restricting and even eliminating the nutrients which are essential for growth the population of microorganisms within the column can be maintained constant. This is quite surprising in view of the expectation that the microorganisms would slowly die off and decrease in population as a result of nonavailability of essential nutrients. However, in most systems, it has been found that the microorganisms which die become a source of essential nutrients for the remaining microorganisms. By having the active microorganisms utilize the essential nutrients contained in the inactive or dead microorganisms, an equilibrium is established which limits the microorganisms population to a stable size.

In some systems, however, it has been found that the microorganisms do not readily assimilate the biomass of the dead microorganisms into their own metabolism. However, it has been found that the addition of a small amount, i.e., 0.001 to 5 wt % of a membrane disruptive detergent to the influent stream will make the nutrients of the dead microorganisms available to the live microorganisms so as to maintain a blance between the death rate and repopulation due to microorganism population expansion. The net result is that the population remains substantially constant.

Example: A column similar to the column shown in the figure was used to convert glucose to ethanol. The column was 1.5" in inside diameter and 24" long. The column was constructed of glass and packed with ¼" Berl saddles and ¼" glass beads. A coating for the packing was produced by forming a mixture of: 25 ml of 1 g/l polyelectrolyte solution (A-23 by Dow), 10 ml formaldehyde solution, 5 ml formamide solution, and 5 g gelatin dissolved in 20 ml water previous to addition.

This was stirred with 1 qt of packing material in a rotary tumble until there was no formaldehyde odor. The resulting packing was dried and baked at 60° to 80°C for 6 to 8 hr. Yeast cream was prepared as follows. To 5 g of active dry ethanol tolerant yeast culture was added: 10 ml water; 10 ml of a 1 g/l solution of Calgon polyelectrolyte CP #8, 3.1 meq/g; and 20 ml saturated ferric chloride solution.

This was mixed vigorously between additions and passed through a Dounce homogenizer using a loose pestle. The mixture was centrifuged and the pellet resuspended in 1 ml saturated sodium propionate, 1 g alum, 1 g magnesium sulfate heptahydrate, 10 ml of a 1 g/l solution of Calgon CP #8, and 20 ml water.

This was again centrifuged and the pellet resuspended in a small amount of water to form a cream. The yeast cream and the coated packing were then placed in a rotary tumbler and tumbled until a good coating was formed. Bottom yeast cultures were mixed with a 1% solution of Nalco 8172 polyelectrolytes prior to being introduced into a sterilized aqueous feed solution consisting of 1% yeast extract and 5 to 30 wt % glucose and was passed upwardly through the column at about 900 ml per day. During a 1 month start-up period, to optimize yeast multiplication and to enhance attachment to the coated packing, aqueous feed solution containing 1% yeast extract and sufficient malt extract to give a specific gravity of 1.02 was passed through the column at about 900 ml per day.

Following the start-up period Difco yeast extract, the water-soluble portion of autolyzed yeast, which was the source of metabolizable nitrogen (an essential growth nutrient) was reduced by a factor of 5 to 0.2% to restrict the growth of the biomass population.

The packed column now having a yeast culture well established on the coated packing was operated continuously on an experimental production basis for 26 weeks. Feed solutions ranging from 20 to 30% glucose were passed upward through the column at several flow rates. The operating characteristics of this yeast column are given in the following table.

The process was stable to variations in feed glucose concentrations between 5 and 30 wt %. During processing, pH values were tested between pH 3 and 8. However, very little effect upon alcohol performance was noted within that range. During this period, the source of metabolizable nitrogen was yeast extract (Difco). This material was maintained at a concentration of 1 wt %. The unit operated stably for a period of several months which is significantly longer than the culture use (1 to 2 weeks) in commercial stirred reactors.

Operating Characteristics of Yeast Column

...... Glucose (%)			Ethanol	Flow	Detention Time
In	Out	Used	(%)	(ml/day)	(hours)
20		20	11	900	8.8
				950	8.3
				1400	5.7
				1450	5.5
				1500	5.3
26	0.2	25.8	15	750	10.6
				1300	6.1
				3550	2.2
30	12	18	10	1200	6.6
	2.5	27.5	15.5	1430	5.5

Control of Limiting Nutrient

F.C. Roesler; U.S. Patent 4,048,017; September 13, 1977; assigned to Imperial Chemical Industries Limited, England describes a method for the continuous fermentation of a culture comprising a nutrient medium, one of the constituents of which is a growth-limiting nutrient, and microorganisms capable of utilizing the medium for growth wherein the growth-limiting nutrient is supplied to the culture at sufficient positions and in such amounts at each position that substantially all the limiting nutrient supplied at each position is consumed by the microorganisms present in the culture in the vicinity of that position before another nutrient available to them in the culture in the vicinity of that position is exhausted and becomes the limiting nutrient.

The fermenter may be any type of fermenter, e.g., a tank wherein circulation of the culture is induced by mechanical stirring or, in the case of an aerobic fermentation, by blowing air thereinto. Preferably it is a fermenter such as those described in United Kingdom Patent 1,353,008, comprising a riser and a downcomer connected at their upper and lower ends and wherein circulation of culture around the system is caused by injecting an oxygen-containing gas such as air into the lower part of the riser.

This process is very suitable for use in aerobic fermentations and may be usefully employed in bacterial fermentations such as the process of United Kingdom Patent 1,370,892 for culturing methanol-utilizing strains of bacteria of the species *Pseudomonas methylotropha, Microcyclus polymorphum, Hyphomicrobium variabile* or *Pseudomonas rosea*, cultures of a number of strains which are available from the National Collection of Industrial Bacteria (NCIB), Scotland, NICB Nos. 10508-17 and 10592-612.

In aerobic fermentations the process has the main advantage that it avoids regions developing in the fermenter wherein the culture grows in effective oxygen limitation, i.e., all the oxygen available to the microorganisms has been used up in a region in which supplies of the limiting nutrient, usually the carbon source, are still available.

The limiting nutrient is suitably supplied to the fermenter at an average of at least one position per cubic meter of the effective volume of the fermenter, preferably at 3 to 6 positions per cubic meter. By the effective volume of the fermenter is meant that volume which is occupied by the culture and in which gasification of, nutrient utilization by, and growth of the microorganisms present in the culture is taking place.

In an apparatus for aerobic fermentation the effective volume for microorganism growth is that part of the volume in which, through mass transfer from gas bubbles, a positive dissolved oxygen tension (partial pressure of oxygen) can be maintained.

Figure 8.3: Apparatus for Continuous Fermentation

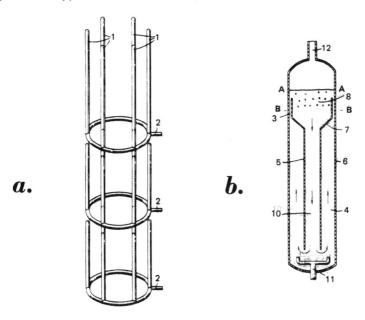

(a)   Perspective view of one arrangement.
(b)   Diagrammatic representation of one fermenter
       in which the apparatus may be installed.

Source: U.S. Patent 4,048,017

The apparatus shown in Figure 8.3a comprises six vertical medium distribution tubes **1** located at the corners of a regular hexagon and connected to three medium supply tubes **2** located at different levels along the heights of distribution tubes **1**, the lowest supply tube being connected to the lower ends of distribution tubes **1**. Each of supply tubes **2** comprises a circular portion connected to distribution tubes **1** with an outwardly extending portion which is connected to a medium supply. Distribution tubes **1** and the circular parts of supply tubes **2** are perforated by a plurality of holes. These holes occur at intervals along the entire height of the distribution tubes **1** being closer together towards the lower ends of the tubes than at the upper ends, the vertical distance between successive holes increasing up the tubes. Successive holes are not located vertically above one another along the tubes **1** but are spaced around the circumference of the tubes forming a suitable pattern, e.g., a spiral. On the circular portions of supply tubes **2** holes are located at substantially equal horizontal distances apart and may also be distributed around the vertical section of the tube, e.g., by pointing alternatively inward and outward relative to the axis of the fermenter vessel.

The fermenter shown in Figure 8.3b has a riser **6** and a downcomer **5**. In each case riser **6** is divided into two sections, one vertically above the other, the lower section or pot **4** being of greater cross-sectional area than the upper section or spout **3**. Pot **4** and spout **3** are linked by connecting piece **7**. In each case downcomer **5** has two sections, one vertically above the other, the upper section or choke **8** being of greater cross-sectional area than the lower section or sink **10**.

Spout **3** opens into choke **8** while the lower end of sink **10** communicates with pot **4**. Air is sparged into the lower part of pot **4** through spargers **11** causing culture contained in the fermenter to rise upwardly in the riser and to flow over into choke **8** and then pass into sink **10**. Culture fills each fermenter up to the level A–A, the region above level B–B in the choke being occupied by bubbly culture. It is from the region above level B–B in choke **8** that gas disengages from the culture to escape through port **12** at the upper end of each fermenter. The spout **3** and choke **8** and also the sink **10** and pot **4** are coaxially located.

During operation of the fermenter, culture medium containing the carbon source and inorganic nutrients is supplied to the culture in riser **6** through supply tube **2** and distribution tubes **1**, passing into the culture through the holes in the supply tube **2** and distribution tubes **1** at substantially the same rate as culture is removed at a point or points not shown in the drawings.

Hydroxylation of 2,6-Methano-3-Benzazocines

H.-J. Vidic and K. Kieslich; U.S. Patent 3,919,047; November 11, 1975; assigned to Schering AG, Germany describe a process whereby compounds of general Formula 1:

(1)

wherein Q' is optionally substituted benzoyl; R_1 and R_2 each are a hydrogen atom or lower alkyl, and Z' is a free, etherified, or esterified hydroxy group, are prepared by subjecting a compound of general Formula 2:

(2)

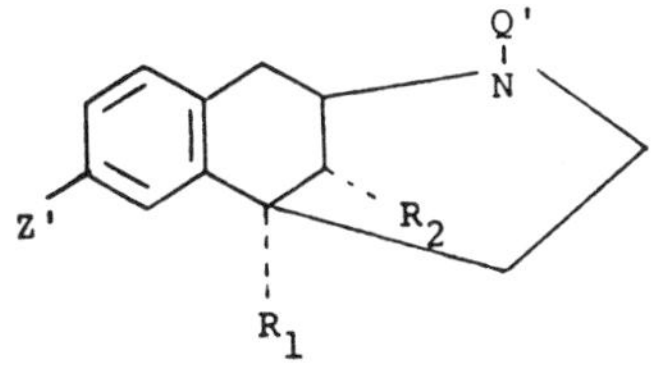

wherein Q', R_1, R_2 and Z' have the values given for Formula 1, to the hydroxylating activity of a hydroxylating microorganism of the family *Agaricaceae.*

Suitable hydroxylating microorganisms of the family *Agaricaceae* are those of the genus *Pellicularia*, especially *P. filamentosa f. sp. sasakii* (ATCC 13289). Preferred starting compounds are those wherein:

(a) R_1 and/or R_2 are lower alkyl, i.e., of 1 to 4 carbon atoms, e.g., methyl, ethyl, propyl, butyl, preferably methyl, especially those wherein either or both R_1 and R_2 are methyl;

(b) Z' is HO−, especially those of (a); and

(c) Q' is unsubstituted benzoyl, especially those of (a) and (b).

The compounds of general Formula 1 are described as central nervous system depressants, especially analgesics and analgesic antagonists, e.g., of meperidine and morphine.

The microbiological hydroxylation is carried out according to conventional methods.

It was found advantageous to employ concentrations of 50 to 1,000 mg, preferably 80 to 250 mg/l, of a conventional nutrient medium. The pH is preferably adjusted to a value in the range from 5 to 7. The culturing temperature is in the range of 20° to 40°C, preferably 25° to 35°C. For aeration purposes, approximately 1 liter of air is fed per minute and per liter of culture broth. The conversion of the substrate is suitably monitored by the analysis of sample extracts by thin-layer chromatography. In general, adequate production of hydroxylated benzazocines is achieved within 20 to 50 hr.

The products of this process are isolated and purified in a conventional manner. For example, the products can be extracted with an organic solvent, e.g., methyl isobutyl ketone, the extract can be evaporated, and the process products can be separated and purified by column chromatography.

The 2,6-methano-3-benzazocines hydroxylated in the 1-position according to the process can then be structurally modified in the usual manner. For example, the 1-hydroxy group can be oxidized to a keto group. The thus obtained keto group is reacted with alkyl magnesium halide to produce 1-alkyl-1-hydroxy compounds. The N-benzoyl side chain can be reduced to a benzyl side chain and, if desired, thereafter removed in a conventional manner. The thus obtained benzazocines with a secondary amino group are amenable to a wide variety of reactions.

Aerobic Cultivation of Microorganisms

Aerobic cultivation of microorganisms such as yeasts, bacteria, etc. must be carried out under the condition of an oxygen-containing atmosphere such as in the presence of air. In the case of microorganisms which assimilate n-paraffins, a very large quantity of oxygen has to be supplied to the cultivating system because these hydrocarbons do not contain an oxygen atom in the molecule. Therefore, the cultivation using n-paraffins needs about three times the amount of oxygen in comparison to the case where organic carbon sources containing oxygen atoms, such as carbohydrates are used for a cultivation medium.

In the conventional process employing the constant aeration condition, the amount of oxygen supplied to the system is about two times the amount that is actually needed for the microorganism during the cultivation. Therefore, this process is uneconomical in view of the operating cost and operating efficiency.

T. Iijima, Y. Odawara and T. Yamaguchi; U.S. Patent 3,912,585; October 14, 1975; assigned to Hitachi, Ltd., Japan describe a continuous process for the aerobic multibatch cultivation of a microorganism which assimilates n-paraffins in a cultivating plant comprising a plurality of cultivating batches, each batch being provided with an agitator for agitating an aqueous suspension comprising the microorganism, n-paraffins having 7 to 20 carbon atoms and water charged thereto, and being connected through controllable valve means to a gas source for supplying an oxygen-containing gas to each of the batches, which comprises the following steps.

(a) Establishing an S-shaped curve representing the relationship between a cultivation period required for performing the aerobic cultivation and the necessary amount of oxygen with respect to the microorganism.

(b) Subdividing the cultivation period into a plurality of steps including a first cultivation step and a final cultivation step to thereby establish a time schedule for the start and finish of each step of cultivation among the batches.

(c) Determining the necessary amount of oxygen at each of the steps in accordance with the S-shaped curve.

(d) Supplying successively to each of the batches the oxygen-containing gas at a constant rate and in amount corresponding to the necessary amount of oxygen at each of the steps from the gas source, with agitation at a constant speed throughout the cultivation.

(e) Regulating successively the amount of gas to be supplied to each of the batches by controlling the valve means so as to increase stepwise the rate of gas supplied from the first step to the final step in accordance with the predetermined time schedule, after initiating the supply and regulation of the gas to a first batch, the supply and regulation of the gas being effected simultaneously to at least two of the batches during the process until termination thereof.

It is well known, as shown in Figure 8.4, that almost all kinds of microorganisms grow in accordance with an S-shaped curve I. Based upon the increase of cell concentration and the state of the microorganism, the process of growth can be subdivided, as shown in Figure 8.4 into a lag phase, a log phase, a decrement phase and a stational phase. Consequently, the amount of oxygen necessary at each of the phases changes nonlinearly with time as shown by curve II in the figure. As stated above, in the conventional methods, the amount of oxygen supplied is set up to the condition that requires the maximum amount of oxygen in the cultivation, that is, the amount of oxygen at the decrement phase and stational phase, as shown in Figure 8.4, so as to avoid a shortage of oxygen throughout the whole cultivation period. However, this method is uneconomical since it always requires a supply of a large amount of oxygen.

Figure 8.4: Time for Cultivation vs Amounts of Microorganism and Oxygen Consumed

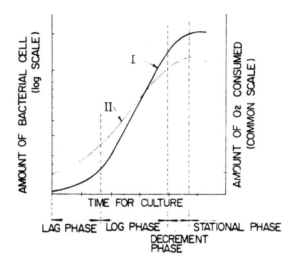

Source: U.S. Patent 3,912,585

This process for aerobic cultivation of microorganisms can be performed within a shorter time for cultivation than the conventional cultivation method. Suitable yeasts include *Torulopsis* and *Saccharomyces* and the bacteria include *Corynebacterium* and *Pseudomonas*.

Blood Control Standard

Blood serum is a complex biological fluid containing numerous components of substantial physiological importance. In the normal or average healthy person the concentrations of these components fall within certain reasonably well-defined limits. When one or more of these components is determined upon analysis to fall outside these acceptable limits, various diseases or pathological conditions of the body system are indicated.

A.L. Louderback and A.J. Fontana; U.S. Patent 3,897,363; July 29, 1975; assigned to Baxter Laboratories, Inc. describe a process relating to a blood control standard and method of preparation thereof.

A principal raw material used in making blood control standards is stored blood plasma obtained from blood donor centers and blood banks. Blood plasma is normally collected and stored in various anticoagulant materials such as, for example, sodium citrate, heparin and sodium ethylenediamine tetraacetate. Certain widely used anticoagulant materials contain, additionally, dextrose (D-glucose). ACD blood (containing citric acid, sodium citrate and dextrose) is a principal example of an anticoagulant stored blood containing elevated levels of dextrose. Another such anticoagulant stored blood is CPD blood (containing citrate, phosphate and dextrose).

Due to the extraneous addition of anticoagulant materials containing dextrose to the stored blood or blood plasma, the stored product will contain an elevated, or abnormally high, level of dextrose. Consequently, such stored blood plasma is not generally suitable for use as a raw material in the preparation of blood control standards except in the case of so-called abnormal control sera where high levels of dextrose are desired.

This process comprises selective destruction of the dextrose in stored blood by aerobic fermentation with yeast in the negative acceleration phase or in the stationary phase of the yeast growth. It is important to use these growth phases in the process, otherwise the yeast feeds on the blood proteins to make more yeast cells and the protein content of the product is undesirably reduced.

These various phases of yeast growth can be controlled or regulated by providing suitable conditions of nutrient, oxygen supply, pH, temperature and inoculant. A description of the kinetics of yeast growth and growth phases is found in *The Chemistry and Biology of Yeasts* edited by A.H. Cook, Academic Press Inc. (1958) pp 252–275. It is also important to use aerobic rather than anerobic fermentation.

In a preferred method of this process, defibrinated plasma or blood serum is incubated with the yeast for about 12 to 24 hr, preferably about 18 hr, at normal room temperature (ca 20° to 25°C).

The commercially important yeasts which can be used in this process are those such as *Saccharomyces cerevisiae, Saccharomyses cerevisiae* variety *ellipsoideus, Saccharomyces carlsbergensis, Saccharomyces fragilis,* and the Torula yeasts, e.g., *Torulopsis spherica, Torulopsis utilis (Candida utilis)* and *Candida pseudotropicalis.*

Examples of suitable commercially available yeast products are Fleishmann's Active Dry Yeast marketed by Standard Brands Inc., and Red Star Active Dry Yeast, marketed by Universal Foods Corp.

In general, from about 0.1 g to about 10 g of active dry yeast per liter of blood plasma is suitable and about 1 g/l is preferred.

Following the foregoing fermentation, the treated plasma is filtered or centrifuged to separate particulate residue and the resulting filtrate or supernant is retained for use as the base blood control standard. Sufficient dextrose can then be added back to this base control standard to provide any desired predetermined dextrose level whereby various normal or abnormal blood control standards can be prepared. For example, the dextrose level can be increased to a range of from 80 to 400 mg/100 ml.

The blood control standard prepared as above can be further treated to reduce the inorganic ion level, particularly sodium, potassium and calcium, and/or to remove the lipoprotein components as described in U.S. Patent 3,682,835.

Example: Pooled human ACD stored blood plasma (10 liters) was obtained from a blood donor center. Upon assay, the plasma was determined to contain 345 mg % (milligram per 100 ml) of glucose and had a total protein content of 6.5 g %. The blood plasma was defibrinated by reaction with 30,000 units of thrombin (Thrombin Topical, Park, Davis & Co.) followed by removal of the clotted material. Upon assay, the defibrinated plasma was determined to contain 350 mg % of glucose.

To 1 liter of the defibrinated plasma was added 1 g of Fleischmann's Active Dry Yeast. The mixture was stirred overnight (about 12 hr) on a magnetic mixer at room temperature (about 20°C). The mixture was then filtered to remove the particulate matter and the filtrate was retained as the desired base blood control standard. Upon assay, it was determined to contain 10 mg % of glucose.

The thus prepared base blood control standard was mixed with 30 g of Dowex-50 ion exchange resin in three increments of 10 g each whereby the Na^+ ion level was reduced from its original level of 163 meq/l to 97 meq/l and the K^+ ion level was reduced from its original level of 11 meq/l to 4.8 meq/l. The resin was removed after treatment with each 10 g increment by filtration through glass wool. Upon assay, the final product was determined to have a total protein content of 6.3 g %.

ε-Caprolactam-Utilizing Microorganism

At the polymer purification step in the production of polyamides, e.g., nylon 6, unreacted monomer is separated in a proportion of about 10%. Hitherto, a major portion of the monomer has been recovered and recycled, and the remainder has been abandoned with the waste liquor. Since the monomer contained in the waste liquor is at the low concentration of about 1,000 ppm, the recovery thereof is not economically feasible. However, it is desirable that the waste liquor not be abandoned to the environment in view of its BOD (biological oxygen demand), so that the effective recovery or removal of the dilute monomer is required.

A. Mimura, S. Hayakawa and T. Iguchi; U.S. Patent 3,880,740; April 29, 1975; assigned to Asahi Kasei Kogyo Kabushiki Kaisha, Japan describe a process for culturing an ε-caprolactam-utilizing strain belonging to the species *Pseudomonas lactamolyticus* (NRRL B-5749) in a medium containing ε-caprolactam as carbon and nitrogen sources. A specific example of the species is *Pseudomonas lactamolyticus* SB-1442 (NRRL No. B-5749).

Conditions utilized for cultivation of the microorganism of this process do not substantially differ from the ordinary growth conditions of other microorganisms. The microorganism is preferably grown in a medium containing ε-caprolactam at a concentration of 0.2 to 4.0 wt %. However, if the concentration of ε-caprolactam in the medium is more than 1.5%, the microorganism is inhibited in growth. In case a high concentration of ε-caprolactam is used, it is necessary to add the compound in small increments. The medium does not require a nitrogen source, but the growth of the microorganism becomes vigorous if a small amount of a nitrogen source such as ammonium sulfate, ammonium chloride, ammonia or urea which is used for ordinary fermentation is added. Other inorganic salts may be added to the medium, such as potassium phosphate, magnesium sulfate, iron sulfate, zinc sulfate and manganese sulfate. Addition of such organo-nitrogen compounds as cane molasses, corn steep liquor and yeast extract helps growth of the microorganism.

The cultivation is ordinarily carried out by means of shaking or stirring the culture under aerobic conditions at a temperature from 18° to 37°C and a pH from 5 to 9. The cultivation time varies depending on the concentration of added carbon source, and is ordinarily from 10 to 50 hr. An example of the medium used in this process is shown below where ingredients are in percent by weight.

ε-Caprolactam	1.0
Monopotassium phosphate	0.075
Magnesium sulfate	0.025
Ferrous sulfate	0.003
Zinc sulfate	0.002
Manganese sulfate	0.002
pH	7.0

Separation of microorganism cells from the cultured medium is easily effected by conventional methods, for example, centrifugation or filtration with or without an agglomerating agent.

Example: The medium shown in the above table was mixed with 2.0% of agar powder to prepare a stock medium. *Pseudomonas lactamolyticus* (NRRL No. B-5749) was inoculated in the stock medium and cultured at 30°C for 24 hr to obtain a stock slant. Subsequently, 50 ml of the medium shown in the above table was charged into a 500 ml shaking flask and then sterilized. The aforementioned stock slant was inoculated in the above medium and shaken reciprocally at 30°C for 24 hr to prepare a seed culture. Each 4% of the seed culture was inoculated into an ε-caprolactam medium, which had been prepared in the same manner as in the case of the seed culture, and then reciprocally shaken at 30°C for 24 hr. After the cultivation, the culture liquor was heat treated at 80°C for 10 min and centrifuged to separate cells, which were then washed with water and dried. The cells were obtained in a proportion of 5.2 g/l of the culture liquor. No ε-caprolactam was observed in the separated liquor.

Production of Yeast Cells

S. Abe and Y. Yokote; U.S. Patent 3,862,006; January 21, 1975; assigned to Kyowa Hakko Kogyo Kabushiki Kaisha, Japan describe a process for producing yeast cells by culturing a strain having an excellent methanol-assimilability, designated *Torulopsis methanosorbosa* KY 12001 (FERM P-1208)(ATCC 20361). In particular, this process comprises producing yeast cells with a high yield by culturing the yeast in a culture medium containing methanol as the main carbon source in the presence of at least one member selected from the group consisting of thiamine and salts thereof, p-aminobenzoic acid and salts thereof, folic acid and salts thereof, riboflavin, biotin, inositol and calcium ion.

Torulopsis methanosorbosa is superior to the known yeasts with respect to the yield of cells to methanol and the specific growth rate. Furthermore, the maximum growth temperature and optimum growth temperature of this yeast are 43°C and 37°C, respectively. Accordingly the yeast is characterized by a higher growth temperature when compared with usual conventional yeasts.

In the culturing of the microorganism, the cooling of fermentation heat is an important problem regarding cooling water and cooling apparatus, etc. Higher culturing temperature is advantageous for promoting a cooling effect since the difference between culturing temperature and temperature of the cooling water becomes larger, and furthermore, the fermentation is accelerated at a higher temperature.

Culturing methods as used with conventional yeasts may in general be applied to this culturing process. Liquid culture, especially submerged culture with agitation is most suitable. It is preferred to carry out the cultivation at a temperature of 20° to 40°C, preferably 30° to 40°C at a pH of 4 to 10.

The concentration of added vitamins and calcium-containing compounds may vary, depending upon the additive. For example, the concentrations in case of biotin and folic acid range preferably from 10 to 100 γ/l and the concentration in case of thiamine hydrochloride is preferably from 50 to 1,000 γ/l; the concentration in the case of p-aminobenzoic acid, folic acid, riboflavin and inositol is preferably from 100 to 1,000 γ/l, and the preferable concentration of calcium chloride, calcium phosphate, etc., is above 10^{-5} M.

Microorganism Immobilization in a Hydrophilic Gel

T. Hino, H. Yamada and S. Okamura; U.S. Patent 4,148,689; April 10, 1979; assigned to Sanraku-Ocean Co., Ltd., Japan describe a process for preparing a hydrophilic complex gel by mixing a water-soluble polymer and an organic silicate, particularly, tetraalkoxysilane, and gelling under mild conditions, whereby is obtained a hydrophilic complex gel having no harmful effect on the activity of biological substances such as microbial cells or enzymes and useful for the immobilization of microbial cells possessing enzyme activity and for use as a pharmaceutical material.

The water-soluble polymer compounds employed in this process have many polar groups such as –OH group, –COOH group, –NH$_2$ group, =NH group and so on, which form strong hydrogen bonds with acidic –OH groups of silicate by hydrolysis. The water-soluble polymer compounds shown below can be used.

  (a) Natural polymer compounds and their derivatives:
       (1) Celluloses; carboxymethylcellulose, methyl-
            cellulose, ethylcellulose, hydroxyethylcellulose;
       (2) Starches; hydroxyethyl starch, carboxymethyl
            starch;
       (3) Other polysaccharides; mannan, dextran, chitosan,
            pullulan, guar gum, locust bean gum, tragacanth,
            xanthan gum, agar, sodium arginate; and
       (4) Proteins; gelatin, albumin.
  (b) Synthetic polymer compounds: Polyvinyl alcohol, polyethyl-
       ene glycol, polyethyleneimine.

The preferred water-soluble polymers may be a polyvinyl alcohol that has an average polymerization degree between 500 and 2,000 and a saponification degree within the range of 70 to 100%; a commercial gelatin of edible grade that has a gel strength of more than 200 g shot per 5 sec in a Bloom Gelometer; or a carboxymethyl cellulose of commercial edible grade that has a carboxylation degree between 0.4 and 0.8 and a sodium content within the range of 7.0 to 8.5%.

The water-soluble polymer compound is solubilized in water at a concentration having a viscosity below 10,000 cp, preferably below 3,000 cp for full agitation. The abovementioned viscosity is measured by No. 3 rotor of a B type viscosity meter at 40°C.

The tetraalkoxysilanes which may be used are those of the structural formula: $Si(OR)_4$ in which R is alkyl group of up to 12 carbon atoms, e.g., methyl, ethyl, propyl, butyl, octyl and lauryl groups. Compounds in which R represents an alkyl group of up to 3 carbon atoms are preferably utilized for the effective formation of the homogenous complex solution; tetraethoxysilane is the most suitable compound.

The water-soluble polymer compound is mixed with the abovementioned tetra-alkoxysilane, and then the pH of the mixture is adjusted below 3 with acid or acidic salt which exhibits no harmful effect on the enzymatic activities of microbial cells. The acids or the acidic salts which may be used are shown below.

    (a)   Inorganic acids: HCl, HNO_3, H_3PO_4, H_2SO_4;

    (b)   Organic acids: Acetic acid, glutamic acid, lactic acid, maleic acid, succinic acid, ascorbic acid, citric acid, tartaric acid; and

    (c)   Inorganic or organic acidic salt: $AlCl_3$, citrate-1-ammonium.

The amount of SiO_2 contained in the tetraalkoxysilane, calculated based on SiO_2 formed by acid-hydrolysis, should be 5 to 300% (w/w) of the dry weight of the water-soluble polymer. The preferable amount of SiO_2 is between 50 and 200%.

Microbial cells possessing enzymatic activities can be immobilized under quite mild conditions by entrapping them inside the gel matrix of the abovementioned hydrophilic complex gel produced from the water-soluble polymer compound and silicate. The microbial cells are added into the abovementioned homogenous sol without any pH adjustment or basic salts.

The pH of the mixture of the complex gel and the microbial cells should be chosen in consideration of the enzyme stability or the properties of the formed gel. Usually, it is adjusted to pH 4 to 8, preferably 5 to 7.

The dry weight of the microbial cells to be added is less than 1,000% (w/w), preferably 20 to 500% (w/w), based on the dry weight of the homogeneous complex sol. After adding the microbial cells to the homogeneous complex sol, the mixture is stirred well to disperse them homogeneously. The gelling reaction of the sol containing the microbial cells takes place at any temperature. However, as the enzymes are unstable at the higher temperature, the gelling reaction is carried out at a temperature between 0° and 70°C, preferably between 10° and 40°C. The gelling reaction is usually completed within 10 to 30 min. For example, the sol is fully gelled at pH 6.0 by continuous stirring at room temperature for 10 to 20 min.

One of the most important advantages of this process is the mild conditions for the immobilization, under which the enzymatic activities can be maintained throughout the whole process without remarkable inactivation.

The formed complex gel entrapping microbial cells is dried and converted to the desired shape. The drying process is carried out below 75°C.

Example: 100 parts of 10% PVA (polyvinyl alcohol) aqueous solution (polymerization degree: 1,700 and saponification degree: 99.5) were mixed with 231.5 parts of distilled water, 28.5 parts of tetraethoxysilane and 1 part of 1 N HCl, and stirred for more than 2 hr at room temperature to give a transparently homogenized sol at about pH 3, containing 5% of solid part (abbreviated as PVA-SiO$_2$ complex sol herein). One part of commercially available dried baker's yeast (*Saccharomyces cerevisiae*) was suspended in 2 parts of water and was then mixed with PVA-SiO$_2$ complex sol. After the yeast was dispersed homogeneously by stirring, the pH was adjusted to 7.0 with 1 N NH$_4$OH solution.

The gel was poured into a petri dish and dried spontaneously at room temperature. A yellowish brown film containing the yeast cells was obtained. A strip of the film immobilizing 1 g of the yeast cells was taken out and incubated in 50 g of 5% glucose aqueous solution at 30°C to carry out the fermentation test. A large amount of gas production began in a few minutes and in 30 min a large amount of gas production was observed from the whole surface of the film. The film swelled a little and decolorized, but the shape showed no change, and the bubbling continued for 30 hr.

The reaction was estimated by the measurement of the weight decrease of the medium by liberation of CO$_2$ gas. The reaction was carried out in proportion to the reaction time. A slight fragrant smell peculiar to alcoholic fermentation was detected. Increase in the turbidity of glucose solution, which is considered to come from the leaking yeast cells from the film, was hardly observed, and hence, the mixture was almost transparent. As a control, the same reaction was conducted by using 1 g of dried yeast cells not immobilized. At the very initial stage of reaction, a slightly large amount of CO$_2$ was generated by using the control cells rather than the immobilized cells but, after 30 hr incubation, the rate of CO$_2$ gas liberation from both media became almost equal.

Accordingly, this fact indicated that the unit activity in both cases, native and immobilized cells, was almost comparable to each other, and hence, inactivation of enzymes in the immobilized cells did not occur during the immobilizing process of yeast cells. As a matter of course, since the nonimmobilized cells were suspended in the medium, mild agitation was required for securing the progress of reaction. From the above results, it was demonstrated that the microbial cells were firmly immobilized by the process while maintaining fermentation activity equal to that of the native cells.

Hydrous Metal Oxide-Chelated Polyhydroxy Support

S.A. Barker, R.N. Greenshields, J.D. Humphreys and J.F. Kennedy; U.S. Patent 4,155,813; May 22, 1979; assigned to Gist-Brocades N.V., Netherlands describe a fermentation process which comprises continuously cultivating a bacterial strain, preferably in a vessel of elongated shape, in a medium containing the substrate to be fermented in the presence of a hydrous oxide of titanium, vanadium, zirconium, iron or tin and a chelated polyhydroxy support.

The hydrous metal oxide-chelated polyhydroxy support is prepared by contacting a hydrolyzable compound of titanium, vanadium, zirconium, iron or tin, optionally in solution, with the polyhydroxy support, optionally drying the mixture obtained and subsequently washing the mixture with water. This may be done, for example, by mixing cellulose powder and titanium tetrachloride in solution, drying the mixture and washing the dried mixture with distilled water until the washings are neutral. The hydrous metal oxide-chelated polyhydroxy support thus obtained should not be heat-dried prior to use.

In any of the fermentations of this process, the substrate giving rise to an oxidation product need not be the immediate precursor of the oxidation product, but could be a metabolic precursor several steps removed from the oxidation product, e.g., glucose or sucrose for lactic acid production. Further, although the fermentations are preferably those taking place in the presence of a gas containing oxygen such as air, the oxidation product need not arise by actual reaction of the substrate or its metabolite with oxygen, but can arise by some other oxidation process. Furthermore, this process can be used in nonoxidative fermentations such as the fermentation of glucose to, e.g., butylene glycol.

The addition of hydrous metal oxide or hydrous metal-polyhydroxy support confers further advantages on the fermentation, for example, (a) it confers greater resistance to disturbance by frothing so that, where desired, greater aeration rates can be employed. Where oxygen is a direct participant in the reaction as in the conversion of ethanol to acetic acid, this is particularly advantageous; (b) better consistency of the production can be attained in the presence of the additive since in addition to (a) it becomes less sensitive to washout.

Example: *Production of Malt Vinegar by a Nonaggregating Strain of Acetobacter in the Presence of Titanium-Chelated Cellulose* — A glass tower fermenter of a cylindrical shape was used which had a high aspect ratio (height to diameter ratio) containing microorganisms suspended in the medium and through which there is a unidirectional flow of medium and/or gases. The fermenter used in this experiment had a volume of 2.6 liters, a height of 57 cm and an average diameter of 7.32 cm so that its aspect ratio was 7.8:1. In operating the fermenter, compressed air was passed through a pressure regulator and filter, then through a flowmeter to measure its flow rate and finally into the fermenter medium via a sintered glass disc fused into the base of the fermenter.

The effluent air from the fermenter was passed through two water-cooled (4°C) condensers to ensure that any volatile components in it were returned to the bulk of the liquid. The temperature of the fermenter was maintained at 32°C by a water jacket surrounding the fermenter which was fed by thermostatically controlled water. The rate of medium flow in the fermenter was controlled by a variable speed peristaltic pump and the flow rate through the fermenter was measured by collecting the vinegar produced over a period of approximately 24 hr and expressed as volumetric efficiency (VE) defined as:

$$\frac{\text{flow rate per day}}{\text{fermenter volume}}$$

The medium used in the experiments is known commercially as charging wort and is a dilute aqueous ethanol solution containing about 5% w/v of ethanol and is produced by a process similar to that used for beer and spirit fermentation from malted barley wort. In this experiment, a culture of a nonaggregating strain of *Acetobacter* was used which was obtained from a tower fermenter of a commercial malt vinegar manufacture and was known not to aggregate or produce extracellular polysaccharide material.

At the beginning of each experiment, the fermenter was half-filled with charging wort, an inoculum (50 to 100 ml) of the culture was added and aeration was commenced at a low rate (0.1 to 0.2 v/v/m = volumes air/fermenter volume/ minute). Samples were taken at intervals and tested for acetic acid and ethanol content. When the acidity reached about 4% w/v, the peristaltic pump was turned on and the medium was metered into the fermenter. The fermentation was then allowed to proceed continuously with the medium flow rate and aeration rate being manipulated to maximize the acetic acid content of the fermenter (7% w/v approximately) and minimize ethanol content (less than 0.57%). Measurements of pH, optical density (600 nm) and titanium content (if present) were also made.

The fermentation was started without addition of the titanium hydroxide or the chelated titanium-cellulose complex. Addition of hydrous titanium hydroxide was started only after the fermenter operated continuously at optimal conditions of medium flow rate and aeration for several consecutive days. Some typical results are given below.

Number of Consecutive Days	Average Aeration (v/v/m)	Hydrous Titanium Oxide .... Added Daily		Performance
		Ti	Cellulose	
		.... (g)........		
2	0.58	—	—	0.95
2	0.58	0.26	—	1.00
2	0.57	0.53	—	1.08
2	0.63	0.53	1.2	1.47
2	0.79	0.53	1.2	1.44

The table shows that the addition of hydrous titanium oxide, either from 0.26 g Ti or from 0.53 g Ti does hardly improve the performance of the tower fermenter, but that a marked increase of the performance is achieved after the addition of the hydrous titanium-cellulose chelate (expressed in the table in terms of its equivalent hydrous titanium oxide and cellulose content).

Mercury Extraction

Mercury poisoning results when mercury salts from industrial effluents deposit in river or lake sediments and are then acted upon by anaerobic bacteria. These bacteria convert mercury salts to monoethyl and dimethyl mercury. These methylated mercury derivatives, particularly the monomethyl mercury become stored in the bodies of fish and later, consumption of the flesh of such fish leads to acute mercury poisoning. Chemical methods of selectively binding mercury from industrial effluents have the major disadvantage of being nonspecific for mercury. Thus, the degree of mercury binding is reduced in the presence of a large excess of divalent metals.

A.M. Chakrabarty, D.A. Friello and J.R. Mylroie; U.S. Patent 3,923,597; December 2, 1975; assigned to General Electric Company describe a process employing genetically-engineered *Pseudomonas* microorganisms, which bind mercury through elaboration of mercury-binding protein, the coding for which is specified by a transmissible plasmid.

The terminology of microbial genetics is sufficiently complicated that certain definitions will be particularly useful in the understanding of this process.

(a)   Plasmid is a hereditary unit that is physically separate from the chromosome of the cell; the terms extra chromosomal element and plasmid are synonymous; when physically separated from the chromosome, some plasmids can be transmitted at high frequency to other cells;

(b)   Transmissible plasmid is a plasmid that carries genetic determinations for its own intercell transfer via conjugation;

(c)   DNA is deoxyribonucleic acid;

(d)   Conjugation is the process by which a bacterium establishes cellular contact with another bacterium and the transfer of genetic materials occurs; and

(e)   (Plasmid)$^+$ indicates that the cells contain the designated plasmid.

Plasmids are believed to consist of double-stranded DNA molecules. The genetic organization of a plasmid is believed to include at least one replication site and a maintenance site for attachment thereof to a structural component of the host cell. Generally, plasmids are not essential for cell viability.

Plasmids may be compatible (i.e., they can reside stably in the same host cell) or incompatible (i.e., they are unable to reside stably in a single cell). Among the known plasmids, for example, are sex factor plasmids and drug-resistance plasmids.

The symbol [OCT] signifies the plasmid aggregate of OCT, factor K and MER. The MER plasmid, in the naturally-occurring [OCT] plasmid, is responsible for conferring resistance to mercury ions.

A culture of microorganisms possessing the requisite MER plasmid is on deposit with the U.S. Department of Agriculture. This culture is identified as follows: *P. putida* AC 28 MER$^+$ (NRRL B-8042)—derived from wild-type *Pseudomonas P. putida* strain PpGI (ATCC No. 17453) by genetic transfer thereto of a mercury resistant plasmid from *Pseudomonas oleovorans* (ATCC No. 17633).

In a batch process the liquid and the requisite concentration of bacteria would be mixed and permitted to interact. After an appropriate period of time, the liquid would be filtered off, the cells heated to the 450° to 500°C range to vaporize most of the mercury and incinerate the cells and then the gases and vapors cooled to recover metallic mercury.

In a continuous process the stream to be treated is directed through a bed of these *Pseudomonas* bacteria, which quickly and effectively take up and store the mercury content. After a given period of time, the microorganism bed is removed from the system, heated to a temperature in the 450° to 500°C range whereby the cell structure is destroyed and most of the mercury is removed by vaporization. The mercury vapors are then condensed and the mercury is recovered as metallic mercury.

The best mode contemplated employs the MER$^+$ *Pseudomonas* culture NRRL B-8042. Thus, for an aqueous system containing about 25 μg of mercury per milliliter of liquid a cell concentration and time of contact would be about 10^{10} cells per milliliter and about 60 min, respectively. A batch process is preferred wherein the requisite amount of centrifuged cells is mixed with the liquid to be treated. After the treatment period, the treated (mercury-depleted) liquid is separated from the cell mass. The cell mass is then heated to a temperature in the 450° to 500°C range. At this temperature most of the mercury is evaporated as metallic mercury (some mercury remains as mercuric oxide) and the cell mass is converted to ash, H_2O and CO_2.

The gases and vapors from this incineration step are conducted to a condenser immersed in a bath, e.g., acetone and solid CO_2. Therein the mercury is condensed as a mirror deposit for separation and recovery, the water freezes and the carbon dioxide will be vented from the system.

Silver Recovery by Photosynthetic Sulfur Bacteria

M. Kitajima; U.S. Patent 4,135,976; January 23, 1979; assigned to Fuji Photo Film Co., Ltd., Japan describes a method of purifying a waste photographic processing solution containing sulfur compounds which comprises applying photosynthetic sulfur bacteria to the waste photographic processing solution under irradiation of light and anaerobic conditions.

Also described is a process of effectively purifying waste-processing solutions or effluents used for processing silver halide photographic materials, in particular, waste fix solutions or a mixture of a waste fix solution and another waste photographic processing solution or solutions and the recovery of silver from these waste solutions by treating the waste solutions with photosynthetic sulfur bacteria.

The process can be applied most successfully when photosynthetic bacteria can utilize thiosulfate. By utilizing these photosynthetic sulfur bacteria, a waste fix solution which contains a high concentration of thiosulfate can be treated directly. Among the photosynthetic sulfur bacteria, the following are suitable for this purpose since they can metabolize thiosulfate as well as sulfide.

Of the following photosynthetic sulfur bacteria, *Chromatium vinosum* is particularly suitable since the bacterium can grow under various environmental conditions and compositions in respect to temperature, pH, salts, organic materials, etc., it weeds out other undesirable bacteria if they enter the system, and it captures thiosulfate ions well.

    (a)   Family *Chromatiaceae:*
- (1) The following species of genus *Chromatium:*
  Chromatium violacens ATCC 17096
  Chromatium vinosum ATCC 17899
  Chromatium gracile, and
  Chromatium minutissimum
- (2) The following species of genus *Thiocystis:*
  Thiocystis violacea SMG 207
- (3) The following species of genus *Thiocapsa:*
  Thiocapsa pfennigii
- (4) The following species of genus *Amoebobacter:*
  Amoebobacter rosens and
  Amoebobacter pendens SMG 236
- (5) The following species of genus *Ectothiorhodospira:*
  Ectothiorhodospira mobils SMG 237
  Ectothiorhodospira shaposhnikovii SMG 243 and
  Ectothiorhodospira halophila SMG 244

    (b)   Family *Chlorobiaceae:*
- (1) The following species of genus *Chlorobium:*
  Chlorobium limicola forma *sp. thiosulfatophilum*
  SMG 249 and
  Chlorobium vibrioforme forma *sp. thiosulfatophilum*
  SMG 265
- (2) The following species of genus *Chloropseudomonas:*
  Chloropseudomonas ethylica

The energy density of the light used for radiation is above about 1 μW/cm^2, preferably from 100 μW/cm^2 to 1 W/cm^2. The process may be carried out in practice by a batch system or using a continuous reaction system. In order to grow the photosynthetic sulfur bacteria well, anaerobic conditions must be used but the extent of oxygen removal depends upon the kind of bacteria and the cultivation conditions. For example, *Chromatium vinosum* has a comparatively high oxygen resistance and the existence of a small amount of oxygen in the system does not hinder the growth of the bacteria when this strain is employed as the photosynthetic sulfur bacteria.

In either the batch system or the continuous system, the removal efficiency of silver from a waste fix solution containing the silver complex salt depends largely on the number of cells of bacteria, the concentration of the silver complex salt, and the treatment conditions. However, by returning the waste solution initially subjected to the silver removal treatment by the photosynthetic sulfur bacteria to the reaction tank again or by adding fresh photosynthetic sulfur bacteria inoculated separately to the waste solution, silver in the waste fix solution can be concentrated and recovered at a yield of substantially 100%.

Example 1: 1 liter of a culture solution prepared by adding to Standard Solution (1) having the composition shown below, 2 g of sodium thiosulfate and 1 g of sodium sulfide was inoculated with purple photosynthetic sulfur bacteria, *Chromatium vinosum,* in a transparent glass bottle with a stopper and was continuously irradiated with light from a tungsten incandescent lamp at a mean light intensity of about 2 mW/cm^2 in a water bath maintained at 30°C.

Standard Solution (1) is composed of the following ingredients.

NaCl	30 g
NH_4Cl	1.0 g
KH_2PO_4	0.5 g
K_2HPO_4	0.5 g
$MgCl_2 \cdot 6H_2O$	0.5 g
$CaCl_2$	0.05 g
$FeCl_3 \cdot 6H_2O$	0.005 g
$NaHCO_3$	2.0 g
City water to make	1 l

After 3 days of light irradiation, the concentration of the cells of the bacteria reached a stationary state. The optical density of the culture suspension thus formed was 0.9 when checked at 373 nm (herein this optical density is referred to as OD_{373}). The cells of the bacteria were recovered from the suspension by centrifugal separation for 10 min at 5,000 g and was redispersed in 100 ml of Standard Solution (1) (not containing sodium thiosulfate and sodium sulfate) to provide a concentrated cell suspension.

Example 2: A solution prepared by adding 2.0 g of $Na_2S_2O_3 \cdot 5H_2O$ to 245 ml of Standard Solution (1) as described in Example 1 was inoculated with 5 ml of the concentrated cell suspension prepared in the same manner as in Example 1 in a 250 ml transparent narrow-mouthed reagent bottle to provide a solution having an OD_{373} of 0.25. The solution thus prepared was irradiated with light from a tungsten incandescent lamp at a light intensity of 1.1 mW/cm² for 3 days at room temperature (about 20° to 30°C), subjected then to centrifugal separation for 10 min at 5,000 g.

The concentration of $S_2O_3^{-2}$ in the supernatant liquid was determined by a titration with a 0.05 N $KI-I_2$ solution. As a control sample, a solution prepared by diluting and treating at the same conditions as above but without light irradiation was subjected to centrifugal separation and the concentration of $S_2O_3^{-2}$ in the supernatant liquid was also measured. The result showed that the concentration in the control sample was 6,700 ppm as $Na_2S_2O_3$, while that of the solution subjected to the treatment of this process was 530 ppm. Thus, the above result shows that about 92% of thiosulfate ions were oxidized by the bacterial treatment of this process.

Example 3: A 0.4 M solution of a silver-thiosulfate complex salt $[Na_3Ag(S_2O_3)_2]$ was prepared by dissolving silver chloride in an aqueous solution of sodium thiosulfate in the dark. In a 120 ml transparent narrow-mouthed bottle were placed 25 ml of Standard Solution (1) as described in Example 1 and 90 ml of the 0.4 M solution of the silver-thiosulfate complex salt prepared above and after adding 5 ml of concentrated cell suspension prepared in the manner as in Example 1 to the mixture in the bottle, cultivation was performed for 3 days at room temperature under irradiation of light from a tungsten lamp at an intensity of 1.1 mW/cm².

By the treatment, the cells of the bacteria largely blackened and precipitated. The precipitate was recovered, redispersed in 300 ml of water, and the suspension was subjected to centrifugal separation for 10 min at 4,000 g followed by washing. After repeating the same centrifugal separation and washing twice, the precipitate thus recovered was dried, weighed, and the total amount thereof was dissolved in concentrated nitric acid. After diluting the solution, the amount of silver in the solution thus prepared was measured using atomic absorption analysis. The amount of silver in the cells of the bacteria cultivated in the solution was 15% as a dry weight ratio of the weight of the cells.

Silver Recovery by Chemosynthetic Sulfur Bacteria

In a similar process, *M. Kitajima and A. Abe; U.S. Patent 4,155,810; May 22, 1979; assigned to Fuji Photo Film Co., Ltd., Japan* describe silver recovery from photographic waste using chemosynthetic sulfur bacteria.

Chemosynthetic sulfur bacteria which are suitable for use are those whose energy source arises from the oxidation of reduced or partially reduced sulfur compounds and are of the genera: *Thiobacillus, Sulfolobus, Thiobacterium, Macromonas, Thiovulum and Thiospira* of which the genus *Thiobacillus* is most suitable for use in this process. Specific examples include the following which are classified as *Thiobacilli.*

> (a) *Thiobacillus thioparus*
> (b) *Thiobacillus novellus*
> (c) *Thiobacillus thiooxidans*
> (d) *Thiobacillus thiocyanooxidans*
> (e) *Thiobacillus ferrooxidans*
> (f) *Thiobacillus denitrificans*

The temperature at which the biological activities of the chemosynthetic sulfur bacteria can occur ranged from 0° to 70°C depending on the bacterial species. The method is more effective at a temperature between 5° and 60°C, more preferably between 10° and 55°C and most preferably between 20° and 50°C. Outside these temperature ranges, the life of the bacteria is only sustained or a very low rate of conversion of the sulfur compounds by organisms tends to occur, thus making this method practically ineffective.

ENZYMES AND ENZYMATIC PROCESSES

ENZYMES

α-Amylase Active in Alkaline Conditions

Various amylases have been isolated from animal and plant bodies since Kohn and Ohlsson's work of the nineteen twenties, and some of the amylases are used industrially. All known amylases are effective in acidic or neutral media, and an amylase effective in an alkaline medium has never been known.

K. Mitsugi, N. Kobayashi, T. Shida and Y. Yokokawa; U.S. Patent 4,022,666; May 10, 1977; assigned to Ajinomoto Company, Incorporated, Japan have found that α-amylases effective in-neutral and alkaline media can be produced by culturing bacteria belonging to the genus *Bacillus* on a nutrient medium under aerobic conditions.

Amylases which can be obtained according to the process include neutral α-amylases of optimum amylolytic activity at 40°C at a pH between 6.5 and 8, and alkaline α-amylases whose optimum pH is between 9.0 and 11.5 at 40°C.

Bacteria which produce these types of amylases include *Bacillus subtilis* AJ-3249, *Bacillus subtilis* AJ-3250 (FERM P-374), *Bacillus subtilis* AJ-3252 (FERM P-375), *Bacillus subtilis* AJ-3255 (FERM P-376), *Bacillus subtilis* AJ-3298 (FERM P-660) and *Bacillus subtilis* AJ-3299 (FERM P-661).

The bacteria are cultured on a medium containing an assimilable carbon source, an assimilable nitrogen source, inorganic salts and organic nutrients. The assimilable carbon sources include carbohydrates such as glucose, starch, dextrin or maltose, and organic acids such as acetic acid. Assimilable nitrogen sources are inorganic ammonium salts such as ammonium chloride, ammonium sulfate or ammonium nitrate, and organic nitrogen compounds such as soybean cake, soybean powder, milk casein, peptone, milk whey, meat extracts and amino acids.

The cultivation is carried out at a temperature between 25° and 45°C, preferably between 30° and 40°C, under aerobic conditions by shaking, stirring and/or aerating The best pH of the culture medium varies with the strain used. *Bacillus subtilis* AJ-3249, AJ-3250 (FERM P-374) and AJ-3252 (FERM P-375) are cultured at pH 6.5 to 8.0, *Bacillus subtilis* AJ-3255 (FERM P-376) is cultured at a pH of 8.0 to 10.0, and *Bacillus subtilis* AJ-3298 (FERM P-660) and AJ-3299 (FERM P-661) are cultured at a pH between 9.0 and 10.5.

In order to isolate the neutral and/or alkaline amylase produced in the culture medium, the bacterial cells are removed by centrifuging or filtration, the amylase is precipitated by adding inorganic salts, such as ammonium sulfate or sodium sulfate, and/or organic solvents such as ethanol or acetone. The precipitate can be recovered by centrifuging or filtration. Ion exchange resins may also be used for purification and isolation.

Amylases produced according to the process are useful for many purposes. The crude amylases are useful in protease-containing laundry compositions. They are also useful for decomposing insoluble starch waste discharged from cardboard factories and consisting of alkaline paste.

Example: A culture medium consisting of 1 g/dl of soybean cake extracts (as dry matter), 1 g/dl polypeptone and 8 g/dl soluble starch, of pH 7.0 was prepared; each 50 ml batch of the medium was placed in a 500 ml shaking flask, and sterilized at 120°C for 20 minutes. The medium was inoculated with a strain of *Bacillus subtilis* AJ-3250 (FERM P-374) which had previously been cultured on an agar slant consisting of 2 g/dl soluble starch, 1 g/dl yeast extracts, 1 g/dl polypeptone and 0.5 g/dl NaCl at 34°C for 24 hours, and cultured at 34°C for 96 hours with shaking.

The culture broth was found to contain 2,500 units/ml amylase at pH 7.0. 930 ml of the culture broth collected from 20 flasks was centrifuged to remove bacterial cells, 365 g solid ammonium sulfate was added to the 850 ml of supernatant liquid, and the solution was left to stand overnight with cooling. Amylase precipitated and was collected by centrifuging at 10,000 rpm for 20 minutes, and dried in vacuo overnight with cooling, a crude enzyme powder containing 360,000 units/g of amylase was obtained in an amount of 51.0 g (isolation yield: 87%).

Lipase and Pyocyanine from n-Paraffins

L.J. Gawel and C-S. Chen; U.S. Patent 4,019,959; April 26, 1977; assigned to Continental Oil Company have found that lipase can be produced by hydrocarbon-oxidizing bacterial cultures in various media that contain substrates such as olive oil or n-paraffins. By rupturing the bacterial cells produced, the amount of intracellular lipase recovery can be significantly increased. Since the enzyme shows good stability in relatively adverse temperature and pH environments, it has good potential for application in industrial uses. Pyocyanine pigments are also obtained during the process.

Although many hydrocarbon substrates can be utilized, substrates such as olive oil and the n-paraffins are preferred. Most preferred are the n-paraffins containing from 10 to 22 carbon atoms.

In the process, the bacterial culture is allowed to grow on the substrate at a temperature of from 20 to 50°C for a period of time ranging from 10 to 100 hours. At the end of the growth cycle, the cells are isolated from the growth medium, the cell walls are ruptured using techniques known in the art, and the lipase and pyocyanine are isolated using techniques well known to those skilled in this art.

The bacterial culture of the process was tentatively identified as a *Pseudomonas* species. The microorganism has been duly deposited with the United States Department of Agriculture and has been assigned designation NRRL B-8110.

α-Galactosidase

α-Galactosidase has heretofore been known to be an enzyme capable of hydrolyzing raffinose which is a substance occurring in beet molasses and tending to obstruct crystallization of sucrose. In the course of beet sugar production, therefore, α-galactosidase is added to the beet molasses including syrup or juice so that the enzyme acts on and hydrolyzes the raffinose contained in the beet molasses into sucrose and galactose and consequently enhances the overall yield of sucrose.

S. Narita, H. Naganishi, C. Izumi, A. Yokouchi and M. Yamada; U.S. Patent 3,867,256; February 18, 1975; assigned to Hokkaido Sugar Co., Ltd., Japan describe a method for the production of an enzyme having high α-galactosidase activity and weak invertase activity by the use of a microorganism belonging to the genus of *Circinelle*, i.e., *Circinella muscae* (Berlese et de Toni) nova typica coreanus (ATCC 20394), the mold of the process.

The culturing conditions under which the production of α-galactosidase within the mycelia of the mold of the genus *Circinella* is effectively induced and the growth of the mold promoted are similar to those employed generally for the aerobic culture of molds. To be specific, a medium is obtained by adding such inducers as lactose, raffinose, melibiose, etc. to a basal substrate formed of carbon sources, nitrogen sources and inorganic salts. Then, the mold of the genus *Circinella* is inoculated to the resulting medium and subjected to shaken culture or aerobic culture for 40 to 72 hours, with the pH value kept in the range between 5 and 8 and the temperature kept at about 30°C.

When the culture is carried out as described above, α-galactosidase is produced in a very high yield within the mycelia. Since the mycelia are in the shape of pellets, they can readily be separated from the culture liquid by filtration. The separated mycelia are washed with water, treated by any known methods such as centrifugation for removal of water, and preserved.

The pellet shaped mycelia thus obtained are packed in a column type reaction tank or horizontal reaction tank. When beet molasses is passed through the tank, the enzyme in the mycelia acts on the raffinose contained in the molasses and hydrolyzes it into sucrose and galactose. In the method of this process, the α-galactosidase activity per weight of the enzyme is by far higher than is obtainable with any of the heretofore known molds. Therefore, this method can provide effective treatment for beet molasses of a higher concentration or a larger amount than by conventional methods.

OTHER PROCESSES

Oxidation of Hydrocarbons

R.I. Leavitt; U.S. Patent 3,880,739; April 29, 1975; assigned to Mobil Oil Corp., describes a process which comprises conducting a highly selective oxidation of hydrophobic paraffins or alcohols with great specificity to aldehydes of corresponding chain lengths by use of an aqueous suspension of a crude enzyme mixture suspended or otherwise dispersed in a continuous hydrocarbon phase.

It should be noted that the enzyme referred to herein as catalyzing the oxidation reactions set forth is not necessarily a single enzyme nor is it necessarily an isolated species. As a practical matter, a given microorganism is grown in the usual way for that particular organism, e.g., on a hydrocarbon substrate in conventional fermentation equipment. The particular organism grown is selected based upon a knowledge, from the literature or otherwise, that one or more of the myriad enzymes contained therein have the ability to catalyze the reaction which is of interest.

In any case, an appropriately chosen microorganism is conventionally grown and cells thereof are suitably harvested in a known manner. These cells are washed with water and suspended in a liquid medium suitably buffered to a pH which is appropriate for the particular organism involved. Three alternative procedures are then available for use of the contained enzymes of these cells: the cells can be ruptured, the enzymes extracted therefrom in a suitable solubilizer such as water and the aqueous solution of enzymes separated from the remaining cell fragments and used in the process; the cells can be ruptured, solubilizer added and the entire mass, unresolved, used in the process; or the whole cells, undisrupted, can be fed into the process suspended in a suitable, perhaps aqueous, medium and used therein in an in situ ruptured or unruptured form.

In any case, it has been found that the activity and the stability of the enzymes so produced is markedly improved by extracting the enzymes from their cells in a solubilizing medium which contains the substrate whose reaction will later be catalyzed by the extracted enzyme or by rupturing the cells in the reaction medium. It has been discovered that only a small proportion of reactant is necessarily present in order to achieve this improvement. Proportions of about 1 volume percent, based on the amount of solubilizer, have been found to result in improved enzyme catalytic activity and stability.

Preferred proportions are about 0.1 to 10 volume percent on the same basis. In particular it has been found that extracting the enzyme content of Pichia yeast cells, which have been grown on an n-tetradecane substrate, with a solubilizer, notably water buffered to a pH of about 8.0, containing a small proportion of n-tetradecane produces an aqueous enzyme product which is about twice as active as the same enzyme product produced in the same way except that no n-tetradecane was used in the extraction step.

One of the problems found by workers in the enzyme catalysis field is the difficulty of maintaining the enzyme in a predefined reaction zone. It is recently beginning to be understood how to make a fixed bed catalyst of enzymes. One of the advantages of this process is that it is possible to effectively encapsulate enzymes and keep them in a relatively fixed reaction zone without attaching them to a solid support. While much work has been and is being done in this field of enzyme encapsulation, most of this work has approached the problem from the point of view of fixing the enzyme or enzymes to a solid substrate in some manner so as to fix their location without denaturing them.

This process has taken a slightly different approach to this problem. According to a preferred aspect the enzymes are effectively encapsulated by dissolving or solubilizing them in an aqueous medium which is emulsified in a continuous oil system. Thus, each droplet of water constitutes an encasement or capsule for the enzymes contained therein. Since the enzymes are not readily soluble in the oil phase to an appreciable extent, if at all, the enzymes are indeed for all practical purposes, fixed in place.

In a batch operation, particularly where the aqueous emulsion is created and maintained at least in large part by agitation, after the reaction has been permitted to proceed for an appropriate length of time, the agitation is removed and the aqueous and oil phases separate as the emulsion breaks. The oil phase contains the products, and this is drawn off by decantation and resolved as the composition of the oil phase dictates.

It has been found, according to this process, that commercial semipermeable membranes which as purchased are intended to and do indeed pass water and aqueous solutions with small solute molecules, can be used in the process in order to separate the active enzymes from the reaction zone and permit the remaining oil phase to be resolved. Much more importantly, it has been found that it is possible to treat these commercial water permeable membranes to convert them to water retaining membranes which are capable of passing oil preferentially. This is most surprising since, if the size of the molecule is determinative of what is passed, one would think that in an oil-water system, water, being smaller in molecular size than oil, would always preferentially pass through such filter membrane.

It is not known exactly why the treatment described above converts a water permeable/oil impermeable filter membrane into an oil permeable/water impermeable filter membrane. It is similarly not known what the mechanism of filtrate passage through the filter is. However, regardless of the mechanism involved, it is a fact that filter membranes treated as described above do indeed behave as described, i.e., allow oil molecules to filter through. This is a very fortunate thing because it permits the process to be carried out continuously while retaining the enzyme catalyst encapsulated in the reaction zone, that is upstream of the filter membrane.

Hydrophobic membranes are, in general, known materials. In recent years, hydraulically permeable membranes have been prepared with very uniform pore structures. Recently, as disclosed, for example, in U.S. Patent 3,567,810, it has been shown possible to prepare highly anisotropic, submicroscopically porous membranes from polymers, which membranes have good mechanical integrity and very high mass transfer capabilities. Such membranes are prepared from crystalline or glassy thermoplastic polymers, having from 5 to 50% crystallinity, and a glass transition temperature of at least about 20°C. Particulary advantageous polymers are those having inherently low water sorptivity, such as those having water absorptivities of less than about 10% by weight of moisture at 25°C and 100% relative humidity.

These highly anisotropic membranes are suitably prepared by:

(1) Forming a casting dope of a suitable polymer in an organic solvent;

(2) Casting a film of the casting dope;

(3) Preferentially contacting one side of the film with a diluent characterized by a high degree of miscibility with the organic solvent and a sufficiently low degree of compatibility with the polymer to effect rapid precipitation of the polymer; and

(4) Maintaining the diluent in contact with the membrane until substantially all the solvent has been replaced with the diluent.

In the above procedure, the side of the film brought into contact with solvent (Step 3) forms an extremely thin microporous skin about 0.1 to 5.0 μ thick, with pore sizes in the mμ range; the balance of the film structure becomes a support layer with very much larger pores.

One particular illustrative embodiment of this process is depicted in the accompanying Figure 9.1. In the depicted embodiment, a mixture of long chain normal paraffin and long chain normal alcohol are fed to a reactor **10** in sufficient quantity to make an oil phase containing about 10 volume percent alcohol. Enzyme is extracted from appropriate microorganism growth in a water solution or suspension and, mixed with a small amount of surfactant, fed to the reactor **10**. Stirring means **12** are provided in the reactor.

The reactor may be provided with heat transfer means (not shown) if desired, as is conventional in chemical reactor construction technology. The downstream end of the reactor **10**, or reaction zone, is a hydrophobic membrane **14** which permits passage of the oil, or organic, phase portion of the contents of the reactor and retains the aqueous portion of the reaction mass in the reaction zone. The oil passing through the filter **14** is composed predominantly of n-paraffin reactant, n-alcohol and n-aldehyde product. A typical product resolution scheme is set forth including a liquid-liquid extraction means **16** for separating any dissolved gases and/or nonorganic components and/or carboxylic acids in the product which might have passed through the filter **14**; a first distillation column **18** for recovering n-paraffin for recycle; and a second distillation column system **20** for recovering and resolving the oxidation products free from heavy ends contamination.

Example: Pichia yeast cells were grown in a fermenter on a predominantly long chain paraffin hydrocarbon substrate, harvested therefrom, washed and suspended in an aqueous medium in a concentration of 5 g in 75 ml buffered to a pH of 8.0. The cells were ruptured by treatment with ultrasonic vibration, centifuged to separate out the cell fragments and an aqueous system of the enzyme portion thereof established at a concentration of 10 mg of protein per ml of aqueous system.

Several runs were made by admixing the aqueous solubilized enzymes and a 10% solution of n-tetradecanol in n-tetradecane in varying proportions to a total of 50 parts by volume. The oxidation reactions were carried out by shaking the reactors for 2 hours at 34°C.

Figure 9.1: Oxidation of Hydrocarbons

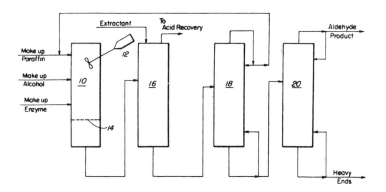

Source: U.S. Patent 3,880,739

Products were recovered as aforesaid and analyzed by gas chromatography. The following correlation of proportion of oil changed to aldehyde formed was observed:

n-Tetradecanol Dissolved in n-Tetradecane (%)	n-Tetradecanal Produced (g/l)
5 vol	0.9
15 vol	1.5
50 vol	2.8

These data substantiate the advantages of operating in a high oil system. Radioactive data demonstrated that most of the aldehyde was derived from the alcohol and that a small portion (2%) was derived from the hydrocarbon.

Hydrolysis of Triglycerides

An effective enzymatic test for the determination of serum triglycerides turns mainly on the development of a rapid and complete process for the hydrolysis of the triglycerides, since various procedures for measuring the glycerol formed by such a reaction are well know.

S.K. Komatsu; U.S. Patent 3,898,130; August 5, 1975; assigned to American Hospital Supply Corporation has found that a complete and surprisingly rapid hydrolysis of triglycerides may be achieved in 3 to 5 minutes using a combination of *Candida* lipase and pancreatic lipase with a bile salt such as sodium taurodeoxycholate. The glycerol so produced may then be assayed by any of a number of known methods.

A variety of pancreatic lipases having activity within the range of about 10 to 100 lipase units per mg, and preferably within the range of 20 to 80 units per mg, are believed suitable, an example being the pancreatic lipase (PL3 Worthington Chemical). The microbial lipase is more specifically a *Candida* lipase which may, for example, be obtained from the cultured broth of *Candida cylindrancea.* Such *Candida* lipase should have activity within the range of 30 to 800 lipase units per mg, and preferably within the range of 200 to 800 lipase units per mg. Other *Candida* lipases are believed to be equally effective when used in combination with the pancreatic lipase and bile salt of the triglyceride-hydrolyzing system.

While both pancreatic and *Candida* lipases must be present, it has been found that the proportional amounts of those constituents, measured in terms of lipase units, may be varied considerably in accordance with selected time requirements for completion of hydrolysis. One lipase unit of activity is the amount sufficient to release one micromol of acid per minute at 25°C from an olive oil emulsion containing gum acacia and 15 mg/ml sodium taurocholate at a pH of about 8.0.

More specifically, the amount of pancreatic lipase in the reaction mixture should be at least 0.14 lipase unit for each µl of body fluid (blood serum or plasma) having a triglyceride value within the range of 0 to 500 mg per 100 ml (mg %) in order to achieve complete hydrolysis within 12 minutes. On the same basis, the amount of *Candida* lipase in the mixture should be at least 0.28 unit, and the amount of bile salt should be at least 0.002 mg, for each µl of body fluid. Where shorter reaction times are required or desired, the amounts of such constituents must be increased. Thus, for complete hydrolysis within 3 to 5 min, at least 1.2 pancreatic lipase units, 0.54 *Candida* lipase units, and 0.02 mg of bile salt, are required for each µl of body fluid.

Example: A reagent suitable for practicing this process may be prepared by making the following 3 ml reaction mixture:

> Pancreatic lipase (Worthington PL3), 180 lipase units
> *Candida* lipase (Worthington), 80 lipase units
> Sodium taurodeoxycholic acid, 3 mg
> LDH, 10 International Units (IU)
> Pyruvate kinase, 10 IU
> NADH, 0.75 micromol
> Phosphoenolpyruvate, 1.5 micromols
> ATP disodium, 0.5 micromol
> Magnesium chloride, 0.0067 M
> Potassium phosphate buffer, 0.1 M, pH 7.0

The assay is carried out by adding an aliquot of liquid containing the triglyceride to be assayed, such as 50 microliters of serum or plasma with triglyceride values of 0 to 500 mg percent, to the above reaction mixture. Following incubation for approximately 5 minutes at a temperature between 25° to 37°C, the optical density is measured at 340 nm. Thereafter, 10 units of glycerol kinase is added and the mixture is again incubated at 25° to 37°C, for another 5 minutes. The optical density is again determined at 340 nm, and the difference in optical densities is proportional to the triglyceride content after appropriate adjustment, using conventional clinical laboratory procedures, for whatever blank reaction is produced.

Gelatin Extraction

Collagen is a major protein constituent of connective tissue in the vertebrate as well as invertebrate animal kingdoms. In practice, meat industry materials rich in collagen such as bone and skin (notably pork skin) are starting materials for the production of collagen or collagen derived products like gelatin.

B.R. Petersen and J.R. Yates; U.S. Patent 4,064,008; December 20, 1977; assigned to Novo Industri A/S, Denmark have found it possible to reduce the conditioning time and to raise the production capacity of a given gelatin producing plant accordingly, if conditioning of the collagen for gelatin extraction is carried out enzymatically under the conditions herein described.

More specifically the alkaline or neutral conditioning of collagen for gelatin extraction comprises treatment of a suitable collagenous raw material, preferably ossein, dehaired hides, or dry- or wet-limed, or salted hides, with a solution of a proteolytic enzyme at a temperature between 20° and 40°C in between 4 and 72 hours and at a pH between 7 and 13, whereafter the proteolytic enzymatic activity is eliminated.

A preferred embodiment of collagen conditioning according to the process comprises use of proteinases produced microbially from microorganisms of the genus *Bacillus* as the proteolytic enzyme, and notably from *B. subtilis* and from *B. licheniformis*.

Other preferred embodiments of collagen conditioning according to the process comprise use of an enzyme concentration in the conditioning float of between 0.05 and 5 Anson units/l, preferably between 0.5 and 1.5 Anson units/l; use of a float ratio (the weight ratio between the float and the collagenous material) between 0.5 and 10; a conditioning time between 6 and 24 hours; the use of the proteolytic enzymes in an amount corresponding to concentration between 10^{-4} and 5×10^{-2} Anson units/g collagenous material; a conditioning temperature between 25° and 35°C; the pH of the float is adjusted to a pH value equal to or substantially equal to the pH optimum of the proteolytic enzyme; and the elimination of the proteolytic enzyme is performed by inactivation of the enzyme activity by means of a pH adjustment to a value in the strongly acid range.

Recovery of Silver from Photographic Wastes

M. Korosi; U.S. Patent 3,982,932; September 28, 1976; assigned to Eastman Kodak Company describes an improved method of precipitating silver out of silver-bearing gelatinous wastes and collecting it in high concentration without the need to centrifuge, to add flocculents (although under certain conditions this can be done advantageously in minute quantities), or to carry out any other operations requiring the handling of relatively large quantities of liquid.

The process in a preferred form includes the steps of adding a proteolytic enzyme to a suspension containing gelatin-bound silver-bearing substances to have the enzyme react with the gelatin molecules to cause them to break down into soluble peptide units. Rapid settling and accumulation of the silver and silver compounds are then accomplished by maintaining suitable conditions for precipitation, and formation of a supernatant liquid.

In a preferred practical embodiment of the process, as shown in Figure 9.2, an aqueous solution or suspension (hereinafter called a liquid) containing emulsified silver metal or silver compounds, for example the waste washing water from photographic emulsion manufacture, or photographic gelatin emulsion removed from scrap film, is treated with proteolytic enzyme material in a reactor tank **11** containing 2,000 gallons of the liquid. Steam is injected into the liquid until the temperature reaches 50°C and the pH is then adjusted to about 8 by introducing an aqueous alkali such as NaOH or KOH. Approximately 5 ppm (parts per million by weight) of a proteolytic enzyme is added, and the mixture is allowed to digest for approximately one half hour at 50°C.

Figure 9.2: Recovery of Silver from Photographic Wastes

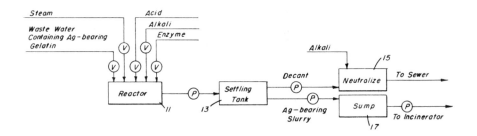

Source: U.S. Patent 3,982,932

The pH is lowered to a value below 4.2, such as between pH 4.2 and pH 2.5, and preferably about 3.5, by acid addition such as 98% undiluted sulfuric acid of which one half to 1 liter may be added per 2,000 gallons in the reactor. During the acid addition the contents of the reactor are continuously stirred by a suitable pump. A fine precipitation is observed within one minute, and coalescing gradually increases the size of the flakes. The time allowed for this operation is about 10 minutes, after which the contents of the tank are pumped into a 10,000 gallon settling tank **13**. The complete cycle for the reactor treatment occupies about 1½ hours.

Several reactor tanks and settling tanks can be used alternately and successively. The sedimentation achieved by the process is found to be quite rapid. After 4 hours of settling, the residual silver content in the supernatant liquid seldom exceeds 4 ppm, and after overnight settling it is generally less than 2 ppm and is often small enough to escape detection. The supernatant liquid is therefore ready for decantation and discharge into the sewer, although before doing so its pH may have to be raised to about 7 by introducing aqueous alkali such as NaOH or KOH in a drip-feed neutralizer **15** to comply with sewerage regulations. In a typical case the discharge to sewer contains about 0.2% gelatin, 0.1% salt and less than 1 ppm of silver.

The wet precipitate slurry is collected in a sump **17** from the bottom of settling tank **13** (and also from reactor **11** when present therein) after a sufficient quantity has accumulated. The wet precipitate is then injected into an incinerator as one way of recovering its silver content.

The solids content of the precipitate before incineration consists of approximately 33% by weight of silver in metal or halide form, and of about 60% of gelatin. Emulsion waste which has been removed from scrap film may be mixed into waste-washing waters in the reactor tank 11 in any desired proportions, e.g., 100 gallons emulsion mixed into the 2,000 gallons in the reactor.

The initial pH adjustment in the reactor may be effected by a 46% aqueous sodium hydroxide solution. The sewer neutralizer added at 15 may be 35% aqueous sodium hydroxide solution diluted with 3 parts H_2O per 1 part solution. Approximately 1 to 2 liters are used for each 2,000 gallon batch in the reactor, and about 3 liters per 10,000 gallons of discharge into the sewer.

Regarding the enzyme, about 5 to 10 ppm may be used for waste-washing waters, and 10 to 50 ppm when 1 part of emulsion has been mixed in 20 parts of waste-washing waters. Preferably the temperature is 48° to 52°C and the pH 7.0 to 10.6 when the enzyme is added. The volume of precipitate sludge may be $1/200$ of the volume of the original waste liquor.

The preferred proteolytic enzyme for performing the process comprises endopeptidases, i.e., proteases which attack the central bonds of protein molecules. A suitable commercially available enzyme of this kind is Bioprase (Nagase & Co., Ltd., Japan). Bioprase is reported by Nagase to be a bacterial proteolytic enzyme preparation which is obtained by establishing a submerged culture of a strain of *Bacillus subtilis*, followed by harvesting, concentrating and purifying of the enzyme.

COMPANY INDEX

The company names listed below are given exactly as they appear in the patents, despite name changes, mergers and acquisitions which have, at times, resulted in the revision of a company name.

INVENTOR INDEX

U.S. PATENT NUMBER INDEX

Copies of U.S. patents are easily obtained
from the U.S. Patent Office at 50¢ a copy.

NOTICE

Nothing contained in this Review shall be construed to constitute a permission or recommendation to practice any invention covered by any patent without a license from the patent owners. Further, neither the author nor the publisher assumes any liability with respect to the use of, or for damages resulting from the use of, any information, apparatus, method or process described in this Review.

VACCINE PREPARATION TECHNIQUES 1980

Edited by J.I. Duffy

Worldwide vaccination programs have the eradication of specific diseases as their goal, and basic immunization with killed or inactivated antigens has become standard practice.

Precipitated or adsorbed preparations, as shown in this book, are more stable and will induce a longer-lasting immune response than pure fluid preparations, although both types are usually administered by injection.

Influenza vaccines, whether killed or live, have not so far been very successful, but new substances, like interferon and the transfer factor, may be instrumental in conferring an immunity that goes far beyond the antigen-antibody response of traditional vaccine technology.

Animal husbandry has received a decided boost with many of the advances in vaccines against animal diseases. Methods of increasing antibody production in the colostrum is a focus of many of the processes described here in detail.

A summarized table of contents follows here with **chapter headings and examples of subtitles**. Numbers in ( ) are numbers of topics.

ISBN 0-8155-0796-8

403 pages